Concepts and Applied Principles of Bioinorganic Chemistry

Volume I

Concepts and Applied Principles of Bioinorganic Chemistry
Volume I

Edited by **Warren Gibbs**

R CALLISTO REFERENCE

New York

Published by Callisto Reference,
106 Park Avenue, Suite 200,
New York, NY 10016, USA
www.callistoreference.com

Concepts and Applied Principles of Bioinorganic Chemistry: Volume I
Edited by Warren Gibbs

International Standard Book Number: 978-1-63239-123-0 (Hardback)

Printed in the United States of America.

Contents

Permissions

List of Contributors

Preface

Bioinorganic chemistry is a field of science that focuses on the examination of the role of metals in biology. This branch of chemistry is about the, function, mechanism and structure dynamics of any and all biologically relevant metal complexes and metal-containing proteins. Bioinorganic chemistry describes the shared relationship between the two sub-disciplines of biochemistry and inorganic chemistry, with emphasis on the function of inorganic substances and chemicals in living systems, as well as processes including the speciation, transport and mineralisation of inorganic materials, along with the use of inorganics in medicinal therapy and diagnosis. This field of study also studies the arena of both natural phenomena such as the behaviour of metallo-proteins as well as artificially introduced metals. These studies also include those that are non-essential metals and chemicals such as in medicine and toxicology. Bioinorganic chemistry is important in elucidating the consequences of substrate bindings and activation, electron-transfer proteins, atom and group transfer chemistry as well as metal properties in biological chemistry. Chemists in this arena use contemporary synthetic methods, theoretical calculations, biological manipulations and advanced characterization techniques that include ultrafast spectroscopy, single-molecule spectroscopy along with synchrotron-based x-ray spectroscopy and diffraction. This is a highly specialized branch of chemistry that requires skilled researchers and scientists to work on new developments and thus the need for newer research is also rapidly increasing.

This book attempts to compile and collate the research and data available in the field of bioinorganic chemistry. I am grateful to those who put in effort and hard work in this field. I am also thankful to those who supported me in this endeavour. I also wish to thank my family for their endless support in my life at every step.

Editor

Synthesis, Spectroscopic Characterization, and In Vitro Antimicrobial Studies of Pyridine-2-Carboxylic Acid N′-(4-Chloro-Benzoyl)-Hydrazide and Its Co(II), Ni(II), and Cu(II) Complexes

Jagvir Singh[1] and Prashant Singh[2]

[1] *Department of Chemistry, Meerut College, Meerut 250001, India*
[2] *Department of Applied Chemistry, B.B.A. University, Lucknow 226025, India*

Correspondence should be addressed to Jagvir Singh, singhjagvir0143@gmail.com

Academic Editor: Spyros Perlepes

N-substituted pyridine hydrazide (pyridine-2-carbonyl chloride and 4-chloro-benzoic acid hydrazide) undergoes hydrazide formation of the iminic carbon nitrogen double bond through its reaction with cobalt(II), nickel(II), and copper(II) metal salts in ethanol which are reported and characterized based on elemental analyses, IR, solid reflectance, magnetic moment, molar conductance, and thermal analysis (TG). From the elemental analyses data, 1 : 2 metal complexes are formed having the general formulae $[MCl_2(HL)_2] \cdot yH_2O$ (where M = Co(II), Ni(II), and Cu(II), y = 1–3). The important infrared (IR) spectral bands corresponding to the active groups in the ligand and the solid complexes under investigation were studied. IR spectra show that ligand is coordinated to the metal ions in a neutral bidentate manner with ON donor sites. The solid complexes have been synthesized and studied by thermogravimetric analysis. All the metal chelates are found to be nonelectrolytes. From the magnetic and solid reflectance spectra, the complexes (cobalt(II), nickel(II), and copper(II)) have octahedral and square planner geometry, respectively. The antibacterial and antifungal activity's data show that the metal complexes have a promising biological activity comparable with the parent ligand against bacterial and fungal species.

1. Introduction

Hydrazones and their metal complexes have been given remarkable attention by the researchers, since they act as the most important stereochemical models in transition metal coordination chemistry, due to their preparative accessibility and structural variety. It has been suggested that the carbonyl linkage (C=O) in Hydrazide is responsible for their biological activities such as antitumor, antibacterial, antifungal, and herbicidal activities [1, 2]. Recently, the developments of new metal drug complexes have received special interests in the field of coordination chemistry [3, 4]. The transition metal-coordinated Hydrazones complexes play a significant role in many catalytic reactions like oxidations, asymmetric cyclopropanation, and polymerisation [5–7]. Hydrazide complexes of transition metal ions are known to provide useful models for elucidation of the mechanisms of enzyme inhibition by hydrazine derivatives [8] and for their possible pharmacological applications [9]. The activities of some such type complexes are very significant against Gram-positive bacteria in vitro. The derivatives of these chelate act as good potential oral drugs to treat the genetic disorders like thalassemia [10]. The structural characterization of these resultant hydrazide complexes revealed some interesting facts, such as their tendency and potency to act as planar pentadentate ligands in most of the complexes [11–14] along with tridentate character [15]. Moreover, these ligands exhibit ketoenol tautomerism and can coordinate in neutral [16], monoanionic [17], dianionic [4, 15, 16], or tetraanionic [18] forms, to the metal ions which have coordination numbers of six and seven [4, 18], generating mononuclear or binuclear species. However, it depends on the reaction

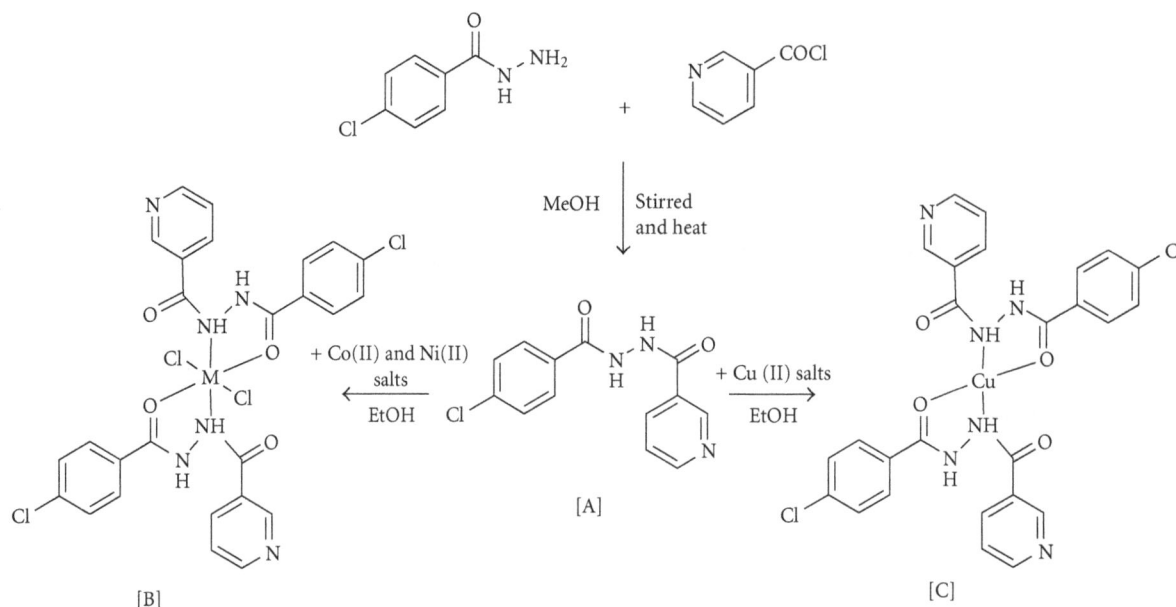

SCHEME 1: Synthesis of ligand (HL) and its metal complexes.

conditions, such as metal ion, its concentration, the pH of the medium, and the nature of the hydrazide ligand [19]. Certain hydrazides and their Cu(II) complexes have antitumor activity. Transition metal complexes having the ability to bind and cleave DNA under physiological conditions are of current interests for their potential applications [20] in nucleic acids chemistry. Such complexes [21] are useful in foot printing studies, as sequence specific DNA binding agents, in genomic research and as diagnostic agents in medicinal applications. Much attention has been paid to biologically active metal complexes in recent years. Oxygen and nitrogen donor ligands [22] have been widely studied due to their high potential to coordinate with transition metals. Compounds containing carbonyl oxygen groups [23] have important position among organic reagents as potential donor ligands for the transition metal ions. Organic compounds and metal complexes [24] both display a wide range of pharmacological activity including anticancer, antibacterial, and fungi static effects. Though some reports are available on the hydrazones metal complexes [21, 22], there is no work focussed in the hydrazide derived from pyridine-2-carboxylic acid N′-(4-chloro-benzoyl)-hydrazide (HL) and its complexes (Scheme 1). In this paper, we report the synthesis, spectroscopic, thermal, and biological studies of the newly synthesized hydrazone complexes of the transition metal ions (Co(II), Ni(II), and Cu(II)).

2. Experimental

2.1. Materials. All the chemicals used in the present investigations were of the analytical reagent grade (AR). Pyridine-2-carbonyl chloride (BDH), 4-chloro-benzoic acid hydrazide (Sigma), metal salts, and solvents were purchased from Qualigens Chemicals Company, India. They were used as received. The elemental analysis (C, H, N) was performed using a Carlo-Erba 1106 Elemental Analyzer, and IR spectra were recorded on a Shimadzu-160 Spectrometer using KBr discs in the range 4000–400 cm^{-1}. Electronic spectra were recorded on a Shimadzu-160 Spectrometer. The ^{1}H and ^{13}CNMR spectra were obtained on a Bruker DPX-400 Spectrometer using DMSO-d_6 solvent and TMS as the internal reference at room temperature. The EPR spectra of the complexes were recorded as polycrystalline sample on a Varian E-4 EPR Spectrometer. The mass losses were measured in nitrogen atmosphere from ambient temperature up to 800°C at a heating rate of 10°C min^{-1}. Molar conductivities in DMF or DMSO at 25°C were measured using a model CM-1 K-TOA company conductivity meter. Magnetic moments at 25°C were determined using the Gouy method with Hg [Co(SCN)$_4$] as calibrant.

2.2. Synthesis of Hydrazine Ligand. In a round bottom flask (100 mL), a methanolic solution (10.0 mL) of pyridine-2-carbonyl chloride (0.02 m mol, 2.62 mL) and an aqueous methanolic solution (10 mL) of 4-chloro-benzoic acid hydrazide (0.01 m mol, 0.20 g) were taken and stirred at room temperature for 30 minutes after then the reaction was refluxed at 50°C for ~6 h. The resulting mixture was left under reflux for 3 h, and the formed solid product was separated by filtration, purified by crystallization from ethanol, washed with diethyl ether, and dried in a vacuum over anhydrous calcium chloride. The yellow product is produced in 52% yield.

2.3. Synthesis of Metal Complexes. The following detailed preparation is given as an example, and the other complexes were obtained similarly. The Cu(II) complex was synthesized by the addition of hot solution (60°C) of the Cu(II) chloride (0.17 g, 1 mmol) in an ethanol-water mixture (1 : 1, 25 mL)

Synthesis, Spectroscopic Characterization, and In Vitro Antimicrobial Studies of Pyridine-2-Carboxylic
Acid N´-(4-Chloro-Benzoyl)-Hydrazide and Its Co(II), Ni(II), and Cu(II) Complexes

3

to the hot solution (60°C) of the ligand (0.550 g, 2 mmol) in the same solvent mixture (25 mL). The resulting mixture was stirred under reflux for one hour whereupon the complex precipitated. It was collected by filtration and washed with a 1 : 1 ethanol: water mixture and diethyl ether.

2.4. Analytical Data of Synthesized Ligand and Its Metal Complexes

2.4.1. Ligand (HL). Yield: 52%; M.P. 195°C, Mol. wt. 275, color: yellow; analytical data for $(C_{13}H_{10}ClN_3O_2)$ found (calc.): C, 56.64 (56.11); H, 3.66 (3.11); N, 15.24 (14.97). IR (KBr, cm^{-1}) 3440 ν_{NH}, 1690 $\nu_{C=O}$, 3015 ν_{C-H}, 2228 ν_{C-N}. ESI-MS, m/z Data found (calc.): 276 (275), ^{1}H NMR (DMSO-d$_6$) δ ppm: 7.1 (m, 8H, HC-Ar), 3.8 (s, 2H, NH–NH). ^{13}C NMR (DMSO-d$_6$) δ ppm: 117.53–121.07 (10C, CH-Ar.), 143.23 (1C, C–N), 153.88 (2C, C=O).

2.4.2. Complex I. Yield: 65%; MP: 230°C; Mol. wt. 681; color: reddish; analytical data for $[C_{26}H_{20}Cl_4CoN_6O_4]$ found (calc.): C, 45.84 (31.15); H, 2.96 (2.85); N, 12.34 (12.17); (KBr, cm^{-1}) 3406 ν_{NH}, 1690 $\nu_{C=O}$, 3015 ν_{C-H}, 2203 ν_{C-N}, 519 ν_{M-N}, 416 ν_{M-O}. ^{1}H NMR (DMSO-d$_6$) δ ppm: 7.1 (m, 8H, HC-Ar), 3.01 (s, 2H, NH–NH). ^{13}C NMR (DMSO-d$_6$) δ ppm: 117.53–121.07 (10C, CH-Ar.), 133.22 (1C, C–N), 153.88 (2C, C=O).

2.4.3. Complex II. Yield: 42%; MP: 235°C; Mol. wt. 680; color: greenish; analytical data for $[C_{26}H_{20}Cl_4N_6NiO_4]$ found (calc.): C, 45.86 (45.55); H, 2.96 (2.55); N, 12.34 (12.12). (KBr, cm^{-1}) 3400 ν_{NH}, 1690 $\nu_{C=O}$, 3015 ν_{C-H}, 2210 ν_{C-N}, 518 ν_{M-N}, 408 ν_{M-O}. ^{1}H NMR (DMSO-d$_6$) δ ppm: 7.1 (m, 8H, HC-Ar), 3.5 (s, 2H, NH–NH). ^{13}C NMR (DMSO-d$_6$) δ ppm: 117.53–121.07 (10C, CH-Ar.), 136.30 (1C, C–N), 153.88 (2C, C=O).

2.4.4. Complex III. Yield: 48%; MP: 220°C; Mol. wt. 615; color: brownish; analytical data for $[C_{26}H_{20}Cl_2CuN_6O_4]$ found (calc.): C, 50.78 (50.30), H, 3.28 (2.90), N, 13.67 (12.17). (KBr, cm^{-1}) 3400 ν_{NH}, 1690 $\nu_{C=O}$, 3015 ν_{C-H}, 2200 ν_{C-N}, 512 ν_{M-N}, 420 ν_{M-O}. ^{1}H NMR (DMSO-d$_6$) δ ppm: 7.1 (m, 8H, HC-Ar), 3.4 (s, 2H, NH–NH). ^{13}C NMR (DMSO-d$_6$) δ ppm: 117.53–121.07 (10C, CH-Ar.), 131.20 (1C, C–N), 153.88 (2C, C=O).

2.5. Biological Activity. Antimicrobial activity of the tested samples was determined using a modified Kirby-Bauer disc diffusion method [25]. One hundred microliters of the tested bacteria or fungi were grown in 10 mL of fresh media until they reached account of approximately 98 cells/mL for bacteria and 95 cells/mL for fungi [26]. One hundred microliters of microbial suspension was spread onto agar plates corresponding to the broth in which they were maintained. Isolated colonies of each organism that might be playing a pathogenic role should be selected from primary agar plates and tested for susceptibility by disc diffusion method [27]. Of the many media available, NCCLS recommends Mueller-Hinton agar due to its results in good

batch-to-batch reproducibility. Disc diffusion method for filamentous fungi tested by using approved standard method (M38-A) developed [28]. For evaluating the susceptibilities of filamentous fungi to antifungal agent, the disc diffusion method was applied for yeast developed by using approved standard method (M44-P) [29]. Plates inoculated with filamentous fungi as *C. albican* MTCC 227 at 25°C for 48 h; Gram (+) bacteria as *Staphylococcus aureus* MTCC 3160; Gram (−) bacteria as *Staphylococcus aureus* MTCC 25923, they were incubated at 35–37°C for 24–28 h and yeast as *S. cerevisiae* MTCC 361 incubated at 30°C for 24–28 h and, and then the diameters of the inhibition zones were measured in millimeters [29]. Standard discs of Gentamycin (antibacterial agent) and amphotericin-B (antifungal agent) served as positive controls for antimicrobial activity, but filter discs impregnated with 10 ML of solvent (distilled water, chloroform, DMSO) were used as a negative control. The agar used is Mueller-Hinton agar that is rigorously tested for composition and pH. Further the depth of the agar in the plate is a factor to be considered in the disc diffusion method. This method is well documented, and standard zones of inhibition have been determined for susceptible and resistant values. Blank paper discs (Schleicher and Schuell, Spain) with a diameter of 8.0 mm were impregnated with 10 ML of tested concentration of the stock solutions. When a filter paper disc impregnated with a tested chemical is placed on agar, the chemical will diffuse from the disc into the agar. This diffusion will place the chemical in the agar only around the disc. The solubility of the chemical and its molecular size will determine the size of the area of chemical infiltration around the disc. If an organism is placed on the agar, it will not grow in the area around the disc if it is susceptible to the chemical. This area of no growth around the disc is known as zone of inhibition or clear zone. For the disc diffusion, the zone diameters were measured with slipping calipers of the national committee for clinical laboratory standards [29]. Agar-based methods such Etest and disc diffusion can be good alternatives because they are simpler and faster than broth-based methods [30].

3. Result and Discussion

The new prepared hydrazide ligand is subjected to elemental analyses, IR, Mass bar and ^{1}H & ^{13}C NMR spectral studies. The results obtained are in good agreement with those calculated for the suggested formula, and the melting point is sharp indicating the purity of the prepared ligand (HL). The structure of HL under study is given in Scheme 1. The IR spectra of hydrazide ligand contain a strong C=O absorption band at 1670 cm^{-1} and N–H absorption band at 3187 cm^{-1}, and after complexation, these bands are slightly shifted to lower wave numbers (1646–1649 cm^{-1} and 3133–3140 cm^{-1}) indicating the involvement of the oxygen and nitrogen atom in chelate formation, respectively. In the FT IR spectra of all the complexes, the nonligand bands observed at 520–586 and 420–466 cm^{-1} regions can be assigned to (M–O) and (M–N), respectively [31–33]. ^{1}H NMR spectrums showed signals in the range δ 8.03 ppm, and these signals

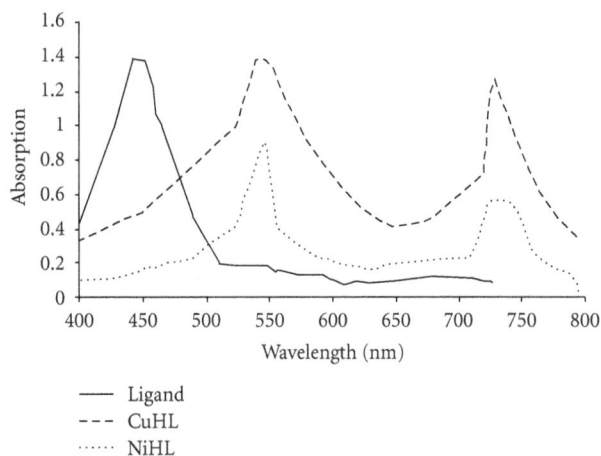

FIGURE 1: UV-visible spectra of free ligand, Cu-complex, and Ni-complex.

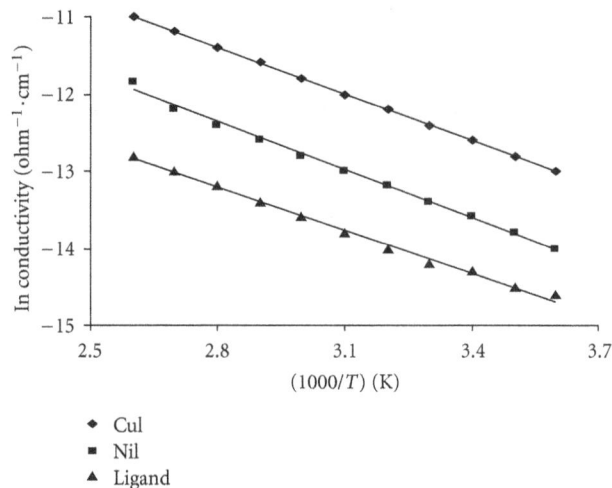

FIGURE 2: The DC conductivities of free ligand, Cu-complex, and Ni-complex before doping.

were the evidence of the secondary amide bonding to the ligand [34]. ^{13}C NMR spectrum of complexes displays a signal at δ 167.45 ppm and δ 163.54 ppm, due to (C–NH) and (C=O), which was indicating that carbon atoms of carbon-amide group and oxygen atom of carbonyl group participate the formation of ligand [28]. The spectrum shows the molecular ion peak at m/z = 275 ($C_{13}H_{10}ClN_3O_2$, calculated atomic mass 276 amu due to ^{13}C and ^{15}N isotopes). The different competitive fragmentation pathways of ligand give the peaks at different mass numbers at 304. The intensity of these peaks reflects the stability and abundance of the ions. The presence of fragments at m/z values 291, 276, 249, and 237 shows the fragmentation. The mass spectrum clearly suggests existence of ligand in the hydrazine form. The molar conductance values of all the complexes lie in the range 122–165 $ohm^{-1}\,cm^2\,mole^{-1}$ corresponding to 1 : 2 nonelectrolytic behaviors [29]. The electronic spectra of the cobalt(II) complex showed three bands at 8780–8810, 17475–17775 and 30235–30270 cm^{-1}, which may be assigned to $^4T_{1g} \rightarrow {}^4T_{2g}$ (F), $^4T_{1g} \rightarrow {}^4T_{1g}$ (P), and $^4T_{1g} \rightarrow {}^3A_{2g}$ (F) transitions and suggested octahedral geometry around the cobalt ion. The magnetic moment 4.80–488 BM is an addition evidence for an octahedral structure [35].

The cobalt complexes showed magnetic moment values 4.70–490 B.M. at room temperature. These high values of the magnetic moments and the stoichiometries suggest a coordination number of six for the central cobalt ions and an octahedral geometry. The magnetic properties of copper complex may be divided into broad classes. First, those are having essentially temperature-independent magnetic moments in rang 2.20 BM. Those exhibiting such moments are mononuclear complexes having no major interaction between the unpaired electrons on different copper ion. The moments in such complex, as in apparent, lie appreciably above the spine-only value (1.73 BM), but as the electronic ground states are nondegenerate, this cannot arise from inherent angular momentum in the ground state. It arises due to mixing of some orbital angular momentum from excited states via spin orbit coupling. The copper (II) complexes exhibit magnetic moments of 1.70–1.75 B.M.,

respectively, at room temperature. The electronic spectra of the copper(II) complexes display a broad band at 14220–14918 cm^{-1} due to $^2B_{1g} \rightarrow {}^2E_g$ and two bands at 16390–16550 and 27250–27350 cm^{-1} assigned to d-d transitions and a charge transfer band, respectively, of square planner environment [31]. The observed magnetic moment of the copper complexes is 1.75–180 BM. The Ni(II) complex is found to have a room temperature magnetic moment value of 3.87 B.M., which is in the normal range observed for octahedral Ni(II) complexes [26, 34, 36]. The electronic spectrum displays three bands in the solid reflectance spectrum at m_1: 14968 cm^{-1}; $^3A_{2g} \rightarrow {}^3T_{2g}$; m_2: 17,788 cm^{-1}; $^3A_{2g} \rightarrow {}^3T_{1g}$ (F) and m_3; 21347 cm^{-1}; $^3A_{2g} \rightarrow {}^3T_{1g}$ (P). The spectrum shows also a band at 24565 cm^{-1} which may be attributed to ligand to metal charge transfer (Figure 1). The maximum conductivity value was 1.81×10^{-4} $ohm^{-1}\,cm^{-1}$ for doped copper(II) complex because of the small size of nickel atom compared with iron atom. Also the prepared hydrazide complexes at different temperatures (303–373) K show that the increased conductivity with increasing of temperatures may be attributed to presence of metals (d-d*) transition. Figures 2 and 3 show that the conductivities of doped and undoped compounds increase with increasing of temperatures which is consistent with semiconductors properties [19, 20].

The dopping compounds have higher conductivities than undoped because iodine doping leads to oxidation of iodine molecules to form I_3^{-1}, I_5^{-3}, and reduction of hydrazide complexes molecules, this effect increase the conductivity by making acceptor bands, and the distance between the energy levels was low. The DC electrical conductivities of doped and undoped prepared compounds over temperature range (0–110°C) and under vacuum were measured by using conductivity apparatus which is consisting of temperature recorder, power supply, voltmeter, resistance and sample cell. The studied samples were as discs covered in two sides by silver paint. The ligand, Cu(II), and Ni(II) complexes were doped with iodine by mixing 1 g (0.004 mole) ligand, 1 g (0.0022 mole) Cu-complex, and 1 g (0.002 mole)

Synthesis, Spectroscopic Characterization, and In Vitro Antimicrobial Studies of Pyridine-2-Carboxylic
Acid N′-(4-Chloro-Benzoyl)-Hydrazide and Its Co(II), Ni(II), and Cu(II) Complexes

5

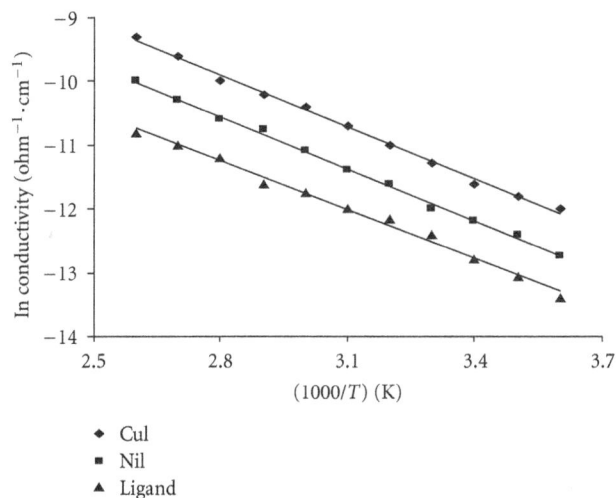

FIGURE 3: The DC conductivity of free ligand, Cu-complex, and Ni-complex after doping.

Ni-complex with 25 mL of iodine solution in CCL$_4$ (4%, w/v), the mixture was refluxed with stirring for 48 hours and then filtered and dried in the vacuum oven at 50°C. The conductivities at different temperatures were calculated according to Arrhenius equation as shown below [19]:

$$\sigma = \sigma o * \exp\left(-\frac{\Delta E}{2K}\right). \tag{1}$$

The resistance of the sample and its electrical conductivity is calculated from the equations:

$$R_x = \frac{R_s * V_x}{V_s}, \quad \sigma = \left(\frac{L}{A}\right) * \frac{1}{R_x}, \tag{2}$$

where R_s is standard resistance (ohm), R_x is sample resistance (ohm), V_s is standard Voltage (volt), V_x is sample voltage (volt), L is Sample thickness (cm), and A is The painted area of the sample surface (cm^2).

3.1. Thermal Analysis (TG).
The thermodynamic activation parameters of decomposition processes of dehydrated complexes, namely, activation energy, enthalpy, entropy, and Gibbs free energy change of the decomposition are evaluated graphically by employing the Coats-Redfern relation. The entropy of activation, enthalpy of activation, and the free energy change of activation were calculated. The high values of the activation energies reflect the thermal stability of the complexes. The entropy of activation is found to have negative values in all the complexes which indicate that the decomposition reactions proceed spontaneously. The thermogram of (CoCl$_2$(HL)) 3H$_2$O and (CuCl$_2$(HL)) 2H$_2$O chelates show five decomposition steps within the temperature range 30–900 and 50–950°C, respectively. The first two steps of decomposition within the temperature range 30–250 and 50–230°C correspond to the loss of three water molecules of hydration, Cl$_2$, one coordinated water molecule, and loss of two water molecules of hydration and coordination and Cl$_2$ (in case of Cu(II) complex, mass loss

of 29.37% (calcd. 28.57%)). The energy of activation was found to be 44.40 and 78.24 (in case of Co(II) complex) and 36.52 and 53.78 kJ mol^{-1} (in case of Cu(II) complex) for the first and second steps, respectively. The subsequent steps (230–950°C) correspond to the removal of the organic part of the ligand leaving metal oxide as a residue. The overall weight loss amounts to 86.02% (calcd. 85.51%) and 84.26 (calcd. 84.11%) for Co(II) and Cu(II) complexes, respectively. The TG curves of the Ni(II) chelate show four stages of decomposition within the temperature range of 30–950°C. The first step at 30–130°C corresponds to the loss of water molecules of hydration. The subsequent three steps (2nd, 3rd, and 4th) involve the loss of Cl$_2$ and ligand molecule. The overall weight loss amounts to 84.97% (calcd. 85.42%), and the activation energy is 102.4–243.4 Ni(II) chelates.

3.2. Biological Activity.
In testing the antibacterial activity of these compounds we used more than one test organism to increase the chance of detecting antibiotic principles in tested materials. The sensitivity of a microorganism to antibiotics and other antimicrobial agents was determined by the assay plates which incubated at 28°C for 2 days for yeasts and at 37°C for 1 day for bacteria. All of the tested compounds showed a remarkable biological activity against different types of Gram-positive and Gram-negative bacteria and against fungi species. The data are listed in Table 1.

It was demonstrated that the newly prepared ligand and its metal complexes showed a higher effect on *C. albican* (Gram-negative bacteria) and *S. aureus* (Gram-positive bacteria). It is known that the membrane of Gram-negative bacteria is surrounded by an outer membrane containing lipopolysaccharides. The newly synthesized hydrazide and its metal complexes seem to be able to combine with the lipophilic layer in order to enhance the membrane permeability of the Gram-negative bacteria. The lipid membrane surrounding the cell favours the passage of only lipid soluble materials; thus the lipophilicity is an important factor that controls the antimicrobial activity. Also the increase in lipophilicity enhances the penetration of hydrazide and its metal complexes into the lipid membranes and thus restricts further growth of the organism. This could be explained by the charge transfer interaction between the studied molecules and the lipopolysaccharide molecules which lead to the loss of permeability barrier activity of the membrane. The hydrazide and its metal complexes could enhance the antimicrobial effect on both strains probably by the hydroxyl group. The Co(II), Ni(II), and Cu(II) complexes were almost the most promising broad spectrum antimicrobial agents due to the presence of coordinated anion with higher antimicrobial activity than the other complexes. From the data the inhibition zone of the metal chelates is higher than that of the ligand. Such increased activity of the metal chelates is due to the lipophilic nature of the metal ion in complexes. Furthermore, the mode of action of hydrazide of the compounds may involve the formation of a hydrogen bond through the azomethine nitrogen atom with the active canters of all the constituents, resulting in interference with normal cell process.

TABLE 1: Antimicrobial activity of ligand and their metal complexes.

Compounds	Time hrs	S. aureus MTCC 3160 Diameter of zone of inhibition (mm)		S. aureus MTCC 25923 Diameter of zone of inhibition (mm)		C. albican MTCC 227 Diameter of zone of inhibition (mm)		S. cerevisiae MTCC 361 Diameter of zone of inhibition (mm)	
		$100\,\mu g$	$50\,\mu g$	$100\,\mu g$	$50\,\mu g$	$100\,\mu g$	$50\,\mu g$	$100\,\mu g$	$50\,\mu g$
HL	24	10	11	—	—	—	—	13	15
	48	12	15	—	—	—	—	13	16
[1]	24	10	12	13	16	19	17	19	14
	48	11	12	14	19	11	10	12	17
[2]	24	14	13	16	18	15	27	21	18
	48	—	—	—	—	11	29	12	19
[3]	24	—	—	—	—	13	10	12	10
	48	14	10	13	12	14	13	14	11
Gentamycin	24	22	24	22	24	—	—	—	—
	48	22	24	22	24	—	—	—	—
Amphotericin-B	24	—	—	—	—	17	21	17	21
	48	—	—	—	—	17	21	17	21

4. Conclusion

The results of this investigation support the suggested structures of the metal complexes. It is obvious from this study that only mononuclear complexes are obtained. The IR spectral studies reveal that HL coordinated to the metal ions via C=O and amide NH–NH group. The chelates are nonelectrolytes. All metal cations have octahedral geometry except copper (II) chelate. The thermal decomposition of the complexes as well as the thermodynamic parameters is studied. The biological activities of the hydrazide under investigation and its complexes against bacterial and fungal organisms are promising which need further and deep studies on animals and humans.

Acknowledgments

The authors are thankful to CDRI, Lucknow, and ACBR, New Delhi, India, for providing the spectral, biological, and analytical facilities.

References

[1] Z. H. Chohan, M. Praveen, and A. Ghaffar, "Structural and biological behaviour of Co(II), Cu(II) and Ni(II) metal complexes of some amino acid derived Schiff-bases," *Metal-Based Drugs*, vol. 4, no. 5, pp. 267–272, 1997.

[2] Z. H. Chohan and S. Kausar, "Synthesis, structural and biological studies of nickel(II), copper(II) and zinc(II) chelates with tridentate Schiff bases having NNO and NNS donor systems," *Chemical and Pharmaceutical Bulletin*, vol. 41, no. 5, pp. 951–953, 1993.

[3] M. J. Seven and L. A. Johnson, *Metal Binding in Medicine*, Lippincott, Philadelphia, Pa, USA, 4th edition, 1960.

[4] R. S. Srivastava, "Studies on some antifungal transition metal chelates of 2-(2-hydroxybenzylideneamino) benzimidazole," *Indian Journal of Chemistry*, vol. 29, p. 1024, 1997.

[5] V. K. Patel, A. M. Vasanwala, and C. N. Jejurkar, "Synthesis of mixed Schiff base complexes of copper(II) and nickel(II) and their spectral, magnetic and antifungal studies," *Indian Journal of Chemistry*, vol. 28, p. 719, 1976.

[6] F. Maggio, A. Pellerito, L. Pellerito, S. Grimaudo, C. Mansueto, and R. Vitturi, "Organometallic complexes with biological molecules—II. Synthesis, solid-state characterization and in vivo cytotoxicity of diorganotin(IV)chloro and triorganotin(IV) chloro derivatives of penicillin G," *Applied Organometallic Chemistry*, vol. 8, no. 1, pp. 71–85, 1994.

[7] R. Vitturi, C. Mansueto, A. Gianguzza, F. Maggio, A. Pellerito, and L. Pellerito, "Organometallic complexes with biological molecules—III: in vivo cytotoxicity of diorganotin(IV)chloro and triorganotin(IV)chloro derivatives of penicillin g on chromosomes of *Aphanius fasciatus*(pisces, cyprinodontiformes)," *Applied Organometallic Chemistry*, vol. 8, no. 6, pp. 509–515, 1994.

[8] L. Pellerito, F. Maggio, A. Consigilo, A. Pellerito, G. C. Stocco, and S. Gremaudo, "Organometallic complexes with biological molecules—4: di- and tri-organotin(IV)amoxicillin derivatives: solid-state and solution-phase spectroscopic investigations," *Applied Organometallic Chemistry*, vol. 9, no. 3, pp. 227–239, 1995.

[9] R. Vitturi, B. Zava, M. S. Colomba, A. Pellerito, F. Maggio, and L. Pellerito, "Organometallic complexes with biological molecules—5: in vivo cytotoxicity of diorganotin(IV)-amoxicillin derivatives in mitotic chromosomes of *Rutilus rubilio*(pisces, Cyprinidae)," *Applied Organometallic Chemistry*, vol. 9, no. 7, pp. 561–566, 1995.

[10] B. Rosenberg and L. VanCamp, "The successful regression of large solid sarcoma 180 tumors by platinum compounds," *Cancer Research*, vol. 30, no. 6, pp. 1799–1802, 1970.

[11] M. J. Cleare and J. D. Hoeschele, "Studies on the antitumor activity of group VIII transition metal complexes—I. Platinum (II) complexes," *Bioinorganic Chemistry*, vol. 2, no. 3, pp. 187–210, 1973.

[12] T. Rosu, E. Pahontu, M. Reka-Stefana et al., "Synthesis, structural and spectral studies of Cu(II) and V(IV) complexes of a novel Schiff base derived from pyridoxal and antimicrobial activity," *Polyhedron*, vol. 31, pp. 352–360, 2012.

[13] A. J. Crowe, "The antitumour activity of tin compounds," in *Metal Based Antitumour Drugs*, vol. 1, pp. 103–149, Freud, London, UK, 1988.

Synthesis, Spectroscopic Characterization, and In Vitro Antimicrobial Studies of Pyridine-2-Carboxylic
Acid N´-(4-Chloro-Benzoyl)-Hydrazide and Its Co(II), Ni(II), and Cu(II) Complexes

7

[14] D. R. Williams, "Thermodynamic considerations in co-ordination—part 10. a potentiometric and calorimetric investigation of copper(II) histidine complexes in solution," *Journal of the Chemical Society, Dalton Transactions*, no. 7, pp. 790–797, 1972.

[15] I. Sakiyan, E. Loğoğlu, S. Arslan, N. Sari, and N. Şakiyan, "Antimicrobial activities of N-(2-hydroxy-1-naphthalidene)-amino acid(glycine, alanine, phenylalanine, histidine, tryptophane) Schiff bases and their manganese(III) complexes," *BioMetals*, vol. 17, no. 2, pp. 115–120, 2004.

[16] N. Sari, S. Arslan, E. Logoglu, and I. Sakiyan, "Antibacterial activities of some Amino acid Schiff bases," *Journal of Animal Science*, vol. 16, no. 2, pp. 283–288, 2003.

[17] R. J. P. Williams, "Role of transition metal ions in biological processes," *Royal Institute of Chemistry, Reviews*, vol. 1, no. 1, pp. 13–38, 1968.

[18] R. C. Maurya, D. D. Mishra, and S. Pillai, "Studies on some mixed-ligand chelates of cobalt(II) involving acetoacetylarylamides and biologically active heterocyclic oxygen donors," *Synthesis and Reactivity in Inorganic and Metal-Organic Chemistry*, vol. 27, no. 10, pp. 1453–1466, 1997.

[19] R. C. Maurya, R. Verma, P. K. Trivedi, and H. Singh, "Synthesis, magnetic and spectral studies of some mixed-ligand chelates of Bis(2-hydroxyacetophenonato)copper(II) with 2- or 3-pyrazoline-5-one derivatives," *Synthesis and Reactivity in Inorganic and Metal-Organic Chemistry*, vol. 28, no. 2, pp. 311–329, 1998.

[20] R. B. Singh, P. Jain, and R. P. Singh, "Hydrazones as analytical reagents: a review," *Talanta*, vol. 29, no. 2, pp. 77–84, 1982.

[21] C. Tan, J. Liu, L. Chen, S. Shi, and L. Ji, "Synthesis, structural characteristics, DNA binding properties and cytotoxicity studies of a series of Ru(III) complexes," *Journal of Inorganic Biochemistry*, vol. 102, no. 8, pp. 1644–1653, 2008.

[22] G. Zuber, J. C. Quada Jr., and S. M. Hecht, "Sequence selective cleavage of a DNA octanucleotide by chlorinated bithiazoles and bleomycins," *Journal of the American Chemical Society*, vol. 120, no. 36, pp. 9368–9369, 1998.

[23] S. M. Hecht, "Bleomycin: new perspectives on the mechanism of action," *Journal of Natural Products*, vol. 63, no. 1, pp. 158–168, 2000.

[24] T. Ghosh, B. G. Maiya, A. Samanta et al., "Mixed-ligand complexes of ruthenium(II) containing new photoactive or electroactive ligands: synthesis, spectral characterization and DNA interactions," *Journal of Biological Inorganic Chemistry*, vol. 10, no. 5, pp. 496–508, 2005.

[25] M. Carcelli, C. Pelizzi, G. Pelizzi, P. Mazza, and F. Zani, "The different behaviour of the di-2-pyridylketone 2-thenoylhydrazone in two organotin compounds. Synthesis, X-ray structure and biological activity," *Journal of Organometallic Chemistry*, vol. 488, no. 1-2, pp. 55–61, 1995.

[26] N. P. Singh, Anu, and J. V. Singh, "Magnetic and spectroscopic studies of the synthesized metal complexes of bis(pyridine-2-carbo) hydrazide and their antimicrobial studies," *E-Journal of Chemistry*, vol. 9, no. 4, pp. 1835–1842, 2012.

[27] S. Saha, D. Dhanasekaran, S. Chandraleka, N. Thajuddin, and A. Panneerselvam, "Synthesis, characterization and antimicrobial activity of cobalt metal complexes against drug resistant bacterial and fungal pathogens," *Advances in Biological Research*, vol. 4, pp. 224–229, 2010.

[28] M. E. Mulligan, K. A. Murray-Leisure, B. S. Ribner et al., "Methicillin-resistant *Staphylococcus aureus*: a consensus review of the microbiology, pathogenesis, and epidemiology with implications for prevention and management," *American Journal of Medicine*, vol. 94, no. 3, pp. 313–328, 1993.

[29] N. Farrell, "Metal complexes as drugs and chemotherapeutic agents," *Comprehensive Coordination Chemistry*, vol. 9, pp. 809–840, 2003.

[30] P. P. Dholakiya and M. N. Patel, "Metal complexes: preparation, magnetic, spectral, and biocidal studies of some mixed-ligand complexes with schiff bases containing NO and NN donor atoms," *Synthesis and Reactivity in Inorganic and Metal-Organic Chemistry*, vol. 34, no. 3, pp. 553–563, 2004.

[31] S. U. Rehman, Z. H. Chohan, F. Gulnaz, and C. T. Supuran, "In-vitro antibacterial, antifungal and cytotoxic activities of some coumarins and their metal complexes," *Journal of Enzyme Inhibition and Medicinal Chemistry*, vol. 20, no. 4, pp. 333–340, 2005.

[32] U. Lee and B. K. Koo, "Synthesis and crystal structures of Mn(II), Co(II), Ni(II), Cu(II), and Zn(II) metal complexes with NNO functionalized ligands," *Bulletin of the Korean Chemical Society*, vol. 26, no. 6, pp. 925–929, 2005.

[33] J. Valdés-Martínez, R. A. Toscano, A. Zentella-Dehesa, M. M. Salberg, G. A. Bain, and D. X. West, "Synthesis, crystal and molecular structure of 5-bromo-salicylaldehyde-2-methylthiosemicarbazonato(nitrato)copper(II) monohydrate," *Polyhedron*, vol. 15, no. 3, pp. 427–431, 1996.

[34] W. J. Geary, "The use of conductivity measurements in organic solvents for the characterisation of coordination compounds," *Coordination Chemistry Reviews*, vol. 7, no. 1, pp. 81–122, 1971.

[35] F. A. Cotton, G. Wilkinson, C. A. Murillo, and M. Bochmann, *Advanced Inorganic Chemistry*, John-Wiley and Sons, New York, NY, USA, 6th edition.

[36] A. I. Vogel, *A Textbook of Quantitative Inorganic Analysis*, ELBS and Longman, London, UK, 4th edition, 1978.

Electrospun Nanostructured Fibers of Collagen-Biomimetic Apatite on Titanium Alloy

Michele Iafisco,[1, 2] Ismaela Foltran,[1, 2] Simona Sabbatini,[3] Giorgio Tosi,[3] and Norberto Roveri[1]

[1] Dipartimento di Chimica "G. Ciamician," Alma Mater Studiorum Università di Bologna, Via Selmi 2, 40126 Bologna, Italy
[2] Dipartimento di Scienze Mediche, Università del Piemonte Orientale, Via Solaroli 17, 28100 Novara, Italy
[3] Dipartimento di Scienze e Tecnologie Chimiche, Università Politecnica delle Marche, Via Brecce Bianche, 60131 Ancona, Italy

Correspondence should be addressed to Michele Iafisco, michele.iafisco@unibo.it

Academic Editor: Giovanni Natile

Titanium and its alloys are currently the mainly used materials to manufacture orthopaedic implants due to their excellent mechanical properties and corrosion resistance. Although these materials are bioinert, the improvement of biological properties (e.g., bone implant contact) can be obtained by the application of a material that mimics the bone extracellular matrix. To this aim, this work describes a new method to produce nanostructured collagen-apatite composites on titanium alloy substrate, by combining electrospinning and biomimetic mineralization. The characterization results showed that the obtained mineralized scaffolds have morphological, structural, and chemical compositional features similar to natural bone extracellular matrix. Finally, the topographic distribution of the chemical composition in the mineralized matrix evaluated by Fourier Transform Infrared microspectroscopy demonstrated that the apatite nanocrystals cover the collagen fibers assembled by the electrospinning.

1. Introduction

Titanium and its alloy are currently the mainly used materials for biomedical and dental implants for their good mechanical properties, high corrosion resistance, and excellent biocompatibility [1]. However, these materials are bioinert since they can nonspecifically downregulate biological responses. For this reason, the research has been directed to the surface modification of these materials to improve bone implant contact and their biological properties. Among the different materials used for the surface modification of Ti substrates, the best way could be the application of a coating that replaces the extracellular matrix (ECM) of bone, basically constituted of biomimetic apatite nanocrystals (HA) and collagen fibers [2].

The mineral phase of bone and teeth consists of plate-like shape nonstoichiometric carbonated-HA crystals with length of about 100 nm, width of 20–30 nm, thickness of 3–6 nm, and different types, and foreign ions substitutions including sodium, magnesium, potassium, carbonate, and fluoride [3, 4]. There are several methods to synthesise hydroxyapatite but mainly it is possible to distinguish two types of hydroxyapatite based on the temperature of synthesis: a high-temperature sintered form (>900°C) and a low-temperature precipitate (~37°C). The compounds synthesized at low-temperature exhibit more solubility, much greater surface area, and they are inherently more active in biological systems in contrast to the biologically inert sintered versions [5]. Sintered hydroxyapatite is most commonly used in load-bearing orthopaedic implants, on the other hand it is not bioresorbable as they are stable over the lifetime of the individual. Since the bone mineralization process occurred at basic pH and physiological temperature (37°C), the synthetic procedure that uses these parameters is called biomimetic mineralization. The final HA formed, in spite of having variable composition, is predominantly carbonated apatite with low crystallinity and it is very similar to the form found in bone, a favourable factor from a tissue engineering perspective [6].

Recent studies concerning collagen coatings on Ti implants have demonstrated their effective role in stimulating cellular responses, increasing bone remodelling, and improving bone growth and bone implant contact [7, 8]. Therefore, hybrids of collagen/HA have greater potential for clinical applications than the pure HA or collagen equivalents, because they benefit from the advantageous properties of both materials. In the composite, the collagen component can act as a matrix to embed the HA particles, alleviating the brittleness of HA and promoting the adhesion of the particles to Ti substrate [9]. At the same time, the HA used as a filler of the protein could improve the mechanical strength and biological performance of the collagen.

Electrospinning process has attracted a great deal of attention in the last years as a way to reproduce the structure of natural ECM by means of producing fibers in the nanometre scale [10]. This technique is used to fabricate nanofibrous structures from natural and synthetic polymers such as collagen, chitosan, silk fibroin, poly(lactide), polyurethane, polycaprolactone, and so forth [11–14]. Several papers reported the production of nanofibrous materials made of collagen mixed with biocompatible synthetic polymers using the electrospinning technique, as well as HA-polymeric fibers [15–18]. Most of these studies have demonstrated enhanced bioactivity of the composite materials, especially in nanofibrous constructs [19, 20]. However, the data on the production of pure nano-structured collagen fibers and HA-collagen composites by electrospining, as well as the data on the deposition of them on titanium alloy or on other orthopaedic implants, are extremely scarce or have not been investigated up to now.

The aim of this study was to develop a method to produce nanostructured collagen/HA composite with characteristics similar to natural-bone ECM on titanium alloy substrates, by combining electrospinning and biomimetic mineralization, and to study the chemical composition distribution of the coating surface by using the Fourier Transform Infrared microspectroscopy.

2. Materials and Methods

2.1. Materials. Common high-purity chemical reagents were supplied from Sigma Aldrich (Milan, Italy). Ultrapure water (0.22 mS, 25°C) was used in all experiments.

2.2. Processing of Pure Collagen Scaffolds. The electrospinning apparatus used in this work consists of three components: (i) spinneret, (ii) collector, and (iii) high-voltage power system. The spinneret is directly connected to a syringe, which acts as a reservoir for the collagen solution. Collagen (type I, extracted from equine Achilles tendon using the manufacturing method of Opocrin S.p.A. (Corlo di Formigine, Italy) [21]) was dissolved into CH_3COOH 0.1 M at the concentration of 2.5 wt%. The solution was fed into a 10 mL glass syringe connected to a stainless steel needle using a Teflon tube having an inner diameter of about 1.0 mm, controlled by a syringe pump (KDS200, KD Scientific Inc., USA), at a feeding rate of 0.01 mL/min. A plate-type titanium alloy (ASTMF136 ELI, Titanium International Group S.r.l., Sala Bolognese, Italy) was used as collector as well as substrate for the electrospun fibers. The distance between the spinneret tip and the collector was adjusted to 18 cm.

A voltage of 15 kV was applied across the spinneret and the collector by a voltage-regulated DC power supply (SL150, Spellman High Voltage Electronics, USA) to generate the electrically charged collagen jet. The titanium coated with the electrospun collagen fibers was vacuum dried overnight before carrying out the characterizations.

2.3. Processing of Mineralized Collagen Scaffolds. The processing steps involved in the preparation of electrospun fibers of collagen-hydroxyapatite are schematically illustrated in Figure 1. The electrospinning of collagen was carried out by the same method described above, but for the preparation of mineralized collagen, the collector was immersed in 10 mL of an aqueous solutions of $Ca(CH_3COO)_2$ 8.5 mM and $NH_4H_2PO_4$ 5.1 mM at pH 3.0, keeping constant the Ca/P molar ratio of 1.67. At the end of collagen deposition, the reaction mixture was titrated with NaOH solution to pH 9. After aging for 15 min, the titanium coated with composite was repeatedly washed with ultrapure water and harvested by vacuum dried before carrying out the characterizations.

2.4. Scaffolds Characterizations. The surface morphology of the electrospun collagen and the collagen-HA scaffolds was examined by using a scanning electron microscope (SEM) (Carl-Zeiss EVO, 40 XVP microscope) using secondary electrons at 25 kV, equipped with an energy-dispersive detector for X-rays (EDAX; Inca 250, Oxford).

The inorganic phase was analyzed by X-ray diffraction (XRD) using a PanAnalytical X'Pert Pro equipped with an X'Celerator detector, using Cu Kα radiation generated at 40 kV and 40 mA. The instrument was configured with 1/2° divergence and receiving slits. The 2θ range was from 20° to 60° with a step size (°2θ) of 0.05 and a counting time of 3 s.

Crystal domain size along the HA axis directions were calculated applying the Scherrer equation:

$$L_{(hkl)} = \frac{0.94\lambda}{\left[\cos\theta\left(\sqrt{\Delta_r^2 - \Delta_0^2}\right)\right]}, \quad (1)$$

where θ is the diffraction angle for (hkl) plane, Δ_r and Δ_0 the widths in radians of reflection (hkl) at half height for the synthesized and pure hydroxyapatite (standard reference material, calcium hydroxyapatite, NIST), respectively, and $\lambda = 1.5405$ Å.

Thermogravimetric analysis (TGA) were carried out on dried samples removed from the titanium alloy plate using a Thermal Analysis SDT Q 600 (TA Instruments, New Castle, DE, USA). Heating was performed in a nitrogen flow (100 mL min^{-1}) using an alumina sample holder. The temperature was increased to 700°C using a heating rate of 10°C/min. The weight of the samples was approximately 3 mg.

Transmission electron microscopy (TEM) investigations were carried out using a CM 100 instrument (80 kV) (Philips, Eindhoven, The Netherlands). The samples were removed from the titanium alloy plate and suspended in ultrapure water. A drop of the coating material suspension was deposited on holey-carbon foils supported on conventional copper microgrids. The samples were observed after staining

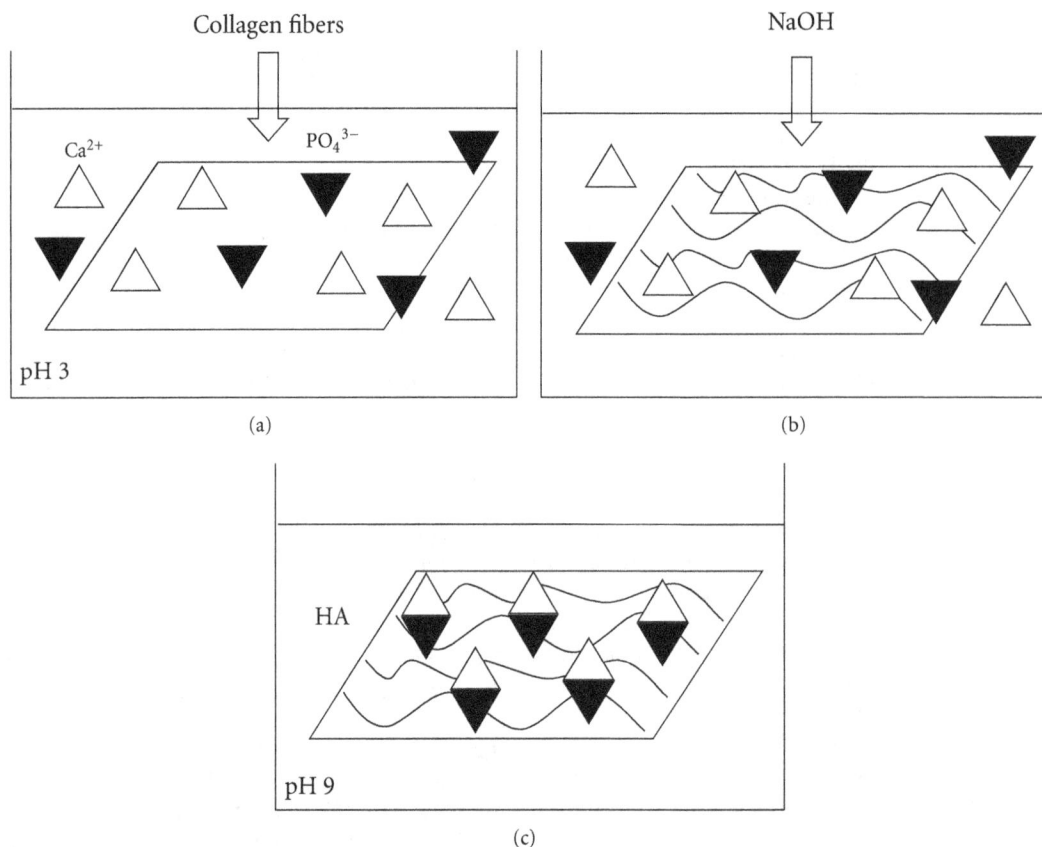

FIGURE 1: Scheme of the processing steps involved in the electrospun nanostructured collagen-hydroxyapatite fibers preparation: (a) electrospinning of collagen in CH_3COOH solution into an aqueous calcium and phosphate precursor solutions ($Ca(CH_3COO)_2$ and $NH_4H_2PO_4$, resp.) at pH 3; (b) hydrolysis with NaOH to pH 9; (c) aging the composite in reaction mixture for 15 min, followed by harvesting and vacuum drying.

at room temperature for 30 minutes with 0.5% aqueous uranyl acetate solution.

Infrared microscopy spectral data were recorded by a Perkin-Elmer Spectrum One Fourier Transform Infrared Spectrometer (FT-IR) equipped with a Perkin-Elmer Autoimage microscope. The spatial resolution was $50 \times 50\,\mu m^2$ and the spectral resolution was $4\,cm^{-1}$. The spatial resolution of $50 \times 50\,\mu m^2$ was chosen in order to optimize the signal-to-noise ratio and considering the high homogeneity of the samples. The spectra were collected in reflectance mode on titanium plate and they are related to the surface of the sample. Specific areas of interest were identified by a microscope television camera and for each one, an IR image was acquired using a liquid nitrogen cooled, 16-pixel mercury cadmium telluride (MCT-A) line detector at a $25\,\mu m$/pixel resolution. Baseline (polynomial line fit), smoothing, and Amide I normalizations were performed in all cases [22]. The correlation maps allow to evidence the chemical topographic distribution of a selected spectrum on the analyzed area. According to a colorimetric scale, the white colour corresponds to a zone with maximum absorption, while the dark colour refers to a zone in which the corresponding absorption band is absent [22].

2.5. *Statistic Analysis.* All experiments were performed at least three times. Determination of HA crystallite domain size along the the c-axis and along the perpendicular to it was carried out 5 times on the same synthesis product.

3. Results and Discussion

In order to maintain the fibrous structure of the polymer, electrospun collagen in the form of nanofibrous scaffolds was produced. The spinning conditions to obtain collagen scaffolds were optimized by manipulating the experimental parameters, such as concentration of solution, distance between the spinneret and the collector, and the flow rate, in order to produce fibrous structure without any bead formation. By optimizing so (see Section 2), uniform fibers with three-dimensional pore structure with high porosity, appropriate for self-assembling mineralization, have been formed. These findings are comparable to the results reported by Boland et al. [23] for electrospun type I collagen fibers with average fiber diameters ranging from about 100 nm to 4.6 μm depending on the concentration. In line with previous reports on other polymeric materials [24], we conclude that electrospinning of collagenous proteins

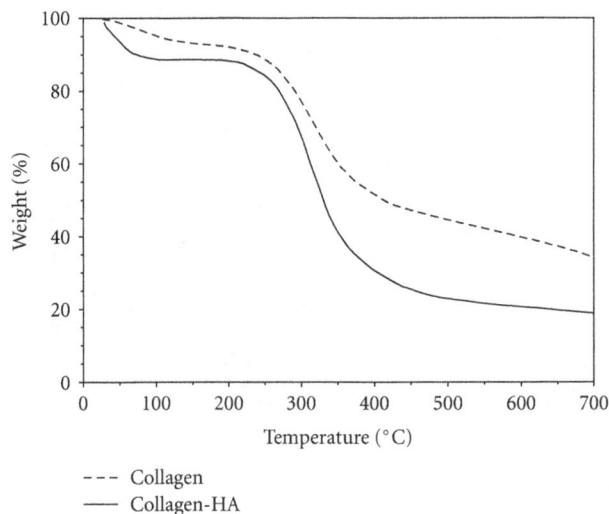

FIGURE 2: TGA curves of electrospun pure collagen (straight line) and collagen-hydroxyapatite composite (dotted line).

at concentrations above 2.5 wt% will yield smooth and uniform fibers of several hundred nanometers in diameters. By contrast, electrospinning at lower concentrations will result in small fibers, but with beads.

The self-assembly process to nucleate HA crystals depends on the negatively charged carboxyl chemical groups of collagen that can bind Ca^{2+} and on the pH of the reaction medium. The final pH value of 9.0 has been chosen in order to induce the HA crystallization in the optimal condition [25]. In order to optimize the HA crystallization, different experiments have been carried out using several calcium and phosphate salts at different concentrations by keeping the Ca/P ratio constant and equal to the HA stoichiometric value of 1.67. The use of $Ca(CH_3COO)_2$ 8.5 mM and $NH_4H_2PO_4$ 5.1 mM was found to be the optimal composition to precipitate HA crystals. Furthermore, these values have been chosen in order to preserve the fiber-like morphology, in fact, at the higher HA amounts, a considerable level of beads was formed instead of the development of a fibrous morphology [26]. It is reasonable that when the HA amount was high, the collagen could not effectively distribute the HA nanocrystals during the precipitation process due to the relatively low density of its amino acid chains. As a result, some of the HA nanocrystals could be precipitated in large clusters without the direct involvement of the amino acids of collagen. In order to quantify the amount of apatite in the mineralized collagen, TGA investigations were carried out (Figure 2). TGA curves of pure collagen and mineralized collagen showed similar profile with loss in the range from room temperature to 200°C due to the evaporation of physisorbed water and weight loss between 200 and 500°C associated with the decomposition of collagen molecules followed by a slight loss between 500 and 700°C resulting from the combustion of the residual organic components [27]. Considering the residue of pure collagen (~17 wt%), the apatitic content in the mineralized composite was determined as about 20 wt%.

The morphology of the electrospun material was studied by SEM, and the micrographs of the pure collagen and mineralized collagen scaffolds are shown in Figure 3. The electrospun fibers of pure collagen have randomly oriented features trough out the matrix, as well as the mineralized fibers. The average diameters of polymers fibers formed during electrospinning with and without HA, determined by SEM, were about 200 nm. These values are well comparable with ECM fibers, which are in the range of 50–500 nm [2]. Finally, the two materials appear very similar in terms of morphology and dimensions of the fibers, and the visualization of the HA crystals was not allowed by SEM probably due to their very small dimensions in the range of 20–30 nm [28].

The presence of deposited HA is evidenced by the EDAX spectrum collected for the mineralized sample (inset in Figure 3(b)). This spectrum indicates that the intensity ratio of Ca and P signals was about 1.5 coherently with the value of biological HA [3].

TEM images of the reconstituted electrospun collagen and collagen-hydroxyapatite are shown in Figures 4(a) and 4(b), respectively. The average diameters of polymers fibers were very similar (about 200 nm) in agreement with the SEM observations. The stained pure collagen fibers display a high degree of fibril assembly, evidencing the characteristic D-band pattern with a regular period of 67 nm along the long axis of the fibril [29]. On the contrary, the mineralized fibers do not show the D-band pattern due to the presence of HA. Figure 4(b) clearly illustrates the presence of HA nanocrystals of about 20–30 nm in size that completely cover the surface of collagen fibers.

To clearly identify the mineral phase developed during the mineralization of the electrospun collagen fibers, the XRD was employed and the pattern is shown in Figure 5. All of the mineral diffraction peaks were indexed to HA pure phase (JCPDS file number 9-432) and no other impurity phases were detected. The most intense peaks at 43.7° and 50.8° 2θ are due to the titanium alloy substrate (marked with * in Figure 5) [30]. The diffraction pattern of the HA exhibits not well-defined diffraction maxima, in fact the peaks corresponding to the crystallographic planes (211), (121), and (300) were all combined into one broad peak centred at about 32° 2θ, indicating a relatively low degree of crystallinity and nanosized dimensions [4]. The crystal domain sizes along the c-axis (D_{002}) and along the perpendicular to it (D_{310}), using the $2\theta = 26°$ (002) and $2\theta = 39°$ (310) diffraction peaks, respectively, were 25 ± 5 nm and 11 ± 3 nm. These values are in agreement with the TEM observations. It is worth to notice that the HA diffraction pattern was very similar to that recorded for the deproteinated bone apatite and this was also confirmed by the similarity of the crystal domain size along the c-axis (21 nm) [3].

The FT-IR spectra of electrospun pure collagen and mineralized collagen are depicted in Figure 6. The spectrum of electrospun collagen shows all the characteristic bands of collagen at 1204, 1240, 1280, and 1338 cm^{-1} arising from C–OH stretching modes and 1400 and 1450 cm^{-1} associated with the asymmetric and symmetric CH_3 bending vibrations

(a)   (b)

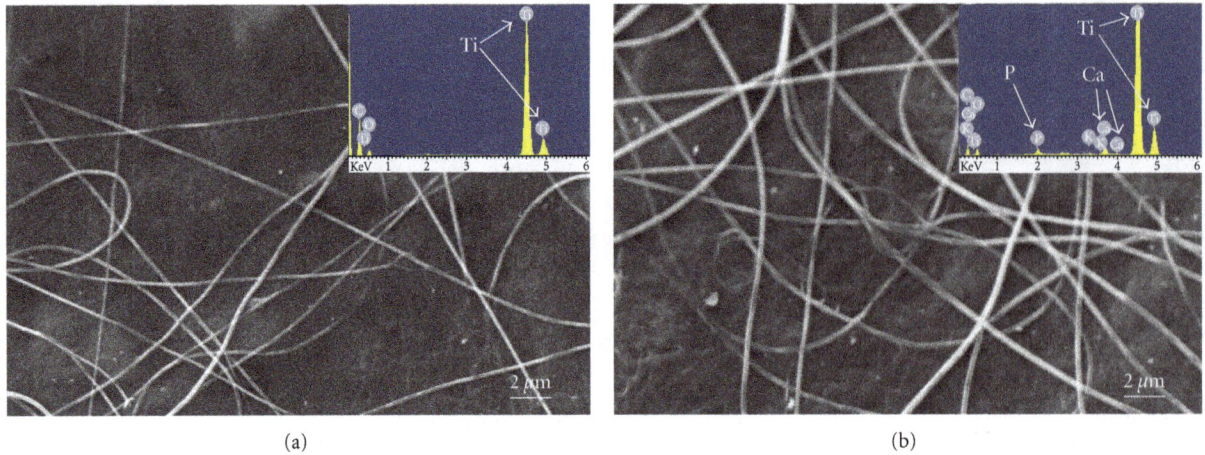

FIGURE 3: Scanning electron microscopy (SEM) images and energy dispersive X-rays (EDAX) analysis of electrospun pure collagen (a) and collagen-hydroxyapatite composite (b).

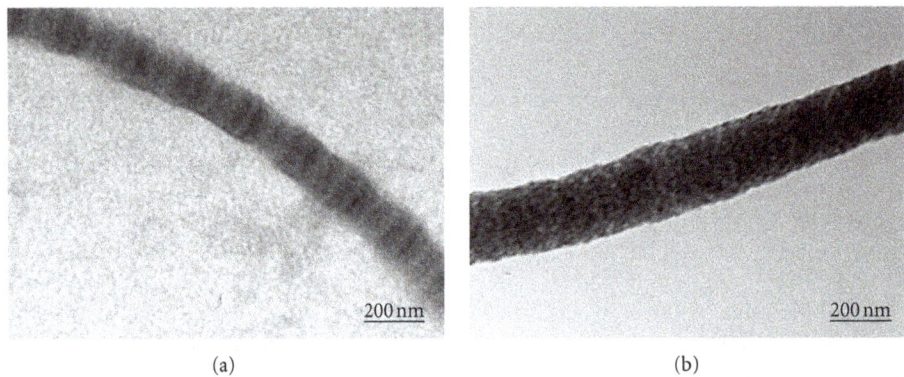

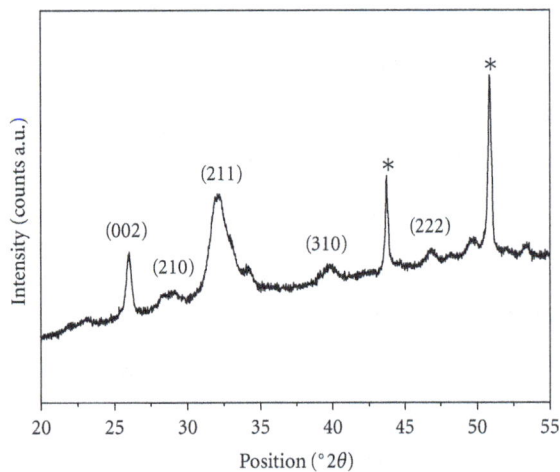

(a)   (b)

FIGURE 4: Transmission electron microscopy (TEM) images of electrospun pure collagen (a) and collagen-hydroxyapatite composite (b).

FIGURE 5: X-Ray diffraction (XRD) pattern of collagen-hydroxyapatite composite; *indicates the Ti alloy diffraction peaks.

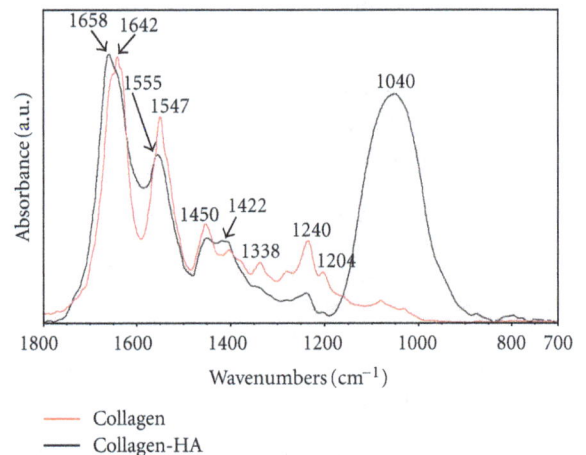

FIGURE 6: FT-IR spectra of electrospun pure collagen (red) and collagen-hydroxyapatite composite (black) in the region 1800–700 cm^{-1}.

[4]. The absorbance ratio of the bands 1240/1450 cm^{-1} which is a measure of the integrity of the triple-helix structure [31] was about 1.0 suggesting that collagen triple helix secondary structure was conserved. The spectra of collagen-HA composite displays the bands of poorly crystalline

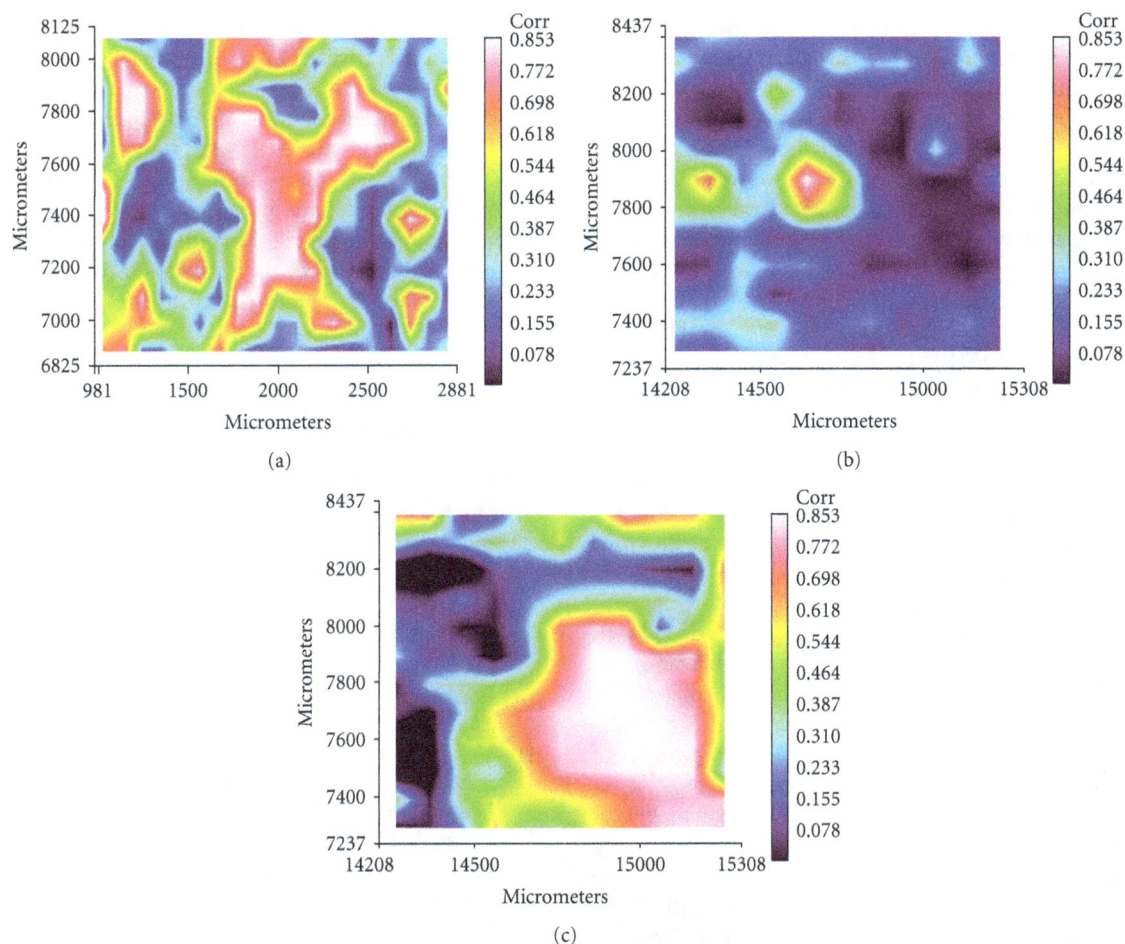

(a)

(b)

(c)

FIGURE 7: Correlation maps obtained by loading the representative FT-IR spectrum of collagen (Figure 6, red spectrum) in the chemical maps of electrospun pure collagen (a) and mineralized collagen (b) and correlation map obtained by loading the FT-IR spectrum of collagen-hydroxyapatite composite (Figure 6, black spectrum) in the chemical map of mineralized collagen (c).

carbonate-HA, a convoluted band centred at $1040 \, cm^{-1}$ (asymmetric stretching modes of the phosphate groups) and a band at $1422 \, cm^{-1}$ (antisymmetric stretching mode of C–O, consistent with a carbonate type-B-substituted apatite, where the carbonate ions replace the phosphate ions in the crystal lattice) [4], while the bands of collagen change their positions and their relative intensities. This finding could indicate that during the mineralization procedure, the collagen secondary structure has been modified by the effective interaction with HA nanocrystals. The limited amount of carbonate in the HA derived from CO_2 dissolved in the preparation media and adsorbed on the surface materials during the previous storage. The presence of carbonate in the structure of apatite was intentionally retained, in order to better mimic the biological ones [3]. In a previous FT-IR study comparing "bulk" collagen with electrospun collagen scaffolds, Stanishevsky et al. [32] reported a shift to lower wavenumbers in electrospun collagen of the major peak in the amide I band from $1667 \, cm^{-1}$ (fibers) to $1642 \, cm^{-1}$ (bulk), as well as the peak of amide II band from $1574 \, cm^{-1}$ (fibers) to $1544 \, cm^{-1}$ (bulk). Similar amide I and amide II peak shifts for our pure collagen scaffolds were observed

(the wavelengths of the bands related to amide I and amide II are 1642 and $1547 \, cm^{-1}$, resp.). It was suggested that changes in the triple-helix structure of collagen fibrils during the fiber drawing in the electrified jet may account for this infrared shift. Other changes in the infrared spectra were reported (such as a blue shift of the amide I band) by the addition of apatite nanoparticles to the electrospun collagen fibers [32]. Similar peak shifts in the amide I and amide II of collagen-apatite scaffolds were clearly observed (1642 to $1658 \, cm^{-1}$ and 1547 to $1555 \, cm^{-1}$, resp.). The observed changes in the infrared spectra are associated with the chemical interaction between the apatite nanoparticles and collagen, in particular they are due to the chemical link between the collagen carboxyl groups and the calcium of HA.

FT-IR microspectroscopy was used to analyze the electrospun coatings with the aim to map and characterize the chemical distribution of the sample. FT-IR microspectroscopy allows to define the structural interactions between HA nanocrystals and reconstituted collagen fibers. Moreover, this technique is able to verify the topographic and quantitative distribution of the components in the sample.

In Figures 7(a) and 7(b), the correlation maps obtained by loading the representative spectrum of collagen (Figure 6, red spectra) in the chemical maps of electrospun pure collagen ($1300 \times 1900\,\mu m$) and mineralized collagen ($1200 \times 1100\,\mu m$) samples, respectively, are shown. The predominance of a dark area in the Figure 7(b) is related to the absence of the pure collagen, due to the mineralization of all the collagen fibers. On the contrary, Figure 7(c) shows the correlation map obtained by loading the spectrum of collagen-HA (Figure 6, black spectra) in the same chemical map of mineralized collagen sample ($1200 \times 1100\,\mu m$) represented in Figure 7(b). In this case, the representative bands of HA prevail on those of pure collagen and the large white area allows to visualize the crystallization of HA over the protein.

It is well accepted that the formation of HA on the surfaces of electrospun fibers was controlled by the charged density of chemical groups on the fiber surface. Previous investigations about the nucleation sites of HA crystals on collagen fibers have suggested that the binding of calcium ions on the negatively charged carboxylate groups of collagen is one of the key factors for the first-step nucleation of HA crystals [33]. Collagen has large number of negatively charged carboxyl chemical groups that can bind Ca^{2+} ions to nucleate the HA crystal growth. The carboxyl groups are present in about 11% of the amino acid residues of collagen molecules. Moreover, in a neutral solution, more than 99% of the carboxyl groups of aspartyl and glutamyl ionize favouring the chelation of calcium ions. The carboxyl groups on the outside of the collagen threefold spiral are one kind of site for collagen mineralization [34]. In this way, higher content and smaller crystal size of HA could be formed on fibers with higher densities of carboxyl groups. Carboxyl groups were initially combined with calcium ions through electrostatic attraction, and then phosphoric ions, which were considered to promote the nucleation of HA, until the whole coverage of the fiber with the primary layer of HA was completed. This hypothesis is well supported in this work by the FTIR and TEM investigations and, in particular, by the blue shift of the amide I and amide II bands in the spectra of mineralized collagen. Moreover, by using FT-IR microspectroscopy, we found that part of the HA crystallized outer the fibers, coherently with the fact that the electrospinning technique allows the formation of the collagen fibers and in this case, the carboxyl groups outside the fibers are responsible for the HA nucleation.

4. Conclusion

In this paper, a new method to produce collagen-HA composites on titanium alloy substrate, by using electrospinning and biomimetic mineralization techniques, has been developed. The results show that the mineralized scaffolds have the most desirable morphological, structural, and chemical compositional features similar to natural bone ECM, which is a good result of the efficacy of the combinational method of this study. The topographic distribution of the coating components has been evaluated by using the FT-IR microspectroscopy, highlighting the fact that with this method the scaffold is composed by collagen fibers, assembled by the electrospinning, covered by poorly crystalline HA nanosized crystals (20–30 nm in size). The spectroscopic investigation has also suggested that the surface functional groups of the collagen-based material, such as carboxyl groups, are crucial for the mineralization *in vitro*. These fibrous nanocomposites should have potential applications as coating materials to improve the biological performances of the titanium medical devices, as scaffolds for tissue engineering, and as fillers for fiber-enforced composites with the aim to improve the bone implant contact and the interaction with the biological environment.

Acknowledgments

The authors would like to thank the Inter University Consortium for Research on Chemistry of Metals in Biological Systems (CIRCMSB) and the Regione Piemonte (Progetto Ricerca Sanitaria Finalizzata 2009, Grant no. 30258/DB2001). M. Iafisco is the recipient of a fellowship from Regione Piemonte.

References

[1] Y. Ramaswamy, C. Wu, and H. Zreiqat, "Orthopedic coating materials: considerations and applications," *Expert Review of Medical Devices*, vol. 6, no. 4, pp. 423–430, 2009.

[2] S. Liao, R. Murugan, C. K. Chan, and S. Ramakrishna, "Processing nanoengineered scaffolds through electrospinning and mineralization suitable for biomimetic bone tissue engineering," *Journal of the Mechanical Behavior of Biomedical Materials*, vol. 1, no. 3, pp. 252–260, 2008.

[3] N. Roveri, B. Palazzo, and M. Iafisco, "The role of biomimetism in developing nanostructured inorganic matrices for drug delivery," *Expert Opinion on Drug Delivery*, vol. 5, no. 8, pp. 861–877, 2008.

[4] M. Iafisco, J. G. Morales, M. A. Hernández-Hernández, J. M. García-Ruiz, and N. Roveri, "Biomimetic carbonate-hydroxyapatite nanocrystals prepared by vapor diffusion," *Advanced Engineering Materials*, vol. 12, no. 7, pp. B218–B223, 2010.

[5] M. Iafisco, E. Varoni, E. Battistella et al., "The cooperative effect of size and crystallinity degree on the resorption of biomimetic hydroxyapatite for soft tissue augmentation," *International Journal of Artificial Organs*, vol. 33, no. 11, pp. 765–774, 2010.

[6] P. A. Madurantakam, I. A. Rodriguez, C. P. Cost et al., "Multiple factor interactions in biomimetic mineralization of electrospun scaffolds," *Biomaterials*, vol. 30, no. 29, pp. 5456–5464, 2009.

[7] H.-W. Kim, L.-H. Li, E.-J. Lee, S.-H. Lee, and H.-E. Kim, "Fibrillar assembly and stability of collagen coating on titanium for improved osteoblast responses," *Journal of Biomedical Materials Research A*, vol. 75, no. 3, pp. 629–638, 2005.

[8] S. Rammelt, T. Illert, S. Bierbaum, D. Scharnweber, H. Zwipp, and W. Schneiders, "Coating of titanium implants with collagen, RGD peptide and chondroitin sulfate," *Biomaterials*, vol. 27, no. 32, pp. 5561–5571, 2006.

[9] S. H. Teng, E. J. Lee, C. S. Park, W. Y. Choi, D. S. Shin, and H. E. Kim, "Bioactive nanocomposite coatings of collagen/hydroxyapatite on titanium substrates," *Journal of Materials Science*, vol. 19, no. 6, pp. 2453–2461, 2008.

[10] W. E. Teo, W. He, and S. Ramakrishna, "Electrospun scaffold tailored for tissue-specific extracellular matrix," *Biotechnology Journal*, vol. 1, no. 9, pp. 918–929, 2006.

[11] A. Nandakumar, H. Fernandes, J. de Boer, L. Moroni, P. Habibovic, and C. A. van Blitterswijk, "Fabrication of bioactive composite scaffolds by electrospinning for bone regeneration," *Macromolecular Bioscience*, vol. 10, no. 11, pp. 1365–1373, 2010.

[12] S.-H. Teng, E.-J. Lee, P. Wang, and H.-E. Kim, "Collagen/hydroxyapatite composite nanofibers by electrospinning," *Materials Letters*, vol. 62, no. 17-18, pp. 3055–3058, 2008.

[13] S. A. Martel-Estrada, C. A. Martínez-Pérez, J. G. Chacón-Nava, P. E. García-Casillas, and I. Olivas-Armendáriz, "*In vitro* bioactivity of chitosan/poly (d,l-lactide-co-glycolide) composites," *Materials Letters*, vol. 65, no. 1, pp. 137–141, 2011.

[14] I. Foltran, E. Foresti, B. Parma, P. Sabatino, and N. Roveri, "Novel biologically inspired collagen nanofibers reconstituted by electrospinning method," *Macromolecular Symposia*, vol. 269, no. 1, pp. 111–118, 2008.

[15] N. Hild, O. D. Schneider, D. Mohn et al., "Two-layer membranes of calcium phosphate/collagen/PLGA nanofibres: *in vitro* biomineralisation and osteogenic differentiation of human mesenchymal stem cells," *Nanoscale*, vol. 3, no. 2, pp. 401–409, 2011.

[16] C. Huang, R. Chen, Q. Ke, Y. Morsi, K. Zhang, and X. Mo, "Electrospun collagen-chitosan-TPU nanofibrous scaffolds for tissue engineered tubular grafts," *Colloids and Surfaces B*, vol. 82, no. 2, pp. 307–315, 2011.

[17] Y. Zhang, J. R. Venugopal, A. El-Turki, S. Ramakrishna, B. Su, and C. T. Lim, "Electrospun biomimetic nanocomposite nanofibers of hydroxyapatite/chitosan for bone tissue engineering," *Biomaterials*, vol. 29, no. 32, pp. 4314–4322, 2008.

[18] J. Zhou, C. Cao, X. Ma, and J. Lin, "Electrospinning of silk fibroin and collagen for vascular tissue engineering," *International Journal of Biological Macromolecules*, vol. 47, no. 4, pp. 514–519, 2010.

[19] M. Yeo, H. Lee, and G. Kim, "Three-dimensional hierarchical composite scaffolds consisting of polycaprolactone, β-tricalcium phosphate, and collagen nanofibers: fabrication, physical properties, and *in vitro* cell activity for bone tissue regeneration," *Biomacromolecules*, vol. 12, no. 2, pp. 502–510, 2011.

[20] H. M. Kim, W.-P. Chae, K.-W. Chang et al., "Composite nanofiber mats consisting of hydroxyapatite and titania for biomedical applications," *Journal of Biomedical Materials Research B*, vol. 94, no. 2, pp. 380–387, 2010.

[21] B. Parma, "Collagen membrane arranged at macromolecular level," European patent EP1307247, 2003.

[22] S. Manara, F. Paolucci, B. Palazzo et al., "Electrochemically-assisted deposition of biomimetic hydroxyapatite-collagen coatings on titanium plate," *Inorganica Chimica Acta*, vol. 361, no. 6, pp. 1634–1645, 2008.

[23] E. D. Boland, J. A. Matthews, K. J. Pawlowski, D. G. Simpson, G. E. Wnek, and G. L. Bowlin, "Electrospinning collagen and elastin: preliminary vascular tissue engineering," *Frontiers in Bioscience*, vol. 9, no. 2, pp. 1422–1432, 2004.

[24] H. Fong and D. H. Reneker, "Elastomeric nanofibers of styrene-butadiene-styrene triblock copolymer," *Journal of Polymer Science B*, vol. 37, no. 24, pp. 3488–3493, 1999.

[25] M. Iafisco, M. Marchetti, J. G. Morales, M. A. Hernández-Hernández, J. M. G. Ruiz, and N. Roveri, "Silica gel template for calcium phosphates crystallization," *Crystal Growth and Design*, vol. 9, no. 11, pp. 4912–4921, 2009.

[26] J.-H. Song, H.-E. Kim, and H.-W. Kim, "Electrospun fibrous web of collagen-apatite precipitated nanocomposite for bone regeneration," *Journal of Materials Science: Materials in Medicine*, vol. 19, no. 8, pp. 2925–2932, 2008.

[27] A. Tampieri, G. Celotti, E. Landi, M. Sandri, N. Roveri, and G. Falini, "Biologically inspired synthesis of bone-like composite: self-assembled collagen fibers/hydroxyapatite nanocrystals," *Journal of Biomedical Materials Research A*, vol. 67, no. 2, pp. 618–625, 2003.

[28] B. Palazzo, M. Iafisco, M. Laforgia et al., "Biomimetic hydroxyapatite-drug nanocrystals as potential bone substitutes with antitumor drug delivery properties," *Advanced Functional Materials*, vol. 17, no. 13, pp. 2180–2188, 2007.

[29] H. M. H. F. Sanders, M. Iafisco, E. M. Pouget et al., "The binding of CNA35 contrast agents to collagen fibrils," *Chemical Communications*, vol. 47, no. 5, pp. 1503–1505, 2011.

[30] H.-H. Sheu, L.-C. Hsiung, and J.-R. Sheu, "Synthesis of multiphase intermetallic compounds by mechanical alloying in Ni-Al-Ti system," *Journal of Alloys and Compounds*, vol. 469, no. 1-2, pp. 483–487, 2009.

[31] R. Sripriya, R. Kumar, S. Balaji, M. Senthil Kumar, and P. K. Sehgal, "Characterizations of polyanionic collagen prepared by linking additional carboxylic groups," *Reactive and Functional Polymers*, vol. 71, no. 1, pp. 62–69, 2011.

[32] A. Stanishevsky, S. Chowdhury, P. Chinoda, and V. Thomas, "Hydroxyapatite nanoparticle loaded collagen fiber composites: microarchitecture and nanoindentation study," *Journal of Biomedical Materials Research A*, vol. 86, no. 4, pp. 873–882, 2008.

[33] W. Cui, X. Li, C. Xie, H. Zhuang, S. Zhou, and J. Weng, "Hydroxyapatite nucleation and growth mechanism on electrospun fibers functionalized with different chemical groups and their combinations," *Biomaterials*, vol. 31, no. 17, pp. 4620–4629, 2010.

[34] W. Zheng, W. Zhang, and X. Jiang, "Biomimetic collagen nanofibrous materials for bone tissue engineering," *Advanced Engineering Materials*, vol. 12, no. 9, pp. B451–B466, 2010.

3

Binding Studies of a New Water-Soluble Iron(III) Schiff Base Complex to DNA Using Multispectroscopic Methods

Nahid Shahabadi, Zeinab Ghasemian, and Saba Hadidi

Department of Inorganic Chemistry, Faculty of Chemistry, Razi University, Kermanshah 74155, Iran

Correspondence should be addressed to Nahid Shahabadi, nahidshahabadi@yahoo.com

Academic Editor: Giovanni Natile

A novel iron(III) complex [Fe(SF)](ClO$_4$)$_3$.2H$_2$O; in which SF = N,N$_0$-bis{5-[(triphenylphosphonium chloride)-methyl] salicylidene }-o-phenylenediamine) has been synthesized and characterized using different physicochemical methods. The binding of this complex with calf thymus (CT) DNA was investigated by circular dichroism, absorption studies, emission spectroscopy, voltammetric studies, and viscosity measurements. The results showed that this complex can bind to DNA via external and groove binding modes.

1. Introduction

Schiff bases, characterized by the azomethine group (–RC=N–), form a significant class of compounds in medicinal and pharmaceutical chemistry and are known to have biological applications due to their antibacterial [1–6], antifungal [3–6], and antitumor [7, 8] activity. Schiff base ligands are considered "privileged ligands" because they are easily prepared by the condensation between aldehydes and imines. The incorporation of transition metals into these compounds leads to the enhancement of their biological activities and decrease in the cytotoxicity of both metal ion and Schiff base ligand [9–11]. Schiff bases with donors (N, O, S, etc.) have structural similarities with neutral biological systems and due to presence of imine group are utilized in elucidating the mechanism of transformation of racemization reaction in biological system [12–14]. Those play an important role in inorganic chemistry as they easily form stable complexes with most transition metal ions.

The structural diversity of transition metal complexes of Schiff base ligands and the structure function relationships of the resulting complexes have been the focus of extensive research in recent years [15–21]. Metal complexes have been widely applied in clinics for centuries, although their molecular mechanism has not yet been entirely understood [22, 23]. The binding and reaction of metal complexes with DNA have been the subjects of intense investigation in relation to the development of new reagents for biotechnology and medicine [24–30]. To understand such functions and to design new DNA-binding metal complexes, investigations of the DNA-binding structures of the complexes are inevitably necessary. Investigations on the interaction between transition metal complexes and DNA has attracted many interests due to their importance in cancer therapy and molecular biology [31–39].

It is well known that some drugs have higher activity when administered as metal complexes than as free ligands [40]. Among them less explored iron-based Schiff base complexes possess many advantages over traditional catalyst due to iron's copious, nontoxic, and inexpensive nature [41]. In this context, we studied the interaction of a new iron(III) complex of a Schiff base ligand (Figure 1) with calf thymus DNA using physicochemical methods.

2. Experimental

2.1. Materials. O-phenylenediamine, salicylaldehyde, p-formaldehyde, triphenylphosphine, absolute ethanol, iron nitrate nona hydrate, diethyl ether, dichloromethane, ethylacetate, NaClO$_4$·H$_2$O, EDTA, Tris-HCl, NaOH, potassium

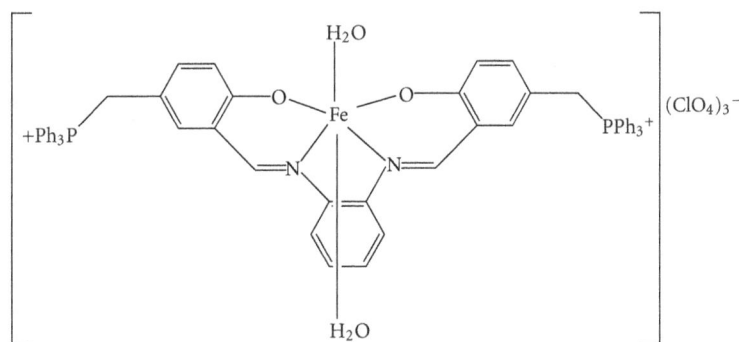

FIGURE 1: The molecular structure of iron(III) complex.

acetate, and chloroform, were purchased from Merck. Doubly distilled deionized water was used throughout. Highly polymerized CT-DNA and Tris-HCl buffer were purchased from Sigma.

Experiments were carried out in Tris-HCl buffer at pH 7.2. Solutions of CT-DNA gave a UV absorbance ratio (260 over 280 nm) of more than 1.8, indicating that the DNA was sufficiently free of protein. The stock solution of CT-DNA was prepared by dissolving approximately 1-2 mg of CT-DNA fibers in 2 mL Tris-HCl (10 mM) by shaking gently and stored for 24 h at 4°C. The DNA concentration (monomer units) of the stock solution (1×10^{-2} M per nucleotide) was determined by UV spectrophotometry in properly diluted samples using a molar absorption coefficient of $6600\,M^{-1}\,cm^{-1}$ at 258 nm. DNA solutions were used after no more than 4 days.

2.2. Instrumentation. The elemental analysis was performed using a Heraeus CHN elemental analyzer. Absorbance spectra were recorded using an HP spectrophotometer (Agilent 8453) equipped with a thermostated bath (Huber polysat cc1). Absorption titration experiments were conducted by keeping the concentration of complex constant (5×10^{-5} M) while varying the DNA concentration from 0 to 2.5×10^{-4} M (ri = [DNA]/[complex] = 0.0, 0.15, 0.35, 0.4, 0.55, 0.65, and 0.7). Absorbance values were recorded after each successive addition of DNA solution, followed by an equilibration period. CD measurements were recorded on a JASCO (J-810) spectropolarimeter, keeping the concentration of DNA constant (8×10^{-5} M) while varying the complex concentration (ri = [complex]/[DNA] = ri = 0, 0.2, 0.4, 0.6, 0.8, and 1). Viscosity measurements were made using a viscosimeter (SCHOT AVS 450) maintained at $25.0 \pm 0.5°C$ using a constant temperature bath. The DNA concentration was fixed at 5×10^{-5} M, and flow time was measured with a digital stopwatch. The mean values of three measurements were used to evaluate the viscosity η of the samples. The values for relative specific viscosity $(\eta/\eta_0)^{1/3}$, where η_0 and η are the specific viscosity contributions of DNA in the absence (η_0) and in the presence of the complex (η), were plotted against ri (ri = [DNA]/[complex] = 0.0, 0.2, 0.4, 0.6, 0.8, and 1.0). All fluorescence measurements were carried out with a JASCO spectrofluorimeter (FP6200) by keeping the concentration of complex constant while varying the DNA concentration from 0 to 5×10^{-5} M (ri = [DNA]/[complex] = 0.0, 0.08, 0.16, 0.32, 0.4, 0.48, 0.55, 0.62, 1, and 1.1) at three different temperatures (298, 310, and 288 K). The cyclic voltammetric (CV) measurements were performed using an μ-AUTOLAB model (PG STAT C), with a three-electrode system: a 0.10 cm diameter Glassy carbon (GC) disc as working electrode, an Ag/AgCl electrode as reference electrode, and a Pt wire as counter electrode. Electrochemical experiments were carried out in a 25 mL voltammetric cell at room temperature. All potentials are referred to the Ag/AgCl reference. Their surfaces were freshly polished with 0.05 mm alumina prior to each experiment and were rinsed using double distilled water between each polishing step. The supporting electrolyte was 0.05 M of Tris-HCl buffer solution (pH 7.4) which was prepared with double distilled water. The current-potential curves and experimental data were recorded on software NOVA 1.6.

2.3. Synthesis of the Iron(III) Schiff Base Complex. To a vigorously stirred solution (10 mL) of 3-formyl-4 hydroxybenzyl-triphenyl phosphonium chloride (0.12 g, 2 mmol) was first added $Fe(NO_3)_3 \cdot 9H_2O$ (0.136 g, 2 mmol) dissolved in water (l0 mL) and then Et_3N (pH 7.5–8). A few minutes later, an ethanolic solution (2 mL) of o-phenylenediamine (0.015 g, 1 mmol) was added drop wise. The solution turned to dark brown and the mixture was vigorously stirred and refluxed for 3 h under an argon atmosphere. After that, $NaClO_4 \cdot H_2O$ (2 mmol) was dissolved in a minimum amount of water (5 mL) and added to the reaction mixture. The resulting brown powder precipitate was collected by filtration, washed with cold ethanol and ether, and dried in air. The complex was dissolved in CH_2Cl_2 (6 mL), Ethylacetate (3 mL) was added and the resulting solution was allowed to evaporate slowly at room temperature. The solid product was filtered off and washed with ethyl acetate and ether yielding the iron complex as a brown powder.

3. Results and Discussion

3.1. Synthesis and Characterization of Iron(III) Complex. The iron(III) complex of SF ligand (SF = N, N_0-bis{5-[(triphenylphosphonium chloride)-methyl] salicylidene}-o-phenylenediamine) synthesized by mixing 3-formyl-4 hydroxybenzyl-triphenyl phosphonium chloride, Fe(III) nitrate and o-phenylenediamine in high yield. FT-IR spectrum of the complex was recorded in KBr pellets from 4000 to 400 cm^{-1}. The FT-IR spectrum of the Schiff base was characterized by the appearance of a band at 3424 cm^{-1} due to the (OH) group [42]. In the FT-IR spectrum of the iron(III) complex, the absence of this band indicates deprotonation of the donor and involvement oxygen in bonding with the metal atom. However, the strong band at ca. 1627 cm^{-1} assigned to (CN) of the Schiff base ligand [43] which shifts approximately 16 cm^{-1} to the lower wave numbers upon coordination to metal [42]. Since the iron(III) is paramagnetic in nature, its NMR spectrum could not be obtained. Elemental analysis also confirms the synthesis of the complex. $FeC_{58}H_{52}N_2P_2O_{16}Cl_3$ anal. calc.: C, 55.39; H, 4.13; N, 2.22. Found: C, 55.21; H, 4.30; N, 2.45.

The electronic spectrum of the complex shows the band at 260 nm is attributable to intramolecular $\pi \rightarrow \pi^*$ transition. Another band at 340 nm could be attributed to metal to ligand charge transfer. The other weaker bands are attributed to $d \rightarrow d$ transitions of metal center. Molar conductance of the complex in DMF is 289 $\Omega^{-1}\, cm^2\, mol^{-1}$ that is indicative of 3 : 1 electrolytic nature of the complex. Conclusive evidence of the bonding was also shown by the observation that new bands in the FT-IR spectra of the metal complex appearing in the low frequency region at 420–518 cm^{-1} characteristic to -(M–O) and -(M–N) stretching vibrations respectively that were not observed in the spectrum of free ligand.

3.2. Viscosity Measurements.

Hydrodynamic methods that are sensitive to length are regarded as one of the least ambiguous and most critical tests of a binding mode in solution in the absence of crystallographic structural data. Intercalating agents are expected to elongate the double helix to accommodate the ligands in between the base leading to an increase in the viscosity of DNA. In contrast, complex that bind exclusively in the DNA grooves by partial and/or non-classical intercalation, under the same conditions, typically cause less pronounced (positive or negative) or no change in DNA solution viscosity [44]. The values of $(\eta/\eta_0)^{1/3}$ were plotted against [complex]/[DNA] (Figure 2). With the increase in the amount of the complex, the relative viscosity of DNA increases steadily which is consistent with DNA groove binding which is also known to enhance DNA viscosity [42].

As intercalation causes a significant increase in viscosity of DNA solutions due to lengthening of the DNA helix as base pairs are separated to accommodate the aromatic chromophore of the bound molecule, it is tempting to ascribe the observed increase in viscosity to intercalative interaction of the complex. However, the effective intercalation of the phenyl rings of SF in iron complex is discouraged by several steric factors. This type of steric clash has been also suggested for the binding of $[Cu(phen)_2]^+$ and $[Ru(phen)_3]^{2+}$ complexes, where intercalation of a phen ligand is inhibited by steric interactions involving the other phen ligands with DNA surface. Therefore, it is obvious that the complex prefers to engage in DNA surface binding with

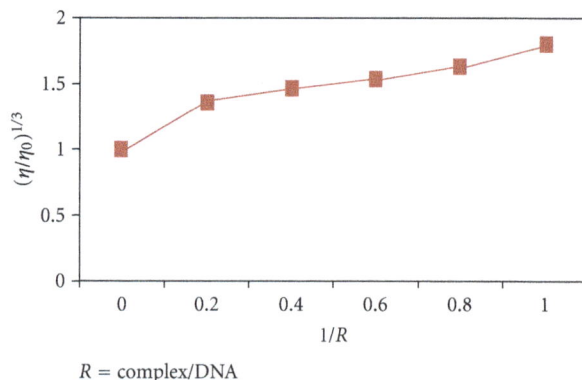

FIGURE 2: Effect of increasing amounts of iron complex on the viscosity of calf thymus DNA (5×10^{-5} M) in 10 mM Tris HCl buffer (ri = 0.2, 0.4, 0.6, 0.8, and 1).

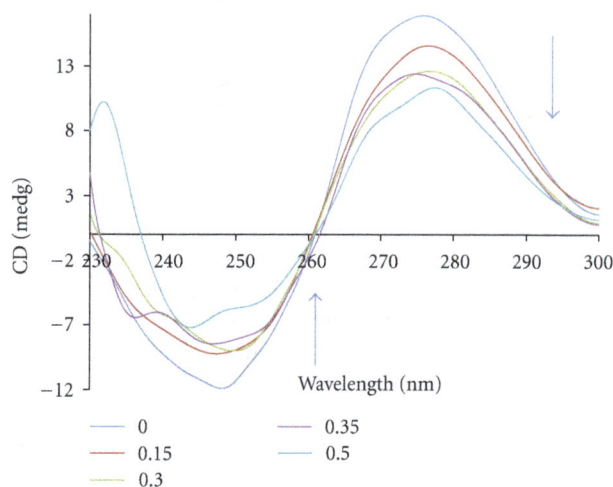

—— 0	—— 0.35
—— 0.15	—— 0.5
—— 0.3	

FIGURE 3

its large size affecting an increase in DNA viscosity, rather than in intercalative DNA interaction. The slow increase in viscosity is an indication of groove binding [45, 46].

3.3. CD Spectral Studies.

Circular dichroic spectral techniques give us useful information on how the conformation of DNA is influenced by the binding of the metal complex to DNA. The observed CD spectrum of calf-thymus DNA consists of a positive band at 277 nm due to base stacking and a negative band at 245 nm due to helicity, which is characteristic of DNA in the right-handed B form. While groove binding and electrostatic interaction of small molecules with DNA show little or no perturbations on the base stacking and helicity bands, intercalation enhances the intensities of both the bands, stabilizing the right-handed B conformation of CT-DNA. The interaction between the complex and CT-DNA was studied by CD spectroscopy (Figure 3).

In this case, the intensities of both the negative and positive bands decrease significantly (shifting to zero levels). This suggests that the DNA binding of the complex induces certain conformational changes, such as the conversion

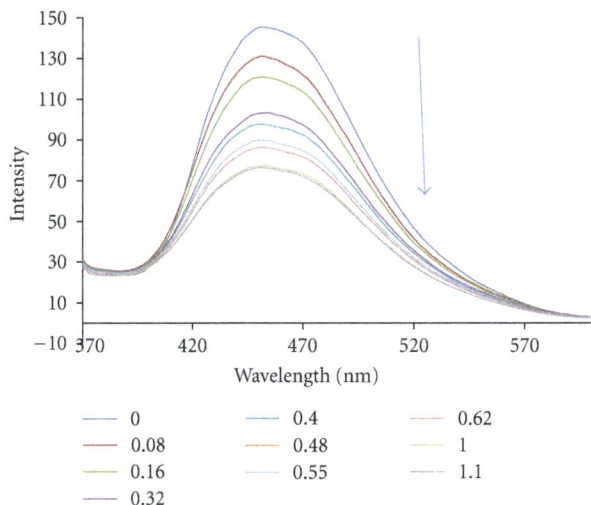

FIGURE 4

TABLE 1: The quenching constants of the Fe(III) complex by CT-DNA at different temperatures.

Temperature (K)	R^2	K_{SV} (L mol^{-1}) $\times 10^4$	K_q (L mol^{-1}) $\times 10^{12}$
288	0.992	7.046	7.046
298	0.971	6.276	6.276
310	0.921	2.683	2.683

TABLE 2: Binding constants (K_f) and the number of binding sites (n) of the complex DNA system.

Temperature (K)	Linear equation	R^2	n	K_f	LogK_f
288	$Y = 1.678X + 8.176$	0.990	1.678	1.49×10^8	8.176
298	$Y = 0.644X + 4.549$	0.987	0.644	3.54×10^4	4.549
310	$Y = 0.527X + 2.293$	0.957	0.527	1.96×10^2	2.293

from a more B-like to a more C-like structure within the DNA molecule [47]. These changes are indicative of a non-intercalative mode of binding of the complex and offer support to its groove binding nature [48].

3.4. Fluorescence Spectroscopy.

Quenching can occur by different mechanisms which are usually classified as dynamic and static quenching. Dynamic quenching refers to a process in which the fluorophore and the quencher come into contact during the transient existence of the exited state, while static quenching refers to fluorophore-quencher complex formation. In general, dynamic and static quenching can be distinguished by their differing dependence on temperature and excited-state lifetime. Since in both cases the fluorescence intensity is related to the concentration of the quencher, the quenched fluorophore can serve as an indicator for the quenching agent [49]. The effect of DNA on the fluorescence intensity of the complex is shown in Figure 4. Upon addition of CT-DNA, a decrease in emission intensity was observed for the complex. This implies that the title complex has a strong interaction with DNA. The quenching of the luminescence of the complex by CT-DNA may be attributed to the photoelectron transfer from the guanine bases of DNA to the excited MLCT state, as reported for other complexes [50, 51].

Fluorescence quenching is described by the Stern-Volmer equation:

$$\frac{F_0}{F} = 1 + Kq\tau_0[Q] = 1 + K_{sv}[Q], \quad (1)$$

where F_0 and F represent the fluorescence intensities in the absence and in the presence of quencher, respectively. Kq is the fluorophore quenching rate constant, K_{sv} is quenching constant, τ_0 is the lifetime of the fluorophore in the absence of a quencher ($\tau_0 = 10^{-8}$), and $[Q]$ is the concentration of quencher [52]. Dynamic and static quenching can be distinguished by their different dependence on temperature [53]. The results in Table 1 indicate that the probable quenching mechanism of this complex by CT-DNA involves static quenching, because K_{sv} decreases with increasing temperature [54].

3.4.1. Binding Constant and the Number of Binding Sites.

The binding constant (K_f) and the binding stoichiometry (n) for the complex formation between iron complex and DNA were measured using (2) [55]

$$\text{Log}\left(\frac{F_0 - F}{F}\right) = \text{Log}K_f + n\text{Log}[Q]. \quad (2)$$

Here, F_0 and F are the fluorescence intensities of the fluorophore in the absence and presence of different concentrations of CT-DNA, respectively. The values of K_f aaa and n were found to be $3.54 \times 10^4 \, \text{M}^{-1}$ and 0.64, respectively, (Table 2).

3.4.2. Binding Mode between the Complex and DNA.

According to the thermodynamic data, interpreted as follows, the model of interaction between a drug and biomolecule can be [56]: (1) $\Delta H > 0$ and $\Delta S > 0$, hydrophobic forces; (2) $\Delta H < 0$ and $\Delta S < 0$, van der Waals interactions and hydrogen bonds; (3) $\Delta H < 0$ and $\Delta S > 0$, electrostatic interactions [57]. In order to elucidate the interaction of our complex with DNA, the thermodynamic parameters were calculated. The plot of ln$K$ versus $1/T$ (Figure 5; (3)) allows the determination of ΔH and ΔS. If the temperature does not vary significantly, the enthalpy change can be regarded as a constant. Based on the binding constants at different temperatures, the free energy changes can be estimated (Table 3; (4)) by the following equations

$$\text{Ln}K = -\frac{\Delta H}{RT} + \frac{\Delta S}{R}, \quad (3)$$

$$\Delta G = \Delta H - T\Delta S = -RT\text{Ln}K, \quad (4)$$

where K is the Stern-Volmer quenching constant at the corresponding temperatures and R is the gas constant. When

TABLE 3: Thermodynamic parameters and binding constants for the binding of Fe(III) complex to calf thymus DNA.

Temperature (K)	ΔG (KJ mol^{-1})	ΔH (KJ mol^{-1})	ΔS (J mol^{-1} K^{-1})
283	−21.99	−197.45	−620.05
299	−12.07	−197.45	−620.05
310	−5.25	−197.45	−620.05

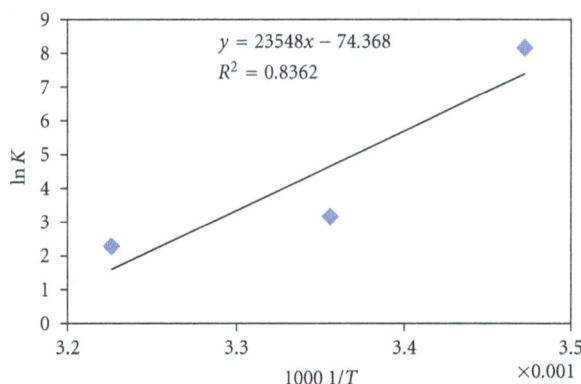

FIGURE 5

we apply this analysis to the binding of the complex with CT-DNA, we find that $\Delta H < 0$ and $\Delta S < 0$. Therefore, van der Waals interactions or hydrogen bonds are probably the main forces in the binding of the titled complex to CT-DNA.

3.5. Absorption Studies. Electronic absorption spectroscopy was an effective method to examine the binding mode of DNA with metal complexes. In general, hypochromism and red shift are associated with the binding of the complex to the helix by an intercalative mode involving strong stacking interaction of the aromatic chromophore of the complex between the DNA base pairs. Figure 6 shows the UV-absorption spectra of the complex in the absence and presence of DNA. In the ultraviolet region from 200 to 400 nm, the complex had strong absorption peaks at 260 nm, besides a shoulder band around 340 nm. The absorption intensity of the complex sample increased (hyperchromism; %H = 37.3%) and no red shift evidently after the addition of DNA, which indicated the interactions between DNA and the complex. The hyperchromism of the complex, on addition of calf-thymus DNA, implies that the binding mode is non-intercalative in nature. This hyperchromism can be attributed to external contact (surface binding) with the duplex. Some similar hyperchromism have been observed [58–60]. But this needs further clarification of the DNA binding mode of the complex by viscosity measurements.

The absorption data were analyzed to evaluate the intrinsic binding constant, K_b, which can be determined from (5) [61],

$$\frac{[DNA]}{(\varepsilon a - \varepsilon f)} = \frac{[DNA]}{(\varepsilon 0 - \varepsilon f)} + \frac{1}{K_b(\varepsilon b - \varepsilon f)}, \quad (5)$$

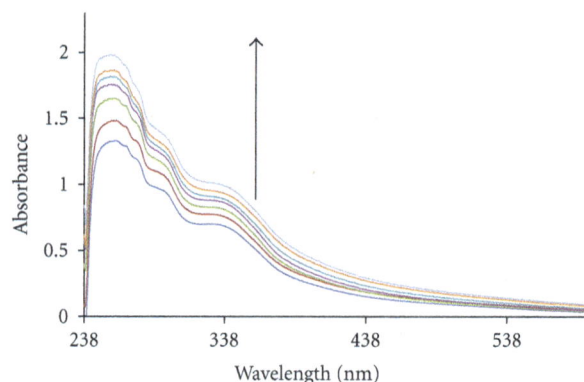

FIGURE 6

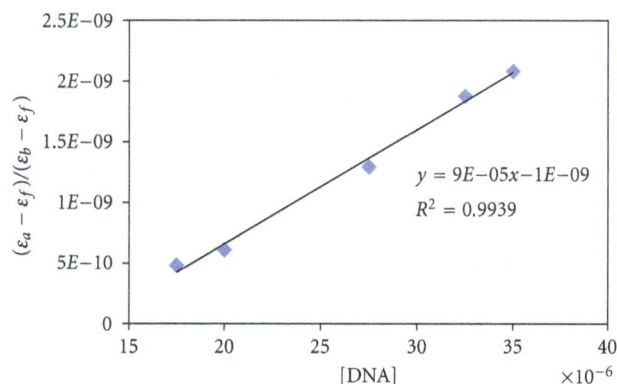

FIGURE 7

where [DNA] is the concentration of DNA in base pairs, the apparent absorption coefficient εa, and εf and εb correspond to $A_{obsd}/[M]$, the extinction coefficient of the free compound and the extinction coefficient of the compound when fully bound to DNA, respectively.

In plots of $[DNA]/(\varepsilon a - \varepsilon f)$ versus [DNA], K_b is given by the ratio of slope to the intercept. The intrinsic bindingconstant, K_b, of complex was 9×10^4 M^{-1} (Figure 7).

The K_b value obtained here is lower than that reported for classical intercalators (for ethidium bromide and [Ru(phen)DPPZ] whose binding constants have been found to be in the order of 10^6-10^7 M) [62, 63]. The observed binding constant is more in keeping with the groove binding with DNA, as observed in the literature [64].

The K_b value for the complex is comparable to that observed for some complexes like, [Co (phen)$_3$]$^{3+}$ (1.6 × 10^4 M^{-1}) [65]; [Ru (phen)$_3$)$^{2+}$] (0.55 × 10^4 M^{-1}) [66]; (Ru (phen)$_2$(bpy)](PF$_6$)$_2$ (0.36 × 10^4 M^{-1}) [67]; [Ru (phen)$_2$L]$^{+4}$ (2.5 × 10^4 M^{-1}) [68] (electrostatic binding mode); [Ru (tpy)(pph$_3$)$_2$Cl]$^+$ (K_b = 4.1 × 10^4 M^{-1}, s = 0.34) [69]; [Ru (tpy)(Asph$_3$)Cl]$^+$ (K_b = 4.5 × 10^4 M^{-1}, s = 0.1) [69]; tricationic Co(III) complexes with asymmetric ligand, [Co(phen)$_2$(pdta)]$^{3+}$ (K_b = 2.8 × 10^4 M^{-1}) [70]; [Co (bpy)$_2$(CNOIP)]$^{3+}$ (K_b = 5 × 10^4 M^{-1}) [71].

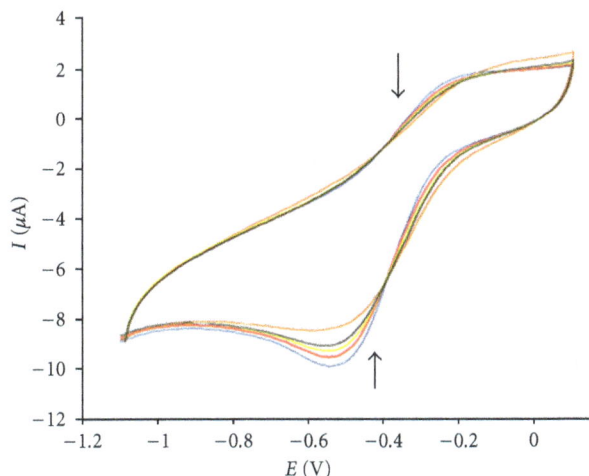

FIGURE 8

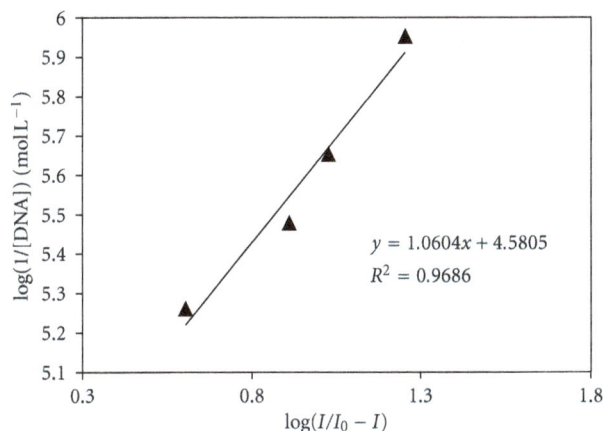

FIGURE 9: Cyclic voltammogram of 0.1 mM Fe(III) complex on a polished glassy carbon electrode in the absence and presence of different concentrations of DNA. v: 100 mVs^{-1}, buffer: 0.05 M Tris-HCl buffer (pH 7.4).

3.6. Voltammetric Studies. The application of electrochemical methods to the study of metallointercalation and coordination of transitional metal complexes to DNA provides a useful complement to the previously used methods of investigation, such as UV–vis spectroscopy [72, 73].

The electrochemical behavior of complex is well known, and was strongly influenced by the electrode material. A well-defined and sensitive peak was observed from the solution of the complex with a GC electrode rather than the Pt one. Therefore, a GC electrode was used in this investigation . When CT-DNA is added to a solution of the complex both the anodic and cathodic peak current heights of the complex decreased in the same manner of increasing additions of DNA, (Figure 8).

The substantial diminution in peak current is attributed to the formation of slowly diffusing complex-DNA supramolecular complex due to which the concentration of the free drug (mainly responsible for the transfer of current) is lowered [74].

The gradual decay in peak current of the complex by increasing of DNA concentration, ranging from 20–100 μM, can be used to quantify the binding constant by the application of the following equation [75]:

$$\log\left(\frac{1}{[\text{DNA}]}\right) = \log K + \log\left(\frac{I}{I_0 - I}\right), \qquad (6)$$

where K is the binding constant, I_0 and I are the peak currents of the free guest and the complex, respectively.

The binding constant with a value of 3.81×10^4 mol^{-1} L was obtained from the intercept of $\log(1/[\text{DNA}])$ versus $\log(I/(I_0 - I))$ plot (Figure 9).

The values of binding constants obtained from UV-vis absorption, fluorescence spectroscopy and CV measurements, 9×10^4, 3.54×10^4 and 3.81×10^4 mol^{-1} L, respectively, were in close agreement.

From the results of the above experiments, the formation of an electrochemically nonactive complex of the complex with DNA resulted in the decrease of the free concentration of the complex in the reaction solution, which caused the decrease of the peak current.

4. Conclusion

One of the most important goals of pharmacological research is the search for new molecular structures which exhibit effective antitumor activities. This has driven inorganic and organometallic chemists to look for new metal compounds with good activities, preferably against tumors that are responsible for high-cancer mortality. In this study, binding interaction of a water soluble iron(III) complex of a Schiff base, SF, (SF 1/4 N, N$_0$-bis{5-[(triphenylphosphonium chloride)-methyl] salicylidene }-o-phenylenediamine) with calf thymus DNA has been investigated and the results show that the complex can bind to DNA via groove binding mode.

Acknowledgment

The authors are grateful for the financial support of Razi University Research Center of Iran.

References

[1] A. A. A. Abu-Hussen, "Synthesis and spectroscopic studies on ternary bis-Schiff-base complexes having oxygen and/or nitrogen donors," *Journal of Coordination Chemistry*, vol. 59, no. 2, pp. 157–176, 2006.

[2] M. S. Karthikeyan, D. J. Prasad, B. Poojary, K. Subrahmanya Bhat, B. S. Holla, and N. S. Kumari, "Synthesis and biological activity of Schiff and Mannich bases bearing 2,4-dichloro-5-fluorophenyl moiety," *Bioorganic and Medicinal Chemistry*, vol. 14, no. 22, pp. 7482–7489, 2006.

[3] K. Singh, M. S. Barwa, and P. Tyagi, "Synthesis, characterization and biological studies of Co(II), Ni(II), Cu(II) and Zn(II) complexes with bidentate Schiff bases derived by heterocyclic ketone," *European Journal of Medicinal Chemistry*, vol. 41, no. 1, pp. 147–153, 2006.

[4] P. Panneerselvam, R. R. Nair, G. Vijayalakshmi, E. H. Subramanian, and S. K. Sridhar, "Synthesis of Schiff bases of 4-(4-aminophenyl)-morpholine as potential antimicrobial agents,"

European Journal of Medicinal Chemistry, vol. 40, no. 2, pp. 225–229, 2005.

[5] S. K. Sridhar, M. Saravanan, and A. Ramesh, "Synthesis and antibacterial screening of hydrazones, Schiff and Mannich bases of isatin derivatives," *European Journal of Medicinal Chemistry*, vol. 36, no. 7-8, pp. 615–625, 2001.

[6] S. N. Pandeya, D. Sriram, G. Nath, and E. Declercq, "Synthesis, antibacterial, antifungal and anti-HIV activities of Schiff and Mannich bases derived from isatin derivatives and *N*-[4-(4′-chlorophenyl)thiazol-2-yl] thiosemicarbazide," *European Journal of Pharmaceutical Sciences*, vol. 9, no. 1, pp. 25–31, 1999.

[7] R. Mladenova, M. Ignatova, N. Manolova, T. Petrova, and I. Rashkov, "Preparation, characterization and biological activity of Schiff base compounds derived from 8-hydroxyquinoline-2-carboxaldehyde and Jeffamines ED®," *European Polymer Journal*, vol. 38, no. 5, pp. 989–999, 2002.

[8] O. M. Walsh, M. J. Meegan, R. M. Prendergast, and T. Al Nakib, "Synthesis of 3-acetoxyazetidin-2-ones and 3-hydroxyazetidin-2-ones with antifungal and antibacterial activity," *European Journal of Medicinal Chemistry*, vol. 31, no. 12, pp. 989–1000, 1996.

[9] Z. Trávníček, M. Malo, Z. Šindelář et al., "Preparation, physicochemical properties and biological activity of copper(II) complexes with 6-(2-chlorobenzylamino)purine (HL₁) or 6-(3-chlorobenzylamino)purine (HL₂). The single-crystal X-ray structure of [Cu(H⁺L₂)₂Cl₃]Cl·2H₂O," *Journal of Inorganic Biochemistry*, vol. 84, no. 1-2, pp. 23–32, 2001.

[10] U. El-Ayaan and A. A. M. Abdel-Aziz, "Synthesis, antimicrobial activity and molecular modeling of cobalt and nickel complexes containing the bulky ligand: *bis*[*N*-(2,6-diisopropylphenyl) imino] acenaphthene," *European Journal of Medicinal Chemistry*, vol. 40, no. 12, pp. 1214–1221, 2005.

[11] M. Sönmez, I. Berber, and E. Akbaş, "Synthesis, antibacterial and antifungal activity of some new pyridazinone metal complexes," *European Journal of Medicinal Chemistry*, vol. 41, no. 1, pp. 101–105, 2006.

[12] E. Keskioğlu, A. B. Gündüzalp, S. Çete, F. Hamurcu, and B. Erk, "Cr(III), Fe(III) and Co(III) complexes of tetradentate (ONNO) Schiff base ligands: synthesis, characterization, properties and biological activity," *Spectrochimica Acta—Part A*, vol. 70, no. 3, pp. 634–640, 2008.

[13] J.-Z. Wu and L. Yuan, "Synthesis and DNA interaction studies of a binuclear ruthenium(II) complex with 2,9-bis(2-imidazo[4,5-*f*][1,10]phenanthroline)-1,10-phenanthroline as bridging and intercalating ligand," *Journal of Inorganic Biochemistry*, vol. 98, no. 1, pp. 41–45, 2004.

[14] K. P. Balasubramanian, K. Parameswari, V. Chinnusamy, R. Prabhakaran, and K. Natarajan, "Synthesis, characterization, electro chemistry, catalytic and biological activities of ruthenium(III) complexes with bidentate N, O/S donor ligands," *Spectrochimica Acta—Part A*, vol. 65, no. 3-4, pp. 678–683, 2006.

[15] S. Yamada, "Advancement in stereochemical aspects of Schiff base metal complexes," *Coordination Chemistry Reviews*, vol. 190–192, pp. 537–555, 1999.

[16] A. Böttcher, T. Takeuchi, K. I. Hardcastle et al., "Spectroscopy and electrochemistry of cobalt(III) schiff base complexes," *Inorganic Chemistry*, vol. 36, no. 12, pp. 2498–2504, 1997.

[17] E. Canpolat and M. Kaya, "Studies on mononuclear chelates derived from substituted Schiff-base ligands (part 2): synthesis and characterization of a new 5-bromosalicyliden-*p*-aminoacetophenoneoxime and its complexes with Co(II),

Ni(II), Cu(II) and Zn(II)," *Journal of Coordination Chemistry*, vol. 57, no. 14, pp. 1217–1223, 2004.

[18] R. L. Lucas, M. K. Zart, J. Murkerjee, T. N. Sorrell, D. R. Powell, and A. S. Borovik, "A modular approach toward regulating the secondary coordination sphere of metal ions: differential dioxygen activation assisted by intramolecular hydrogen bonds," *Journal of the American Chemical Society*, vol. 128, no. 48, pp. 15476–15489, 2006.

[19] P. E. Aranha, M. P. dos Santos, S. Romera, and E. R. Dockal, "Synthesis, characterization, and spectroscopic studies of tetradentate Schiff base chromium(III) complexes," *Polyhedron*, vol. 26, no. 7, pp. 1373–1382, 2007.

[20] A. A. Soliman and W. Linert, "Structural features of ONS-donor salicylidene *Schiff* base complexes," *Monatshefte fur Chemie*, vol. 138, no. 3, pp. 175–189, 2007.

[21] M. Xie, G. Xu, L. Li, W. Liu, Y. Niu, and S. Yan, "In vivo insulin-mimetic activity of [*N,N*-1,3-propyl-bis(salicyladimine)]oxovanadium(IV)," *European Journal of Medicinal Chemistry*, vol. 42, no. 6, pp. 817–822, 2007.

[22] V. Milacic, D. Chen, L. Ronconi, K. R. Landis-Piwowar, D. Fregona, and Q. P. Dou, "A novel anticancer gold(III) dithiocarbamate compound inhibits the activity of a purified 20S proteasome and 26S proteasome in human breast cancer cell cultures and xenografts," *Cancer Research*, vol. 66, no. 21, pp. 10478–10486, 2006.

[23] I. Kostova, "Gold coordination complexes as anticancer agents," *Anti-Cancer Agents in Medicinal Chemistry*, vol. 6, no. 1, pp. 19–32, 2006.

[24] F. Li, W. Chen, C. Tang, and S. Zhang, "Recent development of interaction of transition metal complexes with DNA based on biosensor and its applications," *Talanta*, vol. 77, no. 1, pp. 1–8, 2008.

[25] R. A. Sanchez-Delgado, A. Anzellotti, and L. Suarez, "Metal complexes as chemotherapeutic agents against tropical diseases: malaria, leishmaniasis and trypanosomiasis," in *Metal Ions in Biological Systems*, H. Sigel and A. Sigel, Eds., pp. 379–419, Marcel Dekker, New York, Ny, USA, 2004.

[26] I. Bertini, H. B. Gray, and E. I. Stiefel, *Biological Inorganic Chemistry: Structure and Reactivity*, University Science Books, Sausalito, Calif, USA, 2007.

[27] R. R. Tidwell and D. W. Boykin, "Dicationic DNA minor groove binders as antimicrobial agents," in *DNA and RNA Binders From Small Molecules to Drags*, M. Demeunynck, C. Bailly, and W. D. Wilson, Eds., Wiley-VCH, 2003.

[28] A. K. Patra, M. Nethaji, and A. R. Chakravarty, "Synthesis, crystal structure, DNA binding and photo-induced DNA cleavage activity of (*S*-methyl-ʟ-cysteine)copper(II) complexes of heterocyclic bases," *Journal of Inorganic Biochemistry*, vol. 101, no. 2, pp. 233–244, 2007.

[29] A. K. Patra, T. Bhowmick, S. Roy, S. Ramakumar, and A. R. Chakravarty, "Copper(II) complexes of L-arginine as netropsin mimics showing DNA cleavage activity in red light," *Inorganic Chemistry*, vol. 48, no. 7, pp. 2932–2943, 2009.

[30] S. Ramakrishnan, V. Rajendiran, M. Palaniandavar et al., "Induction of cell death by ternary copper(II) complexes of l-tyrosine and diimines: role of coligands on DNA binding and cleavage and anticancer activity," *Inorganic Chemistry*, vol. 48, no. 4, pp. 1309–1322, 2009.

[31] D. B. Hall, R. E. Holmlin, and J. K. Barton, "Oxidative DNA damage through long-range electron transfer," *Nature*, vol. 382, no. 6593, pp. 731–735, 1996.

[32] K. E. Erkkila, D. T. Odom, and J. K. Barton, "Recognition and reaction of metallointercalators with DNA," *Chemical Reviews*, vol. 99, no. 9, pp. 2777–2795, 1999.

[33] B. Lippert, "Multiplicity of metal ion binding patterns to nucleobases," *Coordination Chemistry Reviews*, vol. 200–202, pp. 487–516, 2000.

[34] C. Liu, M. Wang, T. Zhang, and H. Sun, "DNA hydrolysis promoted by di- and multi-nuclear metal complexes," *Coordination Chemistry Reviews*, vol. 248, no. 1-2, pp. 147–168, 2004.

[35] G. Pratviel, J. Bernadou, and B. Meunier, "DNA and RNA cleavage by metal complexes," *Advances in Inorganic Chemistry*, vol. 45, pp. 251–312, 1998.

[36] M. Shionoya, T. Ikeda, E. Kimura, and M. Shiro, "Novel "multipoint" molecular recognition of nucleobases by a new zinc(II) complex of acridine-pendant cyclen(cyclen = 1,4,7,10-tetraazacyclododecane)," *Journal of the American Chemical Society*, vol. 116, no. 9, pp. 3848–3859, 1994.

[37] D. M. Epstein, L. L. Chappell, H. Khalili, R. M. Supkowski, W. W. De Horrocks, and J. R. Morrow, "Eu(III) macrocyclic complexes promote cleavage of and bind to models for the 5′-cap of mRNA. Effect of pendent group and a second metal ion," *Inorganic Chemistry*, vol. 39, no. 10, pp. 2130–2134, 2000.

[38] X.-Y. Wang, J. Zhang, K. Li et al., "Synthesis and DNA cleavage activities of mononuclear macrocyclic polyamine zinc(II), copper(II), cobalt(II) complexes which linked with uracil," *Bioorganic and Medicinal Chemistry*, vol. 14, no. 19, pp. 6745–6751, 2006.

[39] X. B. Yang, J. Feng, J. Zhang et al., "Synthesis, DNA binding and cleavage activities of the copper (II) complexes of estrogen-macrocyclic polyamine conjugates," *Bioorganic and Medicinal Chemistry*, vol. 16, no. 7, pp. 3871–3877, 2008.

[40] R. Ramesh and M. Sivagamasundari, "Synthesis, spectral, and antifungal activity of Ru(II) mixed-ligand complexes," *Synthesis and Reactivity in Inorganic and Metal-Organic Chemistry*, vol. 33, no. 5, pp. 899–910, 2003.

[41] N. Deligönül, M. Tümer, and S. Serin, "Synthesis, characterization, catalytic, electrochemical and thermal properties of tetradentate Schiff base complexes," *Transition Metal Chemistry*, vol. 31, no. 7, pp. 920–929, 2006.

[42] N. Shahabadi, S. Kashanian, and F. Darabi, "*In Vitro* study of DNA interaction with a water-soluble dinitrogen schiff base," *DNA and Cell Biology*, vol. 28, no. 11, pp. 589–596, 2009.

[43] B. A. Uzoukwu, K. Gloe, and H. Duddeck, "4-acylpyrazolo-neiminjz Schiff bases and their metal complexes: synthesis and spectroscopic studies," *Synthesis and Reactivity in Inorganic and Metal-Organic Chemistry*, vol. 28, no. 5, pp. 819–831, 1998.

[44] J. M. Kelly, A. B. Tossi, D. J. Mcconnell, and C. Ohuigin, "A study of the interactions of some polypyridylruthenium(II) complexes with DNA using fluorescence spectroscopy, topoisomerisation and thermal denaturation," *Nucleic Acids Research*, vol. 13, no. 17, pp. 6017–6034, 1985.

[45] G. L. Eichhorn and Y. A. Shin, "Interaction of metal ions with polynucleotides and related compounds. XII. The relative effect of various metal ions on DNA helicity," *Journal of the American Chemical Society*, vol. 90, no. 26, pp. 7323–7328, 1968.

[46] N. Shahabadi, S. Kashanian, M. Khosravi, and M. Mahdavi, "Multispectroscopic DNA interaction studies of a water-soluble nickel(II) complex containing different dinitrogen aromatic ligands," *Transition Metal Chemistry*, vol. 35, no. 6, pp. 699–705, 2010.

[47] S. Mahadevan and M. Palaniandavar, "Spectroscopic and voltammetric studies on copper complexes of 2,9-dimethyl-1,10-phenanthrolines bound to calf thymus DNA," *Inorganic Chemistry*, vol. 37, no. 4, pp. 693–700, 1998.

[48] P. Uma Maheswari and M. Palaniandavar, "DNA binding and cleavage properties of certain tetrammine ruthenium(II) complexes of modified 1,10-phenanthrolines—effect of hydrogen-bonding on DNA-binding affinity," *Journal of Inorganic Biochemistry*, vol. 98, no. 2, pp. 219–230, 2004.

[49] M. R. Eftink and C. A. Ghiron, "Fluorescence quenching studies with proteins," *Analytical Biochemistry*, vol. 114, no. 2, pp. 199–227, 1981.

[50] B. Peng, H. Chao, B. Sun, H. Li, F. Gao, and L.-N. Ji, "Synthesis, DNA-binding and photocleavage studies of cobalt(III) mixed-polypyridyl complexes: [Co(phen)$_2$(dpta)]$^{3+}$ and [Co(phen)$_2$(amtp)]$^{3+}$," *Journal of Inorganic Biochemistry*, vol. 101, no. 3, pp. 404–411, 2007.

[51] V. G. Vaidyanathan and B. U. Nair, "Photooxidation of DNA by a cobalt(II) tridentate complex," *Journal of Inorganic Biochemistry*, vol. 94, no. 1-2, pp. 121–126, 2003.

[52] Y. Sun, S. Bi, D. Song, C. Qiao, D. Mu, and H. Zhang, "Study on the interaction mechanism between DNA and the main active components in Scutellaria baicalensis Georgi," *Sensors and Actuators, B*, vol. 129, no. 2, pp. 799–810, 2008.

[53] F.-L. Cui, J. Fan, J.-P. Li, and Z.-D. Hu, "Interactions between 1-benzoyl-4-*p*-chlorophenyl thiosemicarbazide and serum albumin: investigation by fluorescence spectroscopy," *Bioorganic and Medicinal Chemistry*, vol. 12, no. 1, pp. 151–157, 2004.

[54] P. B. Kandagal, S. M. T. Shaikh, D. H. Manjunatha, J. Seetharamappa, and B. S. Nagaralli, "Spectroscopic studies on the binding of bioactive phenothiazine compounds to human serum albumin," *Journal of Photochemistry and Photobiology A*, vol. 189, no. 1, pp. 121–127, 2007.

[55] M. Jiang, M.-X. Xie, D. Zheng, Y. Liu, X.-Y. Li, and X. Chen, "Spectroscopic studies on the interaction of cinnamic acid and its hydroxyl derivatives with human serum albumin," *Journal of Molecular Structure*, vol. 692, no. 1–3, pp. 71–80, 2004.

[56] M. Jiang, M.-X. Xie, D. Zheng, Y. Liu, X.-Y. Li, and X. Chen, "Spectroscopic studies on the interaction of cinnamic acid and its hydroxyl derivatives with human serum albumin," *Journal of Molecular Structure*, vol. 692, no. 1–3, pp. 71–80, 2004.

[57] N. Shahabadi and A. Fatahi, "Multispectroscopic DNA-binding studies of a tris-chelate nickel(II) complex containing 4,7-diphenyl 1,10-phenanthroline ligands," *Journal of Molecular Structure*, vol. 970, no. 1-3, pp. 90–95, 2010.

[58] R. F. Pasternack, E. J. Gibbs, and J. J. Villafranca, "Interactions of porphyrins with nucleic acids," *Biochemistry*, vol. 22, no. 10, pp. 2406–2414, 1983.

[59] S. Mahadevan and M. Palaniandavar, "Spectroscopic and voltammetric studies on copper complexes of 2,9-dimethyl-1,10-phenanthrolines bound to calf thymus DNA," *Inorganic Chemistry*, vol. 37, no. 4, pp. 693–700, 1998.

[60] S. Kashanian, M. B. Gholivand, F. Ahmadi, A. Taravati, and A. H. Colagar, "DNA interaction with Al-*N,N*-bis(salicylidene)2,2′-phenylendiamine complex," *Spectrochimica Acta—Part A*, vol. 67, no. 2, pp. 472–478, 2007.

[61] A. Wolfe, G. H. Shimer Jr., and T. Meehan, "Polycyclic aromatic hydrocarbons physically intercalate into duplex regions of denatured DNA," *Biochemistry*, vol. 26, no. 20, pp. 6392–6396, 1987.

[62] M. Cory, D. D. McKee, J. Kagan, D. W. Henry, and J. A. Miller, "Design, synthesis, and DNA binding properties of bifunctional intercalators. Comparison of polymethylene and diphenyl ether chains connecting phenanthridine," *Journal of the American Chemical Society*, vol. 107, no. 8, pp. 2528–2536, 1985.

[63] M. J. Waring, "Complex formation between ethidium bromide and nucleic acids," *Journal of Molecular Biology*, vol. 13, no. 1, pp. 269–282, 1965.

[64] V. G. Vaidyanathan and B. U. Nair, "Photooxidation of DNA by a cobalt(II) tridentate complex," *Journal of Inorganic Biochemistry*, vol. 94, no. 1-2, pp. 121–126, 2003.

[65] S. Arounaguiri, D. Easwaramoorthy, A. Ashokkumar, A. Dattagupta, and B. G. Maiya, "Cobalt(III), nickel(II) and ruthenium(II) complexes of 1,10-phenanthroline family of ligands: DNA binding and photocleavage studies," *Proceedings of the Indian Academy of Sciences: Chemical Sciences*, vol. 112, no. 1, pp. 1–17, 2000.

[66] T.-R. Li, Z.-Y. Yang, B.-D. Wang, and D.-D. Qin, "Synthesis, characterization, antioxidant activity and DNA-binding studies of two rare earth(III) complexes with naringenin-2-hydroxy benzoyl hydrazone ligand," *European Journal of Medicinal Chemistry*, vol. 43, no. 8, pp. 1688–1695, 2008.

[67] P. Uma Maheswari, V. Rajendiran, M. Palaniandavar, R. Thomas, and G. U. Kulkarni, "Mixed ligand ruthenium(II) complexes of 5,6-dimethyl-1,10-phenanthroline: the role of ligand hydrophobicity on DNA binding of the complexes," *Inorganica Chimica Acta*, vol. 359, no. 14, pp. 4601–4612, 2006.

[68] J. Sun, S. Wu, Y. An et al., "Synthesis, crystal structure and DNA-binding properties of ruthenium(II) polypyridyl complexes with dicationic 2,2′-dipyridyl derivatives as ligands," *Polyhedron*, vol. 27, no. 13, pp. 2845–2850, 2008.

[69] S. Sharma, S. K. Singh, M. Chandra, and D. S. Pandey, "DNA-binding behavior of ruthenium(II) complexes containing both group 15 donors and 2,2′:6′,2″-terpyridine," *Journal of Inorganic Biochemistry*, vol. 99, no. 2, pp. 458–466, 2005.

[70] X.-L. Wang, H. Chao, H. Li et al., "DNA interactions of cobalt(III) mixed-polypyridyl complexes containing asymmetric ligands," *Journal of Inorganic Biochemistry*, vol. 98, no. 6, pp. 1143–1150, 2004.

[71] Q.-L. Zhang, J.-G. Liu, J. Liu et al., "DNA-binding and photocleavage studies of cobalt(III) mixed-polypyridyl complexes containing 2-(2-chloro-5-nitrophenyl)imidazo [4,5-*f*][1,10]phenanthroline," *Journal of Inorganic Biochemistry*, vol. 85, no. 4, pp. 291–296, 2001.

[72] D. H. Johnston, K. C. Glasgow, and H. H. Thorp, "Electrochemical measurement of the solvent accessibility of nucleobases using electron transfer between DNA and metal complexes," *Journal of the American Chemical Society*, vol. 117, no. 35, pp. 8933–8938, 1995.

[73] R. Indumathy, S. Radhika, M. Kanthimathi, T. Weyhermuller, and B. Unni Nair, "Cobalt complexes of terpyridine ligand: crystal structure and photocleavage of DNA," *Journal of Inorganic Biochemistry*, vol. 101, no. 3, pp. 434–443, 2007.

[74] Y. M. Song, Q. Wu, P. J. Yang, N. N. Luan, L. F. Wang, and Y. M. Liu, "DNA Binding and cleavage activity of Ni(II) complex with all-trans retinoic acid," *Journal of Inorganic Biochemistry*, vol. 100, no. 10, pp. 1685–1691, 2006.

[75] Q. Feng, N.-Q. Li, and Y.-Y. Jiang, "Electrochemical studies of porphyrin interacting with DNA and determination of DNA," *Analytica Chimica Acta*, vol. 344, no. 1-2, pp. 97–104, 1997.

4

On the Discovery, Biological Effects, and Use of Cisplatin and Metallocenes in Anticancer Chemotherapy

Santiago Gómez-Ruiz,[1] **Danijela Maksimović-Ivanić,**[2]
Sanja Mijatović,[2] **and Goran N. Kaluđerović**[3]

[1] *Departamento de Química Inorgánica y Analítica, E.S.C.E.T., Universidad Rey Juan Carlos, 28933 Móstoles, Spain*
[2] *Institute for Biological Research "Sinisa Stankovic", University of Belgrade, Boulevard of Despot Stefan 142, 11060 Belgrade, Serbia*
[3] *Institut für Chemie, Martin-Luther-Universität Halle-Wittenberg, Kurt-Mothes-Straße 2, 06120 Halle, Germany*

Correspondence should be addressed to Goran N. Kaluđerović, goran.kaluderovic@chemie.uni-halle.de

Academic Editor: Zhe-Sheng Chen

The purpose of this paper is to summarize mode of action of cisplatin on the tumor cells, a brief outlook on the metallocene compounds as antitumor drugs as well as the future tendencies for the use of the latter in anticancer chemotherapy. Molecular mechanisms of cisplatin interaction with DNA, DNA repair mechanisms, and cellular proteins are discussed. Molecular background of the sensitivity and resistance to cisplatin, as well as its influence on the efficacy of the antitumor immune response was evaluated. Furthermore, herein are summarized some metallocenes (titanocene, vanadocene, molybdocene, ferrocene, and zirconocene) with high antitumor activity.

1. Cisplatin

Since 1845, when Italian doctor Peyrone synthesized cisplatin (Figure 1), through Rosenberg's discovery of cisplatin antiproliferative potential [1], and subsequent approval for clinical usage in 1978, this drug is considered as most promising anticancer therapeutic [2, 3]. Cisplatin is highly effective against testicular, ovarian, head and neck, bladder, cervical, oesophageal as well as small cell lung cancer [4].

For more than 150 years, first exaltation about this "drug of the 20th century" was replaced with discouraging data about its toxicity and ineffectiveness got from clinical practice. It was found that cisplatin induced serious side effects such as nephrotoxicity, neurotoxicity, ototoxicity, nausea, and vomiting [5]. General toxicity and low biological availability restricted its therapeutically application. In addition, it is known that some tumors such as colorectal and nonsmall lung cancers are initially resistant to cisplatin while other like ovarian and small cell lung cancers easily acquired resistance to drug [6]. Numerous examples from *in vitro* studies confirmed that exposure to cisplatin often resulted in development of apoptotic resistant phenotype

[7–9]. Following this, development of cisplatin resistant cell lines is found useful for testing the efficacy of future cisplatin modified drugs and on the other hand for evaluation of mechanisms involved in development of resistance. For better understanding of unresponsiveness to cisplatin, it is necessary to define the exact molecular targets of drug action from the moment of entering tumor cell. It is proposed the intact cisplatin which avoided bounding to plasma proteins enter the cell by diffusion or active transport via specific receptors (Figure 2) [10, 11]. Cisplatin is able to use copper-transporting proteins to reach intracellular compartments [12–14]. In addition, regarding to its chemical reactivity, cisplatin can influence cell physiology even through interaction with cell membrane molecules such as different receptors.

1.1. Cisplatin and DNA. Although it is known that DNA is a major target for cisplatin, only 5–10% intracellular concentration of cisplatin is found in DNA fraction while 75–85% binds to nucleophilic sites of intracellular constituents like thiol containing peptides, proteins, replication enzymes, and RNA [6, 15–17]. This preferential binding to non-DNA targets offers the explanation for cisplatin resistance but

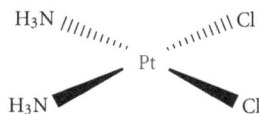

FIGURE 1: Cisplatin.

also its high toxicity. Prerequisite of efficient formation of cisplatin DNA adducts is hydratization of cisplatin enabled by low chloride ions content inside the cells [18]. N7 of guanine and in less extend adenine nucleotide are targeted by platinum [19]. Binding of cisplatin to DNA is irreversible and structurally different adducts are formed. The adducts are classified as intrastrand crosslinking of two nucleobases of single DNA strand, interstrand crosslinking of two different strands of one DNA molecule, chelate formation through N- and O-atoms of one guanine, and DNA-protein crosslinks [20, 21]. Cisplatin forms about 65% pGpG-intrastrand crosslinks, 25% pApG-intrastrand crosslinks, 13% interstrand or intrastrand crosslinks on pGpXpG sequences, and less than 1% of monofunctional adducts (Figure 3) [22]. Crucial role of 1,2-intrastrand crosslinks in antitumor potential of the cisplatin is supported by two facts. First, high mobility group proteins (HMG) specifically recognize this type of cisplatin-DNA interaction and second, these adducts are less efficiently removed by repair enzymes [17]. In addition, important mediators of cisplatin toxicity are ternary DNA-platinum-protein crosslinks (DPCL) whose frequency is dependent from the cell type as well as the type of the treatment. DPCLs inhibited DNA polymerization or their own removal by nucleotide excision repair system more potently than other DNA adducts [17]. In fact, cisplatin DNA adducts can be repaired by nucleotide excision repair proteins (NER), mismatch repair (MMR), and DNA-dependent protein kinases protein [17].

1.2. DNA Repair Mechanism. Nucleotide excision repair proteins are ATP-dependent multiprotein complex able to efficiently repair both inter as well as intrastrand DNA-cisplatin adducts. Successful repair of 1,2-d(GpG) and 1,3-d(GpNpG) intrastrand crosslinks has been found in different human and rodent NER systems [23, 24]. This repair mechanism is able to correct the lesions promoted by chemotherapeutic drugs, UV radiation as well as oxidative stress [17]. Efficacy of NER proteins varying in different type of tumors and is responsible for acquirement of cisplatin resistance. Low level of mentioned proteins is found in testis tumor defining their high sensitivity to cisplatin treatment. Oppositely, ovarian, bladder, prostate, gastric, and cervical cancers are resistant to cisplatin based therapy due to overexpression of several NER genes [25, 26].

Mismatch repair (MMR) proteins are the post replication repair system for correction of mispaired and unpaired bases in DNA caused by DNA Pt adducts. MMR recognized the DNA adducts formed by ligation of cisplatin but not oxaliplatin [27–30]. Defective MMR is behind the resistance of ovarian cancer to cisplatin and responsible for the mutagenicity of cisplatin [31].

DNA dependent protein kinase is a part of eukaryotic DNA double strand repair pathway. This protein is involved in maintaining of genomic stability as well as in repair of double strand breaks induced by radiation [31]. In ovarian cancer presence of cisplatin DNA adducts inhibited translocation of DNA-PK subunit Ku resulting in inhibition of this repair protein [32].

Special attention is focused on recognition of cisplatin-modified DNA by HMG proteins (HMG). It is hypothesized that HMG proteins protected adducts from recognition and reparation [17, 31]. Moreover, it was postulated that these proteins modulate cell cycle events and triggered cell death as a consequence of DNA damage. One of the members from this group marked as HMGB1 is involved in MMR, increased the p53 DNA-binding activity and further stimulated binding of different sequence specific transcription factors [33]. Few studies revealed that cisplatin sensitivity was in correlation with HMGB level, while other studies eliminated its significance in response to cisplatin treatment. Contradictory data about the relevance of HMG proteins in efficacy of cisplatin therapy indicated that this relation is defined by cell specificity.

1.3. Cytotoxicity of Cisplatin. Other non-HMG nuclear proteins are also involved in cytotoxicity of cisplatin. Presence of cisplatin DNA adducts is able to significantly change or even disable the primary function of nuclear proteins essential for transcription of mammalian genes (TATA binding protein, histon-linker protein H1 or 3-methyladenine DNA glycosylase mammalian repair protein) [34–36].

Although cytotoxicity of cisplatin is usually attributed to its reactivity against DNA and subsequent lesions, the fact that more than 80% of internalized drug did not reach DNA indicated the involvement of numerous non-DNA cellular targets in mediation of cisplatin anticancer action [6]. As a consequence of exposure to cisplatin, different signaling pathways are affected. There is no general concept applicable to all types of tumor. It is evident that response to cisplatin is defined by cell specificity. Numerous data revealed changes in activity of most important signaling pathways involved in cell proliferation, differentiation and cell death such as PI3K/Akt, MAPK as well as signaling pathways involved in realization of death signals dependent or independent of death receptors [33]. It is very important to note that alteration in signal transduction upon the cisplatin treatment could be the consequence of both, DNA damage or interaction with exact protein or protein which is relevant for appropriate molecular response. Some of the interactions between protein and cisplatin are already described. Therefore, it was found that cisplatin directly interacts with telomerase, an enzyme that repairs the ends of eukaryotic chromosomes [31, 37]. In parallel, cisplatin-induced damage of telomeres which are not transcribed and therefore hidden from NER. Other important protein targeted by cisplatin is small, tightly folded molecule known as ubiquitin (Ub) [38]. Ub is implicated in selective degradation of short-lived cellular proteins [39]. It has been hypothesized that direct interaction of cisplatin with this protein presented a strong signal for cell death [40]. Two binding sites were identified as target

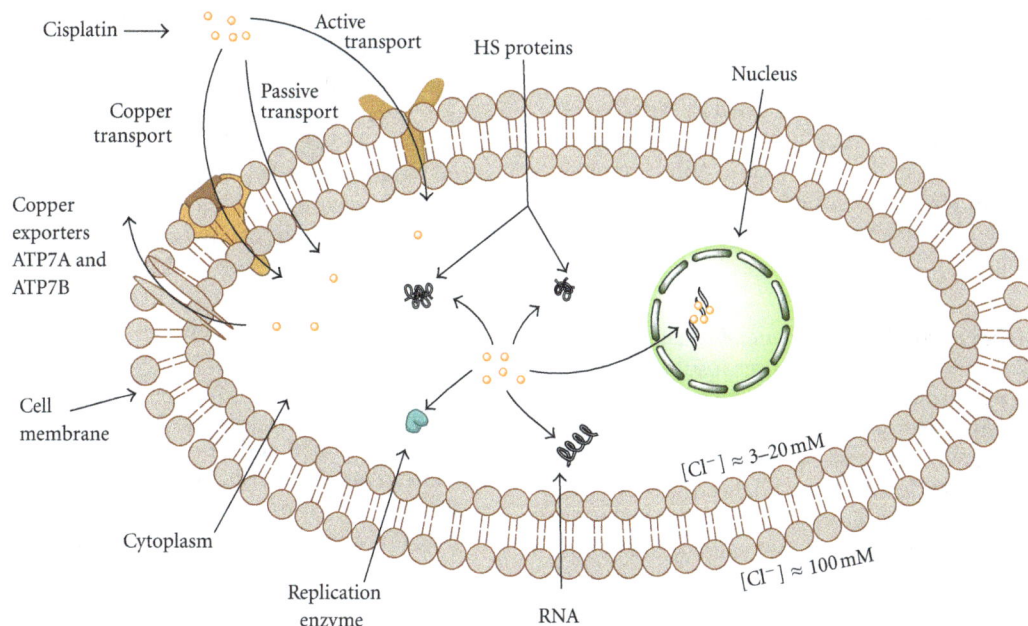

FIGURE 2: Cisplatin and the cell: transport/export and targets.

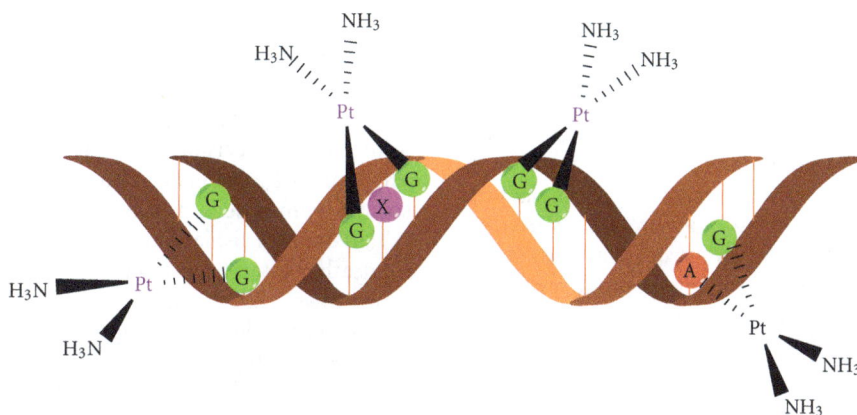

FIGURE 3: DNA adduct formation with cisplatin moiety.

for cisplatin ligation: *N*-terminal methionine (Met1) and histidine at position 68, while the drug makes at least four types of adducts with protein [38]. This resulted in disturbed proteasomal activity and further cell destruction. Having in mind that proteasomal inactivation by specific inhibitors showed promising results in cancer treatment, this aspect of cisplatin reactivity can be leading cytotoxic effect even to be more powerful than DNA damage [41]. One of the crucial molecules involved in propagation of apoptotic signal through depolarization of mitochondrial potential—cytochrome c is also targeted by cisplatin on Met65 [42]. Further, on the list of protein or peptide targets for cisplatin are glutathione and metallothioneins, superoxide dismutase, lysozyme as well as extracellular protein such as albumin, transferrin, and hemoglobin [43]. Some of mentioned interactions served as drug intracellular pool while their biological relevance is still under investigation.

1.4. Activation of Signaling Pathways Induced with Cisplatin. DNA damage induced by cisplatin represent strong stimulus for activation of different signaling pathways. It was found that AKT, c-Abl, p53, MAPK/JNK/ERK/p38 and related pathways respond to presence of DNA lesions [31, 33]. AKT molecule as most important Ser/Thr protein kinase in cell survival protects cells from damage induced by different stimuli as well as cisplatin [44]. Cisplatin downregulated XIAP protein level and promoted AKT cleavage resulting in apoptosis in chemosensitive but not in resistant ovarian cancer cells [45, 46]. Recently published data about synergistic effect of XIAP, c-FLIP, or NFkB inhibition with cisplatin are mainly mediated by AKT pathway [47].

Protein marked as the most important in signaling of the DNA damage is c-Abl which belongs to SRC family of nonreceptor tyrosine kinases [31, 33]. This molecule acts as transmitter of DNA damage triggered by cisplatin from nucleus to cytoplasm [48]. Moreover, sensitivity to cisplatin induced apoptosis is directly related with c-Abl content and could be blocked by c-Abl overexpression [33]. Key role of c-Abl in propagation of cisplatin signals is confirmed in experiments with ABL deficient cells [49]. It was found that cisplatin failed to activate p38 and JNK in the absence of c-Abl. Homology of this kinases with HMGB indicated the possibility that c-Abl recognized and interact with cisplatin DNA lesions like HMGB1 protein [31].

1.5. The Role of the Functional p53 Protein. Evaluation of a 60 cell line conducted by the National Cancer Institute revealed that functional p53 protein is very important for successful response to cisplatin treatment [33]. This tumor suppressor is crucial for many cellular processes and determined the balance between cell cycle arrest as a chance for repair and induction of apoptotic cell death [33]. However, despite extensive NCI study, there are controversial data about correlation between cisplatin sensitivity and p53. For example, it was found that functional p53 was associated with amplified cisplatin sensitivity in SaOS-2 osteosarcoma cells in high serum growth conditions while the opposite relation was observed upon starvation [33]. This phenomenon could be connected to autophagic process triggered in serum deficient conditions, which in turn downregulate cisplatin promoted apoptosis [50]. In some other studies, the response to cisplatin was not influenced by p53. It is indicative that antitumor potential of cisplatin and its interaction with p53 is a question of multiple factors such as tumor cell type, specific signaling involved in cancerogenesis, as well as other genetic alterations. In addition, protein involved or influenced by p53 pathway such as Aurora kinase A, cyclin G, BRCA1 as well as proapoptotic or antiapoptotic mediators are also able to control cisplatin toxicity [33].

1.6. Relation between Cisplatin and Mitogen-Activated Protein (MAP) Kinases. Finally, signaling pathways mediated by mitogen-activated kinases are strongly influenced by cisplatin. These enzymes are highly important in definition of cellular response to applied treatment because they are the major regulators of cell proliferation, differentiation, and cell death. ERK (extracellular signal-related kinase) preferentially responds to growth factor and cytokines but also determines cell reaction to different stress conditions, particularly, oxidative [33]. Cisplatin treatment mainly activated ERK in a dose- and time-dependent manner [33, 51, 52]. However, like as previously described, changes in ERK activity upon the exposure to cisplatin varying from type to type of the malignant cell and is defined by their intrinsic features. Following this, in some circumstances ERK activation antagonized cisplatin toxicity. In cells with significant upregulation of ERK activity in response to cisplatin treatment, exposure to specific MEK1 inhibitor PD98052 abrogated its toxicity. Also, development of the resistance to the cisplatin in HeLa cells is connected with reduced ERK response to the treatment [52]. Moreover, combined treatment with some of the naturally occurring compounds such as aloe-emodin-neutralized cisplatin toxicity through inhibition of

ERK, indicated possible negative outcome of combining of conventional and phytotherapy [53, 54].

Regardless of numerous evidences about its critical role in cisplatin-mediated cell death, ERK is not the only molecule from MAP family which responded to cisplatin. Several studies revealed JNK (c-Jun N-terminal kinase) activation upon the cisplatin addition [55, 56]. However, similarly to other molecules previously mentioned this signal is not the unidirectional and could be responsible for realization but also protection from death triggered by the cisplatin [57, 58]. Finally, there are numerous evidences about highly important role of third member of MAP kinases, p38, in response to cisplatin [59, 60]. Lack of p38 MAPK leads to appearance of resistant phenotype in human cells [55, 60]. Early and short p38 activation is principally described in cells unresponsive to cisplatin while long-term activation was found in sensitive clones. Moreover, in the light of the fact that this kinase has a role in modifying the chromatin environment of target genes, its involvement in cisplatin-induced phosphorilation of histon 3 was determined [61].

1.7. On the Mode of Cell Death Induced by Cisplatin. The net effect of intracellular interaction of cisplatin with DNA and non-DNA targets is the cell cycle arrest and subsequent death in sensitive clones. There are two type of death signals resulting from cellular intoxication by this drug (Figure 4). Fundamentally, the drug concentration presents the critical point for cell decision to undergo apoptotic or necrotic cell death [62]. Primary cultures of proximal tubular cells isolated from mouse died by necrosis if they were exposed to high doses of cisplatin just for a few hours while apoptotic cell death is often triggered by long-term exposure to significantly lower concentrations [63]. However, the presence of necrosis in parallel with apoptosis in tumor-cell population indicated that type of cell death is not just the question of dose but also is defined by cell intrinsic characteristics and energetic status of each cell at the moment of the treatment. In fact, it was considered that intracellular ATP level dictate cell decision to die by necrotic or apoptotic cell death [64, 65]. One of the signals which are provoked with DNA damage is PARP-1 activation and subsequent ATP depletion caused by PARP-1 mediated cleavage of NAD+. This event is a trigger for necrotic cell death. However, activated caspases cleaved the PARP-1, preventing necrotic signal and favor the execution of apoptotic process. On the other hand, the inhibition of caspases by intracellular inhibitors IAP together with continual PARP activity and ATP depletion resulted in necrosis [31]. As numerous biological phenomena, this one is not unidirectional. It was found that failure in PARP cleavage may also serve to apoptosis [66]. This paradox was ascribed to changes in pyridine nucleotide pool as well as in pool of ATP/ADP responsible for regulation of mitochondrial potential [67]. Atypical apoptosis was observed in L1210 leukemia cell line exposed to cisplatin. Different death profiles in cisplatin treated cells confirmed plasticity of signals involved in cell destruction and focus the attention to the molecules responsible for resistance to death as possible targets for the therapy. Having in mind that

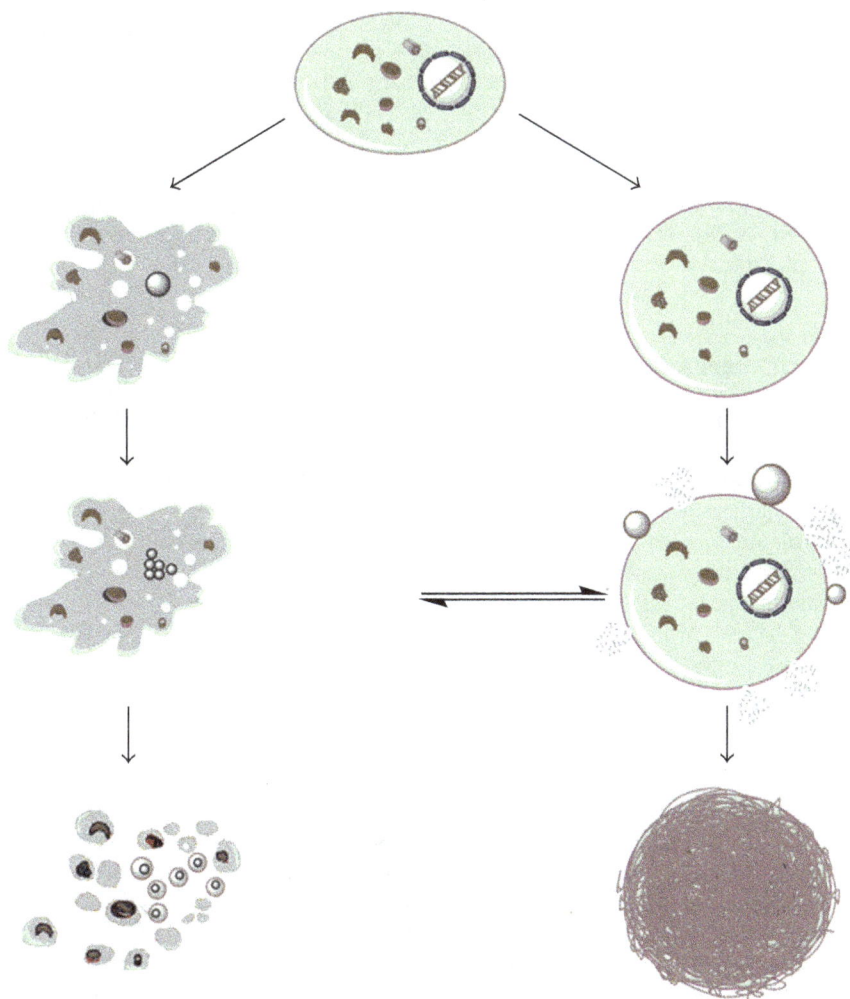

FIGURE 4: Mode of cell death induced by cisplatin: apoptosis (left) and necrosis (right).

cisplatin is toxic agent against whom the cell can activate autophagy as protective process; the specific inhibition of autophagy by certain type of molecules could amplify the effectiveness of cisplatin [50].

1.8. Cisplatin in Immune Senzitization. One of the rarely mentioned but very important aspects of antitumor activity of cisplatin is based on the experimental data about its potential to amplified the sensitivity of malignant cells to one or the most potent and selective antitumor immune response mediated by TNF-related apoptosis inducing ligand TRAIL [68, 69]. This molecule is produced by almost all immune cells involved in nonspecific as well as adoptive immune response. Unfortunately, in the moment when tumor is diagnosed, its sensitivity to natural immunity is debatable. In most of the situations, malignant cells became resistant to TRAIL-mediated cytotoxicity [70]. Moreover, it was confirmed that cisplatin promoted their sensitivity to TRAIL. Nature of its immune sensitizing potential is at least partly due to upregulation of expression of TRAIL receptors—DR4 and DR5 on the cellular membrane glioma,

colon and prostate cell lines as well as downregulation of cellular form of caspase 8 inhibitor FLIP [68, 69]. In addition, presence of cysteine rich domen in the structure of TRAIL specific death receptors indicated possibility that cisplatin directly interact with them.

1.9. Resistance to Cisplatin and How to Surmount It. Resistance to cisplatin could be established at multiple levels, from cellular uptake of the drug through interaction with protein and DNA and finally activation of signals which lead the cell to death. Disturbed drug uptake, drug scavenging by cellular proteins, upregulation of prosurvival signals together with upregulated expression of antiapoptotic molecules such as Bcl-2 and BclXL, overexpressed natural inhibitors of caspases like FLIP and XIAP, diminished MAP signaling pathway or deficiency in proteins involved in signal transferring from damaged DNA to cytoplasm, enhanced activity of repair mechanisms and efficient redox system are features mainly responsible for unsuccessful treatment with cisplatin [33]. Well defined molecular background of the resistance to cisplatin point out the way on how to surmount it. It was already known that some of combined treatments of cis-platin with other chemotherapeutics such as 5-fluorouracil improved therapeutic response rates in patients with head and neck cancer [71, 72]. Furthermore, inhibition of NER DNA repair system, cotreatment with histone deacetylase inhibitors (HDAC) such as trichostatin A (TSA) or suberoy-lanilide hydeoxamic (SAHA) [73], small molecules inhibitors of FLIP and XIAP as well as topoisomerase inhibitors strongly synergized with cisplatin, elevating its therapeutic potential.

2. Metallocenes in Anticancer Chemotherapy

Most of the metallodrugs used currently in chemotherapy treatment are based on platinum (cisplatin analogues), although as side effects are the weakest point in the use of cisplatin-based drugs in chemotherapy are the high number of side effects, many efforts are focused on the search of novel metal complexes with similar antineoplastic activity and less side effects as an alternative for platinum complexes. Transition-metal complexes have shown very useful proper-ties in cancer treatment, and the most important work in chemotherapy with transition metals has been carried out with Group 4, 5, 6, 8, and 11 metal complexes.

From all the studied metal complexes, a wide variety of studies have been carried out for metallocenes which have become an alternative to platinum-based drugs.

According to the IUPAC classification metallocene contains a transition metal and two cyclopentadienyl ligands coordinated in a sandwich structure. These compounds have caused a great interest in chemistry due to their versatility which comes from their interesting physical properties, electronic structure, bonding, and their chemical and spectroscopical properties [74]. Academic and industrial research on metallocene chemistry has led to the utilization of these derivatives in many different applications such as olefin polymerization catalysis, asymmetric catalysis or organic syntheses, preparation of magnetic materials, use as nonlinear optics or molecular recognizers, flame retardants or in medicine [74].

Within medicine, metallocene complexes are being nor-mally used as biosensors or as antitumor agents. Regard-ing their anticancer applicability, titanocene, vanadocene, molybdocene, and ferrocene have been traditionally used with very good results, however, recently also zirconocene derivatives have pointed towards a future potential applica-bility due to the increase of their cytotoxicity. All the other metallocene derivatives have been either not tested or have demonstrated no remarkable applicability in the fight against cancer.

In this part of the paper, we will briefly discuss separately the properties of metallocene derivatives of titanium, zirco-nium, vanadium, molybdenum, and iron.

2.1. Titanocene Derivatives. Titanocene derivatives are together with ferrocene complexes the most studied metal-locenes in the fight against cancer. The pioneering work of Köpf and Köpf-Maier in the early 1980's showed the antipro-liferative properties of titanocene dichloride, [TiCp$_2$Cl$_2$]

(Cp = η^5-C$_5$H$_5$, Figure 5(a)). This compound was studied in phase I clinical trials in 1993 [75–77] using water soluble formulations developed by Medac GmbH (Germany) [78].

Phase I clinical trials pointed towards a dose-limiting side effect associated to nephrotoxicity which together with hypoglycemia, nausea, reversible metallic taste immediately after administration, and pain during infusion, seemed to be the weakest part of the administration of titanocene dichloride in humans. On the other hand, the absence of any effect on proliferative activity of the bone marrow, one of the most common dose-limiting side-effect of nonmetallic drugs, was in interesting result that increased the potential applicability of this compound in humans.

Although phase I clinical trials were not as satisfactory as expected, some phase II clinical trials with patients with breast metastatic carcinoma [79] and advanced renal cell carcinoma [80] have been carried out observing a low activity which discouraged further studies.

However, after the recent work of many groups such as Tacke, Meléndez, McGowan, Baird, and Valentine the interest in this field has been renewed [81–85]. In this con-text a wide variety of titanocene derivatives with amino acids [86, 87], benzyl-substituted titanocene or *ansa*-titanocene derivatives [81], amide functionalized titanocenyls [88, 89], titanocene derivatives with alkylammonium substituents on the cyclopentadienyl rings [90–92], steroid-functionalized titanocenes [93], and alkenyl-substituted titanocene or *ansa*-titanocene derivatives (Figure 5) [94–96], have been reported with very interesting cytotoxic properties which enhance their applicability in humans. In particular, [Ti{η^5-C$_5$H$_4$(CH$_2$C$_6$H$_4$OCH$_3$)}$_2$Cl$_2$] (titanocene Y, Figure 5(b)) and its family, reported by Tacke and coworkers, have demonstrated to have extremely interesting anticancer prop-erties which need to be highlighted.

In general, the cytotoxic activity of titanocene complexes has been correlated to their structure, however, there are still several questions regarding the anticancer mechanism of titanocene(IV) complexes. According to the reported studies in the topic, it seems clear titanium ions reach cells assisted by the major iron transport protein "transferrin" [97–100], and the nucleus in an active transport facilitated probably by ATP. In a final step, binding of titanium ion to DNA leads to cell death (Figure 6) [101, 102]. However, recent experiments have shown interactions of a ligand-bound Ti(IV) complex to other proteins or enzymes [103–105], indicating alterna-tives in cell death mechanisms, which is currently leading to intensive studies by several research groups.

2.2. Zirconocene Derivatives. An alternative to titanium com-plexes may be zirconium(IV) derivatives which are in a very early stage of preclinical experiments. Already in the 1980's Köpf and Köpf-Maier showed the potential of zirconocene derivatives as anticancer agents and very recently, two different studies on zirconocene anticancer chemistry have been reported [106, 107]. These studies by Allen et al. [106] and Wallis et al. [107] have described the cytotoxic activity of different functionalized zirconocene complexes, observing an irregular behavior in the anticancer tests, from which only the complexes [Zr{η^5-C$_5$H$_4$(CH$_2$)$_2$N(CH$_2$)$_5$}$_2$Cl$_2$·2HCl]

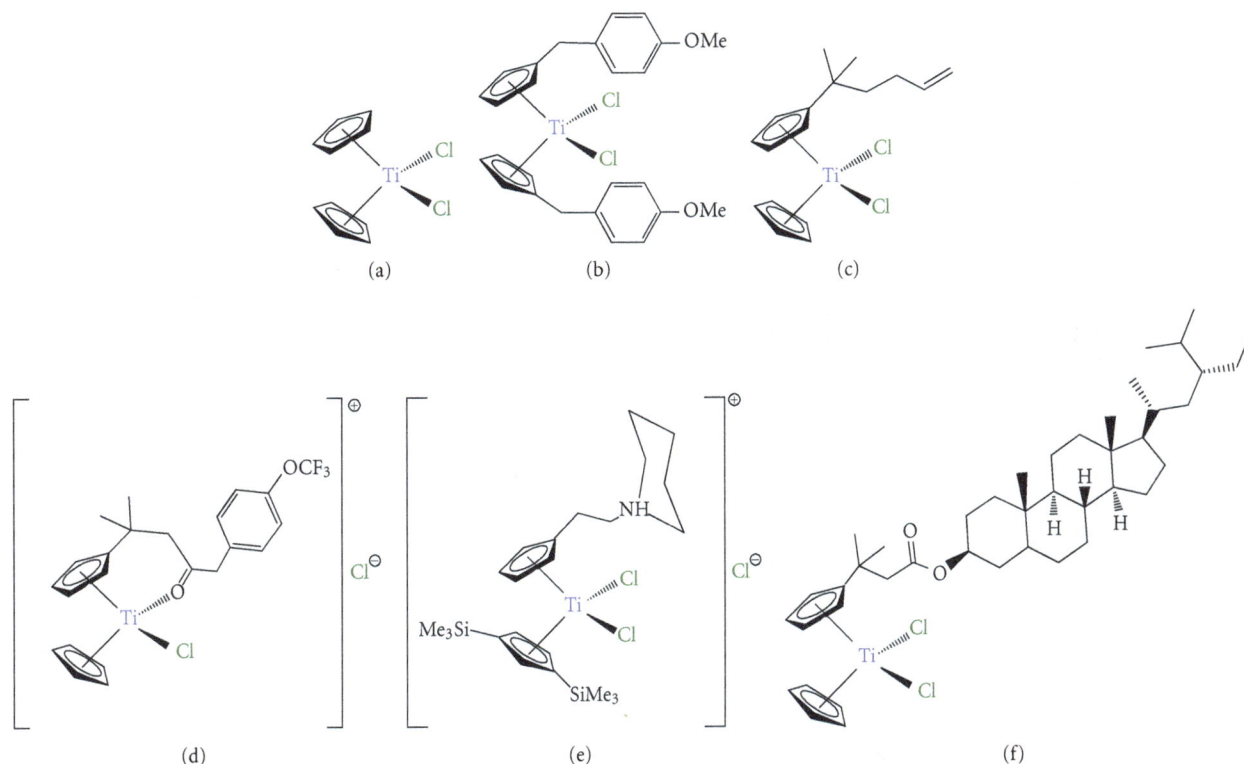

FIGURE 5: Titanocene derivatives used in preclinical and clinical trials: (a) titanocene dichloride; (b) titanocene-Y; (c) alkenyl-substituted titanocene derivative; (d) titanocenyl complex; (e) titanocene derivative with alkylammonium substituents; (f) steroid-functionalized titanocene derivative.

(Figure 7(a)) and [Zr{η^5-C$_5$H$_4$(CH$_2$C$_6$H$_4$OCH$_3$)}$_2$Cl$_2$] (zirconocene Y, Figure 7(b)) have shown promising activity that needs to be improved in order to apply them in anticancer chemotherapy.

In parallel, our research group reported the synthesis, structural characterization, catalytic behavior in the polymerization of ethylene and copolymerization of ethylene and 1-octene and the cytotoxic activity on different human cancer cell lines of a novel alkenyl substituted silicon-bridged *ansa*-zirconocene complex (Figure 7(c)) which proved to be the most active zirconocene complex on human A2780 ovarian cancer cells, reported to date [108].

There is still hard work to do in this field to find a suitable zirconocene complex with increased cytotoxic activity and good applicability in humans.

2.3. Vanadocene Derivatives. Vanadocene dichloride, [VCp$_2$Cl$_2$] (Cp = η^5-C$_5$H$_5$), was extensively studied in preclinical testing against both animal and human cancer cell lines, observing a higher *in vitro* activity of vanadocene(IV) dichloride on direct comparison with titanocene(IV) dichloride [109–111].

These results encouraged further preclinical studies which were restarted around eight years ago [112–114], and have been recently extended [115–118] with the study of the cytotoxic properties of vanadocene Y (Figure 8(a)) and similar derivatives. In addition, a comprehensive study of

the cytotoxic activity of methyl- and methoxy-substituted vanadocene(IV) dichloride toward T-lymphocytic leukemia cells MOLT-4 has also been recently reported [119]. In most cases, vanadocene derivatives are more active than their corresponding titanocene analogues, however, the paramagnetic nature of the vanadium center, which precludes the use of classical NMR tools, makes the characterization of these compounds and their biologically active species more difficult. The need of the use of X-ray crystallography and other methods such as electron-spin resonance (ESR) spectroscopy slows down their analysis and the advances in this topic.

2.4. Molybdocene Derivatives. After the work of Köpf and Köpf-Maier there were some evidences of the potential properties as anticancer agents of molybdocene dichloride derivatives. In recent years, the extensive work carried out by many different research groups confirmed the anticancer properties of molybdocene [120–124]. But not only the cytotoxic properties of these compounds have been reported, the hydrolysis chemistry of [MoCp$_2$Cl$_2$] has been intensively studied [125–127]. In the case of molybdocene derivatives the stability of the Cp ligands at physiological pH has led to the study of many different biological experiments with results which show new insights on the mechanism of antitumor action of [MoCp$_2$Cl$_2$] and some analogous carboxylate derivatives (Figure 8(b)) [117, 128–130].

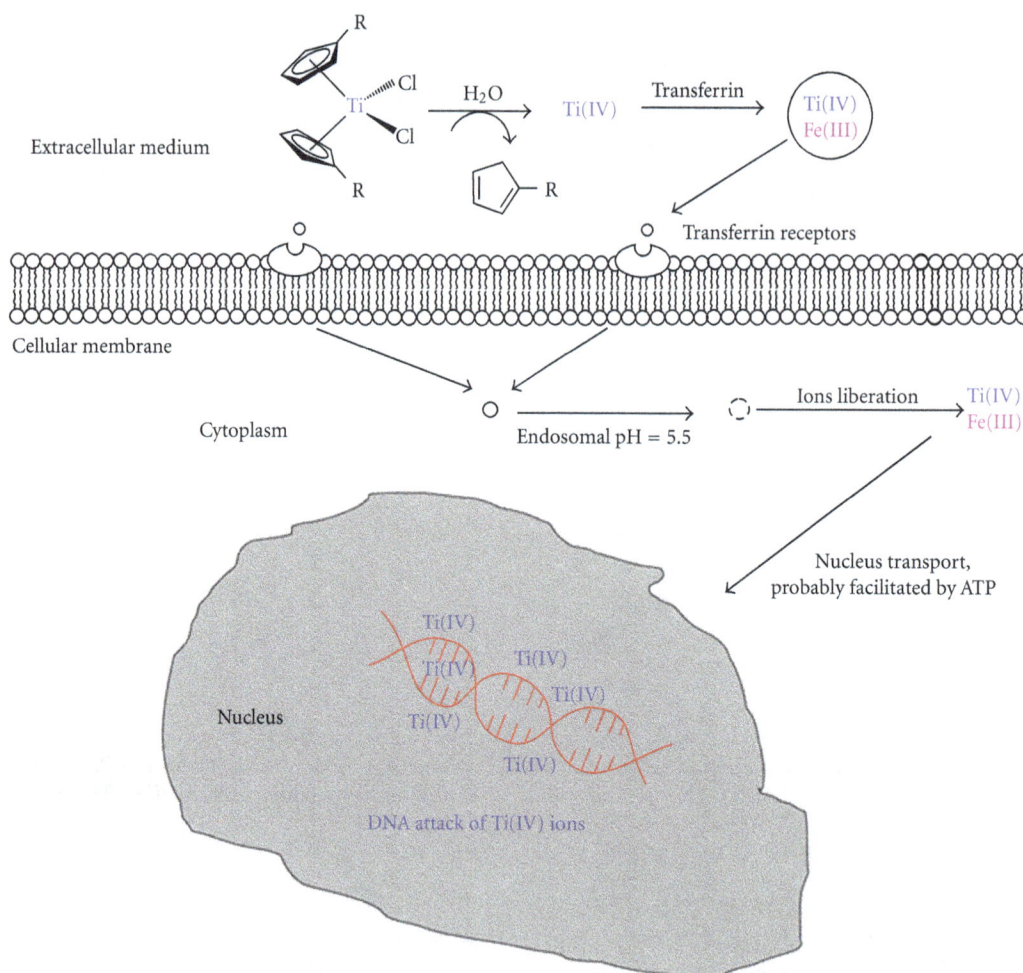

FIGURE 6: Proposed mechanism of action of titanocene derivatives (adapted from Abeysinghe and Harding, Dalton Trans. 32 (2007) 3474).

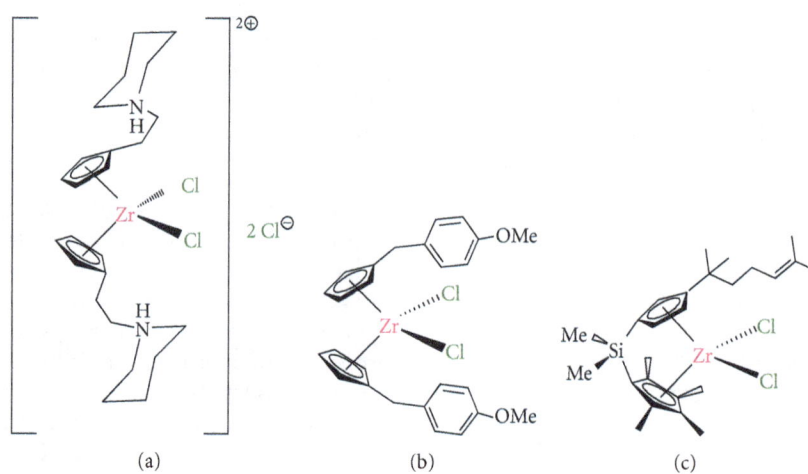

FIGURE 7: Zirconocene derivatives with anticancer activity: (a) zirconocene derivative with alkylammonium substituents; (b) zirconocene-Y; (c) alkenyl-substituted *ansa*-zirconocene complex.

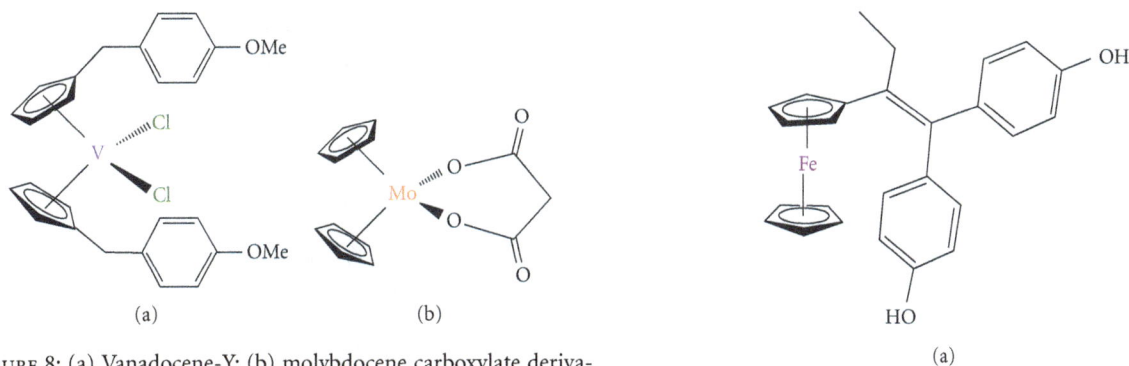

FIGURE 8: (a) Vanadocene-Y; (b) molybdocene carboxylate derivative.

2.5. Ferrocene Derivatives. The discovery of the cytotoxic properties of ferricinium salts on Ehrlich ascite tumors by Köpf and Köpf-Maier [131, 132] were an early breakthrough for the subsequent development of novel preparations of this class of anticancer agents.

There are different groups working in this field, however, to date, the most interesting work in the field of anticancer applications of ferrocene derivatives is being carried out by Jaouen and coworkers.

This group has published several reports on the synthesis of novel functionalized ferrocene derivatives "hydoxyferrocifens" which consist of the linking of the active metabolite of tamoxifen and ferrocene moieties (Figure 9(a)) [133, 134]. This novel class of compounds are able to combine the antioestrogenic properties of tamoxifen with the cytotoxic effects of ferrocene [135–137]. From all these complexes, the outstanding cytotoxicity of a ferrocene complex with a [3] ferrocenophane moiety conjugated to the phenol group (Figure 9(b)) is important to be remarked [138].

In addition, ferrocene-functionalized complexes with steroids or nonsteroidal antiandrogens have also been reported to be very effective to target prostate cancer cells [139].

But not only the design and synthesis of novel ferrocene derivatives with different ligands and cytotoxic properties have been studied, several investigations on the cell death induced mechanism of these anticancer drugs have been reported. Thus, two different action mechanisms have been proposed for ferrocene derivatives, production of electrophilic species, and/or production of ROS species [140].

2.6. Future Tendencies in the Use of Metallocenes in Anticancer Chemotherapy. Almost all metallocene derivatives which have been studied either in preclinical or clinical trials are extremely hydrophobic to be intravenously administered, thus limiting their bioavailability for clinical applications.

Novel formulations of metallocene derivatives in macromolecular systems such as cucurbit(n)urils [140] or cyclodextrins [141] leading to a presumably higher applicability in humans.

In addition, using a different approach, but with the same goal of circumventing the solubility problems of metallocenes in biological media, several metallocene-functionalized MCM-41 or SBA-15 starting from different

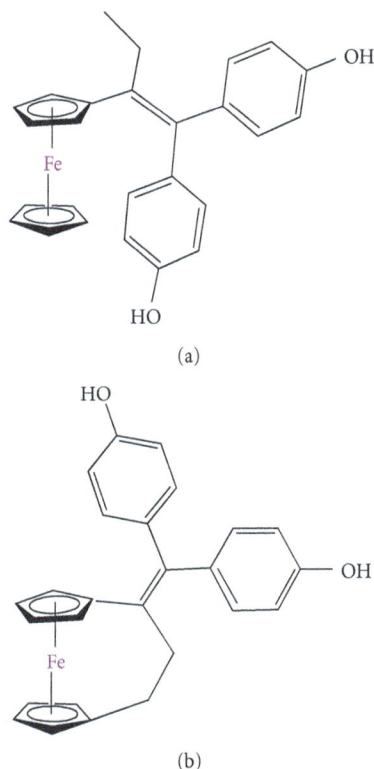

FIGURE 9: Ferrocene derivatives used in preclinical trials: (a) hydroxyferrocifens; (b) ferrocene complex with a [3] ferrocenophane moiety.

titanocene dichloride derivatives with anticancer activity have been reported and may be a good starting point for the development of novel metallocene-based drugs for the treatment of bone tumors [142–145].

3. Conclusions

One of the most potent antitumoral drugs cisplatin deserves special attention as exceptional of few with healing effect. Important role in the action of cisplatin is interaction with nuclear DNA and unfeasibility of the cell response to repair DNA strain containing covalently bonded diammine-platinum(II) moiety (nucleotide excision repair mechanism). Beside DNA, cisplatin might interact with other biomolecules (thioproteins, RNA) and in that way could be deactivated or even may possibly tune different signaling pathways involved in mediation of cell death, which is cell type specific. Namely, cisplatin has intense effects on signaling pathways facilitated by MAPs (e.g., ERK, JNK, p38). In recent years information on the cellular processing of cisplatin has essentially arisen. Knowledge collected from studies about biological effects of cisplatin and development of cisplatin resistant phenotype afford important clues for the design of more efficient and less toxic platinum and nonplatinum metal based drugs in cancer therapy. It is to be expected that nonplatinum metal compounds may demonstrate anticancer activity and toxic side effects noticeably different from that of platinum based drugs. Thus titanocene, vanadocene, molybdocene, ferrocene, and zirconocene revealed encouraging results in *in vitro* studies. These compounds might enter by different transport mechanism through cell membrane and distinctly interact with biomolecules than cisplatin. Notwithstanding the extensive applications of cisplatin in the new investigations will provide us with powerful facts for finding a novel efficient and nontoxic metallotherapeutics in anticancer treatment.

Acknowledgments

The authors would like to acknowledge financial support from Alexander von Humboldt Foundation (GNK, SGR), from the Ministerio de Educación y Ciencia, Spain (Grant no. CTQ2011-24346), and from the Ministry of Science and Technological Development of the Republic of Serbia (Grant no. 173013 DMI, SI).

References

[1] F. Arnesano and G. Natile, "Mechanistic insight into the cellular uptake and processing of cisplatin 30 years after its approval by FDA," *Coordination Chemistry Reviews*, vol. 253, no. 15-16, pp. 2070–2081, 2009.

[2] D. J. Higby, H. J. Wallace, D. J. Albert, and J. F. Holland, "Diaminodichloroplatinum: a phase I study showing responses in testicular and other tumors," *Cancer*, vol. 33, no. 5, pp. 1219–1225, 1974.

[3] T. W. Hambley, "Developing new metal-based therapeutics: challenges and opportunities," *Dalton Transactions*, vol. 21, no. 43, pp. 4929–4937, 2007.

[4] E. R. Jamieson and S. J. Lippard, "Structure, recognition, and processing of cisplatin-DNA adducts," *Chemical Reviews*, vol. 99, no. 9, pp. 2467–2498, 1999.

[5] G. Giaccone, "Clinical perspectives on platinum resistance," *Drugs*, vol. 59, no. 4, pp. 9–17, 2000, discussion 37-38.

[6] M. A. Fuertes, C. Alonso, and J. M. Pérez, "Biochemical modulation of cisplatin mechanisms of action: enhancement of antitumor activity and circumvention of drug resistance," *Chemical Reviews*, vol. 103, no. 3, pp. 645–662, 2003.

[7] B. Köberle, J. R. W. Masters, J. A. Hartley, and R. D. Wood, "Defective repair of cisplatin-induced DNA damage caused by reduced XPA protein in testicular germ cell tumours," *Current Biology*, vol. 9, no. 5, pp. 273–276, 1999.

[8] E. L. Mamenta, E. E. Poma, W. K. Kaufmann, D. A. Delmastro, H. L. Grady, and S. G. Chaney, "Enhanced replicative bypass of platinum-DNA adducts in cisplatin-resistant human ovarian carcinoma cell lines," *Cancer Research*, vol. 54, no. 13, pp. 3500–3505, 1994.

[9] A. G. Eliopoulos, D. J. Kerr, J. Herod et al., "The control of apoptosis and drug resistance in ovarian cancer: influence of p53 and Bcl-2," *Oncogene*, vol. 11, no. 7, pp. 1217–1228, 1995.

[10] A. I. Ivanov, J. Christodoulou, J. A. Parkinson et al., "Cisplatin binding sites on human albumin," *The Journal of Biological Chemistry*, vol. 273, no. 24, pp. 14721–14730, 1998.

[11] R. C. DeConti, B. R. Toftness, R. C. Lange, and W. A. Creasey, "Clinical and pharmacological studies with cis diammineedichloroplatinum(II)," *Cancer Research*, vol. 33, no. 6, pp. 1310–1315, 1973.

[12] D. P. Gately and S. B. Howell, "Cellular accumulation of the anticancer agent cisplatin: a review," *British Journal of Cancer*, vol. 67, no. 6, pp. 1171–1176, 1993.

[13] S. Ishida, J. Lee, D. J. Thiele, and I. Herskowitz, "Uptake of the anticancer drug cisplatin mediated by the copper transporter Ctr1 in yeast and mammals," *Proceedings of the National Academy of Sciences of the United States of America*, vol. 99, no. 22, pp. 14298–14302, 2002.

[14] R. A. Alderden, M. D. Hall, and T. W. Hambley, "The discovery and development of cisplatin," *Journal of Chemical Education*, vol. 83, no. 5, pp. 728–734, 2006.

[15] A. R. Timerbaev, C. G. Hartinger, S. S. Aleksenko, and B. K. Keppler, "Interactions of antitumor metallodrugs with serum proteins: advances in characterization using modern analytical methodology," *Chemical Reviews*, vol. 106, no. 6, pp. 2224–2248, 2006.

[16] E. Volckova, F. Evanics, W. W. Yang, and R. N. Bose, "Unwinding of DNA polymerases by the antitumor drug, cis-diamminedichloroplatinum(II)," *Chemical Communications*, vol. 9, no. 10, pp. 1128–1129, 2003.

[17] S. Ahmad, "Platinum-DNA interactions and subsequent cellular processes controlling sensitivity to anticancer platinum complexes," *Chemistry and Biodiversity*, vol. 7, no. 3, pp. 543–566, 2010.

[18] R. J. Knox, F. Friedlos, D. A. Lydall, and J. J. Roberts, "Mechanism of cytotoxicity of anticancer platinum drugs: evidence that cis-diamminedichloroplatinum(II) and cis-diammine-(1,1-cyclobutanedicarboxylato)platinum(II) differ only in the kinetics of their interaction with DNA," *Cancer Research*, vol. 46, no. 4, pp. 1972–1979, 1986.

[19] L. L. Munchausen and R. O. Rahn, "Physical studies on the binding of cis dichlorodiamine platinum(II) to DNA and homopolynucleotides," *Biochimica et Biophysica Acta*, vol. 414, no. 3, pp. 242–255, 1975.

[20] A. Eastman, "Characterization of the adducts produced in DNA by cis-diamminedichloroplatinum(II) and cis-dichloro(ethylenediamine)platinum(II)," *Biochemistry*, vol. 22, no. 16, pp. 3927–3933, 1983.

[21] A. C. M. Plooy, A. M. J. Fichtinger-Schepman, and H. H. Schutte, "The quantitative detection of various Pt-DNA-adducts in Chinese hamster ovary cells treated with cisplatin: application of immunochemical techniques," *Carcinogenesis*, vol. 6, no. 4, pp. 561–566, 1985.

[22] A. E. Egger, C. G. Hartinger, H. B. Hamidane, Y. O. Tsybin, B. K. Keppler, and P. J. Dyson, "High resolution mass spectrometry for studying the interactions of cisplatin with oligonucleotides," *Inorganic Chemistry*, vol. 47, no. 22, pp. 10626–10633, 2008.

[23] D. B. Zamble, D. Mu, J. T. Reardon, A. Sancar, and S. J. Lippard, "Repair of cisplatin-DNA adducts by the mammalian excision nuclease," *Biochemistry*, vol. 35, no. 31, pp. 10004–10013, 1996.

[24] J. T. Reardon, A. Vaisman, S. G. Chaney, and A. Sancar, "Efficient nucleotide excision repair of cisplatin, oxaliplatin, and bis-acetoammine-dichloro-cyclohexylamine-platinum(IV) (JM216) platinum intrastrand DNA diadducts," *Cancer Research*, vol. 59, no. 16, pp. 3968–3971, 1999.

[25] S. W. Johnson, R. P. Perez, A. K. Godwin et al., "Role of platinum-DNA adduct formation and removal in cisplatin resistance in human ovarian cancer cell lines," *Biochemical Pharmacology*, vol. 47, no. 4, pp. 689–697, 1994.

[26] K. V. Ferry, T. C. Hamilton, and S. W. Johnson, "Increased nucleotide excision repair in cisplatin-resistant ovarian

cancer cells: role of ERCC1-XPF," *Biochemical Pharmacology*, vol. 60, no. 9, pp. 1305–1313, 2000.

[27] D. Fink, H. Zheng, S. Nebel et al., "In vitro and in vivo resistance to cisplatin in cells that have lost DNA mismatch repair," *Cancer Research*, vol. 57, no. 10, pp. 1841–1845, 1997.

[28] E. D. Scheeff, J. M. Briggs, and S. B. Howell, "Molecular modeling of the intrastrand guanine-guanine DNA adducts produced by cisplatin and oxaliplatin," *Molecular Pharmacology*, vol. 56, no. 3, pp. 633–643, 1999.

[29] A. Vaisman, M. Varchenko, A. Umar et al., "The role of hMLH1, hMSH3, and hMSH6 defects in cisplatin and oxaliplatin resistance: correlation with replicative bypass of platinum-DNA adducts," *Cancer Research*, vol. 58, no. 16, pp. 3579–3585, 1998.

[30] S. G. Chaney, S. L. Campbell, E. Bassett, and Y. Wu, "Recognition and processing of cisplatin- and oxaliplatin-DNA adducts," *Critical Reviews in Oncology/Hematology*, vol. 53, no. 1, pp. 3–11, 2005.

[31] V. Cepeda, M. A. Fuertes, J. Castilla, C. Alonso, C. Quevedo, and J. M. Pérez, "Biochemical mechanisms of cisplatin cytotoxicity," *Anti-Cancer Agents in Medicinal Chemistry*, vol. 7, no. 1, pp. 3–18, 2007.

[32] K. M. Henkels and J. J. Turchi, "Induction of apoptosis in cisplatin-sensitive and -resistant human ovarian cancer cell lines," *Cancer Research*, vol. 57, no. 20, pp. 4488–4492, 1997.

[33] D. Wang and S. J. Lippard, "Cellular processing of platinum anticancer drugs," *Nature Reviews Drug Discovery*, vol. 4, no. 4, pp. 307–320, 2005.

[34] Z. S. Juo, T. K. Chiu, P. M. Leiberman, I. Baikalov, A. J. Berk, and R. E. Dickerson, "How proteins recognize the TATA box," *Journal of Molecular Biology*, vol. 261, no. 2, pp. 239–254, 1996.

[35] J. Yaneva, S. H. Leuba, K. Van Holde, and J. Zlatanova, "The major chromatin protein his tone H1 binds preferentially to cis-platinum-damaged DNA," *Proceedings of the National Academy of Sciences of the United States of America*, vol. 94, no. 25, pp. 13448–13451, 1997.

[36] M. Kartalou, L. D. Samson, and J. M. Essigmann, "Cisplatin adducts inhibit 1,N6-ethenoadenine repair by interacting with the human 3-methyladenine DNA glycosylase," *Biochemistry*, vol. 39, no. 27, pp. 8032–8038, 2000.

[37] C. Söti, A. Rácz, and P. Csermely, "A nucleotide-dependent molecular switch controls ATP binding at the C-terminal domain of Hsp90. N-terminal nucleotide binding unmasks a C-terminal binding pocket," *The Journal of Biological Chemistry*, vol. 277, no. 9, pp. 7066–7075, 2002.

[38] T. Peleg-Shulman and D. Gibson, "Cisplatin-protein adducts are efficiently removed by glutathione but not by 5′-guanosine monophosphate," *Journal of the American Chemical Society*, vol. 123, no. 13, pp. 3171–3172, 2001.

[39] H. Daino, I. Matsumura, K. Takada et al., "Induction of apoptosis by extracellular ubiquitin in human hematopoietic cells: possible involvement of STAT3 degradation by proteasome pathway in interleukin 6-dependent hematopoietic cells," *Blood*, vol. 95, no. 8, pp. 2577–2585, 2000.

[40] P. A. Nguewa, M. A. Fuertes, V. Cepeda et al., "Pentamidine is an antiparasitic and apoptotic drug that selectively modifies ubiquitin," *Chemistry and Biodiversity*, vol. 2, no. 10, pp. 1387–1400, 2005.

[41] D. Maksimovic-Ivanic, S. Mijatovic, D. Miljkovic et al., "The antitumor properties of a nontoxic, nitric oxide-modified version of saquinavir are independent of Akt," *Molecular Cancer Therapeutics*, vol. 8, no. 5, pp. 1169–1178, 2009.

[42] A. Casini, C. Gabbiani, G. Mastrobuoni et al., "Insights into the molecular mechanisms of protein platination from a case study: the reaction of anticancer platinum(II) iminoethers with horse heart cytochrome C," *Biochemistry*, vol. 46, no. 43, pp. 12220–12230, 2007.

[43] F. Arnesano and G. Natile, ""Platinum on the road": interactions of antitumoral cisplatin with proteins," *Pure and Applied Chemistry*, vol. 80, no. 12, pp. 2715–2725, 2008.

[44] S. R. Datta, A. Brunet, and M. E. Greenberg, "Cellular survival: a play in three akts," *Genes and Development*, vol. 13, no. 22, pp. 2905–2927, 1999.

[45] M. Fraser, B. M. Leung, X. Yan, H. C. Dan, J. Q. Cheng, and B. K. Tsang, "p53 is a determinant of X-linked inhibitor of apoptosis protein/Akt-mediated chemoresistance in human ovarian cancer cells," *Cancer Research*, vol. 63, no. 21, pp. 7081–7088, 2003.

[46] H. C. Dan, M. Sun, S. Kaneko et al., "Akt phosphorylation and stabilization of X-linked inhibitor of apoptosis protein (XIAP)," *The Journal of Biological Chemistry*, vol. 279, no. 7, pp. 5405–5412, 2004.

[47] J. G. Viniegra, J. H. Losa, V. J. Sánchez-Arévalo et al., "Modulation of PI3K/Akt pathway by E1a mediates sensitivity to cisplatin," *Oncogene*, vol. 21, no. 46, pp. 7131–7136, 2002.

[48] Y. Shaul, "c-Abl: activation and nuclear targets," *Cell Death and Differentiation*, vol. 7, no. 1, pp. 10–16, 2000.

[49] J. Gong, A. Costanzo, H. Q. Yang et al., "The tyrosine kinase c-Abl regulates p73 in apoptotic response to cisplatin-induced DNA damage," *Nature*, vol. 399, no. 6738, pp. 806–809, 1999.

[50] L. Harhaji-Trajkovic, U. Vilimanovich, T. Kravic-Stevovic, V. Bumbasirevic, and V. Trajkovic, "AMPK-mediated autophagy inhibits apoptosis in cisplatin-treated tumour cells," *Journal of Cellular and Molecular Medicine*, vol. 13, no. 9, pp. 3644–3654, 2009.

[51] X. Wang, J. L. Martindale, and N. J. Holbrook, "Requirement for ERK activation in cisplatin-induced apoptosis," *The Journal of Biological Chemistry*, vol. 275, no. 50, pp. 39435–39443, 2000.

[52] W. Cui, E. M. Yazlovitskaya, M. S. Mayo et al., "Cisplatin-induced response of c-jun N-terminal kinase 1 and extracellular signal—regulated protein kinases 1 and 2 in a series of cisplatin-resistant ovarian carcinoma cell lines," *Molecular Carcinogenesis*, vol. 29, pp. 219–228, 2000.

[53] S. Mijatovic, D. Maksimovic-Ivanic, J. Radovic et al., "Aloe emodin decreases the ERK-dependent anticancer activity of cisplatin," *Cellular and Molecular Life Sciences*, vol. 62, no. 11, pp. 1275–1282, 2005.

[54] S. Mijatovic, D. Maksimovic-Ivanic, J. Radovic et al., "Antiglioma action of aloe emodin: the role of ERK inhibition," *Cellular and Molecular Life Sciences*, vol. 62, no. 5, pp. 589–598, 2005.

[55] A. Mansouri, L. D. Ridgway, A. L. Korapati et al., "Sustained activation of JNK/p38 MAPK pathways in response to cisplatin leads to Fas ligand induction and cell death in ovarian carcinoma cells," *The Journal of Biological Chemistry*, vol. 278, no. 21, pp. 19245–19256, 2003.

[56] I. Sánchez-Perez, J. R. Murguía, and R. Perona, "Cisplatin induces a persistent activation of JNK that is related to cell death," *Oncogene*, vol. 16, no. 4, pp. 533–540, 1998.

[57] J. Hayakawa, M. Ohmichi, H. Kurachi et al., "Inhibition of extracellular signal-regulated protein kinase or c-Jun N-terminal protein kinase cascade, differentially activated by

cisplatin, sensitizes human ovarian cancer cell line," *The Journal of Biological Chemistry*, vol. 274, no. 44, pp. 31648–31654, 1999.

[58] R. J. Davis, "Signal transduction by the JNK group of MAP kinases," *Cell*, vol. 103, no. 2, pp. 239–252, 2000.

[59] P. Pandey, J. Raingeaud, M. Kaneki et al., "Activation of p38 mitogen-activated protein kinase by c-Abl-dependent and -independent mechanisms," *The Journal of Biological Chemistry*, vol. 271, no. 39, pp. 23775–23779, 1996.

[60] J. Hernández Losa, C. Parada Cobo, J. Guinea Viniegra, V. J. Sánchez-Arevalo Lobo, S. Ramón y Cajal, and R. Sánchez-Prieto, "Role of the p38 MAPK pathway in cisplatin-based therapy," *Oncogene*, vol. 22, no. 26, pp. 3998–4006, 2003.

[61] D. Wang and S. J. Lippard, "Cisplatin-induced post-translational modification of histones H3 and H4," *The Journal of Biological Chemistry*, vol. 279, no. 20, pp. 20622–20625, 2004.

[62] V. M. Gonzalez, M. A. Fuertes, C. Alonso, and J. M. Perez, "Is cisplatin-induced cell death always produced by apoptosis?" *Molecular Pharmacology*, vol. 59, no. 4, pp. 657–663, 2001.

[63] W. Lieberthal, V. Triaca, and J. Levine, "Mechanisms of death induced by cisplatin in proximal tubular epithelial cells: apoptosis vs. necrosis," *American Journal of Physiology*, vol. 270, no. 4, pp. F700–F708, 1996.

[64] Y. Eguchi, S. Shimizu, and Y. Tsujimoto, "Intracellular ATP levels determine cell death fate by apoptosis or necrosis," *Cancer Research*, vol. 57, no. 10, pp. 1835–1840, 1997.

[65] R. Zhou, M. G. Vander Heiden, and C. M. Rudin, "Genotoxic exposure is associated with alterations in glucose uptake and metabolism," *Cancer Research*, vol. 62, no. 12, pp. 3515–3520, 2002.

[66] Z. Herceg and Z. Q. Wang, "Failure of poly(ADP-ribose) polymerase cleavage by caspases leads to induction of necrosis and enhanced apoptosis," *Molecular and Cellular Biology*, vol. 19, no. 7, pp. 5124–5133, 1999.

[67] T. Hirsch, P. Marchetti, S. A. Susin et al., "The apoptosis-necrosis paradox. Apoptogenic proteases activated after mitochondrial permeability transition determine the mode of cell death," *Oncogene*, vol. 15, no. 13, pp. 1573–1581, 1997.

[68] L. Ding, C. Yuan, F. Wei et al., "Cisplatin restores TRAIL apoptotic pathway in glioblastoma-derived stem cells through up-regulation of DR5 and down-regulation of c-FLIP," *Cancer Investigation*, vol. 29, pp. 511–520, 2011.

[69] O. Vondálová Blanárová, I. Jelínková, A. Szöor et al., "Cisplatin and a potent platinum(IV) complex-mediated enhancement of TRAIL-induced cancer cells killing is associated with modulation of upstream events in the extrinsic apoptotic pathway," *Carcinogenesis*, vol. 32, no. 1, pp. 42–51, 2011.

[70] D. Maksimovic-Ivanic, S. Stosic-Grujicic, F. Nicoletti, and S. Mijatovic, "Resistance to TRAIl and how to surmount it," *Immunology Research*, vol. 52, no. 1-2, pp. 157–168, 2012.

[71] M. P. Decatris, S. Sundar, and K. J. O'Byrne, "Platinum-based chemotherapy in metastatic breast cancer: current status," *Cancer Treatment Reviews*, vol. 30, no. 1, pp. 53–81, 2004.

[72] M. D. Shelley, K. Burgon, and M. D. Mason, "Treatment of testicular germ-cell cancer: a cochrane evidence-based systematic review," *Cancer Treatment Reviews*, vol. 28, no. 5, pp. 237–253, 2002.

[73] M. S. Kim, M. Blake, J. H. Baek, G. Kohlhagen, Y. Pommier, and F. Carrier, "Inhibition of histone deacetylase increases cytotoxicity to anticancer drugs targeting DNA," *Cancer Research*, vol. 63, no. 21, pp. 7291–7300, 2003.

[74] N. J. Long, *Metallocenes*, Blackwell Science, Oxford, UK, 1998.

[75] A. Korfel, M. E. Scheulen, H. J. Schmoll et al., "Phase I clinical and pharmacokinetic study of titanocene dichloride in adults with advanced solid tumors," *Clinical Cancer Research*, vol. 4, no. 11, pp. 2701–2708, 1998.

[76] C. V. Christodoulou, D. R. Ferry, D. W. Fyfe et al., "Phase I trial of weekly scheduling and pharmacokinetics of titanocene dichloride in patients with advanced cancer," *Journal of Clinical Oncology*, vol. 16, no. 8, pp. 2761–2769, 1998.

[77] K. Mross, P. Robben-Bathe, L. Edler et al., "Phase I clinical trial of a day-1, -3, -5 every 3 weeks schedule with titanocene dichloride (MKT 5) in patients with advanced cancer: a study of the phase I study group of the association for medical oncology (AIO) of the German Cancer Society," *Onkologie*, vol. 23, no. 6, pp. 576–579, 2000.

[78] B. W. Müller, R. Müller, S. Lucks, and W. Mohr, "Medac Gesellschaft fur Klinische Spzeilpräparate GmbH," US Patent 5, 296, 237, 1994.

[79] N. Kröger, U. R. Kleeberg, K. Mross, L. Edler, G. Saß, and D. K. Hossfeld, "Phase II clinical trial of titanocene dichloride in patients with metastatic breast cancer," *Onkologie*, vol. 23, no. 1, pp. 60–62, 2000.

[80] G. Lümmen, H. Sperling, H. Luboldt, T. Otto, and H. Rübben, "Phase II trial of titanocene dichloride in advanced renal-cell carcinoma," *Cancer Chemotherapy and Pharmacology*, vol. 42, no. 5, pp. 415–417, 1998.

[81] E. Meléndez, "Titanium complexes in cancer treatment," *Critical Reviews in Oncology/Hematology*, vol. 42, no. 3, pp. 309–315, 2002.

[82] F. Caruso and M. Rossi, "Antitumor titanium compounds and related metallocenes," in *Metal Ions in Biological System*, A. Sigel and H. Sigel, Eds., vol. 42 of *Metal Complexes in Tumor Diagnostics and as Anticancer Agents*, Marcel Dekker, New York, NY, USA, 2004.

[83] J. C. Dabrowiak, *Metals in Medicine*, John Wiley & Sons, West Sussex, UK, 2009.

[84] U. Olszewski and G. Hamilton, "Mechanisms of cytotoxicity of anticancer titanocenes," *Anti-Cancer Agents in Medicinal Chemistry*, vol. 10, no. 4, pp. 302–311, 2010.

[85] K. Strohfeldt and M. Tacke, "Bioorganometallic fulvene-derived titanocene anti-cancer drugs," *Chemical Society Reviews*, vol. 37, no. 6, pp. 1174–1187, 2008.

[86] R. Hernández, J. Lamboy, L. M. Gao, J. Matta, F. R. Román, and E. Meléndez, "Structure-activity studies of Ti(IV) complexes: aqueous stability and cytotoxic properties in colon cancer HT-29 cells," *Journal of Biological Inorganic Chemistry*, vol. 13, no. 5, pp. 685–692, 2008.

[87] R. Hernández, J. Méndez, J. Lamboy, M. Torres, F. R. Román, and E. Meléndez, "Titanium(IV) complexes: cytotoxicity and cellular uptake of titanium(IV) complexes on caco-2 cell line," *Toxicology in Vitro*, vol. 24, no. 1, pp. 178–183, 2010.

[88] L. M. Gao, J. Matta, A. L. Rheingold, and E. Meléndez, "Synthesis, structure and biological activity of amide-functionalized titanocenyls: improving their cytotoxic properties," *Journal of Organometallic Chemistry*, vol. 694, no. 26, pp. 4134–4139, 2009.

[89] A. Gansäuer, I. Winkler, D. Worgull et al., "Carbonyl-substituted titanocenes: a novel class of cytostatic compounds with high antitumor and antileukemic activity," *Chemistry*, vol. 14, no. 14, pp. 4160–4163, 2008.

[90] O. R. Allen, L. Croll, A. L. Gott, R. J. Knox, and P. C. McGowan, "Functionalized cyclopentadienyl titanium organometallic compounds as new antitumor drugs," *Organometallics*, vol. 23, no. 2, pp. 288–292, 2004.

[91] O. R. Allen, A. L. Gott, J. A. Hartley, J. M. Hartley, R. J. Knox, and P. C. McGowan, "Functionalised cyclopentadienyl titanium compounds as potential anticancer drugs," *Dalton Transactions*, no. 43, pp. 5082–5090, 2007.

[92] G. D. Potter, M. C. Baird, and S. P. C. Cole, "A new series of titanocene dichloride derivatives bearing chiral alkylammonium groups; Assessment of their cytotoxic properties," *Inorganica Chimica Acta*, vol. 364, no. 1, pp. 16–22, 2010.

[93] L. M. Gao, J. L. Vera, J. Matta, and E. Meléndez, "Synthesis and cytotoxicity studies of steroid-functionalized titanocenes as potential anticancer drugs: sex steroids as potential vectors for titanocenes," *Journal of Biological Inorganic Chemistry*, vol. 15, no. 6, pp. 851–859, 2010.

[94] S. Gómez-Ruiz, G. N. Kaluđerović, S. Prashar et al., "Cytotoxic studies of substituted titanocene and ansa-titanocene anticancer drugs," *Journal of Inorganic Biochemistry*, vol. 102, no. 8, pp. 1558–1570, 2008.

[95] S. Gómez-Ruiz, G. N. Kaluđerović, Ž. Žižak et al., "Anticancer drugs based on alkenyl and boryl substituted titanocene complexes," *Journal of Organometallic Chemistry*, vol. 694, no. 13, pp. 1981–1987, 2009.

[96] G. N. Kaluđerović, V. Tayurskaya, R. Paschke, S. Prashar, M. Fajardo, and S. Gómez-Ruiz, "Synthesis, characterization and biological studies of alkenyl-substituted titanocene(IV) carboxylate complexes," *Applied Organometallic Chemistry*, vol. 24, no. 9, pp. 656–662, 2010.

[97] H. Sun, H. Li, R. A. Weir, and P. J. Sadler, "You have full text access to this content the first specific Ti^{IV}-protein complex: potential relevance to anticancer activity of titanocenes," *Angewandte Chemie International Edition*, vol. 37, no. 11, pp. 1577–1579, 1998.

[98] M. Guo and P. J. Sadler, "Competitive binding of the anticancer drug titanocene dichloride to N,N′-ethylenebis(o-hydroxyphenylglycine) and adenosine triphosphate: a model for Ti^{IV} uptake and release by transferrin," *Journal of the Chemical Society, Dalton Transactions*, vol. 1, pp. 7–9, 2000.

[99] M. Guo, H. Sun, S. Bihari et al., "Stereoselective formation of seven-coordinate titanium(IV) monomer and dimer complexes of ethylenebis(o-hydroxyphenyl)glycine," *Inorganic Chemistry*, vol. 39, no. 2, pp. 206–215, 2000.

[100] M. Guo, H. Sun, H. J. McArdle, L. Gambling, and P. J. Sadler, "Ti(IV) uptake and release by human serum transferrin and recognition of Ti(IV)-transferrin by cancer cells: understanding the mechanism of action of the anticancer drug titanocene dichloride," *Biochemistry*, vol. 39, no. 33, pp. 10023–10033, 2000.

[101] P. Köpf-Maier and D. Krahl, "Tumor inhibition by metallogenes: ultrastructural localization of titanium and vanadium in treated tumor cells by electron energy loss spectroscopy," *Chemico-Biological Interactions*, vol. 44, no. 3, pp. 317–328, 1983.

[102] P. Köpf-Maier, "Intracellular localization of titanium within xenografted sensitive human tumors after treatment with the antitumor agent titanocene dichloride," *Journal of Structural Biology*, vol. 105, no. 1–3, pp. 35–45, 1990.

[103] A. D. Tinoco, C. D. Incarvito, and A. M. Valentine, "Calorimetric, spectroscopic, and model studies provide insight into the transport of Ti(IV) by human serum transferrin," *Journal of the American Chemical Society*, vol. 129, no. 11, pp. 3444–3454, 2007.

[104] A. D. Tinoco, E. V. Eames, and A. M. Valentine, "Reconsideration of serum Ti(IV) transport: albumin and transferrin trafficking of Ti(IV) and its complexes," *Journal of the American Chemical Society*, vol. 130, no. 7, pp. 2262–2270, 2008.

[105] M. Pavlaki, K. Debeli, I. E. Triantaphyllidou, N. Klouras, E. Giannopoulou, and A. J. Aletras, "A proposed mechanism for the inhibitory effect of the anticancer agent titanocene dichloride on tumour gelatinases and other proteolytic enzymes," *Journal of Biological Inorganic Chemistry*, vol. 14, no. 6, pp. 947–957, 2009.

[106] O. R. Allen, R. J. Knox, and P. C. McGowan, "Functionalised cyclopentadienyl zirconium compounds as potential anticancer drugs," *Dalton Transactions*, no. 39, pp. 5293–5295, 2008.

[107] D. Wallis, J. Claffey, B. Gleeson, M. Hogan, H. Müller-Bunz, and M. Tacke, "Novel zirconocene anticancer drugs?" *Journal of Organometallic Chemistry*, vol. 694, no. 6, pp. 828–833, 2009.

[108] S. Gómez-Ruiz, G. N. Kaluđerović, D. Polo-Cerón et al., "A novel alkenyl-substituted ansa-zirconocene complex with dual application as olefin polymerization catalyst and anticancer drug," *Journal of Organometallic Chemistry*, vol. 694, no. 18, pp. 3032–3038, 2009.

[109] P. Köpf-Maier, "Antitumor bis(cyclopentadienyl) metal complexes," in *Metal Complexes in Cancer Chemotherapy*, B. K. Keppler, Ed., pp. 259–296, VCH Verlagsgesellschaft, Weinheim, Germany, 1993.

[110] C. S. Navara, A. Benyumov, A. Vassilev, R. K. Narla, P. Ghosh, and F. M. Uckun, "Vanadocenes as potent anti-proliferative agents disrupting mitotic spindle formation in cancer cells," *Anti-Cancer Drugs*, vol. 12, no. 4, pp. 369–376, 2001.

[111] P. Ghosh, O. J. D'Cruz, R. K. Narla, and F. M. Uckun, "Apoptosis-inducing vanadocene compounds against human testicular cancer," *Clinical Cancer Research*, vol. 6, no. 4, pp. 1536–1545, 2000.

[112] H. Paláčková, J. Vinklárek, J. Holubová, I. Císařová, and M. Erben, "The interaction of antitumor active vanadocene dichloride with sulfur-containing amino acids," *Journal of Organometallic Chemistry*, vol. 692, no. 17, pp. 3758–3764, 2007.

[113] J. Vinklárek, J. Honzíček, and J. Holubová, "Interaction of the antitumor agent vanadocene dichloride with phosphate buffered saline," *Inorganica Chimica Acta*, vol. 357, no. 12, pp. 3765–3769, 2004.

[114] J. Vinklárek, H. Paláčková, J. Honzíček, J. Holubová, M. Holčapek, and I. Císařová, "Investigation of vanadocene(IV) α-amino acid complexes: synthesis, structure, and behavior in physiological solutions, human plasma, and blood," *Inorganic Chemistry*, vol. 45, no. 5, pp. 2156–2162, 2006.

[115] B. Gleeson, J. Claffey, A. Deally et al., "Synthesis and cytotoxicity studies of fluorinated derivatives of vanadocene Y," *European Journal of Inorganic Chemistry*, no. 19, pp. 2804–2810, 2009.

[116] B. Gleeson, J. Claffey, M. Hogan, H. Müller-Bunz, D. Wallis, and M. Tacke, "Novel benzyl-substituted vanadocene anticancer drugs," *Journal of Organometallic Chemistry*, vol. 694, no. 9-10, pp. 1369–1374, 2009.

[117] B. Gleeson, J. Claffey, A. Deally et al., "Novel benzyl-substituted molybdocene anticancer drugs," *Inorganica Chimica Acta*, vol. 363, no. 8, pp. 1831–1836, 2010.

[118] B. Gleeson, M. Hogan, H. Müller-Bunz, and M. Tacke, "Synthesis and preliminary cytotoxicity studies of indole-substituted vanadocenes," *Transition Metal Chemistry*, vol. 35, no. 8, pp. 973–983, 2010.

[119] J. Honzíček, I. Klepalová, J. Vinklárek et al., "Synthesis, characterization and cytotoxic effect of ring-substituted and ansa-bridged vanadocene complexes," *Inorganica Chimica Acta*, vol. 373, no. 1, pp. 1–7, 2011.

[120] J. B. Waern and M. M. Harding, "Bioorganometallic chemistry of molybdocene dichloride," *Journal of Organometallic Chemistry*, vol. 689, no. 25, pp. 4655–4668, 2004.

[121] J. B. Waern, C. T. Dillon, and M. M. Harding, "Organometallic anticancer agents: cellular uptake and cytotoxicity studies on thiol derivatives of the antitumor agent molybdocene dichloride," *Journal of Medicinal Chemistry*, vol. 48, no. 6, pp. 2093–2099, 2005.

[122] J. B. Waern, H. H. Harris, B. Lai, Z. Cai, M. M. Harding, and C. T. Dillon, "Intracellular mapping of the distribution of metals derived from the antitumor metallocenes," *Journal of Biological Inorganic Chemistry*, vol. 10, no. 5, pp. 443–452, 2005.

[123] J. B. Waern, P. Turner, and M. M. Harding, "Synthesis and hydrolysis of thiol derivatives of molybdocene dichloride incorporating electron-withdrawing substituents," *Organometallics*, vol. 25, no. 14, pp. 3417–3421, 2006.

[124] K. S. Campbell, A. J. Foster, C. T. Dillon, and M. M. Harding, "Genotoxicity and transmission electron microscopy studies of molybdocene dichloride," *Journal of Inorganic Biochemistry*, vol. 100, no. 7, pp. 1194–1198, 2006.

[125] J. H. Toney and T. J. Marks, "Hydrolysis chemistry of the metallocene dichlorides $M(\eta^5\text{-}C_5H_5)_2Cl_2$, M = Ti, V, Zr. Aqueous kinetics, equilibria, and mechanistic implications for a new class of antitumor agents," *Journal of the American Chemical Society*, vol. 107, no. 4, pp. 947–953, 1985.

[126] C. Balzarek, T. J. R. Weakley, L. Y. Kuo, and D. R. Tyler, "Investigation of the monomer-dimer equilibria of molybdocenes in water," *Organometallics*, vol. 19, no. 15, pp. 2927–2931, 2000.

[127] L. Y. Kuo, M. G. Kanatzidis, M. Sabat, A. L. Tipton, and T. J. Marks, "Metallocene antitumor agents. Solution and solid-state molybdenocene coordination chemistry of DNA constituents," *Journal of the American Chemical Society*, vol. 113, no. 24, pp. 9027–9045, 1991.

[128] P. M. Abeysinghe and M. M. Harding, "Antitumour bis(cyclopentadienyl) metal complexes: titanocene and molybdocene dichloride and derivatives," *Dalton Transactions*, no. 32, pp. 3474–3482, 2007.

[129] M. M. Harding and G. Mokdsi, "Antitumour metallocenes: structure-activity studies and interactions with biomolecules," *Current Medicinal Chemistry*, vol. 7, no. 12, pp. 1289–1303, 2000.

[130] K. S. Campbell, C. T. Dillon, S. V. Smith, and M. M. Harding, "Radiotracer studies of the antitumor metallocene molybdocene dichloride with biomolecules," *Polyhedron*, vol. 26, no. 2, pp. 456–459, 2007.

[131] P. Köpf-Maier, H. Köpf, and E. W. Neuse, "Ferrocenium salts—the first antineoplastic iron compounds," *Angewandte Chemie—International Edition in English*, vol. 23, no. 6, pp. 456–457, 1984.

[132] P. Köpf-Maier, H. Kopf, and E. W. Neuse, "Ferricenium complexes: a new type of water-soluble antitumor agent," *Journal of Cancer Research and Clinical Oncology*, vol. 108, no. 3, pp. 336–340, 1984.

[133] S. Top, J. Tang, A. Vessières, D. Carrez, C. Provot, and G. Jaouen, "Ferrocenyl hydroxytamoxifen: a prototype for a new range of oestradiol receptor site-directed cytotoxics," *Chemical Communications*, vol. 8, pp. 955–956, 1996.

[134] S. Top, B. Dauer, J. Vaissermann, and G. Jaouen, "Facile route to ferrocifen, 1-[4-(2-dimethylaminoethoxy)]-1-(phenyl-2-ferrocenyl-but-1-ene), first organometallic analogue of tamoxifen, by the McMurry reaction," *Journal of Organometallic Chemistry*, vol. 541, no. 1-2, pp. 355–361, 1997.

[135] S. Top, A. Vessières, C. Cabestaing et al., "Studies on organometallic selective estrogen receptor modulators. (SERMs) Dual activity in the hydroxy-ferrocifen series," *Journal of Organometallic Chemistry*, vol. 637–639, no. 1, pp. 500–506, 2001.

[136] S. Top, A. Vessières, G. Leclercq et al., "Synthesis, biochemical properties and molecular modelling studies of organometallic Specific Estrogen Receptor Modulators (SERMs), the ferrocifens and hydroxyferrocifens: evidence for an antiproliferative effect of hydroxyferrocifens on both hormone-dependent and hormone-independent breast cancer cell lines," *Chemistry*, vol. 9, no. 21, pp. 5223–5236, 2003.

[137] G. Jaouen, S. Top, A. Vessières, G. Leclercq, and M. J. McGlinchey, "The first organometallic selective estrogen receptor modulators (SERMs) and their relevance to breast cancer," *Current Medicinal Chemistry*, vol. 11, no. 18, pp. 2505–2517, 2004.

[138] D. Plazuk, A. Vessières, E. A. Hillard et al., "A [3]ferrocenophane polyphenol showing a remarkable antiproliferative activity on breast and prostate cancer cell lines," *Journal of Medicinal Chemistry*, vol. 52, no. 15, pp. 4964–4967, 2009.

[139] E. A. Hillard, A. Vessières, and G. Jaouen, "Ferrocene functionalized endocrine modulators as anticancer agents," *Topics in Organometallic Chemistry*, vol. 32, pp. 81–117, 2010.

[140] D. P. Buck, P. M. Abeysinghe, C. Cullinane, A. I. Day, J. G. Collins, and M. M. Harding, "Inclusion complexes of the antitumour metallocenes Cp_2MCl_2 (M = Mo, Ti) with cucurbit[n]urils," *Dalton Transactions*, no. 17, pp. 2328–2334, 2008.

[141] C. C. L. Pereira, C. V. Diogo, A. Burgeiro et al., "Complex formation between heptakis(2,6-di-O-methyl)-β-cyclodextrin and cyclopentadienyl molybdenum(II) dicarbonyl complexes: structural studies and cytotoxicity evaluations," *Organometallics*, vol. 27, no. 19, pp. 4948–4956, 2008.

[142] D. Pérez-Quintanilla, S. Gomez-Ruiz, Ž. Žižak et al., "A new generation of anticancer drugs: mesoporous materials modified with titanocene complexes," *Chemistry*, vol. 15, no. 22, pp. 5588–5597, 2009.

[143] G. N. Kaluđerović, D. Pérez-Quintanilla, I. Sierra et al., "Study of the influence of the metal complex on the cytotoxic activity of titanocene-functionalized mesoporous materials," *Journal of Materials Chemistry*, vol. 20, no. 4, pp. 806–814, 2010.

[144] G. N. Kaluđerović, D. Pérez-Quintanilla, Ž. Žižak, Z. D. Juranić, and S. Gómez-Ruiz, "Improvement of cytotoxicity of titanocene-functionalized mesoporous materials by the increase of the titanium content," *Dalton Transactions*, vol. 39, no. 10, pp. 2597–2608, 2010.

[145] A. García-Peñas, S. Gómez-Ruiz, D. Pérez-Quintanilla et al., "Study of the cytotoxicity and particle action in human cancer cells of titanocene-functionalized materials," *Journal of Inorganic Biochemistry*, vol. 106, no. 2, pp. 100–110, 2012.

5

Biosorption of Mercury (II) from Aqueous Solutions onto Fungal Biomass

Víctor M. Martínez-Juárez,[1] **Juan F. Cárdenas-González,**[2] **María Eugenia Torre-Bouscoulet,**[2] **and Ismael Acosta-Rodríguez**[2]

[1] *Área Académica de Medicina Veterinaria y Zootecnia, Instituto de Ciencias Agropecuarias, Universidad Autónoma del Estado de Hidalgo, Zona Universitaria, Rancho Universitario Km 1. C.P. 43600, Tulancingo de Bravo Hidalgo, Mexico*
[2] *Laboratorio de Micología Experimental, Centro de Investigación y de Estudios de Posgrado, Facultad de Ciencias Químicas, Universidad Autónoma de San Luis Potosí, Avenida Dr. Manuel Nava No. 6, Zona Universitaria, 78320 San Luis Potosí, SLP, Mexico*

Correspondence should be addressed to Ismael Acosta-Rodríguez, iacosta@uaslp.mx

Academic Editor: Concepción López

The biosorption of mercury (II) on 14 fungal biomasses, *Aspergillus flavus* I–V, *Aspergillus fumigatus* I-II, *Helminthosporium* sp., *Cladosporium* sp., *Mucor rouxii* mutant, *M. rouxii* IM-80, *Mucor* sp 1 and 2, and *Candida albicans*, was studied in this work. It was found that the biomasses of the fungus *M. rouxii* IM-80, *M. rouxii* mutant, *Mucor* sp1, and *Mucor* sp 2 were very efficient removing the metal in solution, using dithizone, reaching the next percentage of removals: 95.3%, 88.7%, 80.4%, and 78.3%, respectively. The highest adsorption was obtained at pH 5.5, at 30°C after 24 hours of incubation, with 1 g/100 mL of fungal biomass.

1. Introduction

Heavy metal ion pollution has become wide spread throughout the world as a result of industrialization, which significantly threats the ecosystem, especially the people's health due to their severe toxicity. In order to minimize the impacts of metals contaminated, wastewaters need to be treated before discharge to water bodies. Environmental mercury levels have increased considerably in recent years. The direct anthropogenic sources of mercury in water bodies are related to numerous industrial applications (e.g., chloroalkali productions, pharmaceutical and cosmetic preparations, electrical instruments, and pulp and paper industries) and many products of common use (e.g., thermometers, batteries, and medical drugs) [1].

Mercury is one of the priority pollutants listed by the USEPA as it can easily pass the blood-brain barrier and affect the fetal brain [2]. High concentrations of Hg (II) cause impairment of pulmonary function and kidney, chest pain, and dyspnea [3–6]. The illness, which came to be known as Minamata disease, was caused by mercury poisoning gas as a result of eating contaminated fish. Mercury has very high tendency for binding to proteins and it mainly affects the renal and nervous systems [7]. Mercury removal from wastewaters needs to achieve very low levels for all these reasons. Metal sorption by different types of biomaterials such as inactive dried biomass of algae, bacteria, and fungi can serve for removing metals from solution because of their unique chemical composition [8–10] investigated the metal binding capacity of the thermophilic bacteria *Geobacillus thermodenitrificans*. According to this study, bacterial biomass reduced the concentration of Fe^{3+} (91.31%), Cr^{3+} (80.80%), Co^{2+} (79.71%), Cu^{2+} (57.14%), Zn^{2+} (55.14%), Cd^{2+} (49.02%), Ag^+ (43.25%), and Pb^{2+} (36.86%) at different optimum pH at 720 min [11], also it was investigated the biosorption of cadmium (II) from aqueous solutions by industrial fungus *Rhizopus cohnii*, for this, some researchers reported the maximum uptake of cadmium by fungal biomass at 40.5 mg/g in optimal conditions, which was higher than many other adsorbents, including activated carbon. Some other researches also indicated that biosorption is a very effective method to remove metals from the water and wastewater [12–18]. The objective of this work was to study the removal of mercury (II) in solution by 14 species of fungi isolated from different areas of mine waste and resistant to various heavy metals.

2. Experimental

2.1. Biosorbents. The biosorbents utilized were 14 fungal biomasses of *Aspergillus flavus* I–V, *Aspergillus fumigatus* I-II isolated from a mining waste in Zimapan, HGO, Mexico; *Helminthosporium* sp., *Cladosporium* sp., *Mucor* sp. 1 and 2 resistant to zinc, lead, and copper isolated from the air collected near a zinc smelting plant in San Luis Potosi, S.L.P., Mexico; *Mucor rouxii* mutant resistant to copper and lead, obtained by mutagenesis with ethylmethanesulfonate; *Mucor rouxii* IM-80 (wild type), and *Candida albicans* isolated from a leather works, located in Leon, GTO Mexico.

2.2. Microorganism and Mercury (II) Solutions. The fungi were grown at 28°C in an agitated and aerated liquid mediUM containing thioglycolate broth, 8 g/L. After 4-5 days of incubation for *A. flavus* I–V, *A. fumigates* I-II, *Helminthosporium* sp., *Cladosporium* sp., *Mucor* sp 1-2, *M. rouxii* mutant, *M. rouxii*, IM-80, and *C. albicans*, the cells were centrifuged at 3000 rpm for 5 min, washed twice with trideionized water, and then dried at 80°C for 4 h in an oven. Finally, the fungal biomass was milled and stored in an amber bottle in the refrigerator until their use.

For analysis were prepared a series of solutions of mercury of 100 mg/L, pH was adjusted with nitric acid, and the quantity of biomass added to each flask was of 1 g/100 mL for the mercury's solution. It taken samples at different times, the biomass is removed for centrifugation (3000 rpm/5 min), and the supernatant is analyzed to define the ion metal concentration.

2.3. Determination of Mercury (II). The concentration of mercury ions in solution was determined spectrophotometrically at 492 nm using Dithizone (1,5-Diphenylthiocarbazone) as the complexing agent, by the formation of orange colored solution. The minimum detectable mercury concentration was 1.0 μg/10 mL of dithizone solution [19].

3. Results and Discussion

3.1. The Effect of Incubation Time and pH. Figure 1 shows the effect of contact time and pH on biosorption of Hg (II) ions (100 mg/L) to the dried *M. rouxii* IM-80 biomass, it was found that the highest removal occurred at 24 h of incubation and pH 5.5 (95.4%) (Figure 1), and these results resemble those reported by *Aspergillus versicolor* [20] and *Rhizopus oligosporus* [21]. Structural properties of the biosorbent including the cellular support and other several factors are known to affect the biosorption rate [22]. The pH is a critical parameter in biosorption because it influences the equilibrium by affecting the speciation of the metal ion(s) in solution, the concentration of competing hydrogen ions,

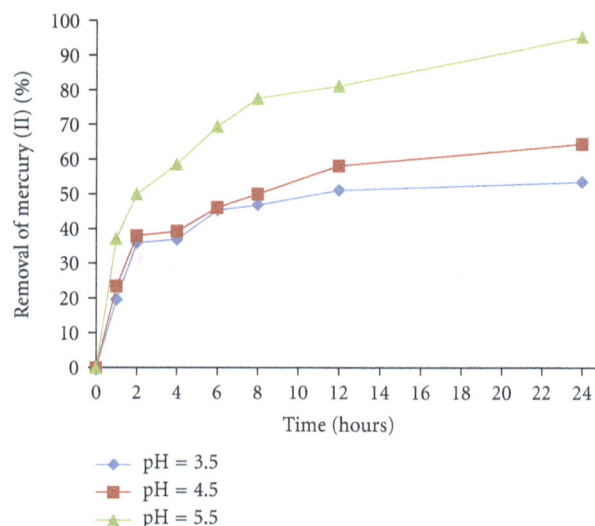

FIGURE 1: The effect of pH and incubation time on the biosorption of mercury (II). 100 mg/L Hg (II), 30°C, 100 rpm, 1 g of fungal biomass of *M. rouxii* IM-80.

and the chemistry of the active binding sites on the biomass. The fungal cell wall contains amino, carboxyl, and phosphate functional reactive groups. The carboxyl and phosphate groups carry negative charges that allow the fungal cell wall components to be potential detainer of metal ions [23]. The maximum biosorption of Hg (II) was observed at pH 5.5 (95.3%, Figure 1). At acidic pH (3.0), protonation of the cell wall components adversely affected the biosorption capacity of the fungal biomass, but its effect became minor with increasing pH in the medium. With an increase in pH, the negative charge density on the cell surface increases due to the deprotonation of the metal binding sites and thus increases biosorption [23]. Several researchers investigated the effect of pH on biosorption of mercury (II) by using different kinds of microbial biomasses. For example, *A. versicolor* [20], *R. oligosporus* [21], *Penicillium purpurogenum* [23], and the maximum biosorption were obtained in the pH range of 5.0 to 7.0.

3.2. Effect of Temperature. Figure 2 shows the effect of varying temperatures (30°C, 35°C, and 40°C), the maximal adsorption capacity was found at 30 ± 1°C, (95.3%), and the adsorption capacity of dried *M. rouxii* IM-80 biomass decreased with temperatures higher than 30 ± 1°C (83.2% at 35°C, and 71.4% at 40°C). This is like to the report for *A. versicolor*, *R. oligosporus*, and *Bacillus subtilis* [20, 21, 24]. The temperature of the adsorption medium could be important for energy-dependent mechanisms in metal biosorption by microorganisms. Energy-independent mechanisms are less likely to be affected by temperature since the process responsible for biosorption is largely physicochemical in nature. The biosorption of Hg (II) by *M. rouxii* IM-80 fungus appears to be temperature-dependent over the temperature range tested (30–40°C).

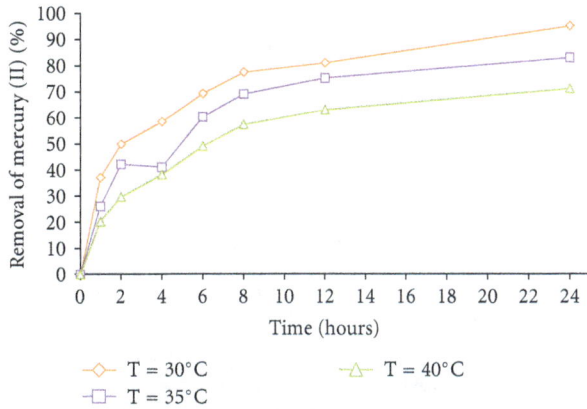

FIGURE 2: The effect of the temperature on mercury (II) removal. 100 mg/L Hg (II), 100 rpm. pH 5.5, 1 g of fungal biomass of *M. rouxii* IM-80.

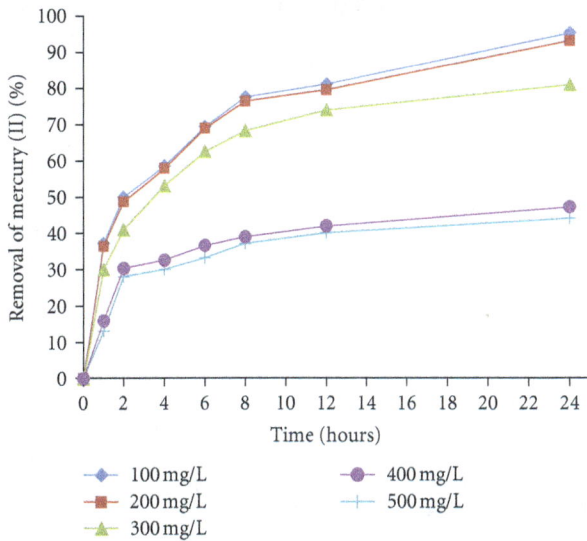

FIGURE 3: The effect of the concentration of mercury (II) in solution on the removal of Hg (II) ions. 100 rpm, 30°C, pH 5.5.1 g of fungal biomass of *M. rouxii* IM-80.

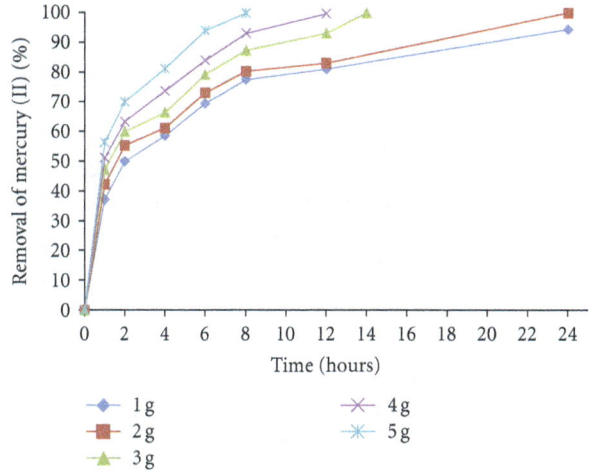

FIGURE 4: The effect of fungal biomass concentration on the removal of mercury (II). 100 mg/L, mercury (II), 100 rpm, 30°C, pH 5.5.

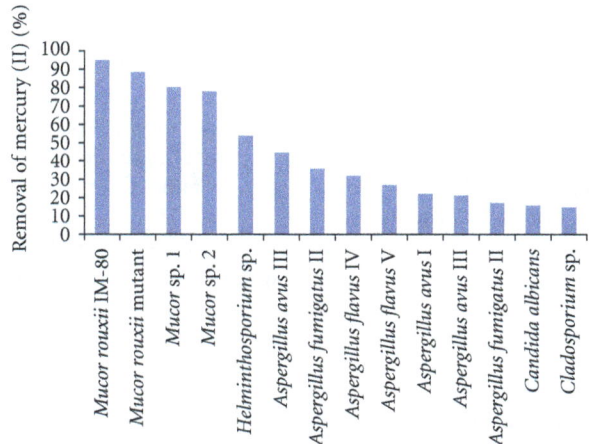

FIGURE 5: Biosorption of mercury (II) on different fungal biomasses. 100 mg/L Hg (II), 100 rpm, 30°C, pH 5.5, 1 g of fungal biomass.

3.3. Effect of Initial Mercury (II) Concentration.

Biosorption capacities of the *M. rouxii* IM-80 biomass for the mercury (II) ions were studied as a function of the initial Hg (II) ions concentration between 100 and 500 mg/L in the biosorption medium (Figure 3). Although the percentage of adsorption decreased, when ions concentration increased. A similar type of trend was reported for the removal of Hg (II) from aqueous solution by sorption on *R. oligosporus* [21], *B. subtilis* [24], *Pleurotus sapidus* [25], biogenic silica modified with L-cysteine [26], and activated carbon prepared from agricultural byproduct/waste [6]. These results may be explained to be due to the increase in the number of ions competing for the available binding sites and also because of the lack of active sites on solution biomass at higher concentrations [6].

3.4. Effect of Initial Biomass Concentration.

The influence of biomass on the removal capacity of mercury (II) was depicted in Figure 4. If we increase the amount of biomass also increases the removal of the metal in solution (100% of removal, with 5 g of fungal biomass, at 8 hours), with more biosorption sites of the same, because the amount of added biosorbent determines the number of binding sites available for metal biosorption [27]. Similar results have been reported for *Acetobacter xylinum* cellulose [27], *Mucor racemosus* biomass [28], and *Saccharomyces cerevisiae* [29].

3.4.1. Biosorption of Mercury (II) By Different Fungal Biomasses.

In Figure 5, we show the biosorption of mercury (II) by the different biomasses analyzed. It was found that the biomass of the fungus *M. rouxii* IM-80, *M. rouxii* mutant, *Mucor* sp1, and *Mucor* sp2 were very efficient at removing the metal in solution (95.3%, 88.7%, 80.4%, and 78.3%, resp.).

We do not know why the fungal biomasses of the mucorales were the most efficient at removing mercury (II) in solution. However, this difference may be because the polysaccharides of the cell wall could provide binding groups including amino, carboxyl groups and the nitrogen and oxygen of the peptide bonds could be accompanied by displacement of protons, dependent in part upon the extent of protonation as determined by the pH [21–23, 30].

Otherwise, in mercury detoxification process, work is still necessary to illustrate the distribution and diversity of the microbial communities under heavy metals stress in order to employ them for the bioremediation of these toxic pollutants, singly or in combination for greater efficiency [31]. Moreover, some mercury biosorbent fungi cannot only detoxify mercury but also remove other metals such as cadmium, chromium (VI), and lead [32].

4. Conclusion

In this study, mercury uptake by different fungal biomasses was investigated. The performance of the biosorbents was examined as a function of the operating conditions, in particular incubation time, pH and initial metal ion concentration, and fungal biomass. The experimental evidence shows a strong effect of the experimental conditions. Maximum biosorption capacity values showed that some biosorbents used are very effective in recovery or removal of mercury ion from aquatic systems. When the ease of production and economical parameters are concerned, it was observed that *M. rouxii* IM-80, *M. rouxii* mutant, *Mucor* sp. 1, and *Mucor* sp. 2 are a very promising biomaterial for removal or recovery of the metal ion studied.

References

[1] N. Li and R. Bai, "Copper adsorption on chitosan-cellulose hydrogel beads: behaviors and mechanisms," *Separation and Purification Technology*, vol. 42, no. 3, pp. 237–247, 2005.

[2] M. Zabihi, A. Haghighi Asl, and A. Ahmadpour, "Studies on adsorption of mercury from aqueous solution on activated carbons prepared from walnut shell," *Journal of Hazardous Materials*, vol. 174, no. 1–3, pp. 251–256, 2010.

[3] B. S. Inbaraj and N. Sulochana, "Mercury adsorption on a carbon sorbent derived from fruit shell of *Terminalia catappa*," *Journal of Hazardous Materials*, vol. 133, no. 1–3, pp. 283–290, 2006.

[4] H. Yavuz, A. Denizli, H. Güngünes, M. Safarikova, and I. Safarik, "Biosorption of mercury on magnetically modified yeast cells," *Separation and Purification Technology*, vol. 52, no. 2, pp. 253–260, 2006.

[5] B. S. Inbaraj, J. S. Wang, J. F. Lu, F. Y. Siao, and B. H. Chen, "Adsorption of toxic mercury(II) by an extracellular biopolymer poly(glutamic acid)," *Bioresource Technology*, vol. 100, no. 1, pp. 200–207, 2009.

[6] M. M. Rao, D. H. K. K. Reddy, P. Venkateswarlu, and K. Seshaiah, "Removal of mercury from aqueous solutions using activated carbon prepared from agricultural by-product/waste," *Journal of Environmental Management*, vol. 90, no. 1, pp. 634–643, 2009.

[7] H. E. Byrne and D. W. Mazyck, "Removal of trace level aqueous mercury by adsorption and photocatalysis on silica-titania composites," *Journal of Hazardous Materials*, vol. 170, no. 2-3, pp. 915–919, 2009.

[8] C. Chen and J. Wang, "Removal of Pb^{2+}, Ag^+, Cs^+ and $Sr2^+$ from aqueous solution by brewery's waste biomass," *Journal of Hazardous Materials*, vol. 151, no. 1, pp. 65–70, 2008.

[9] H. D. Ozsoy, H. Kumbur, B. Saha, and J. H. van Leeuwen, "Use of *Rhizopus oligosporus* produced from food processing wastewater as a biosorbent for Cu(II) ions removal from the aqueous solutions," *Bioresource Technology*, vol. 99, no. 11, pp. 4943–4948, 2008.

[10] S. K. Chatterjee, I. Bhattacharjee, and G. Chandra, "Biosorption of heavy metals from industrial waste water by *Geobacillus thermodenitrificans*," *Journal of Hazardous Materials*, vol. 175, no. 1-3, pp. 117–125, 2010.

[11] J. M. Luo, X. Xiao, and S. L. Luo, "Biosorption of cadmium(II) from aqueous solutions by industrial fungus *Rhizopus cohnii*," *Transactions of Nonferrous Metals Society of China*, vol. 20, no. 6, pp. 1104–1111, 2010.

[12] K. C. Gavilana, A. V. Pestovc, H. M. Garcia, Y. Yatlukc, J. Roussya, and E. Guibala, "Mercury sorption on a thiocarbamoyl derivative of chitosan," *Journal of Hazardous Materials*, vol. 165, pp. 415–426, 2009.

[13] H. D. Ozsoy, H. Kumbur, and Z. Özer, "Adsorption of copper (II) ions to peanut hulls and Pinus brutia sawdust," *International Journal of Environment and Pollution*, vol. 31, no. 1-2, pp. 125–134, 2007.

[14] W. Ma and J. M. Tobin, "Determination and modelling of effects of pH on peat biosorption of chromium, copper and cadmium," *Biochemical Engineering Journal*, vol. 18, no. 1, pp. 33–40, 2004.

[15] J. Hanzlik, J. Jehlicka, O. Sebek, Z. Weishauptova, and V. Machovic, "Multi-component adsorption of Ag(I), Cd(II) and Cu(II) by natural carbonaceous materials," *Water Research*, vol. 38, pp. 2178–2184, 2004.

[16] V. K. Garg, R. Gupta, R. Kumar, and R. K. Gupta, "Adsorption of chromium from aqueous solution on treated sawdust," *Bioresource Technology*, vol. 92, no. 1, pp. 79–81, 2004.

[17] H. D. Ozsoy and H. Kumbur, "Adsorption of Cu(II) ions on cotton boll," *Journal of Hazardous Materials*, vol. 136, no. 3, pp. 911–916, 2006.

[18] A. Sari and M. Tuzen, "Removal of mercury(II) from aqueous solution using moss (*Drepanocladus revolvens*) biomass: equilibrium, thermodynamic and kinetic studies," *Journal of Hazardous Materials*, vol. 171, no. 1–3, pp. 500–507, 2009.

[19] A. E. Greenberg, L. S. Clesceri, and A. D. Eaton, *Standard Methods For the Examination of Water and wasteWater*, vol. 3, American Public Health Association, Washington, DC, USA, 18 edition, 1992.

[20] S. K. Das, A. R. Das, and A. K. Guha, "A study on the adsorption mechanism of mercury on *Aspergillus versicolor* biomass," *Environmental Science and Technology*, vol. 41, no. 24, pp. 8281–8287, 2007.

[21] H. D. Ozsoy, "Biosorptive removal of Hg(II) ions by *Rhizopus oligosporus* produced from corn-processing wastewater," *African Journal of Biotechnology*, vol. 9, no. 51, pp. 8791–8799, 2010.

[22] G. Bayramoğlu, S. Bektaş, and M. Y. Arica, "Biosorption of heavy metal ions on immobilized white-rot fungus *Trametes versicolor*," *Journal of Hazardous Materials*, vol. 101, no. 3, pp. 285–300, 2003.

[23] R. Say, N. Yilmaz, and A. Denizli, "Biosorption of cadmium, lead, mercury, and arsenic ions by the fungus *Penicillium*

purpurogenum," *Separation Science and Technology*, vol. 38, no. 9, pp. 2039–2053, 2003.

[24] X. S. Wang, F. Y. Li, W. He, and H. H. Miao, "Hg(II) removal from aqueous solutions by *Bacillus subtilis* biomass," *Clean*, vol. 38, no. 1, pp. 44–48, 2010.

[25] Y. Yalcinkaya, M. Y. Arica, L. Soysal, A. Denizli, O. Genc, and S. Bektaş, "Cadmium and mercury uptake by immobilized *Pleurotus sapidus*," *Turkish Journal of Chemistry*, vol. 26, no. 3, pp. 441–452, 2002.

[26] M. R. M. Chaves, K. T. Valsaraj, R. D. De Laune, R. P. Gambrell, and P. M. Buchler, "Mercury uptake by biogenic silica modified with L-cysteine," *Environmental Technology*, vol. 32, no. 14, pp. 1615–1625, 2011.

[27] A. Rezaee, J. Derayat, S. B. Mortazavi, Y. Yamini, and M. T. Jafarzadeth, "Removal of mercury from Chlor-alkali industry wastewater using *Acetobacter xylinum* cellulose," *American Journal of Environmental Sciences*, vol. 1, no. 2, pp. 102–105, 2005.

[28] T. Liu, H. Li, Z. Li, X. Xiao, L. Chen, and L. Deng, "Removal of hexavalent chromium by fungal biomass of *Mucor racemosus*: influencing factors and removal mechanism," *World Journal of Microbiology and Biotechnology*, vol. 23, no. 12, pp. 1685–1693, 2007.

[29] M. E. Rodríguez, R. C. Miranda, R. Olivas, and C. A. Sosa, "Efectos de las condiciones de operación sobre la biosorción de Pb^{++}, Cd^{++} and Cr^{+++}, en solución por Saccharomyces cerevisiae residual," *Información Tecnológica*, vol. 19, no. 6, pp. 47–55, 2008.

[30] S. Bartnichi-García, "Chemistry of fungal cell wall," *Annual Review of Microbiology*, vol. 22, pp. 13–26, 1968.

[31] M. Ashraf and M. Essa, "The effect of a continuous mercury stress on mercury reducing community of some characterized bacterial strains," *African Journal of Microbiology Research*, vol. 6, no. 18, pp. 4006–4012, 2012.

[32] J. De, N. Ramaiah, and L. Vardanyan, "Detoxification of toxic heavy metals by marine bacteria highly resistant to mercury," *Marine Biotechnology*, vol. 10, no. 4, pp. 471–477, 2008.

6

Behavior and Distribution of Heavy Metals Including Rare Earth Elements, Thorium, and Uranium in Sludge from Industry Water Treatment Plant and Recovery Method of Metals by Biosurfactants Application

Lidi Gao,[1] **Naoki Kano,**[2] **Yuichi Sato,**[1] **Chong Li,**[1] **Shuang Zhang,**[1] **and Hiroshi Imaizumi**[2]

[1] *Graduate School of Science and Technology, Niigata University, Niigata 950-2181, Japan*
[2] *Department of Chemistry and Chemical Engineering, Faculty of Engineering, Niigata University, Niigata 950-2181, Japan*

Correspondence should be addressed to Naoki Kano, kano@eng.niigata-u.ac.jp

Academic Editor: Virtudes Moreno

In order to investigate the behavior, distribution, and characteristics of heavy metals including rare earth elements (REEs), thorium (Th), and uranium (U) in sludge, the total and fractional concentrations of these elements in sludge collected from an industry water treatment plant were determined and compared with those in natural soil. In addition, the removal/recovery process of heavy metals (Pb, Cr, and Ni) from the polluted sludge was studied with biosurfactant (saponin and sophorolipid) elution by batch and column experiments to evaluate the efficiency of biosurfactant for the removal of heavy metals. Consequently, the following matters have been largely clarified. (1) Heavy metallic elements in sludge have generally larger concentrations and exist as more unstable fraction than those in natural soil. (2) Nonionic saponin including carboxyl group is more efficient than sophorolipid for the removal of heavy metals in polluted sludge. Saponin has selectivity for the mobilization of heavy metals and mainly reacts with heavy metals in F3 (the fraction bound to carbonates) and F5 (the fraction bound to Fe-Mn oxides). (3) The recovery efficiency of heavy metals (Pb, Ni, and Cr) reached about 90–100% using a precipitation method with alkaline solution.

1. Introduction

With the rapid development of industry, a large quality of industrial sludge is settled down in wastewater treatment plants (WWTPs) every year. The sludge must be treated and disposed in a safe and effective manner because it may be contaminated with toxic organic and inorganic compounds. Much of this sludge is treated using a variety of digestion techniques to reduce the amount of organic matter and the number of disease-causing microorganisms, then the nutrient-rich sludge is provided to use as agricultural soil for landscaping and garden planting or as natural fertilizer [1–3]. These techniques have reduced the amount of landfill and changed waste into resource [4, 5]. However, the digested sludge cannot be directly used for practical use because it may contain hazardous inorganic substances such as heavy metals and radioactive elements. For this reason, it is of

significant importance to investigate the removal of these metals by eco-friendly methods and to study the behavior and distribution of heavy metals in sludge from an environmental protection and human health perspective. On the other hand, the demand for trace metals such as rare earth elements (REEs) in modern society has increased markedly in recent years. The shortage of trace metals including REEs and uranium (U) has been of concern, and the investigation of new sources of these trace metals is important from a resources recovery point of view.

In recent years, the concentrations and distribution of heavy metals in sludge has been extensively studied [6–10]. Furthermore, the investigations of methods for removing of heavy metals from sludge have been widely carried out [11–14].

Total concentrations and fractions of heavy metals in sewage sludge from municipal and industrial wastewater treatment plants have been studied [6]. The results showed that the total concentrations of heavy metals in sludge varied greatly and that there was no significant difference in total metal concentration between municipal and industrial wastewater treatment plants. Chen et al. [7] reported the bioavailability and eco-toxicity of heavy metals in municipal sludge by taking into consideration both the speciation of metals and the local environmental characteristics. From this work, it was found that only the sludge from Xia Wan sewage treatment plant showed elevated concentrations of heavy metals and that the sludge from other plants showed low total concentrations of heavy metals except for a slightly higher concentration of Cd. The results of the sequential extraction procedure showed that Cu and Zn were principally distributed in the oxidize fraction and that Pb was mainly in the residual fraction. Furthermore, the different types of sludge and the distribution of the heavy metals in sludge have been studied [8]. It was confirmed that the total concentrations of heavy metals did not exceed the limits set out by the European legislation and that the stabilization method undergone by the sludge strongly influenced the distribution and the associated phases of heavy metals. The extractable forms of heavy metals in sludge from wastewater treatment plants have been determined to obtain suitable information about their bioavailability or toxicity [9]. In regard to current international legislation on the use of sludge for agricultural purposes, the concentrations of any metal did not exceed permitted levels. For most of the subject metallic elements, the increase of the concentrations was clearly found in two less-available fractions (oxidizable fraction and residual fraction) with the sludge treatment. In contrast, Ščančar et al. [10] determined the total and fractional concentrations of Cd, Cr, Cu, Fe, Ni, and Zn in sewage sludge samples from an urban wastewater treatment plant and showed that the sludge could not be used in agriculture due to the high total Ni concentration and its high mobility.

Currently, the removal of ultrasond-assisted metals from sludge is applied widely. For example, Deng et al. [11] and Li et al. [12] investigated the removal or recovery of heavy metals from sludge using ultrasound-assisted acid. The results showed that ultrasonic treatment is a necessary and effective method for assisting the improvement of heavy metal removal. However, ultrasonic treatment has an effect on the physical and chemical properties of sludge to some extent and is energy-consuming. In another study, Babel and Mundo Dacera [13] reviewed various methods for the removal of heavy metal from sewage sludge, including chemical extraction, bioleaching, electroreclamation, and supercritical fluid extraction (SFE). They compared the advantages and limitations of each and gave a detailed analysis of their findings. A combination of two methods (i.e., bioleaching and electrokinetic remediation technology) for removing heavy metals from sludge has been also reported [14]. The combined technology can not only remove the heavy metals in the sludge but also make them be recycled, although it is energy-consuming to some extent.

As mentioned above, most research has been mainly focused on toxic heavy metallic elements such as Cd, Pb, Cu, and Cr as subject elements and on the differences and characteristics of these elements according to different types of sludge or different treatment processes. However, few reports have been published about the behavior and distribution of REEs, Th, and U. Moreover, there have been very few comparisons between concentrations of heavy metals in sludge and those in natural soil carried out. It is important to compare the concentration and distribution of metals in sludge with those in natural soil when considering the utilization of sludge as agricultural soil in the future. The purposes of this paper are (1) to investigate the behavior, distribution, and characteristics of heavy metals including REEs, Th, and U in sludge compared with those in natural soil and (2) to study the removal/recovery process of heavy metals from polluted sludge with biosurfactant elution by batch and column experiments.

2. Experimental

2.1. Apparatus and Reagents. An inductively coupled plasma mass spectrum (ICP-MS) instrument (Thermo scientific X-Series) was used to determine the concentrations of REEs, Th, and U, and an inductively coupled plasma atomic emission spectrophotometer (ICP-AES) instrument (SPS1500, Seiko Instruments Inc.) was employed to determine the concentrations of heavy metals (Zn, Cd, Pb, Cr, Ni, and Cu). The operating conditions of the ICP-MS are the same as shown in our previous paper [15] and those of ICP-AES are based on our other previous paper [16].

Heavy metal standard solutions, including REEs, Th, and U, were purchased from SPEX CertiPrep, Inc. (USA). Each working standard solution was prepared by diluting the original standard solution.

In this work two kinds of biosurfactants were used, saponin and sophorolipid. Saponin was purchased commercially from Sigma-Aldrich, Inc. (Germany). It is a nonionic biosurfactant but includes the carboxyl group (–COOH) as shown in Figure 1(a) based on the analysis of *quillaja bark* by Guo and Kenne [17]. Sophorolipid was supplied by State Key Laboratory for Microbial Technology (Shandong University, China). It is also a nonionic biosurfactant, and one possible structure of sophorolipid from *Wickerhamiella domercqiae* analyzed by Chen et al. is given in Figure 1(b) [18].

All other chemical reagents, purchased from Kanto Chemical Co., Inc. (Japan) were of analytical grade. Water (>18.2 MΩ), which was treated using an ultrapure water system (Advantec aquarius: RFU 424TA), was employed throughout the work.

2.2. Samples. The original sludge sample was collected in May 2010 from an industrial water treatment plant. The sample was air-dried and removed coarse sand and stone, then ground and sieved through 120 mesh (0.125 mm).

The polluted sludge sample was prepared by adding the solution containing three kinds of metallic salts ($NiCl_2 \cdot 6H_2O$ (1000 ppm), $Pb(NO_3)_2$ (1360 ppm), and $Cr(NO_3)_3 \cdot 9H_2O$ (1000 ppm)) to the air-dried original sludge sample. The polluted sample was shaken for 3 days on a shaker at room temperature ($25.0 \pm 0.2°C$) and subsequently

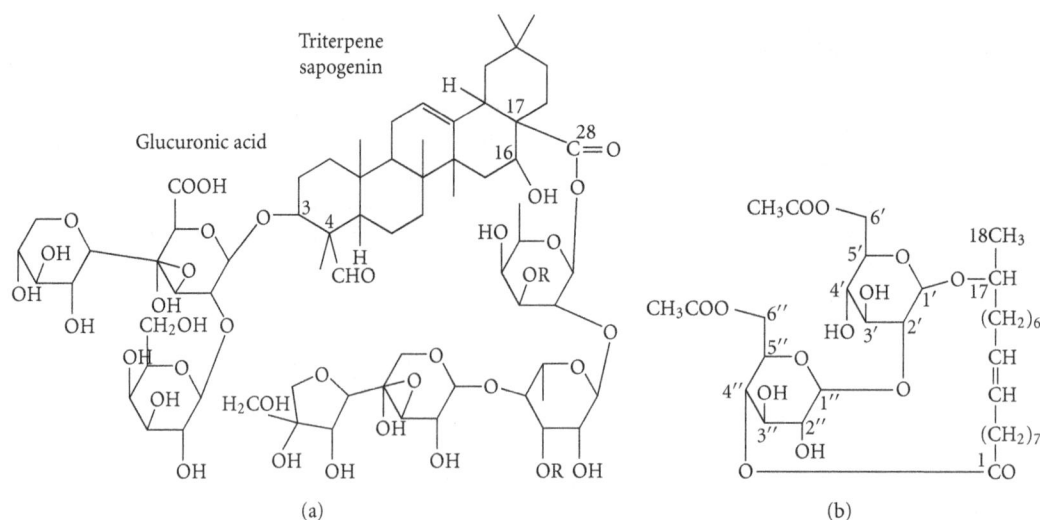

FIGURE 1: One structure of saponin (15) and sophorolipid (16).

TABLE 1: Sequential extraction procedure for fractional determination of metallic elements.

Step	Fraction	Extraction reagents	Ratio of sludge : reagent	Extraction condition
1	F1	Extrapure water	7 : 70	Shake 6 h, 30°C
2	F2	0.05 mol·dm^{-3} Ca(NO$_3$)$_2$	7 : 70	Shake 24 h, 30°C
3	F3	2.5% CH$_3$COOH	7 : 70	Shake 24 h, 30°C
4	F4	6% H$_2$O$_2$	7 : 120	Water bath, 95°C (Evaporate)
		2.5% CH$_3$COOH	7 : 70	Shake 24 h, 30°C
5	F5	0.1 mol·dm^{-3} (COOH)$_2$ + 0.175 mol·dm^{-3} (COONH$_4$)$_2$ + Ascorbic acid	7 : 210	Water bath, 6 h, 95°C Occasional shaking
6	F6	HNO$_3$ + HF		Microwave digestion

centrifuged at 3000 rpm for 30 min using a centrifugal separator (Kubota Co. 5200). The supernatant was discarded, and the polluted sludge was air-dried and sieved through 120 mesh (0.125 mm).

Basic characteristics of the sludge samples, such as pH, EC, moisture content, and cation exchange capacity (CEC) were measured, based on the method for soil testing recommended by The Japanese Geotechnical Society [19]. For measuring organic matter content, 10 g air-dried sludge samples were heated for 2 h at 105°C and then burned at 550°C in a furnace for 6 h. Organic matter content was estimated from the weight differences of the sludge before and after burning divided by the sludge weight before burning. Permeability is an important physical parameter to determine the feasibility of the soil flushing process. Therefore, the permeability of the sludge was also determined using the Unfirmed Water Head Test [20]. The specific surface area of each sample was measured using Micrometritics TriStar3000. The BET method and Langmuir method, as well as methods used in our previous work, were applied to determine the surface area [21, 22].

For measuring total metal concentration, the sludge was digested with HNO$_3$-HF by using the microwave digestion method as well as the case of digesting soil [23]. After this,

the analysis of metallic elements was performed using ICP-AES or ICP-MS.

2.3. Distribution of Metallic Elements in Sludge.

All heavy metals including REEs, Th, and U in sludge samples were partitioned into six fractions with sequential extraction procedures mainly based on Sadamoto et al. [24] and Tessier et al. [25]. In this paper, these six fractions, (1) water soluble, (2) exchangeable, (3) bound to carbonates, (4) bound to organic matter, (5) bound to Fe-Mn oxides, and (6) residual were, denoted as F1, F2, F3, F4, F5, and F6, respectively. The sequential extraction procedure is outlined in Table 1. For the initial step in this sequential extraction procedure, 7 g of dried sludge sample in 100 cm^3 polypropylene centrifuge tube was used. Following extraction in each step, the mixture of sludge sample and each extraction reagent was centrifuged (3000 rpm × 30 min) using a centrifugal separator (Kubota Co. 5200). This procedure is the same used in our previous work on soil [26]. The concentrations of metallic elements in each fraction were determined with ICP-AES or ICP-MS.

2.4. Batch Test and Column Test.

The effect of the concentration and pH value of biosurfactant solution on the removal of heavy metals in polluted sludge was investigated

Behavior and Distribution of Heavy Metals Including Rare Earth Elements, Thorium, and Uranium in Sludge from Industry Water Treatment Plant and Recovery Method of Metals by Biosurfactants Application

47

in batch experiments at room temperature (25°C). Each 1.0 g of contaminated sludge was weighed into a centrifuge tube, and 25 cm^3 of biosurfactant solution, varying in initial concentrations from 1 to 50 g·dm^{-3}, was added to each tube. The tubes were then shaken in a reciprocating shaker for 24 hours to attain equilibrium. The suspension was centrifuged (3000 rpm × 30 min) using a centrifugal separator (Kubota Co. 5200). The supernatant solutions were separated and dissolved with 1 mol·dm^{-3} HNO$_3$ after digestion for analysis. Subsequently, the effect of pH value (varied from 2.5 to 6.5) was investigated with the same procedure as above.

Column tests were also conducted to remove heavy metals from polluted sludge at the optimum concentration and pH of the biosurfactant solution. Permeability is a useful parameter for the flushing techniques like column experiments. Silica was mixed with polluted sludge (mass ratio of 4 : 1) to be compacted in column to improve the low permeability of sludge. 30 g of each mixture was packed in a glass column (internal diameter of 1.5 cm and length of 30 cm) and two in total. Saponin solution and ultrapure water were prepared for mobilization and leaching of heavy metals from sludge at less than a rate of 0.2 cm^3·min^{-1} using multichannel peristaltic pump. Each leachate was collected per one flush volume and then was digested and dissolved in 1 mol·dm^{-3} HNO$_3$ for analysis.

2.5. Recovery of Heavy Metals from Leachates. The precipitation method was applied by using 3 mol dm^{-3} hydroxide sodium (NaOH) [27]. The leachate from polluted sludge from washing with saponin was used as a sample after determining the concentration of heavy metals. The pH value of the leachate was gradually increased because heavy metals were precipitated as hydroxide. The solution was allowed to stand for 24 hours before being centrifuged with a refrigerated centrifuge, after which, the concentration of heavy metal was measured with ICP spectrometry.

3. Result and Discussion

3.1. Characteristics of Sludge. Some physical-chemical characteristics of sludge have been determined (Table 2). As shown in this table, the pH value of the polluted sludge is lower than that of the original sludge, whereas EC is remarkably large. These results may be attributable to the fact that the polluted sludge was prepared by adding a solution containing three kinds of metallic elements (Pb, Ni, and Cr). Furthermore, CEC of the sludge became small after the introduction of heavy metals. It is known that soil flushing proves effective only for permeable soil ($K > 1.0 \times 10^{-3}$ cm·s^{-1}) or, to a lesser extent, slightly permeable soil (1.0×10^{-5} cm·s^{-1} $< K < 1.0 \times 10^{-3}$ cm·s^{-1}) [28]. The permeability of the sludge studied in this work ($K \approx 1.7 \times 10^{-5}$ cm·s^{-1}) is much lower than the value of permeable soil, so quartz sand was added into the sludge to improve its permeability in column washing experiments. From the above mentioned, it is perhaps obvious that the pH, EC, and CEC values were changed by the introduction of metals.

TABLE 2: Physical-chemical characteristics of original sludge and polluted sludge.

Parameters	Original sludge	Polluted sludge
pH(H$_2$O)	5.78	4.47
pH(KCl)	5.08	4.42
EC, μS·cm^{-1}	360	1926
Moisture content, %	6.10	7.13
Permeability, cm·s^{-1}	1.75×10^{-5}	1.72×10^{-5}
Organic matter content, %	14.0	14.0
Cation exchange capacity (CEC), cmol·kg^{-1}	33.6	30.5
BET-specific surface area (SSA), m^2·g^{-1}	36.0	29.4
Langmuir-specific surface area (SSA), m^2·g^{-1}	56.2	45.8

TABLE 3: The concentrations of heavy metals in sludge (mg·kg^{-1}).

	Zn	Pb	Cd	Ni	Cr	Cu
Original sludge	232	33.5	11.2	32.7	53.0	50.7
Polluted sludge	233	1.21×10^3	11.5	716	984	51.2

3.2. Concentrations and Distribution of Heavy Metals in Sludge. The concentrations of heavy metals (Zn, Pb, Cd, Ni, Cr, and Cu) found in the sludge are listed in Table 3. The relative standard deviation (RSD) of the triplicated analyses of each sample was less than 5%. From Table 3, the concentration of Zn was the highest, and the total concentration of Pb, Ni, Cr, and Cu did not exceed the limits in "The Criterion about the Waste Including Metals" [29], but Cd is relatively high in the original sludge. For polluted sludge, the concentrations of uncontaminated elements (Zn, Cd, and Cu) were almost unchanged; however, the concentrations of contaminated element increased remarkably. These results indicate that most of lead, chromium, and nickel in solution have been introduced successfully into the sludge. The absorption rates of Pb, Ni, and Cr in this study were 86.5%, 68.4%, and 93.1%, respectively, which may be mainly attributed to competitive sorption onto the sludge. These results concur with those reported by Juwarkar et al. [30].

For reference, the concentrations of heavy metals in natural soil are also shown in Figure 2 along with those of the original sludge. It was found that the concentrations of heavy metals in sludge are higher than those in natural soil [26] (natural soil used in this work is no plow soil from Ueno, Sekikawa village in Niigata Prefecture, Toyasato, Sakata town and Tateoka, Murayama town in Yamagata Prefecture, resp.). One possible reason for high concentrations in sludge is that the sludge was mainly precipitated from wastewater (containing many kinds of heavy metals), which is discharged from the industries such as paper manufacturing, petrochemical engineering, glass production, textiles, and transportation. In particular, the concentrations of Cd and Ni in sludge are markedly higher (up to double) than those

TABLE 4: The concentrations of Fe and Mn in original sludge and natural soil (mg·kg^{-1}).

	Original sludge	Natural soil A	Natural soil B	Natural soil C
Fe	4.73×10^4	2.69×10^4	3.71×10^4	3.20×10^4
Mn	769	694	936	618

in natural soil. This suggests that heavy metals in sludge may tend to accumulate in agriculture soil if the sludge is used repeatedly.

The relative distribution of heavy metals is shown in Figure 3. The results in Figure 3(a) suggested that, in addition to the residual fraction, Pb, Cd, Ni, and Cr mainly exist as Fe-Mn oxides fraction, Zn exists as carbonate fraction, and Cu exists as organic fraction in original sludge. These results are in accordance with the distribution characteristics of heavy metals in soil [26, 31, 32].

Comparing Figure 3(a) with Figure 3(b), the following results can be obtained. (1) The concentration of Ni in F2 (bound to exchangeable fraction) sharply increased, which suggests that Ni may be very harmful to the environment at the beginning period of pollution. (2) Heavy metals (Pb, Ni, and Cr) were in relatively unstable fractions (from F1 to F5) at the early stage of pollution and generally moved to the stable residual fraction (F6) with time and become difficult to remove from soil. Considering this, the remediation of polluted soil by heavy metals should be carried out as soon as possible. (3) The dominant fraction (except residual fraction, F6) is different among elements; that is, the order of the relative distributions is "oxide fraction" > "carbonate fraction" > "organic fraction" for Pb, and "organic fraction" > "carbonate fraction" > "oxide fraction" for Cr. Ni does not show any dominant chemical fraction although the general tendency is "organic fraction" > "oxide fraction" > "carbonate fraction" > "exchangeable fraction." It indicates that Pb is easily adsorbed to Fe-Mn oxides and that Cr is drawn to organic matter in sludge. (4) The dominant fraction of uncontaminated elements (i.e., Zn, Cd and Cu) hardly changed, although the relative distribution of F3 (bound to carbonate fraction) decreased in case of Zn.

Comparing Figures 3(a) and 3(c), the proportion of residual fraction (F6) in natural soil is relatively higher than that in original sludge. In contrast, the proportions of oxide fraction (F5) and carbonate fraction (F3) in original sludge are higher than those in natural soil. From these results, it is found that heavy metals in natural soil usually exist in a more stable state than those in sludge. It indicates that it may be hard for heavy metals in natural soil to permeate into groundwater or to be absorbed by crops. On the other hand, the high proportion of oxide fraction in sludge may be due to relative large contents of Fe and Mn in sludge (Table 4).

In brief, the concentrations of heavy metals in sludge are larger than those in natural soils. The relative distribution of residual fraction in natural soil is higher than that in original sludge, while the ratio of oxide fraction in natural soils is lower than that in original sludge.

3.3. Concentrations and Distribution of REEs, Th, and U in Sludge.
REEs, Th, and U were also extracted from the sludge along with heavy metals and determined with ICP-MS. The concentrations are shown in Table 5, and the relative standard deviation (RSD) of the triplicated analyses of each sample was less than 10%. The relative distribution of REEs, Th, and U is shown in Figure 4 (distribution characteristics of REEs, Th, and U in natural soil A and B are similar to those in natural soil C, so the data for soil A and B are not shown in this figure.). Judging from Table 5, the concentrations of REEs in the sludge are similar to those in natural soil C, while the concentrations of Th and U are smaller than those in natural soil C. On the other hand, the concentrations of metallic elements (except for HREE, i.e., heavy rare earth elements) in sludge are higher than those in natural soil A and B.

Figures 4(a) and 4(b) show that the distribution characteristics of REEs, Th, and U are generally similar to those of heavy metals. It is noted that, except for U which is higher, the proportion of carbonate fraction (F3) in sludge is lower than that in natural soil. These results show that REEs, Th, and U in natural soil exist in more stable states than those in sludge. In addition, high carbonate fraction of U in sludge is noticeable because the available content in crops is generally considered to be reflected, to some extent, by metal content in carbonate fraction [33].

As mentioned above, the concentrations of REEs, Th, and U in sludge are variable when compared with those in natural soils. Furthermore, heavy metallic elements including REEs, Th, and U in sludge exist as more unstable fraction than those in natural soil.

3.4. Removal of Heavy Metals in Sludge.
The removal of heavy metals (Pb, Ni, and Cr) in polluted sludge was investigated with elution technology by using biosurfactant (nonionic biosurfactant sophorolipid and saponin). To quantify the factors influencing the removal efficiency of biosurfactant, the effects of the concentration and pH value of the bio-surfactant solution in batch experiments and the washing volume of the biosurfactant solution in column experiments were tested in this work.

3.4.1. Batch Experiments.
The effects of the concentrations of the biosurfactants solution on the removal efficiency of heavy metals are shown in Figure 5, and the effects of pH value of the biosurfactants solution on the removal efficiency of heavy metals are shown in Figure 6. For both biosurfactants, the concentration ranged from 1 to 50 g·dm^{-3}, and pH ranged from 2.5 to 6.5. The removal efficiency of heavy metals by both biosurfactants generally ascended with increasing concentration and decreasing pH value; however it was clear that saponin is more efficient than sophorolipid. Although both biosurfactants are nonionic, the saponin used in this work contained the carboxyl group in sapogenin moiety [17]. For this reason, saponin reacts more easily with metallic elements and to make metallic elements depart from the sludge surface into the soil solution. Because of this only the results using saponin are discussed in the following.

Behavior and Distribution of Heavy Metals Including Rare Earth Elements, Thorium, and Uranium in Sludge from Industry
Water Treatment Plant and Recovery Method of Metals by Biosurfactants Application

49

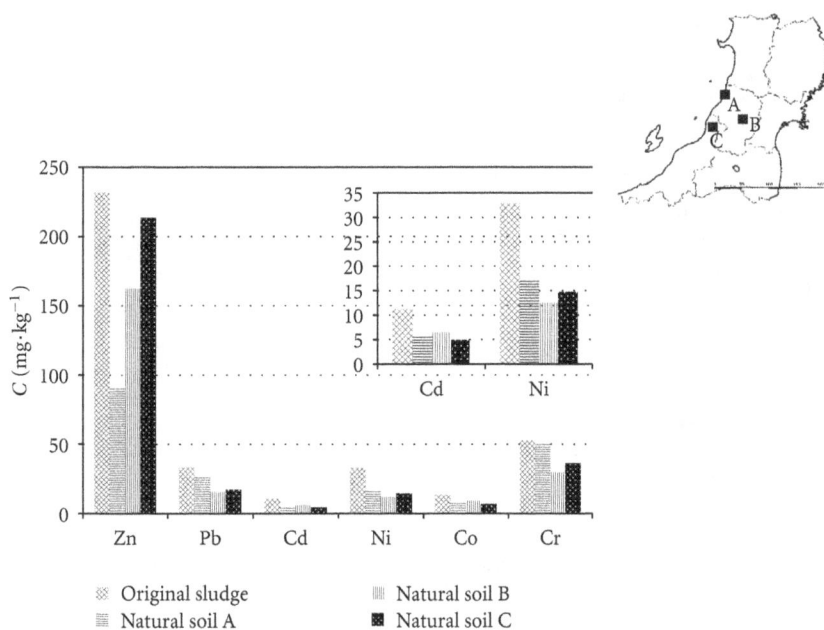

FIGURE 2: The concentrations of some heavy metals (Zn, Pb, Cd, Ni, Co, and Cr) in original sludge and natural soil ((A) Sakata City and (B) Murayama City in Yamagata Prefecture in Japan and (C) Sekikawa Village in Niigata Prefecture in Japan).

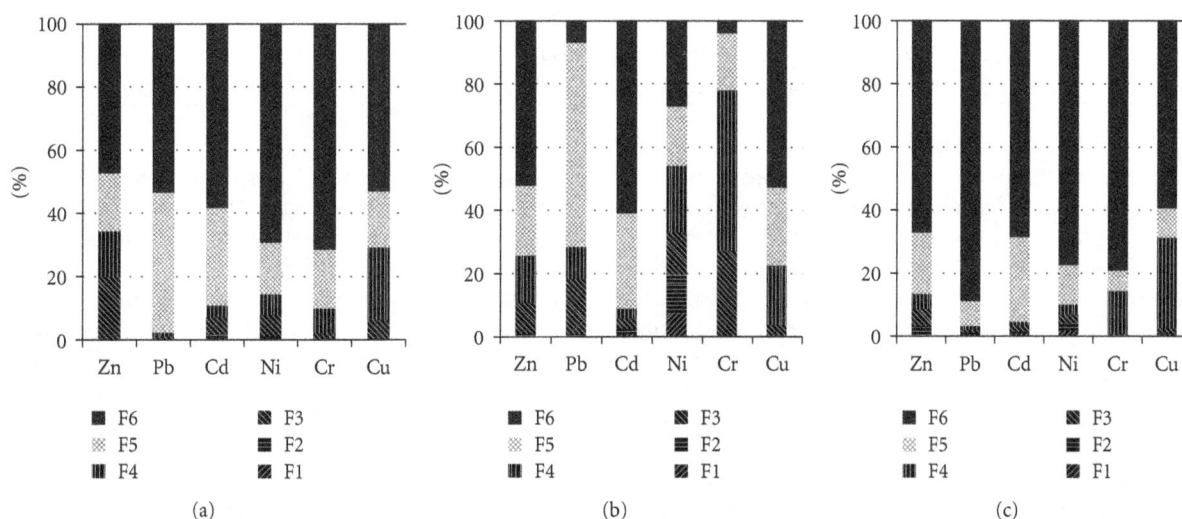

FIGURE 3: The relative distribution of some heavy metals (Zn, Pb, Cd, Ni, Co, and Cr). (a) Original sludge. (b) Polluted sludge. (c) Natural soil C.

The removal efficiency is the greatest when the concentration of saponin solution (i.e., $50 \, g \cdot dm^{-3}$) is the highest (Figure 5(b)). However, the sludge is apt to produce colloidal precipitation due to the adsorption of biosurfactant molecules when the concentration is $50 \, g \cdot dm^{-3}$. Because of this an optimum saponin solution concentration of $30 \, g \cdot dm^{-3}$ was selected for the following column experiments.

Figure 6(b) shows that the removal efficiency is dependent on pH. When the pH value of saponin solution is higher than its pKa (4.6), the removal efficiency is low. It may be considered that sodium ions, which increased by adding NaOH to adjust pH of saponin solution, compete with heavy metals for saponin. In contrast, when the pH value was lower than its pKa, the removal efficiency abruptly increased. However, when the pH value was less than 3.0, the removal efficiency of Ni and Cr was reduced. This may be due to the amount of saponin adsorbed onto sludge, which increased with decreasing pH because electrostatic attraction between saponin and sludge surface increases at low pH [27]. For this reason a pH of 3.0 was applied in the following column experiments.

3.4.2. Column Experiments. The concentrations of heavy metals removed from the polluted sludge with washing

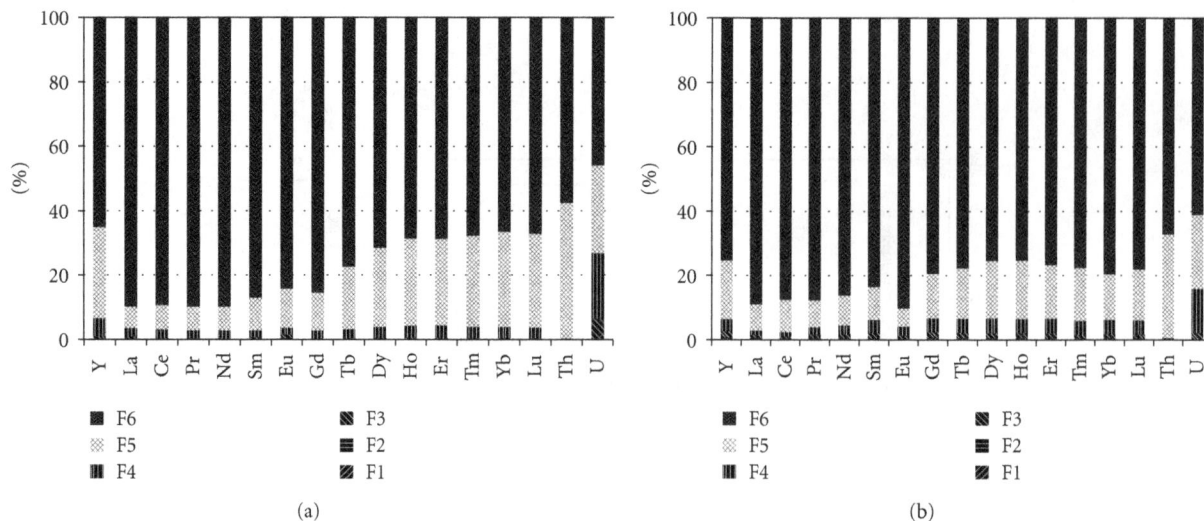

(a)

(b)

FIGURE 4: The relative distribution of REEs, Th, and U. (a) Original sludge. (b) Natural soil C.

TABLE 5: The concentrations of REEs, Th, and U in original sludge and natural soil (mg·kg^{-1}).

	La	Ce	Pr	Nd	Sm	Eu	Gd	Tb	Dy	Ho	Er	Tm	Yb	Lu	Th	U
Original sludge	22.3	48.6	5.54	28.6	5.13	1.28	6.40	0.764	3.75	0.728	2.18	0.297	1.90	0.283	7.90	2.33
Natural soil A	12.9	26.5	2.89	11.5	2.44	0.86	2.51	0.386	2.31	0.463	1.42	0.195	1.33	0.192	4.46	0.911
Natural soil B	14.5	31.8	3.67	15.0	3.46	1.21	3.94	0.644	3.46	0.708	2.19	0.308	2.14	0.321	4.09	1.31
Natural soil C	21.5	47.1	5.16	19.4	4.06	0.718	3.68	0.566	3.02	0.578	1.74	0.241	1.71	0.248	23.4	4.18

volume through column are illustrated in Figure 7. In addition to biosurfactant solution, ultrapure water was used as eluent for the control. As seen in Figure 7(a), the removal of each metal showed a peak with the increasing of washing volume. In the case of Pb, Ni, and Cr, 884, 460, and 552 ppm, respectively, were removed overall from total loaded concentration. Figure 7(b) shows that, except some removal of Ni (58 ppm), hardly any metals were removed with ultrapure water. From Figures 7(a) and 7(b), it is found that saponin has high potential for the removal of heavy metals from polluted sludge compared to ultrapure water.

After 12 washing volumes (one washing volume is about 6.2 cm^3), the total removal efficiency reached 73.2%, 64.2%, and 56.1% for Pb, Ni, and Cr, respectively. The results indicate that saponin facilitates mobilization of metals selectively and that the leaching behavior of biosurfactant is dependent on the characteristics of the metals. This may be due to the specificity of biosurfactant for each metal and the coexistence of metals in the sludge.

3.4.3. Confirmation of Fraction Removed by Saponin Solution.
To confirm the fractions of heavy metals removed by column flushing with saponin solution, sequential extraction was conducted after the column washing. The concentrations of heavy metals (Pb, Ni, and Cr) in polluted sludge before and after the column washing are shown in Figure 8(a), and the relative distribution of each heavy metal is shown in Figure 8(b). Figure 8(a) shows that a remarkable decrease of total concentration was found for each heavy metal and

that the concentration in each fraction was also changed regardless of the kind of metal. The concentration of F1 for Pb and Cr slightly increased due to the residual saponin in the sludge, which can further react with heavy metals in the extraction process. The concentrations of three elements in F3, F4, and F5 all decreased. Of the three fractions, however, F4 showed the smallest decrease. It may be that heavy metals in F3 and F5 could be more easily released than those in F4 under acidic conditions (pH 3). For the same reason, it is suggested that the removal efficiency of Cr (the proportion of this element in F4 was over 50% of total concentration) was the lowest among the three kinds of metals. From Figure 8(b), it is found that the proportion of the relative stable fraction of heavy metals became higher after column washing and that the relative distribution characteristics of heavy metals was closer to that in natural soil.

From the above mentioned, saponin is more efficient than sophorolipid for the removal of heavy metals from sludge in this work. Saponin has selectivity for the mobilization of heavy metals and mainly reacts with the F3 and F5 fractions of heavy metals.

3.5. Recovery of Heavy Metals from Sludge.
In order to recover heavy metals from the sludge leachates, the precipitation method by adding NaOH was firstly considered. Figure 9 shows the recovery efficiency of heavy metals from the sludge leachate at pH 9.2–12.9 using the precipitation method. At pH 10.9, the recovery efficiency of each heavy metal almost reached the maximum possible and was 89.7%, 91.1%, and

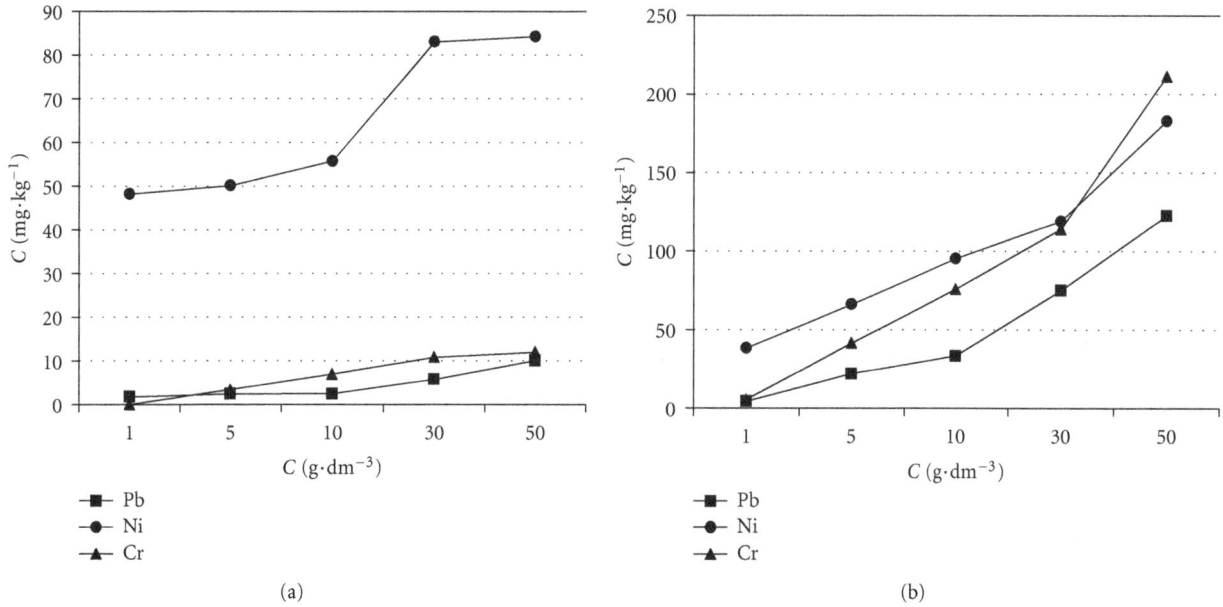

FIGURE 5: Effect of concentrations on removal of heavy metals by batch experiments with biosurfactant as washing agent: (a) sophorolipid and (b) saponin.

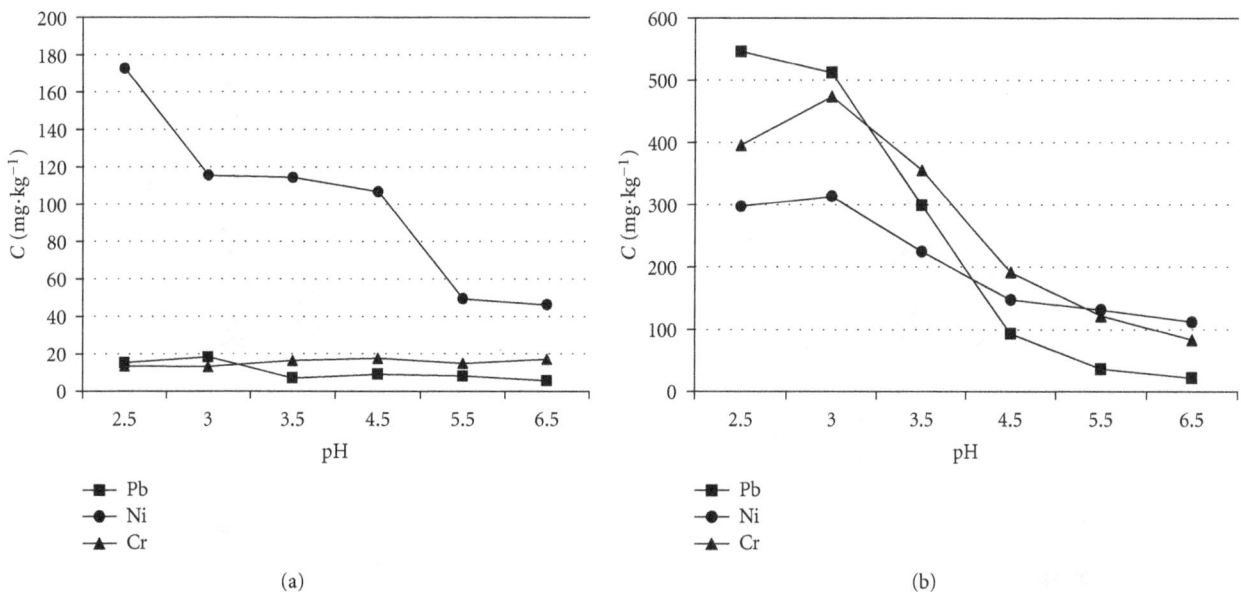

FIGURE 6: Effect of pH value on removal of heavy metals by batch experiments with biosurfactant as washing agent: (a) sophorolipid and (b) saponin.

99.1% for Pb, Ni, and Cr, respectively. Due to the amphoteric nature of lead and chromium, their hydroxide compounds (i.e., precipitate) are redissolved, and this decreased the recovery efficiency (in addition to other issues such as the waste of alkaline solution) at excessively high pH (i.e., >11.5). Therefore, the optimal pH for recovery was considered to be about pH 10.9. That is to say, it is an effective method to use an alkaline solution for obtaining high recovery efficiency of heavy metals such as Pb, Ni, and Cr.

Although most of Pb, Ni, and Cr in the polluted sludge was removed with saponin by washing in the column, the residual concentrations are still higher than those in agricultural soil. Even still, this work has quantitatively shown that, to some extent, saponin could be an efficient sorbent for the removal of heavy metals from sludge. However, further investigations to survey the method for improving the removal efficiency of heavy metals, to elucidate the mechanism of the removal of heavy metals by surfactant,

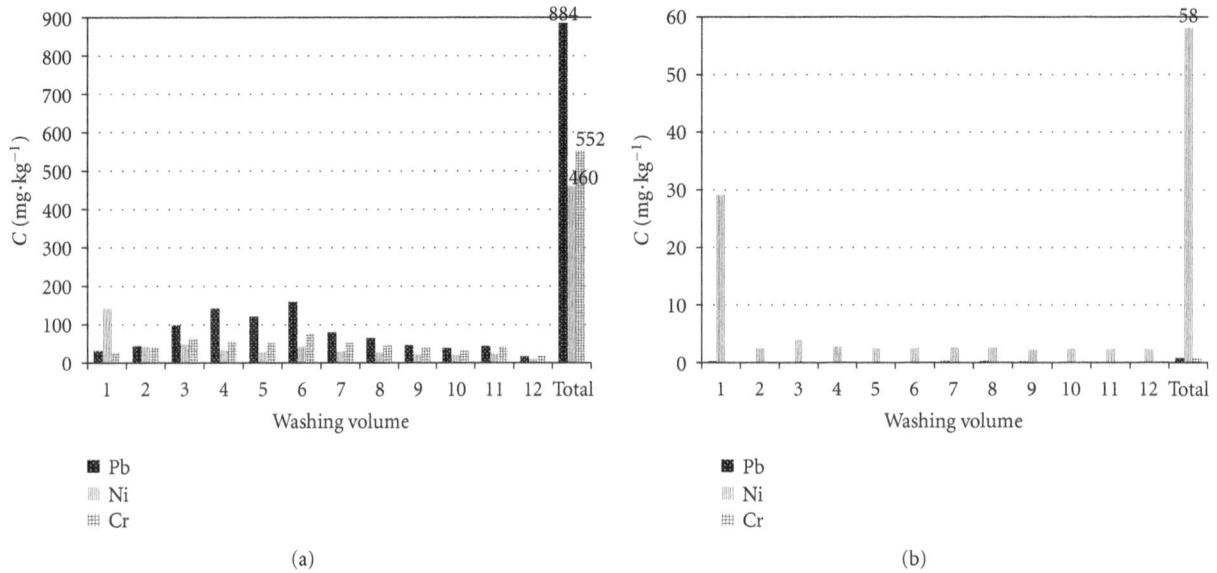

(a)   (b)

FIGURE 7: Effect of washing volume on the removal of heavy metals by column experiments: (a) saponin and (b) ultrapure water (1 w.v. (washing volume) = 6.2 dm³).

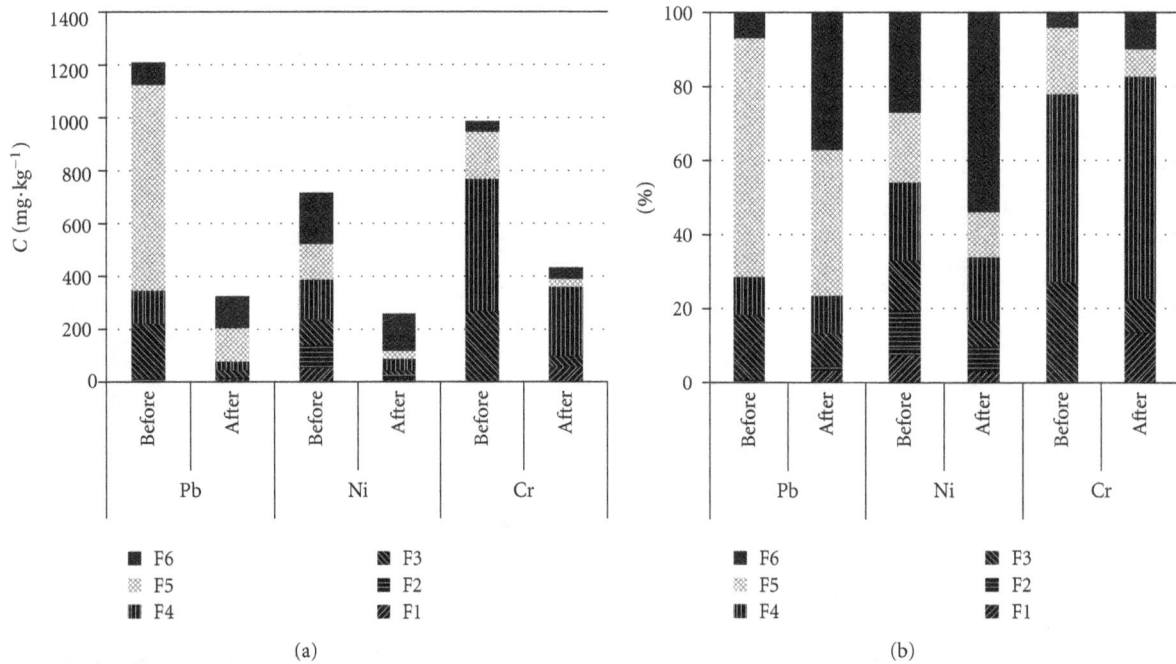

(a)   (b)

FIGURE 8: The concentrations and the relative distribution of heavy metals in polluted sludge before and after the column experiments: (a) concentration and (b) relative distribution.

and to survey the selection of the optimum surfactant and the optimum conditions for the removal of heavy metals are needed in future research.

4. Conclusion

The behavior, distribution, and characteristics of heavy metals including REEs, Th, and U in sludge from an industry water treatment plant were investigated and compared with

those in natural soil. Furthermore, the removal/recovery process of heavy metals (Pb, Cr, and Ni) from the polluted sludge was studied with biosurfactant elution by batch and column experiments. Consequently, the following conclusions have been obtained.

(1) The concentrations of heavy metals in sludge are greater than those in natural soils, and the concentrations of REEs, Th, and U in sludge are variable when compared with those in natural soils. The relative

Behavior and Distribution of Heavy Metals Including Rare Earth Elements, Thorium, and Uranium in Sludge from Industry
Water Treatment Plant and Recovery Method of Metals by Biosurfactants Application

53

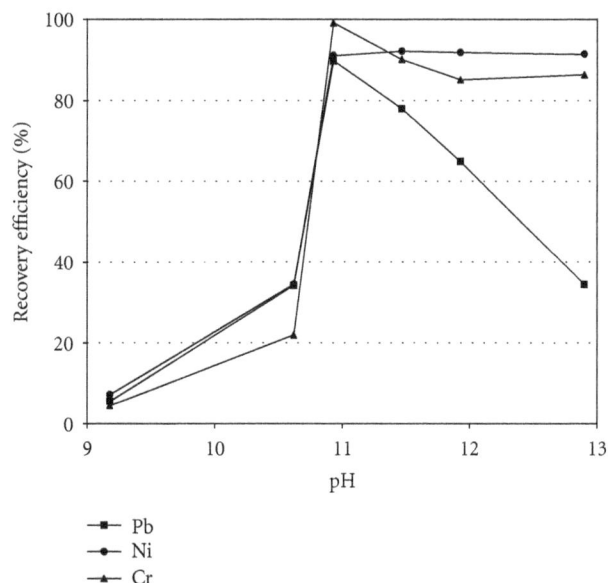

FIGURE 9: The recovery efficiency of heavy metals from the sludge leachate.

distribution of the residual fraction in natural soil is higher than that in original sludge. On the other hand, the relative distribution of oxide fraction in original sludge is higher than that in natural soils. That is, heavy metallic elements in sludge have generally greater concentrations and exist as more unstable fraction than those in natural soil.

(2) Nonionic saponin is more efficient than sophorolipid for the removal of heavy metals from sludge. Saponin has selectivity for the mobilization of heavy metals and mainly reacts with F3 and F5 fractions of heavy metals. In other words, nonionic biosurfactants including the carboxyl group have high potential for the removal of heavy metals in sludge.

(3) The recovery efficiency of heavy metals (Pb, Ni, and Cr) reached about 90–100% by the precipitation method with alkaline solution.

Acknowledgments

The present work was partially supported by a Grant-in-Aid for Scientific Research of the Japan Society for the Promotion of Science. The authors are grateful to Dr. K. Satoh of Faculty of Science, Dr. K. Fujii and Mr. M. Ohizumi of Waste Fluid Treatment Facilities, and Mr. T. Hatamachi of Faculty of Engineering in Niigata University for permitting the use of ICP-MS, ICP-AES, and Surface Area Analyzer and for giving helpful advice in measurement.

References

[1] H. Carrère, C. Dumas, A. Battimelli et al., "Pretreatment methods to improve sludge anaerobic degradability: a review," *Journal of Hazardous Materials*, vol. 183, no. 1–3, pp. 1–15, 2010.

[2] L. Appels, J. Baeyens, J. Degrève, and R. Dewil, "Principles and potential of the anaerobic digestion of waste-activated sludge," *Progress in Energy and Combustion Science*, vol. 34, no. 6, pp. 755–781, 2008.

[3] M. J. Wang, "Land application of sewage sludge in China," *Science of the Total Environment*, vol. 197, no. 1–3, pp. 149–160, 1997.

[4] D. Fytili and A. Zabaniotou, "Utilization of sewage sludge in EU application of old and new methods: a review," *Renewable and Sustainable Energy Reviews*, vol. 12, no. 1, pp. 116–140, 2008.

[5] K. M. Smith, G. D. Fowler, S. Pullket, and N. J. D. Graham, "Sewage sludge-based adsorbents: a review of their production, properties and use in water treatment applications," *Water Research*, vol. 43, no. 10, pp. 2569–2594, 2009.

[6] C. Wang, X. Hu, M. L. Chen, and Y. H. Wu, "Total concentrations and fractions of Cd, Cr, Pb, Cu, Ni and Zn in sewage sludge from municipal and industrial wastewater treatment plants," *Journal of Hazardous Materials*, vol. 119, no. 1–3, pp. 245–249, 2005.

[7] M. Chen, X. M. Li, Q. Yang et al., "Total concentrations and speciation of heavy metals in municipal sludge from Changsha, Zhuzhou and Xiangtan in middle-south region of China," *Journal of Hazardous Materials*, vol. 160, no. 2-3, pp. 324–329, 2008.

[8] A. Fuentes, M. Lloréns, J. Sáez et al., "Simple and sequential extractions of heavy metals from different sewage sludges," *Chemosphere*, vol. 54, no. 8, pp. 1039–1047, 2004.

[9] E. A. Alvarez, M. C. Mochón, J. C. J. Sanchez, and M. T. Rodríguez, "Heavy metal extractable forms in sludge from wastewater treatment plants," *Chemosphere*, vol. 47, no. 7, pp. 765–775, 2002.

[10] J. Ščancar, R. Milačič, M. Stražar, and O. Burica, "Total metal concentrations and partitioning of Cd, Cr, Cu, Fe, Ni and Zn in sewage sludge," *Science of the Total Environment*, vol. 250, no. 1–3, pp. 9–19, 2000.

[11] J. Deng, X. Feng, and X. Qiu, "Extraction of heavy metal from sewage sludge using ultrasound-assisted nitric acid," *Chemical Engineering Journal*, vol. 152, no. 1, pp. 177–182, 2009.

[12] C. Li, F. Xie, Y. Ma et al., "Multiple heavy metals extraction and recovery from hazardous electroplating sludge waste via ultrasonically enhanced two-stage acid leaching," *Journal of Hazardous Materials*, vol. 178, no. 1–3, pp. 823–833, 2010.

[13] S. Babel and D. del Mundo Dacera, "Heavy metal removal from contaminated sludge for land application: a review," *Waste Management*, vol. 26, no. 9, pp. 988–1004, 2006.

[14] G. Peng, G. Tian, J. Liu, Q. Bao, and L. Zang, "Removal of heavy metals from sewage sludge with a combination of bioleaching and electrokinetic remediation technology," *Desalination*, vol. 271, no. 1–3, pp. 100–104, 2011.

[15] N. Sakamoto, N. Kano, and H. Imaizumi, "Determination of rare earth elements, thorium and uranium in seaweed samples on the coast in Niigata Prefecture by inductively coupled plasma mass spectrometry," *Applied Geochemistry*, vol. 23, no. 10, pp. 2955–2960, 2008.

[16] N. Kano, Y. Otsuki, H. Lu, and H. Imaizumi, "Study on reduction of chromium using humic substances and clay minerals," *Journal of Ecotechnology Research*, vol. 13, no. 2, pp. 79–84, 2007.

[17] S. Guo and L. Kenne, "Structural studies of triterpenoid saponins with new acyl components from *Quillaja saponaria* Molina," *Phytochemistry*, vol. 55, no. 5, pp. 419–428, 2000.

[18] J. Chen, X. Song, H. Zhang, and Y. Qu, "Production, structure elucidation and anticancer properties of sophorolipid from *Wickerhamiella domercqiae*," *Enzyme and Microbial Technology*, vol. 39, no. 3, pp. 501–506, 2006.

[19] The Japanese Geotechnical Society, "The method and explanation for soil testing," pp. 125-130 and pp. 160-173, 1990.

[20] Environmental Science Division (EVS), U.S. Department of Energy, http://web.ead.anl.gov/resrad/datacoll/conuct.htm.

[21] Y. D. Wang, Y. Koto, N. Sakamoto, N. Kano, and H. Imaizumi, "Biosorption of rare earth elements, thorium and uranium using *Buccinum tenuissimum* shell biomass," *Radioisotopes*, vol. 59, no. 9, pp. 549–558, 2010.

[22] Y. Koto, N. Kano, Y. D. Wang, N. Sakamoto, and H. Imaizumi, "Biosorption of lanthanides from aqueous solutions using pretreated *Buccinum tenuissimum* shell biomass," *Bioinorganic Chemistry and Applications*, vol. 2010, Article ID 804854, 10 pages, 2010.

[23] N. Kano, T. Tsuchida, N. Sakamoto et al., "Behavior and distribution of rare earth elements, thorium and uranium in soil environment," *Radioisotopes*, vol. 58, no. 11, pp. 727–741, 2009.

[24] H. Sadamoto, K. Iimura, T. Honna, and S. Yamamoto, "Examination of fractionation of heavy metals in soils," *Japanese Journal of Soil Science and Plant Nutrition*, vol. 65, no. 6, pp. 645–653, 1994.

[25] A. Tessier, P. G. C. Campbell, and M. Blsson, "Sequential extraction procedure for the speciation of particulate trace metals," *Analytical Chemistry*, vol. 51, no. 7, pp. 844–851, 1979.

[26] L. D. Gao, N. Kano, Y. Higashidaira, Y. Nishimura, R. Ito, and H. Imaizumi, "Fractional determination of some metallic elements including rare earth elements, thorium and uranium in agriculture soil by sequential extraction procedure," *Radioisotopes*, vol. 60, no. 11, pp. 443–459, 2011.

[27] K. J. Hong, S. Tokunaga, and T. Kajiuchi, "Evaluation of remediation process with plant-derived biosurfactant for recovery of heavy metals from contaminated soils," *Chemosphere*, vol. 49, no. 4, pp. 379–387, 2002.

[28] L. Di Palma and F. Medici, "Recovery of copper from contaminated soil by flushing," *Waste Management*, vol. 22, no. 8, pp. 883–886, 2002.

[29] Administrative Management Bureau, Ministry of Internal Affairs and Communications (Japan), http://law.e-gov.go.jp/cgi-bin/strsearch.cgi.

[30] A. A. Juwarkar, K. V. Dubey, A. Nair, and S. K. Singh, "Bioremediation of multi-metal contaminated soil using biosurfactant—a novel approach," *Indian Journal of Microbiology*, vol. 48, no. 1, pp. 142–146, 2008.

[31] C. N. Mulligan, R. N. Yong, and B. F. Gibbs, "On the use of biosurfactants for the removal of heavy metals from oil-contaminated soil," *Environmental Progress*, vol. 18, no. 1, pp. 50–54, 1999.

[32] G. Z. Zheng, *Theory and Practice of Research on Heavy Metal Pollution in Agricultural Soil*, Chinese Environment Science Press, Beijing, China, 2007.

[33] A. Tsumura and S. Yamasaki, "Behavior of uranium, thorium, and lanthanoids in paddy fields," *Radioisotopes*, vol. 42, pp. 265–272, 1993.

7

Preparation and Characterization of Hydroxyapatite Coating on AZ31 Mg Alloy for Implant Applications

S. A. Salman,[1,2] K. Kuroda,[1] and M. Okido[1]

[1] EcoTopia Science Institute, Nagoya University, Furo-cho, Chikusa, Nagoya 464-8603, Japan
[2] Graduate School of Engineering, Al-Azhar University, Nasr City, Cairo 11371, Egypt

Correspondence should be addressed to S. A. Salman; sa.salman@yahoo.com

Academic Editor: Giovanni Natile

Magnesium alloys as biodegradable metal implants in orthopaedic research received a lot of interest in recent years. They have attractive biological properties including being essential to human metabolism, biocompatibility, and biodegradability. However, magnesium can corrode too rapidly in the high-chloride environment of the physiological system, loosing mechanical integrity before the tissue has sufficiently healed. Hydroxyapatite (HAp) coating was proposed to decrease the corrosion rate and improve the bioactivity of magnesium alloy. Apatite has been cathodically deposited on the surface of Mg alloy from solution that composed of 3 mM $Ca(H_2PO_4)_2$ and 7 mM $CaCl_2$ at various applied potentials. The growing of HAp was confirmed on the surface of the coatings after immersion in SBF solution for 7 days. The coating obtained at −1.4 V showed higher corrosion resistance with bioactive behaviors.

1. Introduction

Metal materials, including stainless steels, titanium, and cobalt-chromium-based alloys, are commonly used for implant devices due to their high strength, ductility, and good anticorrosion properties [1]. It is more suitable for load-bearing applications compared with ceramics or polymeric materials due to their combination of high mechanical strength and fracture toughness [2]. The release of toxic metallic ions or particles by corrosion or wear processes leads to undesirable effects on cell and bone tissues [3]. Moreover, these metallic materials are not biodegradable in the human body and can cause long-term complication (infection) [4]. The elastic modules of current metallic biomaterials are not well matched with that of natural bone tissue, resulting in stress shielding effects that can lead to reduced stimulation of new bone growth and remodeling which decreases implant stability [5]. Comparing to commonly approved metallic bio-materials, magnesium alloys have many outstanding advantages due to their attractive biological property including being essential to human metabolism, biocompatibility, and biodegradability [6].

The mechanical properties of magnesium alloys are similar to those of natural bone (40–57 GPa) [7]. Moreover, magnesium is one of the most important bivalent ions associated with the formation of biological appetites and plays an important role in the changes in the bone matrix that determines bone fragility [8]. On the other hand, implants made of magnesium alloys were degraded in vivo, eliminating the need for a second operation for implant removal. Good biocompatibility was observed in clinical studies [9]. Unfortunately, magnesium can corrode too rapidly in the physiological pH (7.4–7.6) and high-chloride environment of the physiological system, loosing mechanical integrity before the tissue has sufficiently healed and producing hydrogen gas in the corrosion process at a rate that is too fast to be dealt with by the host tissue [10].

Recently, some researches have been done to slow down the biodegradation rate of magnesium alloys, including fluoride conversion coating [11], alkali-heat treatment [12], and plasma immersion ion implantation [13]. Besides improving the biodegradation rate of magnesium alloys, the biocompatibility should also be considered. The hydroxyapatite $[Ca_{10}(PO_4)_6(OH)_2]$, hereafter (HAp), coating can satisfy

TABLE 1: Chemical composition (mass%) of AZ31 Mg alloy.

Al	Zn	Mn	Si	Cu	Ni	Fe	Mg
3.0	1.0	0.43	0.01	<0.01	<0.001	0.003	Bal.

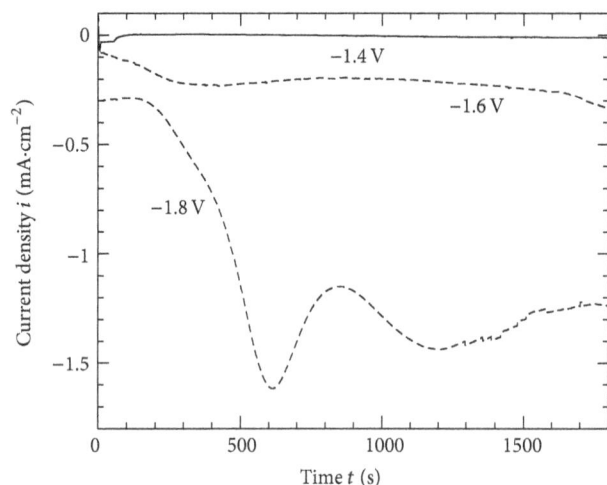

FIGURE 1: Current change during cathodic electrodeposition on AZ31 Mg alloy in 3 mM $Ca(H_2PO_4)_2$, 7 mM $CaCl_2$ for 30 min.

the dual properties. Synthetic HAp ceramics was routinely used as porous implants, powders, and coatings on metallic prostheses to provide bioactive fixation. The presence of sparingly soluble HAp coatings led to bone tissue response (osteoconduction) in which bone grew along the coating and formed a mechanically strong interface [14]. Therefore, HAp coating was deposited on the surface of metallic implants to improve the biocompatibility property and to decrease the degradation rate in the physiological environment. Deposition of HAp coatings has been achieved by a number of methods, including plasma spraying [15], ion implantation [16], sputtering [17], sol-gel coating [18], biomimetic methods [19], electrophoretic deposition (EPD) [20], electrochemical deposition [21], and electrospray deposition (ESD) [22, 23].

Among these methods, electrochemical deposition is known to be simple, flexible, and inexpensive technique for fabricating the metallic thin films and does not require complex and expensive vacuum apparatus. In addition, the deposition processing can be conducted at room temperature, and the morphology of coating can be controlled easily by varying the electrochemical potential and electrolyte concentration [24].

In this study, electrodeposition method was employed to produce HAp coating on AZ31 magnesium alloy in order to improve the anticorrosion properties and to enhance its bioactivity. The electrodeposition process was performed in calcium phosphate electrolyte with other additives at optimized pH and temperature. The effect of applied potential on the structure and the anticorrosion properties was investigated. The anticorrosion properties of the developed coatings were evaluated by the anodic polarization and electrochemical impedance spectroscopy techniques. The surface morphology and phase structure of coated films were also

examined using scanning electron microscope (SEM), X-ray diffraction (XRD), and Energy Dispersive Spectroscopy (EDS). The HAp coating could enhance the anticorrosion property and effectively improved in vitro bioactivity in simulated body fluid.

2. Materials and Methods

2.1. Specimens Preparation. Commercially available AZ31 Mg alloy (3 mass% Al, 1 mass% Zn) was used as the substrate. The chemical composition of the alloy is listed in Table 1. The surface of the alloys was polished up to no. 2000 emery paper followed by 0.05 μm alumina powders. The specimens were carefully cleaned with water, rinsed with acetone, and dried under air. All of the experiment specimens were mounted using polytetrafluoroethylene (PTFE) resin tape leaving and exposed surface area of 1 cm^2.

2.2. The Film Formation. The specimens were immersed in 3 mM $Ca(H_2PO_4)_2$ + 7 mM $CaCl_2$ solution at 37°C and pH 6 for 30 min. The electrodeposition was performed at −1.4, −1.6, and −1.8 V applied potentials using a conventional electrochemical cell equipped with three electrodes. Platinum, Ag/AgCl sat. KCl, and magnesium alloy specimen served as the counter, the reference, and the working electrode, respectively.

2.3. Characterization Methods. The morphology and microstructure of the films were observed with Hitachi S-800 scanning electron microscope (SEM). The crystal structure of the coated films was identified using thin film X-ray diffraction (XRD) analysis using a Shimadzu XRD-6000 X-ray diffractometer with Cu Kα radiation with scan step 0.02° and scan speed 2°/min. from diffraction angle of 10 to 60 at 30 kV and 20 mA. The elemental composition of the coating was identified using Energy Dispersive Spectroscopy (EDS).

2.4. Anticorrosion Measurements. The anodic polarization and electrochemical impedance spectroscopy tests were carried out using a Solartron 1285 Potentiostat from Solartron Analytical, Farnborough, United Kingdom. The measurements were controlled by Scribner Associates Corrware and Z polt electrochemical experiment software, respectively. The polarization curves were measured in phosphate buffered saline PBS(−) solutions with a scanning rate of 1 mV/s.

2.5. Bioactivity Experiment in Simulated Body Fluid. The coating was immersed for 7 days in the SBF (Kokubo solution) with ion concentrations nearly the same as those of the body blood plasma. The pH of the SBF solution was adjusted to 7.4. The phase structure, surface morphology, and the anticorrosion property were examined every day by removing the specimens from SBF solution, rinsed with distilled water, and air-dried.

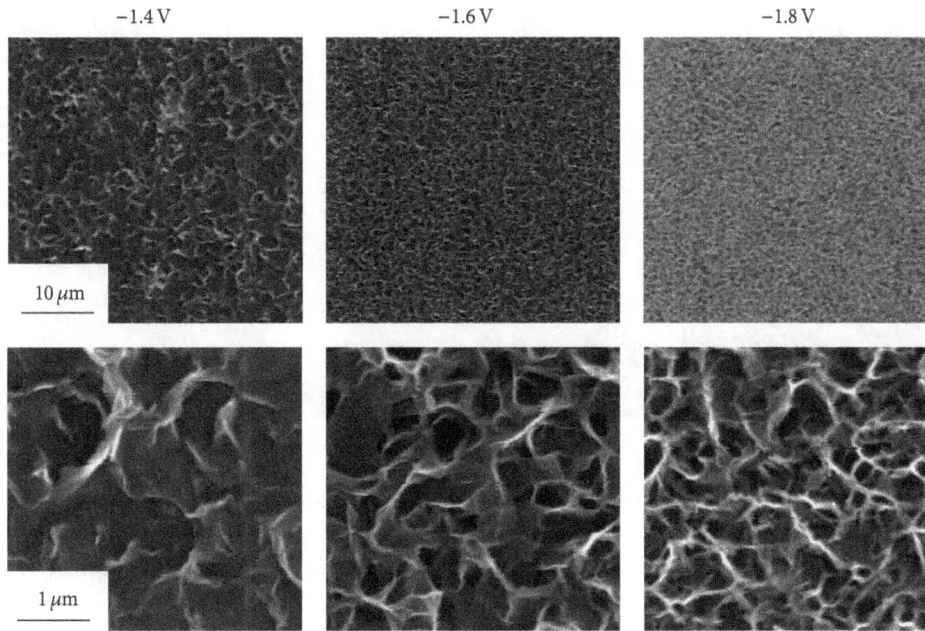

FIGURE 2: SEM images of coating films formed on AZ31 Mg alloy in 3 mM $Ca(H_2PO_4)_2$, 7 mM $CaCl_2$ for 30 min. at -1.4 V, -1.6 V, and -1.8 V.

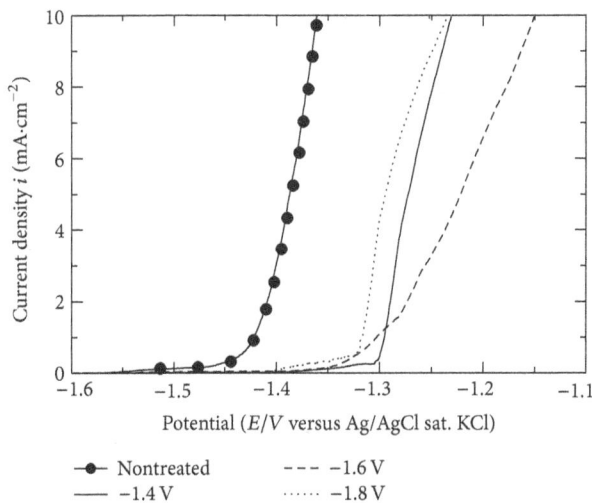

FIGURE 3: Anodic polarization test in PBS($-$) for AZ31 magnesium alloys before and after treatment at -1.4 V, -1.6 V, and -1.8 V.

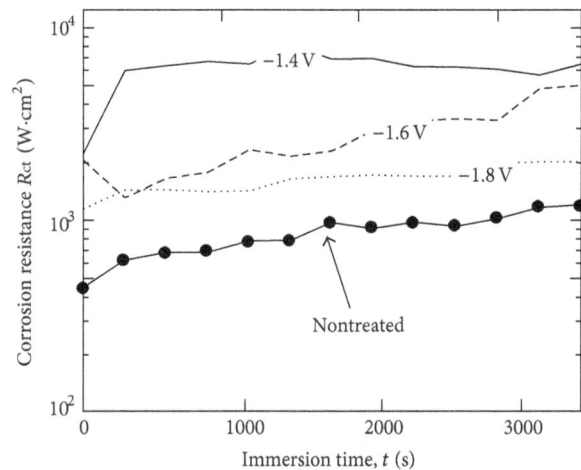

FIGURE 4: Change in Rct in PBS($-$) for AZ31 magnesium alloys before and after treatment at -1.4 V, -1.6 V, and -1.8 V.

3. Results and Discussion

High bioactivity can be achieved by performing the coating in the condition that is near to the human internal environment. Some research groups have applied various ranges of temperatures (room temperature; 45, 50, and 80°C) and initial pH (5.8, 6.5, and neutral) [25].

To achieve pure apatite coating on magnesium alloy, the initial pH and temperature should be carefully chosen. In our previous work, the optimum solution pH and temperature were defined [26]. The HAp peaks could be observed only at the temperature of 37°C or higher. It has been reported that the growth rate of apatite layer increased with the increasing

temperature due to the decreasing in the solubility product of HAp [27].

HAp coating was deposited on the AZ31 alloy using the electrodeposition method. Fixed potential was set to -1.4, -1.6, and -1.8 V, so that the Mg alloy substrate became negatively charged (as a cathode). Magnesium dissolution takes place immediately after immersion of the specimen in 3 mM $Ca(H_2PO_4)_2$ and 7 mM $CaCl_2$ solution for 30 min at pH 6. according to reaction (1). Water is reduced at the cathode surface to produce hydrogen gas and hydroxide ions,

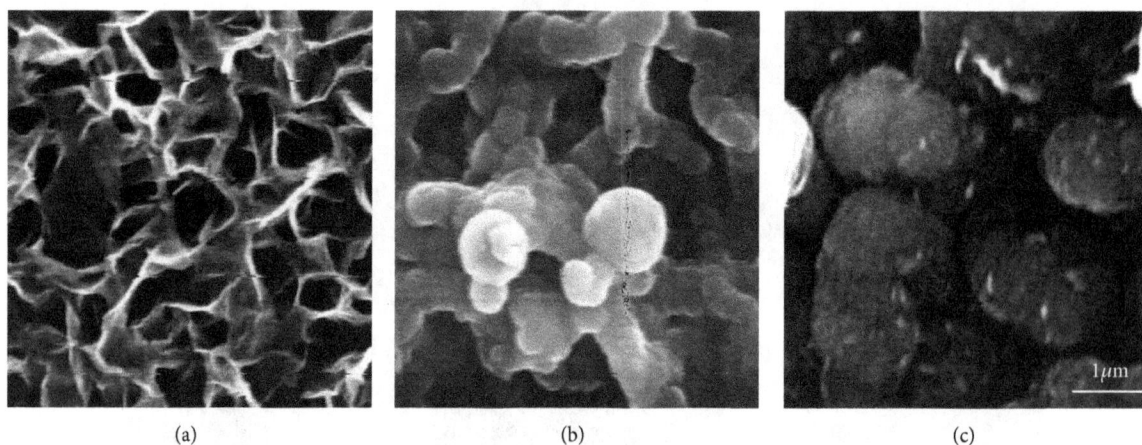

FIGURE 5: SEM images of coating obtained at −1.4 V before and after immersion in SBF solution for 1 day and 7 days.

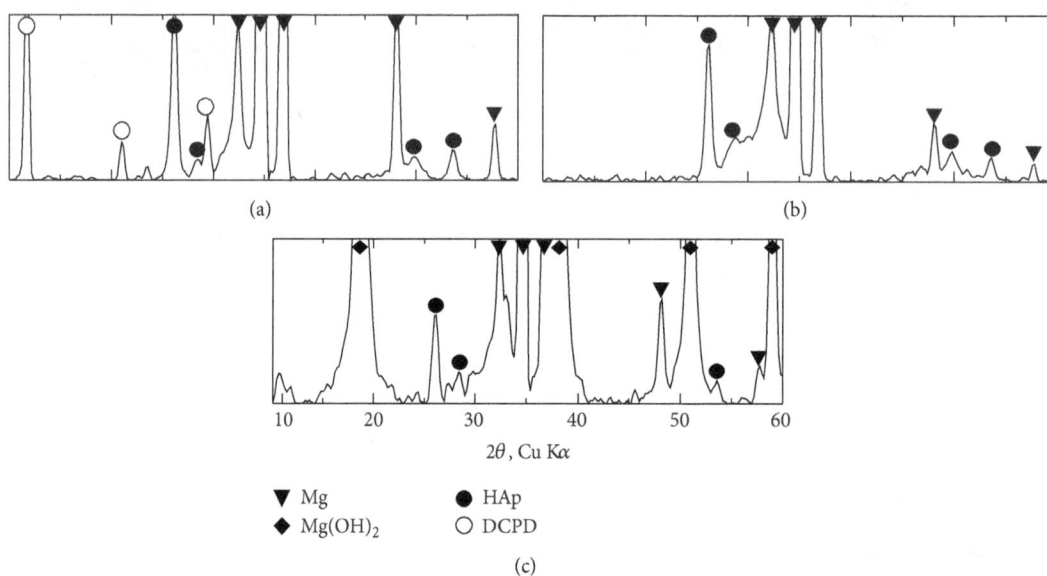

▼ Mg   ● HAp
◆ Mg(OH)$_2$ ○ DCPD

(c)

FIGURE 6: xrd patterns of coatings obtained at −1.4 V before and after immersion in SBF solution for 1 day and 7 days.

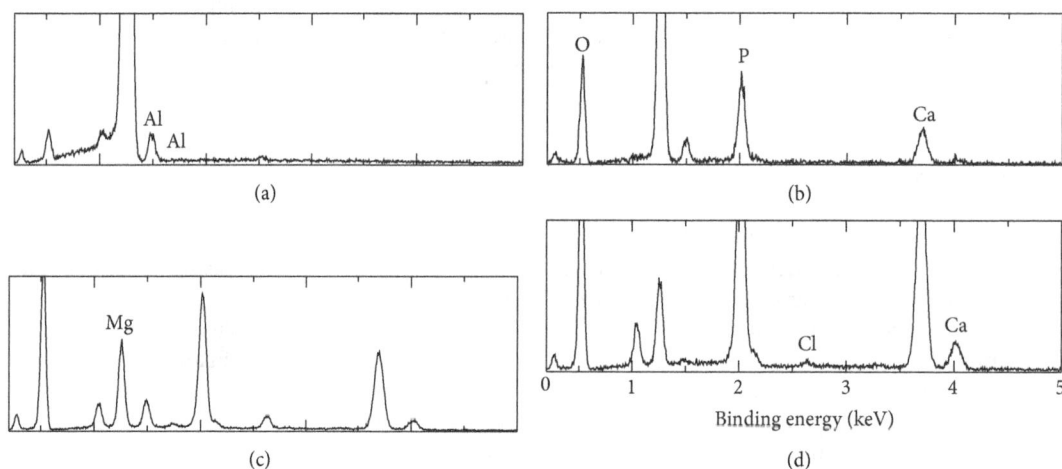

FIGURE 7: EDS patterns of coatings obtained at −1.4 V before and after immersion in SBF solution for 1 day and 7 days.

which will lead to an increase in the solution pH according to reaction (2):

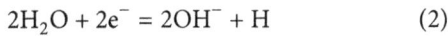

$$Mg = Mg^{2+} + 2e^- \qquad (1)$$

$$2H_2O + 2e^- = 2OH^- + H \qquad (2)$$

Figure 1 shows the current-time curves during the electrodeposition treatment at various potentials. At −1.8 V, the current density sharply decreased indicating the early formation of the coating film. On the other hand, the current density has a slight decrease at −1.6 V and has nearly a stable rate at −1.4. The local increase in the pH value and availability of Ca and P ions in the solution are expected to encourage the deposition of HAp and dicalcium phosphate dihydrate [CaHPO$_4 \cdot$2H$_2$O], hereafter (DCPD) on the magnesium alloy surface as shown in reactions (3) and (4):

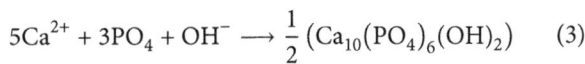

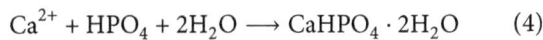

$$5Ca^{2+} + 3PO_4 + OH^- \longrightarrow \frac{1}{2}\left(Ca_{10}(PO_4)_6(OH)_2\right) \qquad (3)$$

$$Ca^{2+} + HPO_4 + 2H_2O \longrightarrow CaHPO_4 \cdot 2H_2O \qquad (4)$$

Figure 2 shows the SEM images of coating films formed at −1.4 V, −1.6 V, and −1.8 V. Flake-like particles were observed on all surfaces after cathodic deposition, while the size of crystallites becomes finer with increasing the cathodic potential to more negative side.

The corrosion resistance of the coating films formed at −1.4, −1.6, and −1.8 V was examined in PBS(−) solution using anodic polarization tests as shown in Figure 3. The results show that the corrosion reaction does not take place easily as it indicated by the decrease of corrosion currents and the increase of polarization resistances. The pitting potential of all Hap-coated specimens shifted toward the noble direction in comparison to the nontreated specimen. This indicates that the coating films offered a corrosion protection, and Mg alloy was able to be prevented from being dissolved. Moreover, the coating obtained at −1.4 V has the nobler pitting potential.

EIS test was performed to ensure the best corrosion resistance film. Corrosion rate determination is associated with the charge transfer resistance (Rct) by using EIS technique. The charge transfer resistance is equal to the diameter of the semicircle in the complex plane graph (Nyquist diagram). Figure 4 shows the time dependence of the charge transfer resistance, which corresponds to corrosion rate. The charge transfer resistance increases by the treatment in 3 mM Ca(H$_2$PO$_4$)$_2$ and 7 mM CaCl$_2$ solution at various applied potentials. The film formed at −1.4 V has much higher Rct value than those formed at −1.8 and −1.6 V. This result is in agreement with anodic polarization test which indicates that the film formed at −1.4 V has the best anticorrosion properties.

Improving corrosion resistance and decreasing dissolution rate of implant materials are highly recommended to allow the bone to heal. Therefore, we performed the bioactivity experiments on the coating obtained at −1.4 V. The in vitro examination was carried out at 37°C for 7 days in order to evaluate the bioactive behaviors of HAp coating in SBF solution, which has the ion element equal to or close to that of blood plasma.

Figure 5 shows the surface morphologies of the coating obtained at −1.4 before and after immersion in SBF solution for 1 day and 7 days. The growing of HAp was confirmed on the surface of the coating after immersion for 1 day in SBF solution. The particles sizes were finer upon immersion in SBF solution. The growing rate of HAp on the surface of the coating is notably increased with increasing the immersion time to 7 days.

Figure 6 shows the XRD results of the coating obtained at −1.4 before and after immersion in SBF solution for 1 day and 7 days. The DCPD and HAp peaks were seen before immersion in SBF solution.

With immersion in SBF solution, DCPD was converted into more stable HAp. With increasing of immersion time to 7 days, HAp peaks have increased which confirms the growth of HAp

Figure 7 shows the EDS results of the coating obtained at −1.4 before and after immersion in SBF solution for 1 day and 7 days. Strong peaks of calcium (Ca), phosphorous (P), oxygen (O), magnesium (Mg), aluminum (Al), and several small peaks were observed before and after immersion in SBF solution. Figure 6(a) shows that the coating film composed of Mg, Al, and calcium phosphate compounds which are identified as HAp and DCPD from previous XRD test. With immersion in SBF solution for 1 day, the peaks of Ca and P were increased and Mg peaks were decreased which indicate growing of HAp as shown in Figure 6(b). With increasing of immersion time to 7 days, the peaks of Ca and P continued to increase and Mg peaks continued to decrease as shown in Figure 6(b). This continuous growing of HAp with immersion for 7 days, obviously confirms the bioactivity of this coating

4. Conclusions

The hydroxyapatite coating was successfully electrodeposited on magnesium alloy surface by immersing in 3 mM Ca(H$_2$PO$_4$)$_2$ and 7 mM CaCl$_2$ solution for 30 min and at temperature of 37°C. The anticorrosion property of AZ31 magnesium alloy was improved with hydroxyapatite coating, and the best anticorrosion properties were obtained at −1.4 V. Hydroxyapatite was grown on the coating with immersion in SBF solution for 7 days, which proves its bioactivity.

Acknowledgments

The authors gratefully acknowledge the Ministry of the Education, Culture, Sports, Science and Technology, Japan, and the Ministry of Higher Education, Egypt. They wish also to acknowledge Professor Ryoichi Ichino of EcoTopia Science Institute, Nagoya University, Japan, for valuable discussion during this research.

References

[1] S. Ono, A. Kodama, and H. Asoh, "Electrodeposition behavior of hydroxyapatite on porous electrode composed of sintered titanium spheres," *Journal of Japan Institute of Light Metals*, vol. 58, no. 11, pp. 593–598, 2008.

[2] M. Sumita, T. Hanawa, and S. H. Teoh, "Development of nitrogen-containing nickel-free austenitic stainless steels for metallic biomaterials—review," *Materials Science and Engineering C*, vol. 24, no. 6–8, pp. 753–760, 2004.

[3] B. D. Hahn, D. S. Park, J. J. Choi et al., "Aerosol deposition of hydroxyapatite-chitosan composite coatings on biodegradable magnesium alloy," *Surface and Coatings Technology*, vol. 205, no. 8-9, pp. 3112–3118, 2011.

[4] C. Lhotka, T. Szekeres, I. Steffan, K. Zhuber, and K. Zweymüller, "Four-year study of cobalt and chromium blood levels in patients managed with two different metal-on-metal total hip replacements," *Journal of Orthopaedic Research*, vol. 21, no. 2, pp. 189–195, 2003.

[5] J. Nagels, M. Stokdijk, and P. M. Rozing, "Stress shielding and bone resorption in shoulder arthroplasty," *Journal of Shoulder and Elbow Surgery*, vol. 12, no. 1, pp. 35–39, 2003.

[6] P. Staiger, A. Pietak, J. Huadmai, and G. Dias, "Magnesium and its alloys as orthopedic biomaterials: a review," *Biomaterials*, vol. 27, no. 9, pp. 1728–1734, 2006.

[7] J. Vormann, "Magnesium: nutrition and metabolism," *Molecular Aspects of Medicine*, vol. 24, no. 1–3, pp. 27–37, 2003.

[8] C. R. Howlett, H. Zreiqat, R. O'Dell et al., "The effect of magnesium ion implantation into alumina upon the adhesion of human bone derived cells," *Journal of Materials Science*, vol. 5, no. 9-10, pp. 715–722, 1994.

[9] F. Witte, V. Kaese, H. Haferkamp et al., "In vivo corrosion of four magnesium alloys and the associated bone response," *Biomaterials*, vol. 26, no. 17, pp. 3557–3563, 2005.

[10] C. E. Wen, M. Mabuchi, Y. Yamada, K. Shimojima, Y. Chino, and T. Asahina, "Processing of biocompatible porous Ti and Mg," *Scripta Materialia*, vol. 45, no. 10, pp. 1147–1153, 2001.

[11] K. Y. Chiu, M. H. Wong, F. T. Cheng, and H. C. Man, "Characterization and corrosion studies of fluoride conversion coating on degradable Mg implants," *Surface and Coatings Technology*, vol. 202, no. 3, pp. 590–598, 2007.

[12] X. N. Gu, W. Zheng, Y. Cheng, and Y. F. Zheng, "A study on alkaline heat treated Mg-Ca alloy for the control of the biocorrosion rate," *Acta Biomaterialia*, vol. 5, no. 7, pp. 2790–2799, 2009.

[13] C. Liu, Y. Xin, X. Tian, and P. K. Chu, "Corrosion behavior of AZ91 magnesium alloy treated by plasma immersion ion implantation and deposition in artificial physiological fluids," *Thin Solid Films*, vol. 516, no. 2–4, pp. 422–427, 2007.

[14] L. L. Hench and J. M. Polak, "Third-generation biomedical materials," *Science*, vol. 295, no. 5557, pp. 1014–1017, 2002.

[15] L. Yan, Y. Leng, and L. T. Weng, "Characterization of chemical inhomogeneity in plasma-sprayed hydroxyapatite coatings," *Biomaterials*, vol. 24, no. 15, pp. 2585–2592, 2003.

[16] M. Hamdi and A. Ide-Ektessabi, "Preparation of hydroxyapatite layer by ion beam assisted simultaneous vapor deposition," *Surface and Coatings Technology*, vol. 163-164, pp. 362–367, 2003.

[17] S. J. Ding, "Properties and immersion behavior of magnetron-sputtered multi-layered hydroxyapatite/titanium composite coatings," *Biomaterials*, vol. 24, no. 23, pp. 4233–4238, 2003.

[18] W. Weng, S. Zhang, K. Cheng et al., "Sol-gel preparation of bioactive apatite films," *Surface and Coatings Technology*, vol. 167, no. 2-3, pp. 292–296, 2003.

[19] W. H. Song, Y. K. Jun, Y. Han, and S. H. Hong, "Biomimetic apatite coatings on micro-arc oxidized titania," *Biomaterials*, vol. 25, no. 17, pp. 3341–3349, 2004.

[20] X. Meng, T. Y. Kwon, and K. H. Kim, "Hydroxyapatite coating by electrophoretic deposition at dynamic voltage," *Dental Materials Journal*, vol. 27, no. 5, pp. 666–671, 2008.

[21] R. Chiesa, E. Sandrini, M. Santin, G. Rondelli, and A. Cigada, "Osteointegration of titanium and its alloys by anodic spark deposition and other electrochemical techniques: a review," *The Journal of Applied Biomaterials & Biomechanics*, vol. 1, no. 2, pp. 91–107, 2003.

[22] W. Siefert, "Corona spray pyrolysis: a new coating technique with an extremely enhanced deposition efficiency," *Thin Solid Films*, vol. 120, no. 4, pp. 267–274, 1984.

[23] M. Iafisco, R. Bosco, S. C. G. Leeuwenburgh et al., "Electrostatic spray deposition of biomimetic nanocrystalline apatite coatings onto titanium," *Advanced Engineering Materials*, vol. 14, pp. B13–B20, 2012.

[24] M. C. Kuo and S. K. Yen, "The process of electrochemical deposited hydroxyapatite coatings on biomedical titanium at room temperature," *Materials Science and Engineering C*, vol. 20, no. 1-2, pp. 153–160, 2002.

[25] H. Qu and M. Wei, "The effect of temperature and initial pH on biomimetic apatite coating," *Journal of Biomedical Materials*, vol. 87, no. 1, pp. 204–212, 2008.

[26] S. A. Salman, *Corrosion protection of magnesium alloys with anodizing and conversion coating [Ph.D. thesis]*, Nagoya University, 2010.

[27] T. Kokubo and H. Takadama, "How useful is SBF in predicting in vivo bone bioactivity?" *Biomaterials*, vol. 27, no. 15, pp. 2907–2915, 2006.

8

Pharmacokinetic Study of Di-Phenyl-Di-(2,4-Difluobenzohydroxamato)Tin(IV): Novel Metal-Based Complex with Promising Antitumor Potential

Yunlan Li, Zhuyan Gao, Pu Guo, and Qingshan Li

School of Pharmaceutical Science, Shanxi Medical University, Taiyuan 030001, China

Correspondence should be addressed to Qingshan Li, qingshanl@yahoo.com

Academic Editor: Sanja Mijatović

Di-phenyl-di-(2,4-difluobenzohydroxamato)tin(IV)(DPDFT), a new metal-based arylhydroxamate antitumor complex, showed high *in vivo* and *in vitro* antitumor activity with relative low toxicity, but no data was reported regarding its pharmacokinetics and dependent toxicity. In this paper, a rapid, sensitive, and reproducible HPLC method *in vivo* using Diamonsil ODS column with a mixture of methanol and phosphoric acid in water (30 : 70, V/V, pH 3.0) as mobile phase was developed and validated for the determination of DPDFT. The plasma was deproteinized with methanol that contained acetanilide as the internal standard (I.S.). The photodiode array detector was set at a wavelength of 228 nm at room temperature and a linear curve over the concentration range $0.1 \sim 25\,\mu g \cdot mL^{-1}$ ($r = 0.9993$) was obtained. The method was used to determine the concentration-time profiles for DPDFT in the plasma after single intravenous administration with doses of 5, 10, 15 mg·kg^{-1} to rats. The pharmacokinetics parameter calculations and modeling were carried out using the 3p97 software. The results showed that the concentration-time curves of DPDFT in rat plasma could be fitted to two-compartment model.

1. Introduction

Metals offer potential advantages over the more common organic-based drugs, including a wide range of coordination numbers and geometries, accessible redox states, "tuneability" of the kinetics of ligand substitution, as well as a wide structural diversity. Medicinal inorganic chemistry is a thriving area of research [1, 2], initially fueled by the discovery of cisplatin, a metal-based antitumor drug about 40 years ago. Since the discovery of cisplatin and its introduction in the clinics, metal compounds have been intensely investigated in view of their possible application in cancer therapy. Platinum anticancer agents, such as cisplatin, have been highly successful but there are several disadvantages associated with their clinical use. What needs to be recognized is that there are many other nonplatin metal-based antitumor drugs in the periodic table with therapeutic potential. Diorganotin(IV) complexes are potential antitumor agents mainly active against P388 lymphocytic leukemia and other tumors [3–5].

Lately, these antitumor agents are actually being studied widely. Among diorganotin(IV) compounds, dibutyltin(IV) derivatives of hydroxamic acid have received more attention due to their structural and biological importance [6–8].

Recently, we reported a series of diorganotin(IV) aryl-hydroxamates which exhibit *in vitro* antitumor activities (against a series of human tumor cell lines) which, in some case, are identical to, or even higher than, that of *cisplatin* [3, 5]. Di-phenyl-di-(2,4- difluobenzohydroxamato)tin(IV) (DPDFT, its structure was shown in Figure 1), a kind of efficient diorganotin(IV) patent compound (number: ZL01135148 and 102826Z) with lower toxicity, is a potential antitumor candidate for the clinical application due to its high *in vivo and in vitro* activity mainly against hepatoma, gastric cancer, nasopharyngeal carcinoma and other tumors (data not shown here). However, its antitumor molecular action mechanism is still unclear. In order to study the precise action mechanism and toxicity of this metal-based antitumor diorganotin(IV) compound, its fate should be first elucidated *in vivo*, including their absorption, distribution, metabolism, and elimination. However, nothing is known about the pharmacokinetic behavior of DPDFT in body. Therefore, it is essential to establish a rapid, sensitive, and accurate method to determine the pharmacokinetic of organotin compound DPDFT in a body.

For medical care, practical, robust, simple, and efficient analytical methods are needed. Several methods have been reported for the determination of organotin compounds, including HPLC-MS [9, 10], HPLC-ICP-MS [10, 11], GC [12], GC-MIP AED [12–14], GC-MS [15–17], GC-ICP MS [18, 19], and so on. However, these methods have obvious disadvantages. For examples, GC methods have complicated pretreatments for the samples and seem to be unsuitable for quantitative determination in rat plasma because only a small amount of blood is normally used in pharmacokinetic studies. HPLC-MS is superior method and should be used whenever is possible. MS methods were used for mentioned analysis in order to increase selectivity of detection of DPDFT from complex matrixes. However, in this study, the HPLC-MS method was not used because the mass spectrometry conditions were still not ripe for DPDFT detection, and the high analysis cost and the expensive apparatus required were other considerations. So far, to the best of our knowledge, no method has been reported for determination of the diorganotin(IV) patent compound DPDFT by HPLC method with UV detection in the pharmacokinetic studies in rat plasma. Therefore, in this research, we chose DPDFT as a typical antitumor diorganotin(IV) arylhydroxamate to develop a simple, sensitive, and specific HPLC assay for its quantitative determination in rat plasma and to investigate its preliminary pharmacokinetic properties.

2. Experimental

2.1. Materials and Reagents. DPDFT used in analysis was synthesized by the same method as described in [20]. Its purity was 99.9%. Acetanilide used as an internal standard (I.S.) was purchased from Sigma Laboratories (St. Louis, MO). The chromatographic solvents and reagents such as methanol and phosphoric acid were obtained from the National Institute for the Control of Pharmaceutical and Biologic Products (Beijing, China). All substances were of chromatographic grade. Deionized water was prepared using a Milli-Q water purifying system from Millipore Corp. (Bedford, MA).

2.2. Animal Treatment. Laboratory bred adult Wistar albino rats (200–250 g), which were supplied by the Animal Research Center at Shanxi Medical University (Taiyuan, Shanxi Province, China), were housed at $25 \pm 2°C$ in a well-ventilated animal house under 12:12 h light dark cycle. The animals drank sterilized drinking water, and standard chow diet was supplied *ad libitum* to each cage. The animal experiments were performed in accordance with the ARVO Statement for the Use of Animals in Ophthalmic and Vision Research and were approved by the Animal Ethics Committee of Shanxi Medical University.

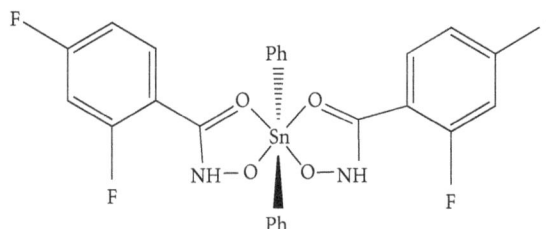

FIGURE 1: Chemical structure of DPDFT.

2.3. Instrumentation and Chromatographic Conditions. HPLC analysis was carried out using a Waters 2695 HPLC system (Waters Associates, Milford, MA) which consisted of a photodiode array detector, an autosampler, and a degasser. The apparatus was interfaced to a DELL PC compatible computer using Empower Pro software for data acquisition. The sensitivity was 0.2 AUFS. The autosampler was cooled to 10°C. The column was maintained at room temperature. Chromatographic separation of DPDFT and the I.S. was achieved on a Diamonsil C_{18} column (250 mm $\times$ 4.6 mm, 5 μm) from Dikma Technologies (Beijing, China) protected by a SHIMADZU LC guard column at 25°C. The mobile phase for HPLC analysis consisted of phosphoric acid in water (solvent A)/methanol (solvent B) (30 : 70, V/V, pH 3.0) with a flow rate of 0.8 mL·min^{-1}. A sample volume of 20 μL was injected. Prior to use, the mobile phase was filtered through a 0.45 μm hydrophilic membrane filter. The photodiode array detector was set at a wavelength of 228 nm at room temperature. Under the chromatographic condition mentioned above, DPDFT and the I.S. acetanilide could be separated completely in the chromatograms ($R > 1.5$), and there was no endogenous interference with the chromatographic peak of DPDFT and the I.S. acetanilide. Besides, the retention time of acetanilide ($t_R = 15.67$ min) was very suitable as the I.S. compared to that of DPDFT ($t_R = 8.34$ min).

2.4. Preparation of Plasma Samples. Blood samples collected from rat blood plasma were immediately transferred to 1.5 mL heparinized microcentrifuge tubes from fossa orbitalis of rats, and then processed for plasma by centrifugation. The supernatant plasma (0.2 mL) was then vortex-mixed with methanol (0.4 mL) containing acetanilide (0.2 mL, 50.0 μg·mL^{-1}) as internal standard (I.S.) for 30 s. After vortex-mixing, the mixture was centrifuged at 13000 rpm for 10 min 4°C to separate precipitated proteins. The supernatant solution of methanol layer was filtered through a 0.45 μm membrane filter. Twenty microliters of filtrate were injected into the chromatography. The same sample processing was also applied to the recovery and to the precision study in plasma. All harvested samples stored at −4°C were brought to room temperature before use and analyzed within one week. No significant differences were found between the samples stored at −4°C and those stored at −20°C (data not shown).

Pharmacokinetic Study of Di-Phenyl-Di-(2,4-Difluobenzohydroxamato)Tin(IV): Novel Metal-Based Complex
with Promising Antitumor Potential

63

2.5. Bioanalytical Method Validation

2.5.1. Preparation of Stocks, Calibration Standards, and Quality Control Samples.
A stock solution of DPDFT was prepared in methanol at the concentration of $100\,mg \cdot mL^{-1}$ and was further diluted in HPLC mobile phase to make working standards. The I.S. stock solution was prepared and diluted to $50.0\,\mu g \cdot mL^{-1}$ working solution with HPLC mobile phase. All the stock solutions were maintained at $4°C$ until use.

The linearity of HPLC method for the determination of DPDFT was evaluated by a calibration curve in the range of $0.1\sim25\,\mu g \cdot mL^{-1}$. Calibration standard samples were prepared by adding different concentrations of the standards of DPDFT and 0.2 mL of the I.S. working solution to the blank plasma. The final concentrations of DPDFT standard samples of plasma (0.1, 0.5, 1.0, 2.5, 5.0, 10.0, and $25.0\,\mu g \cdot mL^{-1}$, resp.) were prepared by spiking control rat plasma with appropriate amounts of the standard stock solution prepared above. The I.S. was added to each standard sample immediately before sample processing. For the evaluation of the linearity of the standard calibration curve, the analyses of DPDFT in plasma samples were performed on three independent days using fresh preparations. Each calibration curve consisted of a double blank sample (without internal standard), a blank sample (with internal standard), and seven calibrator concentrations. Each calibration curve was constructed by plotting the analyte to internal standard peak area ratio (y) against analyte concentrations (x). The calibration curves were fitted using a least-square linear regression model $y = ax + b$. The resulting a, b parameters were used to determine back-calculated concentrations, which were then statistically evaluated. All calibration curves of DPDFT were constructed before the experiments with correlation coefficient (r^2) of 0.99 or better.

2.5.2. Bioanalytical Method Validation.
The specificity was defined as non-interference when DPDFT was being retained from the endogenous plasma components, and no crossinterference between DPDFT and the I.S. using the proposed extraction procedure. Six different lots of blank (DPDFT-free plasma) were evaluated both with and without internal standard to assess the specificity of the method.

Quality control (QC) samples were used to determine the accuracy and precision of method and were independently prepared at low ($0.8\,\mu g \cdot mL^{-1}$), medium ($4\,\mu g \cdot mL^{-1}$), and high ($20\,\mu g \cdot mL^{-1}$) concentrations. To evaluate the accuracy and precision, we used at least five QC samples of three different concentrations of DPDFT. The intraday and interday accuracies were expressed as the percentage difference between the measured concentration and the nominal concentration in rat plasma. The intraday precision and accuracy were calculated using replicate ($n = 6$) determinations for each concentration of the spiked plasma sample during a single analytical run. The interassay precision and accuracy were calculated using replicate ($n = 6$) determinations of each concentration made on three separate days. The variability of determination was expressed as the relative standard deviation (RSD) which should be $\leq15\%$, covering the range of actual experimental concentrations.

The extraction efficiency of DPDFT was determined by analyzing replicate sets ($n = 6$) of QC samples: 0.4, 8, $20\,\mu g \cdot mL^{-1}$ for rat plasma, representing low, medium, and high QCs, respectively. The recoveries were calculated by comparing the peak areas of DPDFT added into blank samples and extracted using the protein precipitation procedure, with those obtained from DPDFT spiked directly into postprotein precipitation solvent at three QC concentration levels.

The stability of DPDFT in rat plasma was assessed by analyzing replicates ($n = 6$) of QC samples at concentrations of 0.8, 4, $20\,\mu g \cdot mL^{-1}$, respectively. The investigation covered the expected conditions during all of the sample storage and process periods, which included the stability data from freeze/thaw cycle and long-term stability tests. The concentrations obtained from stability studies were compared with the freshly prepared QC samples, and the percentage concentration deviation was calculated. In each freeze–thaw cycle, the samples were frozen and stored at $-20°C$ for 10 days, then thawed at room temperature. The stability of the fresh plasma samples was tested after keeping the samples at -4, $-20°C$ and room temperature for 72 h. The stability of deproteinized samples at $10°C$ in the autosampler was evaluated up to 24 h.

To determine the limit and quantification of detection, we prepared the dilutions of 1, 2.5, 5, 10, 15, 20, and $30\,ng \cdot mL^{-1}$ DPDFT in plasma. The results were evaluated by analyzing each plasma sample spiked with the analyte at a final concentration at which the signal-to-noise ratio (S/N) was 10 and 3.

2.5.3. Pharmacokinetic Study.
To evaluate the suitability of the assay for pharmacokinetic studies, 7.5, 15, and $30\,mg \cdot kg^{-1}$ of DPDFT were intravenously administered to rats (half males and half females, resp.). Six animals were used in each dosage by direct injection into a lateral tail vein, with a duration of infusion less than 1 min. Heparinized blood samples (0.5 mL) were collected at 0, 1, 3, 5, 10, 30, 60, and 120 min after injection. Eighteen rats were used for each time point. After each sampling, the removed volume of blood was supplemented with an equal volume of sodium chloride. The blood samples were immediately centrifuged and the resulting plasma was prepared according to the procedure given for the calibrators. Pharmacokinetic calculations were performed using the observed data. Pharmacokinetic analysis of DPDFT concentrations in plasma was performed using two-compartment model methods via the 3p97 software package (Chinese Pharmacology Society). All values obtained were expressed in mean ± standard deviation.

3. Results and Discussion

A deproteinized method to detect DPDFT, an typical antitumor diorganotin(IV) compound in plasma, was first developed in this paper. Diverse proportional solvents (methanol and acidified water) were selected in the deproteinizing process. The plasma deproteinized with double volume of methanol could produce the minimal dilution, optimum peak shape, and an increase of detector sensitivity along with the satisfactory recovery.

3.1. Wavelength Selection Results.

3.1. Wavelength Selection Results. The Waters 2695 HPLC–DAD measurement was performed under the chromatographic condition mentioned above. Chromatographic separation of DPDFT was achieved on a Diamonsil C_{18} column (250 mm × 4.6 mm, 5 μm) protected by a SHIMADZU guard column at 25°C. The mobile phase for HPLC analysis consisted of phosphoric acid in water (solvent A)/methanol (solvent B) (30:70, V/V, pH 3.0) with a flow rate of 0.8 mL·min^{-1}. A sample volume of 20 μL was injected. The DAD wavelength range was set on 190~400 nm with a slid width of 1 nm and a response time of 2.0 s at room temperature. As shown in Figures 2(a), 2(b), and 2(c), the peaks were detected with good baseline separation. Peak identification was confirmed by comparison of UV spectra. The maximum absorption wavelength of DPDFT was 228 nm, and there was no endogenous interference with the chromatographic peak of DPDFT.

3.2. Validation Data of Bioanalytical Method. The method was validated using the criteria described above. The data were found to be linear over a concentration range of 0.1~25 μg·mL^{-1} in blood samples. The regression equation was $y = 32.001x + 0.31$, with the correlation coefficient $r = 0.9993$ ($n = 7$), where y represented the peak-area ratio of DPDFT to the I.S. in rat plasma and x was the concentration of DPDFT. The limit of quantitation (LOQ) was 10 ng, which can be determined with a relative error (RE) and precision (RSD) of <15% at a signal to-noise ratio of 10. The limits of detection (LOD) were 3.5 ng, based on a signal-to-noise ratio of 3.

Under the chromatographic condition, the number of theoretical plates was 5000. The degree of interference by endogenous plasma with DPDFT and the I.S. was assessed by inspection of chromatograms derived from a processed blank plasma sample. The results show that there were no endogenous interfering peaks with the I.S. and DPDFT in the rat plasma. Typical chromatograms of blank plasma, blank plasma spiked with DPDFT QC sample (3 μg·mL^{-1}) and the I.S., and a rat plasma sample after dosing with 15 mg·kg^{-1} DPDFT are presented (Figure 3). DPDFT and the I.S. were eluted at 15.67 and 8.34 min, respectively. The total run time was less than 30 min. A good separation of the I.S. and DPDFT was obtained under the specified chromatographic conditions.

The recoveries of the assay were assessed by comparing the peak-area ratios (analyte/the I.S.) obtained from spiked plasma samples of three DPDFT standard concentrations (0.4, 8, 20 mg/mL) with the peak-area ratios (analyte/the I.S.) for the samples containing the equivalent analyte and the I.S. which were directly dissolved in methanol. The recoveries were approximately 90.0–97.0% in the rat plasma, as shown in Table 1, the mean extraction recovery and the coefficient of variation RSD of DPDFT at three various concentrations

TABLE 1: Recoveries of the assay for determining DPDFT in rat plasma ($n = 6$).

Spiked concentration (μg/mL)	Recovery (%, mean ± SD)	RSD (%)
0.8	90.8 ± 7.0	7.7
4	95.7 ± 12.9	13.5
20	96.2 ± 10.8	11.3

from the rat plasma were 94.2%± 10.3% and 10.9% ($n = 18$), respectively.

The precision and accuracy of this method were evaluated by assaying each low, middle, and high concentration QC sample. The reproducibility of the method was assessed by examining both intraday and interday variance. Accuracy (%) = [(Cobs − Cnom)/Cnom] × 100. The precision (%RSD) was calculated from the observed concentrations as follows: RSD = [standard deviation (SD)/Cobs] ×100. As shown in Table 2, the data showed that the intraday and interday precisions (% RSD) of the three QC samples in rat plasma were <15%. The RSD values of the intraday and the interday for rat plasma samples ranged from 3.8% ~9.0% and 4.0% ~9.0%, respectively. These validations demonstrated the reliability of the assay.

Stability of DPDFT during storage and processing was checked using quality control samples. The DPDFT and I.S. stock solutions were stable for at least 2 months when stored at 4°C. The deviation of the mean test responses was within ±10% of appropriate controls in all stability tests of DPDFT in rat plasma. After three freeze-thaw cycles, the concentration changes of DPDFT were less than 7%. The analyte was stable in the matrices at 4, −20°C and room temperature for 72 h without significant degradation (<8%). The run-time stability study showed that DPDFT in deproteinized rat plasma was stable at 10°C for up to 24 h (<6%). These results suggested that the rat plasma samples containing DPDFT can be handled under normal laboratory conditions without significant loss of the compound. All stability results are summarized in Table 3.

3.3. Pharmacokinetic Applicability. The developed RP-HPLC analytical method has been successfully used for the pharmacokinetic study after a single intravenous administration of DPDFT within 120 min. The mean plasma concentration-time curves of DPDFT after administration of 7.5, 15, and 30 mg·kg^{-1} in rats are shown in Figure 4, the concentration-time data conformed to a two-compartment model and the major mean pharmacokinetic parameters (mean ± SD) are summarized in Table 4. The RP-HPLC method satisfied the requirement of this study and demonstrated its general suitability for pharmacokinetics studies of DPDFT in rats.

3.4. The Pharmacokinetics Features of DPDFT in Rat Plasma. HPLC analysis for amphoteric, polar substances with low wavelength ultraviolet absorption is always associated with some difficulties, especially when high sensitivity is required. The results showed that this classic liquid extraction HPLC

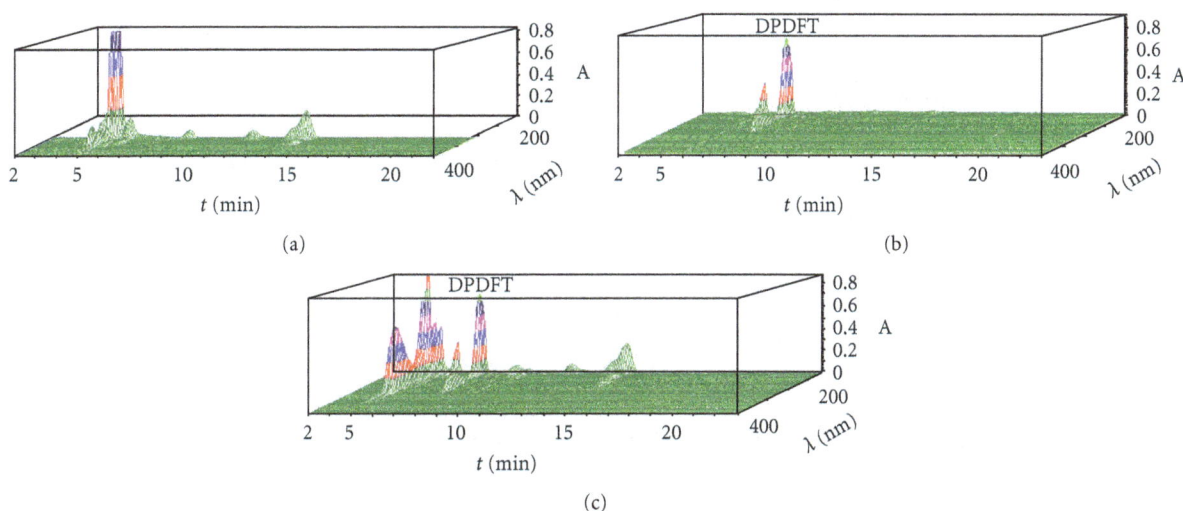

FIGURE 2: HPLC-DAD profiles of DPDFT in plasma sample. (a) Blank plasma; (b) DPDFT ($2\,\mu g \cdot mL^{-1}$) standard; (c) blood sample containing DPDFT ($2.4\,\mu g \cdot mL^{-1}$) collected at 3 min after administration of DPDFT ($15\,mg \cdot kg^{-1}$, i.v.).

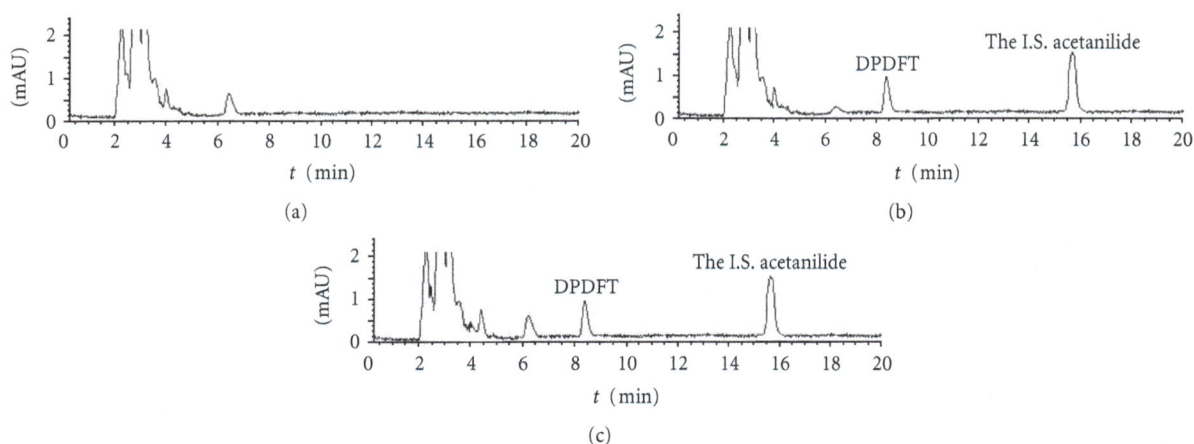

FIGURE 3: Chromatograms of DPDFT in plasma sample. Separation was performed using Waters 2695 HPLC system. The mobile phase consisted of phosphoric acid in water (solvent A)/methanol (solvent B) ($30 : 70$, V/V, pH 3.0) using Diamonsil C_{18} column at 25°C with a flow rate of $0.8\,mL \cdot min^{-1}$. (a) Blank plasma; (b) blank plasma spiked with DPDFT ($3\,\mu g \cdot mL^{-1}$) and the I.S. ($8\,\mu g \cdot mL^{-1}$); (c) blood sample containing DPDFT ($2.4\,\mu g \cdot mL^{-1}$) collected at 3 min after administration of DPDFT ($15\,mg \cdot kg^{-1}$, i.v.).

with UV detecting method was sensitive (the limit of detection was 3.5 ng) enough for pharmacokinetics studies of DPDFT in rats and did not require any forms of analyte derivatization or special columns or instruments. The most important factor for achievement of high sensitivity was the clean baseline owing to appropriate sample preparation on the chromatograms. Thus, a high signal/noise ratio was achieved, which constituted the base for the high sensitivity. The second factor was the high recovery of DPDFT from plasma during the extraction process. Methanol was used not only for deproteination, but also as extractive solvent for assay. So, the high recoveries of DFDPT (90.8, 95.7, and 96.2% for three determinations, resp.) were obtained. During the process of method development, it was discovered that DPDFT spiked with aqueous I.S. showed a symmetric single peak in the chromatograms at a pH around 3.0.

With high sensitivity, small sample requirement, and simple sample treatment procedures, this method was successfully applied to the analysis of rat plasma samples and the pharmacokinetic study of DPDFT in rat.

The mean pharmacokinetic parameters (mean ± SD) are summarized in Table 4. There were no significant differences in all pharmacokinetic parameters between male and female rats at dose of 7.5, 15, and $30\,mg \cdot kg^{-1}$. The results showed that there was significant difference for AUC(0–t), they were calculated to be 7.56, 37.15, 81.25 $mg \cdot kg^{-1} \cdot min^{-1}$ at doses of 7.5, 15, and $30\,mg \cdot kg^{-1}$, respectively, and for the value of V_d after three dosages, 2.13, 1.13, $0.62\,L \cdot kg^{-1}$, respectively. These results suggested that the pharmacokinetics of the complex is a nonlinear process from 7.5 to 30 mg/kg. Otherwise, the distribution half-life $t_{1/2a}$ (1.04, 1.01, 1.12 min, resp.) and elimination half-life $t_{1/2\beta}$ (17.68,

TABLE 2: Intra- and interday precision and accuracy for DPDFT in rat plasma ($n = 6$).

Matrix	Nominal concentration ($\mu g \cdot mL^{-1}$)	Observed concentration ($\mu g \cdot mL^{-1}$) ± SD	Precision (%RSD)	Accuracy (%)
	0.8	0.73 ± 0.05	6.8	−8.75
Intra-day	4	3.73 ± 0.14	3.8	−6.75
	20	19.21 ± 1.02	5.3	−3.95
	0.8	0.69 ± 0.06	8.7	−13.75
Inter-day	4	3.80 ± 0.15	4.0	−5.00
	20	19.02 ± 1.11	5.9	−4.90

Notes: accuracy (%) = [(Cobs − Cnom)/Cnom] × 100, RSD = [standard deviation (SD)/Cobs] × 100.

TABLE 3: Stability results of DPDFT at different conditions in rat plasma ($n = 6$).

Storage period and storage condition	Nominal concentration ($\mu g \cdot mL^{-1}$)	Observed concentration ($\mu g \cdot mL^{-1}$) ± SD	Accuracy (%)	RSD (%)
	0.8	0.85 ± 0.07	6.25	8.2
Concentration of fresh preparation	4	3.91 ± 0.12	−2.25	3.1
	20	21.06 ± 1.33	5.30	6.3
	0.8	0.76 ± 0.05	−5.00	6.6
Three freeze and thaw cycles	4	3.79 ± 0.13	−5.25	3.5
	20	18.97 ± 1.17	−5.15	6.2
	0.8	0.87 ± 0.04	8.75	4.6
Stability for 72 h at 4°C	4	3.70 ± 0.16	−7.50	4.3
	20	21.14 ± 1.27	5.70	6.0
	0.8	0.74 ± 0.05	−7.50	6.8
Stability for 72 h at −20°C	4	4.22 ± 0.12	5.50	2.9
	20	19.13 ± 1.53	−4.35	7.9
	0.8	0.78 ± 0.05	−2.50	6.4
Stability for 72 h at room temperature	4	3.87 ± 0.19	−3.25	4.9
	20	19.46 ± 1.19	−2.75	6.1
	0.8	0.83 ± 0.04	3.75	4.8
Autosampler stability for 24 h at 10°C	4	3.83 ± 0.18	−4.25	4.7
	20	19.19 ± 1.08	−4.05	5.6

Notes: accuracy (%) = [(Cobs − Cnom)/Cnom] × 100, RSD = [standard deviation (SD)/Cobs] × 100.

TABLE 4: Mean pharmacokinetic parameters in rats after intravenous administration of 7.5, 15, and 30 $mg \cdot kg^{-1}$ of DPDFT (mean ± SD, $n = 6$).

Parameter (unit)	Dosage/mg kg^{-1}		
	7.5 (low)	15 (middle)	30 (high)
A/min^{-1}	2.21 ± 0.03	7.94 ± 3.12	22.36 ± 8.12
$B/mg \cdot kg^{-1}$	0.15 ± 0.04	0.92 ± 0.04	1.85 ± 0.44
β/min^{-1}	0.04 ± 0.003	0.04 ± 0.003	0.04 ± 0.003
$V_d/L \cdot kg^{-1}$	2.13 ± 0.07	1.13 ± 0.08	0.62 ± 0.08
$t_{1/2a}/min$	1.04 ± 0.01	1.01 ± 0.01	1.12 ± 0.1
$t_{1/2\beta}/min$	17.68 ± 2.6	19.38 ± 3.6	16.81 ± 3.6
K_{21}/min^{-1}	0.08 ± 0.004	0.11 ± 0.01	0.09 ± 0.01
K_{10}/min^{-1}	0.32 ± 0.05	0.24 ± 0.08	0.30 ± 0.18
K_{12}/min^{-1}	0.26 ± 0.04	0.38 ± 0.01	0.28 ± 0.11
$AUC/mg \cdot kg^{-1} min^{-1}$	7.56 ± 1.02	37.15 ± 3.06	81.25 ± 15.3
$CL(s)/mg \cdot mL^{-1}$, $mg\,kg^{-1}$	0.66 ± 0.05	0.27 ± 0.02	0.018 ± 0.02

Pharmacokinetic Study of Di-Phenyl-Di-(2,4-Difluobenzohydroxamato)Tin(IV): Novel Metal-Based Complex with Promising Antitumor Potential

67

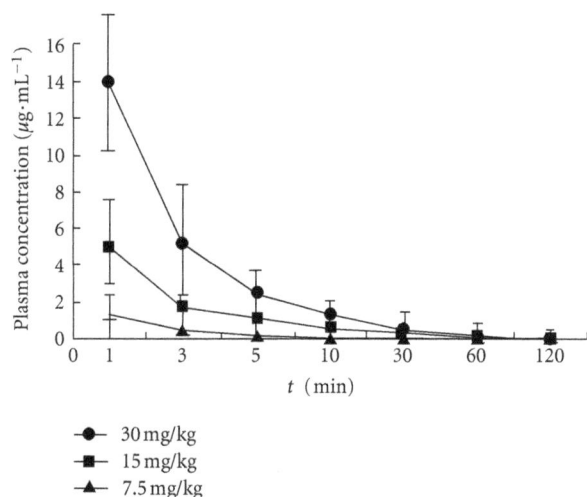

FIGURE 4: Mean plasma concentration-time profiles of DPDFT in rats after intravenous administration of 7.5, 15, and 30 mg·kg^{-1}. Each point represents the mean concentration of six rats.

19.38, 16.81 min, resp.) have no significant difference when the administration dosage of DPDFT was increased from 7.5 to 30 mg/kg, indicating that DPDFT distributed and eliminated very quickly.

4. Conclusions

In this paper, a simple, economical, sensitive, and specific method for the determination of DPDFT, a typical antitumor diorganotin(IV) compound in rat plasma, was first reported. The assay was validated for linearity, specificity, accuracy, precision, recovery, and stability, and good results were obtained. The results of preliminary pharmacokinetic studies indicated that DPDFT showed nonlinear pharmacokinetics in the studied dose ranges in rats and the concentration-time curves of DPDFT in rat plasma could be fitted to two-compartment model. These results hinted that DPDFT might accumulate in certain organs, thus produce the toxicity or could be quickly metabolized in the plasma into active constituent for antitumor. In order to study the precise toxicity mechanism of this metal-based antitumor diorganotin(IV) compounds, we should further elucidate their *in vivo* absorption, distribution, metabolism, and elimination. Meanwhile, based on the structure of DPDFT as a lead compound, structure reconstitution and optimization should be carried out to explore the better antitumor diorganotin(IV) compounds with higher activity, relative lower toxicity, and good pharmacokinetics features in the future.

Abbreviations

DPDFT: Di-phenyl-di-(2,4-difluobenzohydroxamato)tin(IV)
I.S.: Internal standard
HPLC: High performance liquid chromatography
QC: Quality control
S/N: The signal-to-noise ratio
LOQ: The limit of quantitation
LOD: The limits of detection
RE: Relative error
Cobs: Observed concentration
Cnom: Nominal concentration
SD: Standard deviation
RSD: Relative standard deviation
$t_{1/2a}$: Distribution half-life
$t_{1/2\beta}$: Elimination half-life
AUC: Area under the plasma concentration-time curve
K_{21}: First order transfer from compartment 2 to compartment 1
K_{10}: First order elimination from compartment 1
K_{12}: First order transfer from compartment 1 to compartment 2
Cl: Clearance.

Acknowledgments

Financial supports from the support of the State "Innovative Drug Development" Key Science and Technology Projects of China (no. 2009ZX09103-104), the National Natural Science Foundation of China (nos. 30973603 and 30772682), the Shanxi Foundation for overseas returned (no. 2010-54), the Program for the Top Young and Middle-aged Innovative Talents of Higher Learning Institutions of Shanxi Province and by Shanxi Province Foundation Science for Youths, the Foundation for the Younth Doctor from Shanxi Medical University in 2008, and Program for the Top Science and Technology Innovation Teams of Higher Learning Institutions of Shanxi province (2011) are gratefully acknowledged.

References

[1] T. W. Hambley, "Chemistry: metal-based therapeutics," *Science*, vol. 318, no. 5855, pp. 1392–1393, 2007.

[2] T. Storr, K. H. Thompson, and C. Orvig, "Design of targeting ligands in medicinal inorganic chemistry," *Chemical Society Reviews*, vol. 35, no. 6, pp. 534–544, 2006.

[3] L. Yunlan, L. Jinjie, and L. Qingshan, "Mechanisms by which the antitumor compound di-n-butyl-di-(4- chlorobenzohydroxamato)tin(IV) induces apoptosis and the mitochondrial-mediated signaling pathway in human cancer SGC-7901 cells," *Molecular Carcinogenesis*, vol. 49, no. 6, pp. 566–581, 2010.

[4] Q. Li, M. F. C. Guedes Da Silva, and A. J. L. Pombeiro, "Diorganotin(IV) derivatives of substituted benzohydroxamic acids with high antitumor activity," *Chemistry—A European Journal*, vol. 10, no. 6, pp. 1456–1462, 2004.

[5] L. Yunlan, L. Yang, N. Xiaoqiang et al., "Synthesis and antitumor activity of a new mixed-ligand complex di-n-butyl-(4-chlorobenzohydroxamato)tin(IV) chloride," *Journal of Inorganic Biochemistry*, vol. 102, no. 9, pp. 1731–1735, 2008.

[6] L. Pellerito, C. Prinzivalli, G. Casella et al., "Diorganotin(IV) N-acetyl-l-cysteinate complexes: synthesis, solid state, solution phase, DFT and biological investigations," *Journal of Inorganic Biochemistry*, vol. 104, no. 7, pp. 750–758, 2010.

[7] T. S. Basu Baul, A. Paul, L. Pellerito et al., "Dibutyltin(IV) complexes containing arylazobenzoate ligands: chemistry, in vitro cytotoxic effects on human tumor cell lines and mode of interaction with some enzymes," *Investigational New Drugs*, vol. 29, no. 2, pp. 285–289, 2011.

[8] W. Li, Z. W. Zhang, S. M. Ren, Y. Sibiril, D. Parent Massin, and T. Jiang, "Synthesis and antitumor activity of novel dibutyltin carboxylates of aminoglucosyl derivatives," *Chemical Biology and Drug Design*, vol. 73, no. 6, pp. 682–686, 2009.

[9] Y. Suzuki, Y. Endo, M. Ogawa, Y. Kim, N. Onda, and K. Yamanaka, "Development of an analytical method to confirm toxic trimethylated tin in human urine," *Journal of Chromatography B*, vol. 868, no. 1-2, pp. 116–119, 2008.

[10] Z. H. Yu, M. Jing, X. R. Wang, D. Y. Chen, and Y. L. Huang, "Simultaneous determination of multi-organotin compounds in seawater by liquid-liquid extraction-high performance liquid chromatography-inductively coupled plasma mass spectrometry," *Guang Pu Xue Yu Guang Pu Fen Xi*, vol. 29, no. 10, pp. 2855–2859, 2009.

[11] G. Zhai, J. Liu, L. Li et al., "Rapid and direct speciation of methyltins in seawater by an on-line coupled high performance liquid chromatography-hydride generation-ICP/MS system," *Talanta*, vol. 77, no. 4, pp. 1273–1278, 2009.

[12] G. A. Zachariadis and E. Rosenberg, "Speciation of organotin compounds in urine by GC-MIP-AED and GC-MS after ethylation and liquid-liquid extraction," *Journal of Chromatography B*, vol. 877, no. 11-12, pp. 1140–1144, 2009.

[13] D. Sakati and T. Teddy, "Seperation of organotin and organolead conpounds in drinking water by GC-MIP ADE," *Chemosphere*, vol. 32, no. 10, pp. 1983–1992, 1996.

[14] J. Carpinteiro, I. Rodríguez, and R. Cela, "Applicability of solid-phase microextraction combined with gas chromatography atomic emission detection (GC-MIP AED) for the determination of butyltin compounds in sediment samples," *Analytical and Bioanalytical Chemistry*, vol. 380, no. 5-6, pp. 853–857, 2004.

[15] Y. Li, Y. Hu, J. Liu, Y. Guo, and G. Wang, "Determination of organotin compounds in textile auxiliaries by gas chromatography-mass spectrometry," *Se Pu*, vol. 29, no. 4, pp. 353–357, 2011.

[16] K. Inagaki, A. Takatsu, T. Watanabe et al., "Certification of butyltins and phenyltins in marine sediment certified reference material by species-specific isotope-dilution mass spectrometric analysis using synthesized 118Sn-enriched organotin compounds," *Analytical and Bioanalytical Chemistry*, vol. 387, no. 7, pp. 2325–2334, 2007.

[17] R. B. Rajendran, H. Tao, T. Nakazato, and A. Miyazaki, "A quantitative extraction method for the determination of trace amounts of both butyl- and phenyltin compounds in sediments by gas chromatography-inductively coupled plasma mass spectrometry," *The Analyst*, vol. 125, no. 10, pp. 1757–1763, 2000.

[18] H. Nsengimana, E. M. Cukrowska, A. Dinsmore, E. Tessier, and D. Amouroux, "*In situ* ethylation of organolead, organotin and organomercury species by bromomagnesium tetraethylborate prior to GC-ICP-MS analysis," *Journal of Separation Science*, vol. 32, no. 14, pp. 2426–2433, 2009.

[19] M. Vahčič, R. Milačič, and J. Ščančar, "Development of analytical procedure for the determination of methyltin, butyltin, phenyltin and octyltin compounds in landfill leachates by gas chromatography-inductively coupled plasma mass spectrometry," *Analytica Chimica Acta*, vol. 694, no. 1-2, pp. 21–30, 2011.

[20] S. Xianmen, C. Jingrong, and L. Qingshan, "The preliminary structure-activity relationship of aromatic hydroxamic acid organotin compounds," *Science China Chemistry Series B*, vol. 38, no. 5, pp. 429–440, 2008.

9

Fine Tuning of Redox Networks on Multiheme Cytochromes from *Geobacter sulfurreducens* Drives Physiological Electron/Proton Energy Transduction

Leonor Morgado,[1] Joana M. Dantas,[1] Marta Bruix,[2]
Yuri Y. Londer,[3,4] and Carlos A. Salgueiro[1]

[1] Requimte-CQFB, Departamento de Química, Faculdade de Ciências e Tecnologia, Universidade Nova de Lisboa, Campus Caparica, 2829-516 Caparica, Portugal
[2] Departamento de Espectroscopía y Estructura Molecular, Instituto de Química-Física "Rocasolano", CSIC, Serrano 119, 28006 Madrid, Spain
[3] Biosciences Division, Argonne National Laboratory, Argonne, IL 60439, USA
[4] New England Biolabs, 240 County Road, Ipswich, MA 01938, USA

Correspondence should be addressed to Carlos A. Salgueiro, csalgueiro@fct.unl.pt

Academic Editor: Takao Yagi

The bacterium *Geobacter sulfurreducens (Gs)* can grow in the presence of extracellular terminal acceptors, a property that is currently explored to harvest electricity from aquatic sediments and waste organic matter into microbial fuel cells. A family composed of five triheme cytochromes (PpcA-E) was identified in *Gs*. These cytochromes play a crucial role by bridging the electron transfer from oxidation of cytoplasmic donors to the cell exterior and assisting the reduction of extracellular terminal acceptors. The detailed thermodynamic characterization of such proteins showed that PpcA and PpcD have an important redox-Bohr effect that might implicate these proteins in the e^-/H^+ coupling mechanisms to sustain cellular growth. The physiological relevance of the redox-Bohr effect in these proteins was studied by determining the fractional contribution of each individual redox-microstate at different pH values. For both proteins, oxidation progresses from a particular protonated microstate to a particular deprotonated one, over specific pH ranges. The preferred e^-/H^+ transfer pathway established by the selected microstates indicates that both proteins are functionally designed to couple e^-/H^+ transfer at the physiological pH range for cellular growth.

1. Introduction

The ability to use extracellular terminal electron acceptors (e.g., Fe(III), U(VI) oxides, or electrode surfaces) in addition to the more common cytoplasmic acceptors, such as fumarate, spreads the bacterium *Geobacter sulfurreducens* (*Gs*) environmental versatility [1–3]. However, the use of extracellular acceptors sets to the microorganism additional challenges. The first of such is the efficient electron delivery to cell exterior, and the second one is the net production of metabolic energy to support cellular growth. To address them, *Gs* respiratory chains are designed to permit an effective flow of electrons from the oxidation of cytoplasmic

organic compounds to the outer membrane. In fact, the topology of the electron transfer proteins involved in such electron transfer is quite unusual in comparison with other Gram-negative bacteria. In addition to the localization at the cytoplasmic membrane, several electron transfer proteins have also been identified at the outer membrane of *Gs* cells, which constitutes an efficient interface between the cell surface and the extracellular acceptors [4–8]. Albeit much remains to be known on the *Geobacter*'s electron transfer chains, it is consensual nowadays that electrons are transferred at the cytoplasmic membrane from a NAD(P)H dehydrogenase to a quinone pool, then to periplasmic *c*-type cytochromes, and, finally, to at least a metal reductase at the outer membrane [9]. The use of extracellular electron acceptors by *Gs* leads to a decrease in the biomass production in comparison with the soluble acceptor fumarate [10]. The discrepancy observed in the cellular growth yield was attributed to the cytoplasmic acidification, which is circumvented when the acceptor is fumarate. In this case, the cytoplasmic protons produced from acetate oxidation are consumed favouring the H^+ electrochemical potential gradient across the periplasmic membrane that drives ATP synthesis [11]. In comparison to the studies on fumarate respiration, metabolic modelling studies can only simulate the experimental results obtained in presence of extracellular electron acceptors if additional e^-/H^+ coupling mechanisms are considered [10]. To date, these mechanisms remain to be identified, and the present work aims to contribute to their clarification.

Evidence for functional energy transduction in the absence of a membrane confinement was reported for the hydrogenase/tetraheme cytochrome c_3 system from *Desulfovibrio vulgaris* [12, 13]. In this case, the electrons and protons from H_2 are provided by the periplasmic hydrogenase to the periplasmic cytochrome c_3, which couples the deenergization of electrons to the lowering of proton pK_a values to favour their release in the periplasm. This hydrogenase/cytochrome c_3 system works as a proton thruster device, which provides the required diffusion control and thermodynamic drive in the water-protein interface (a two-phase system) in agreement with the Williams localized theory [14].

A family composed of five triheme cytochromes (PpcA-E) was identified in *Gs* [15]. These five cytochromes are small soluble proteins, each with approximately 10 kDa. Cytochromes PpcB, PpcC, PpcD, and PpcE share 77%, 62%, 57%, and 65% amino acid sequence identity with PpcA, respectively [15]. The three hemes form the protein heme core and are covalently linked to cysteine residues in the CXXCH binding motifs, where X designates any amino acid. All the hemes are axially coordinated by two histidine residues and are low spin, both in the reduced (Fe^{2+}, S = 0) and in the oxidized (Fe^{3+}, S = 1/2) forms. The structures of the five cytochromes have been determined, showing that they have a high level of structural homology [16, 17]. Knockout studies on *Gs* cells with the genes encoding PpcA and PpcD deleted showed that cellular Fe^{3+} reduction was impaired [18]. In addition, proteomics studies by Ding and coworkers [19] showed that PpcD is more abundant when *Gs* is grown on insoluble iron oxides compared to ferric citrate. Like cytochromes c_3, PpcA and PpcD can be reduced by hydrogenase and their thermodynamic characterization showed that they have the adequate properties to couple e^-/H^+ transfer, which might contribute for the H^+ electrochemical potential gradient necessary to support bacterial growth in presence of extracellular acceptors [20].

In this work, we show that the e^-/H^+ coupling in PpcA and PpcD driven by the protonation/deprotonation of the redox-Bohr center is observed only within the cellular optimal pH range for growth, reinforcing the physiologically significance of the redox-Bohr effect observed for these bacteria.

2. Methodology

The coexistence of several microstates in solution, connecting the fully reduced and oxidized states, makes the study of the properties of the redox centers in multiheme proteins particularly complex. In the case of *Gs* triheme cytochromes, a single pK_a dominates the pH dependence of the reduction potentials of the heme groups [20], and thus 16 possible microstates can coexist in solution (Figure 1). The several microstates can be grouped in four macroscopic oxidation stages (S_{0-3}) connected by three one-electron redox steps and containing the microstates with the same number of oxidized hemes. Each pair of microstates (protonated or deprotonated) is connected by single electron redox steps and can be described by a total of 24 Nernst equations. In each case, three Nernst equations relate the microstates in oxidation stage 1 to that of stage 0; six equations relate microstates in stage 2 and 1; finally, three equations describe the redox equilibria between the fully oxidized microstate and those in stage 2. Taking as reference the fully reduced and protonated microstate and considering a sequential oxidation of hemes 1, 3, and 4 along the four oxidation stages as an example, the relevant Nernst equations are:

$$E = e_1 + \frac{RT}{F} \ln \frac{P_{1H}}{P_{0H}},$$

$$E = e_3 + I_{13} + \frac{RT}{F} \ln \frac{P_{13H}}{P_{1H}}, \tag{1}$$

$$E = e_4 + I_{13} + I_{14} + I_{34} + \frac{RT}{F} \ln \frac{P_{134H}}{P_{13H}},$$

where the term e_i represents the heme reduction potentials and I_{ij} the heme redox interactions, which account for the effect of the oxidation state of one heme in the reduction potential of its neighbors.

These equations (1) can be rewritten and each microstate can be expressed as a function of P_{0H} (2):

$$P_{1H} = P_{0H} \exp^{(E-e_1)F/RT},$$

$$P_{13H} = P_{0H} \exp^{(2E-e_1-e_3-I_{13})F/RT}, \tag{2}$$

$$P_{134H} = P_{0H} \exp^{(3E-e_1-e_3-e_4-I_{13}-I_{14}-I_{34})F/RT}.$$

The fully reduced and protonated microstate P_{0H} and the correspondent deprotonated microstate P_0 are related by the Handerson-Hasselbach equation:

$$P_{0H} \rightleftharpoons P_0 + H^+,$$

$$P_0 = P_{0H} \exp^{[\ln 10(pH - pK_{red})]}. \tag{3}$$

Fine Tuning of Redox Networks on Multiheme Cytochromes from Geobacter sulfurreducens Drives Physiological Electron/Proton Energy Transduction

71

TABLE 1: Thermodynamic parameters of the fully reduced and protonated forms of PpcA and PpcD obtained at 288 K and 250 mM ionic strength [20]. Redox potentials are relative to standard hydrogen electrode (SHE). Standard errors are given in parenthesis. The pK_1, pK_2, and pK_{ox} values were determined from the values obtained for pK_{red} and redox-Bohr interactions.

		PpcA	PpcD
Heme redox potentials (mV)	e_1	−154(5)	−156(6)
	e_3	−138(5)	−139(6)
	e_4	−125(5)	−149(6)
Heme-heme redox interactions (mV)	I_{13}	27(2)	46(3)
	I_{14}	16(3)	3(4)
	I_{34}	41(3)	14(4)
Redox-Bohr interactions (mV)	I_{1H}	−32(4)	−28(6)
	I_{3H}	−31(4)	−23(6)
	I_{4H}	−58(4)	−53(6)
pK_a	Stage 0 (pK_{red})	8.6 (0.1)	8.7 (0.1)
	Stage 1 (pK_1)	8.0 (0.1)	8.1 (0.1)
	Stage 2 (pK_2)	7.2 (0.1)	7.4 (0.1)
	Stage 3 (pK_{ox})	6.5 (0.1)	6.9 (0.1)

Taking as an example the sequential oxidation of hemes 1, 3, and 4 along the four oxidation stages but for the deprotonated microstates, the Nernst equations are:

$$E = e_1 + I_{1H} + \frac{RT}{F} \ln \frac{P_1}{P_0}$$

$$E = e_3 + I_{13} + I_{1H} + I_{3H} + \frac{RT}{F} \ln \frac{P_{13}}{P_1} \quad (4)$$

$$E = e_4 + I_{13} + I_{14} + I_{34} + I_{1H} + I_{3H} + I_{4H} + \frac{RT}{F} \ln \frac{P_{134}}{P_{13}},$$

where the terms I_{iH} represent the redox-Bohr interactions and account for the effect of the pH on the heme reduction potentials. Equation (4) can also be rewritten by expressing each deprotonated microstate as a function of P_{0H} (5):

$$P_1 = P_{0H} \exp^{[(\ln 10(pH - pK_{red})) + ((E - e_1 - I_{1H})(F/RT))]}$$

$$P_{13} = P_{0H} \exp^{[(\ln 10(pH - pK_{red})) + ((2E - e_1 - e_3 - I_{13} - I_{1H} - I_{3H})(F/RT))]} \quad (5)$$

$$P_{134} = P_{0H} \exp^{[(\ln 10(pH - pK_{red})) + ((3E - e_1 - e_3 - e_4 - I_{13} - I_{14} - I_{34} - I_{1H} - I_{3H} - I_{4H})(F/RT))]}.$$

Thus, in the particular case of a triheme cytochrome, three reduction potentials, one pK_{red} plus six two-center interactions (three heme and three redox-Bohr interactions), are sufficient to determine the fractional contribution of each microstate, across the full range of pH and solution potential.

These parameters can be obtained combining data from NMR and visible spectroscopy and were determined for PpcA and PpcD in a previous work (Table 1) [20].

3. Results and Discussion

In this work, the effect of pH on the functional mechanism of the triheme cytochromes PpcA and PpcD from *Gs* was evaluated. The thermodynamic parameters listed on Table 1 for the fully reduced and protonated proteins, which include the heme reduction potentials, the pK_a values of the redox-Bohr center, heme-heme redox interactions, and redox-Bohr interactions, were used to determine the heme oxidation profiles and the fractional contribution of the 16 microstates at different pH values. This allows to compare the individual heme oxidation profiles inside and outside the physiological pH range and to demonstrate a correlation between the functional properties of these cytochromes and the physiological pH range for cellular growth.

From the heme reduction potential values listed in Table 1, the order of oxidation of the heme groups can be established for the fully reduced and protonated protein,

which is 1-3-4 for PpcA and 1-4-3 for PpcD. However, both proteins display important redox-Bohr interactions, being the largest one observed for heme 4. All the redox-Bohr interactions are negative, which is expected in electrostatic terms, since the removal of proton(s), upon deprotonation of the redox-Bohr center, lowers the affinity for electrons (lower reduction potential) by the heme groups. This is reflected in the heme reduction potentials of the fully reduced deprotonated proteins, which can be obtained by the simple sum of the heme reduction potentials of the fully reduced and protonated proteins with their respective redox-Bohr interactions. In the case of PpcA, the heme reduction potentials for the fully reduced and deprotonated protein are −186, −169, and −183 mV for hemes 1, 3, and 4, respectively. Comparison of these values with those obtained for the correspondent protonated form (Table 1) clearly shows that the order of oxidation is different in both situations: 1-4-3 (deprotonated protein) and 1-3-4 (protonated protein). For PpcD, the same scenario is observed: the heme reduction potentials of the reduced and deprotonated protein are more negative (−184, −162, and −202 mV for hemes 1, 3, and 4, resp.), and the order of oxidation is affected by the deprotonation of the redox-Bohr center (4-1-3 *versus* 1-4-3). This analysis shows that the redox properties of PpcA and PpcD are modulated by the solution pH (redox-Bohr effect). However, the redox-Bohr effect is functionally relevant only if observed at physiological pH range for cellular growth.

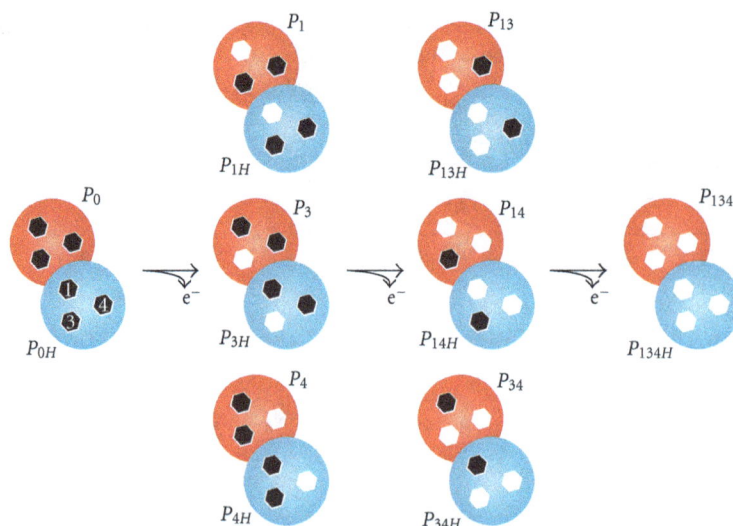

FIGURE 1: Electronic distribution scheme for a triheme cytochrome with a proton-linked equilibrium showing the 16 possible microstates. PpcA and PpcD are structurally similar to tetraheme cytochromes c_3, with the exception that heme 2 and the corresponding fragment of the polypeptide chain are absent. Thus, to be consistent with the literature, the heme groups are numbered 1, 3, and 4 according to the order of attachment to the CXXCH motif in the polypeptide chain. The blue and red circles correspond to the protonated and deprotonated microstates, respectively. Hexagons represent heme groups, which can be either reduced (black) or oxidized (white). The microstates are grouped, according to the number of oxidized hemes, in four oxidation stages connected by three one-electron redox steps. P_{0H} and P_0 represent the reduced protonated and deprotonated microstates, respectively. P_{ijkH} and P_{ijk} indicate, respectively, the protonated and deprotonated microstates, where i, j, and k represent the heme(s) that are oxidized in that particular microstate.

To best of our knowledge, the optimal pH for *Gs* growth has not yet been determined. Kim and Lee [21] studied the effect of the initial pH on the growth rates of *Gs* cultures utilizing fumarate as electron acceptor in the pH range of 5.5–6.8. Within this range, higher rates were observed at pH 6.8, which were reduced at pH 6.4 and completely inhibited at pH 5.5 [21]. In addition to this non-electrode-respiring conditions, several electrode-respiring studies have shown a pH drop inside *Gs* biofilms with a concomitant decrease in the measured current [22–25]. Experiments carried out with anode-respiring bacteria with high presence of *Gs* cells have shown that in the pH range 6–8, maximum current density was achieved at pH 8 and dropped continually to pH 6 [22–24]. These studies also suggested that at pH < 6 *Gs* metabolism is inhibited.

In a previous work, we have used pH 7.5 as representative value for *Gs* physiological pH [20]. In the present work, we aimed to evaluate the significance of the redox-Bohr effect on the functional mechanism of PpcA and PpcD by studying the individual heme oxidation profiles and the fractional contribution of the microstates at a broader pH range. Thus, in order not to confuse literature, in the present work, the individual heme oxidation profiles and fractional contribution of microstates of both proteins were determined at pH 5.5 and 9.5 and compared with the previous analysis carried out at pH 7.5 (Figures 2 and 3). Then, a detailed analysis of the dominant PpcA and PpcD microstates was carried out in the pH range 5.5–9.5 (Figure 4).

The heme oxidation profiles described in Figure 2 show that, at each pH value, the shape of the redox curves is substantially different from a pure Nernst curve and the several crossovers clearly indicate that the electron affinity of each heme is modulated by the heme-heme redox interactions, as protein oxidation progresses. The redox interactions are all positive, which is expected in electrostatic terms, and reflect the stabilization of the reduced state of one heme by the removal of one electron from a neighboring one. From the comparison of the individual heme oxidation profiles at different pH values, it is clear that the heme apparent midpoint reduction potentials e_{app} (i.e., the point at which the oxidized and reduced fractions of each heme group are equally populated) are different due to redox-Bohr interactions (Figure 2). The shape of each heme oxidation curve is therefore a result of the interplay of both heme-heme and redox-Bohr interactions.

In the case of PpcA, at pH 5.5, the e_{app} values of hemes 3 and 4 are similar. However, at high pH, the deprotonation of the redox-Bohr center lowers considerable the e_{app} value of heme 4 (largest redox-Bohr interaction) bringing it closer to that of heme 1. The modulation of the individual heme oxidation profiles is also observed for PpcD, though yielding a distinct result. In fact, due to the similarity of the e_{app} values of hemes 1 and 4 at low pH, the deprotonation of the redox-Bohr center yields a more notorious separation of the three curves being heme 4 the one with smaller e_{app} value.

The effect of the protonation/deprotonation of the redox-Bohr center can be further rationalised from the fractional contribution of each microstate (Figure 1). The fractional contribution of these microstates for PpcA and PpcD at pH 5.5, 7.5, and 9.5 provides functional mechanistic insights on the electron transfer pathways of the proteins (Figure 3). The analysis of this figure shows that the relevant

Fine Tuning of Redox Networks on Multiheme Cytochromes from Geobacter sulfurreducens Drives Physiological
Electron/Proton Energy Transduction

73

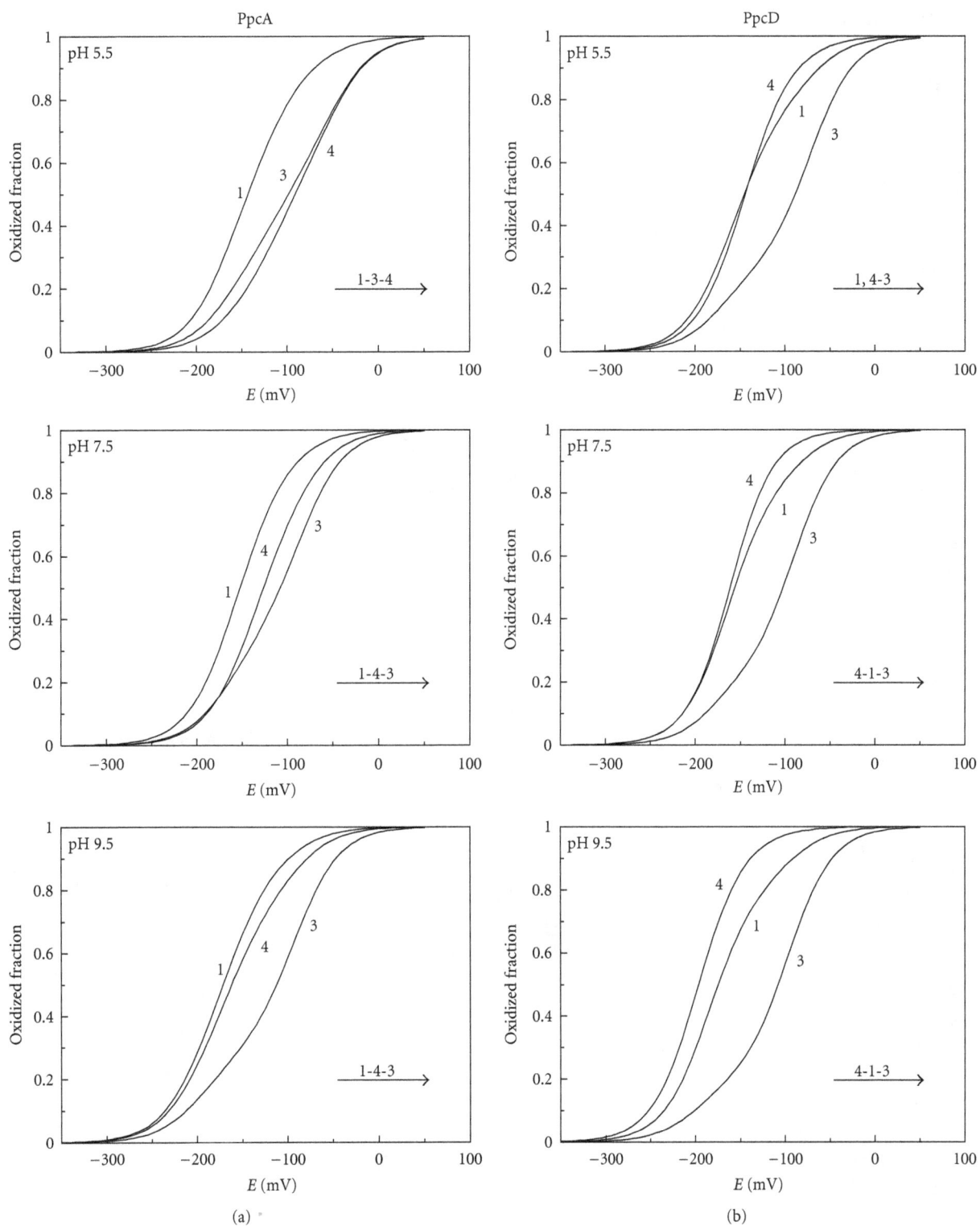

FIGURE 2: Heme oxidation fractions for PpcA (a) and PpcD (b) at different pH values. The curves were calculated as a function of the solution
reduction potential (*versus* SHE) using the parameters listed in Table 1. The order of oxidation of the hemes is indicated by the arrow.

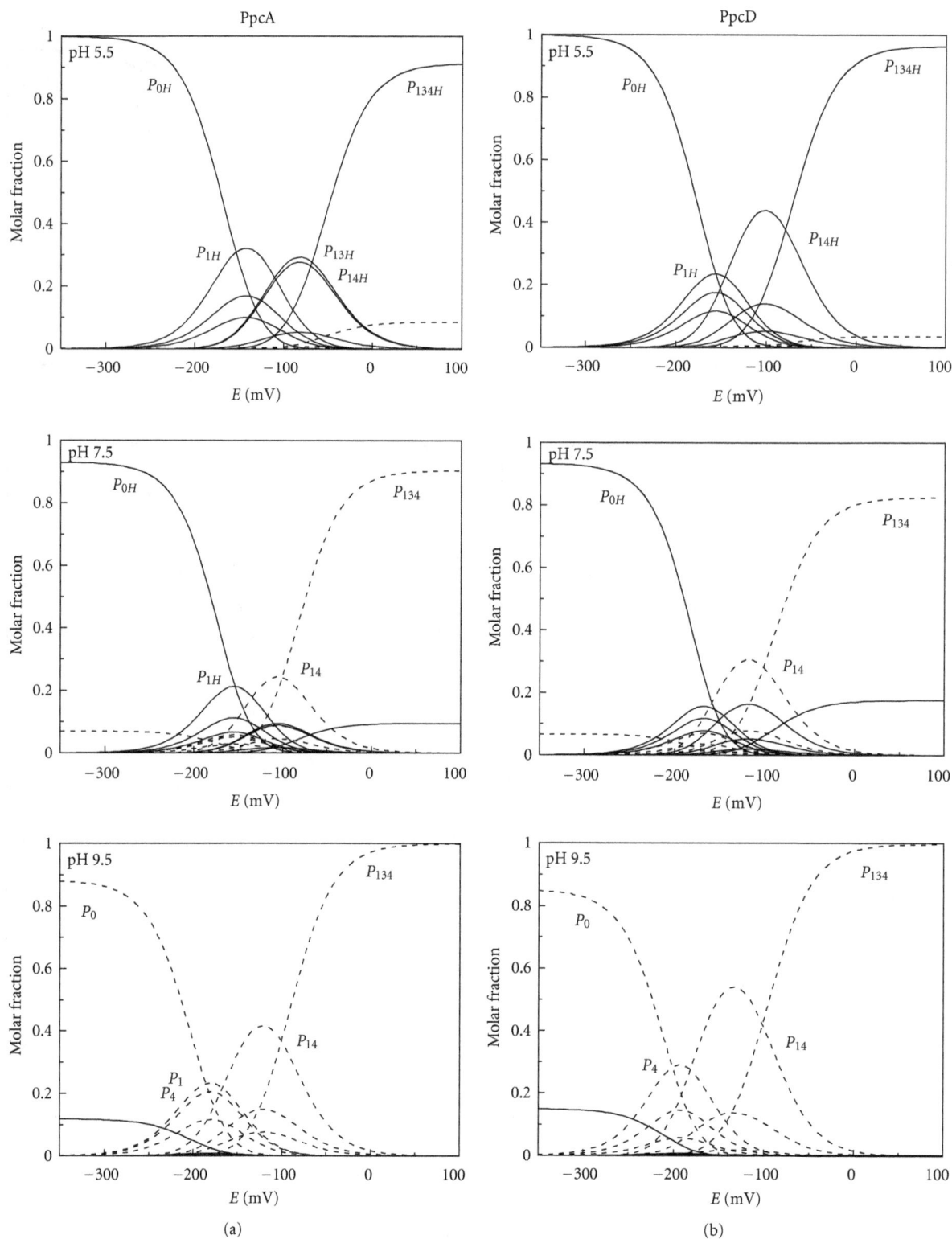

FIGURE 3: Molar fraction of the 16 individual microstates (see Figure 1) of PpcA (a) and PpcD (b) at different pH values. The curves were calculated as a function of the solution reduction potential (*versus* SHE) using the parameters listed in Table 1. Solid and dashed lines indicate the protonated and deprotonated microstates, respectively. For clarity, only the relevant microstates are labeled.

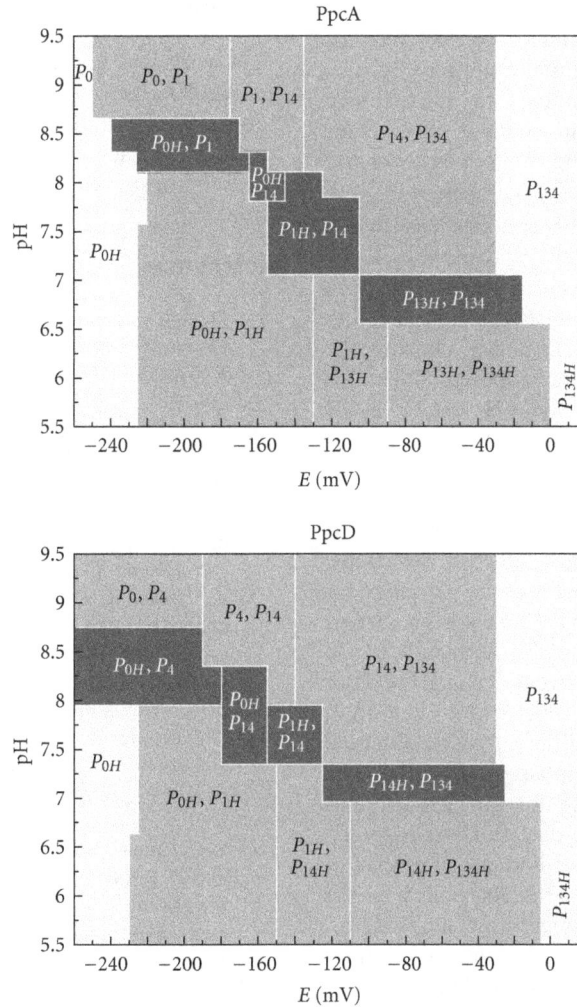

FIGURE 4: Dependence of the molar fractions of PpcA and PpcD microstates with pH and solution potential (*versus* SHE) at 288 K and 250 mM ionic strength. The molar fractions of the individual microstates were determined using the parameters listed in Table 1, and those showing the largest molar fraction are represented. The regions where proteins can perform e^-/H^+ transfer and e^- transfer are highlighted in dark gray and gray, respectively. Regions where the protein is fully reduced or fully oxidized are indicated in light gray.

microstates are quite distinct at different pH values. In the case of PpcA, several microstates dominate the intermediate stages of oxidation either at pH 5.5 or pH 9.5. At pH 7.5, stage 0 is dominated by the protonated form P_{0H} and stage 1 is dominated by the oxidation of heme 1 (P_{1H}) while keeping the acid-base center protonated. Stage 2 is dominated by the oxidation of heme 4 and deprotonation of the acid-base center (P_{14}), which remains deprotonated in stage 3 (P_{134}). Therefore, at pH 7.5, a route is defined for the electrons within PpcA: $P_{0H} \rightarrow P_{1H} \rightarrow P_{14} \rightarrow P_{134}$, whereas at the other pH values there is no coherent path. Moreover, it is also clear that, associated with the favoured electron transfer pathway at physiologic pH, a deprotonation occurs as one electron is transferred between oxidation stages 1 and 2 (Figure 3) suggesting that microstates P_{14} and P_{1H} are the physiological forms of PpcA. This mechanistic information, which can only be obtained from a detailed microscopic analysis shows how selected microstates could confer directionality of events: PpcA microstate P_{14} can

uptake electrons and a weakly acidic proton from the donor associated with the cytoplasmic membrane, originating the microstate P_{1H}. The pK_a value of this proton (pK_a 8) corresponds to the pK_a value of oxidation stage 1 (Table 1). As discussed above, upon protonation of the redox-Bohr center, the reduction potential of the hemes becomes less negative. The loss of electron reducing power is used to lower the pK_a of the proton so that when meeting the physiological downstream redox partner, microstate P_{1H} donates less reducing electrons and a more acidic proton (pK_a 7.2, which corresponds to the pK_a value of oxidation stage 2—see Table 1). This is now sufficiently acidic to be released in the periplasm, and microstate P_{14} is now ready to initiate a new energy transduction cycle.

PpcD can also couple the transfer of electrons and protons at physiologic pH, though by a different pathway (Figure 3). The oxidation stage 0 is dominated by the protonated form P_{0H}. However, the microstates of oxidation stage 1 are overcomed by the P_{0H} curve, which intercepts first the P_{14} curve. This microstate (P_{14}) dominates the oxidation stage 2, whereas P_{134} dominates stage 3. Thus, for this cytochrome, a different preferential route for electrons is established, favouring a proton-coupled $2\,e^-$ transfer step between oxidation stages 0 and 2: $P_{0H} \rightarrow P_{14} \rightarrow P_{134}$. As for PpcA, at pH 5.5 and 9.5, no coupling between electron and proton transfer is observable for PpcD (Figure 3).

Several studies using biofilms on electrodes have shown that electron transfer out of Gs cells is detectable above potentials of about $-200\,$mV *versus* standard hydrogen electrode [22, 26, 27]. Thus, in the present work, we further determined the microstates of PpcA and PpcD showing the largest molar fraction values in the redox potential window $-200\,$mV to $-30\,$mV and also in the pH range 5.5–9.5 (Figure 4). This analysis shows that PpcA and PpcD are not fully reduced nor fully oxidized and thus are functionally active (i.e., capable of receiving and donating electrons) in that redox potential and pH range. Additionally, it is interesting to note that the microstates pairs that can couple e^-/H^+ transfer are essentially detectable at the pH range for which most studies with *Geobacter* cells have been carried out. In particular, PpcA, which is highly abundant in Gs, being most likely the reservoir of electrons destined for outer surface [28, 29], can perform e^-/H^+ energy transduction in the pH range 6.5 to 8.5 albeit at different redox potential values (Figure 4). It should be emphasized that the data presented in Figure 4 was obtained for purified proteins and that the transposition of these data to Gs cells grown in cultures or biofilms should take into account that in these more complex systems, other variables such as the presence of the other cellular components, metabolic status, and environmental conditions (e.g., ionic strength) can alter slightly the working range of these proteins. Studies on Gs biofilms on electrodes showed that cells have a Nernstian response around $-150\,$mV [26, 30, 31]. Since PpcA is functionally active at this redox potential (Figure 4), it can be hypothesized that this response is mainly PpcA driven. Under this hypothesis and since PpcA is also able to perform e^-/H^+ energy transduction at pH values around 7, a possible functional mechanism would involve the transfer of one electron and one proton from the quinone oxidoreductase being the second electron transfer back to the quinone pool.

Overall, the work presented here shows that PpcA and PpcD display the adequate functional properties to perform e^-/H^+ energy transduction but only within the pH range 6.5–8.5. This might represent additional mechanisms contributing to the H^+ electrochemical potential gradient across the periplasmic membrane that drives ATP synthesis and might also explain why Gs cells become metabolically inactive at pH 6 or lower [22–25].

Abbreviations

Gs: *Geobacter sulfurreducens* bacterium
Ppc: Periplasmic *c*-type cytochrome
SHE: Standard hydrogen electrode.

Acknowledgments

This work was supported by Research Grant PTDC/QUI/70182/2006 from Fundação para a Ciência e a Tecnologia (Portugal). The authors thank the anonymous referee for the valuable comments and constructive suggestions that helped them to improve the paper.

References

[1] D. R. Bond and D. R. Lovley, "Electricity production by *Geobacter sulfurreducens* attached to electrodes," *Applied and Environmental Microbiology*, vol. 69, no. 3, pp. 1548–1555, 2003.

[2] D. R. Lovley, "Fe(III) and Mn(IV) reduction," in *Environmental Microbe-Metal Interactions*, A. Press, Ed., ASM Press, Washington, DC, USA, 2000.

[3] D. R. Lovley, T. Ueki, T. Zhang et al., "*Geobacter*. The microbe electric's physiology, ecology, and practical applications," *Advances in Microbial Physiology*, vol. 59, pp. 1–100, 2011.

[4] K. Inoue, X. Qian, L. Morgado et al., "Purification and characterization of OmcZ, an outer-surface, octaheme *c*-type cytochrome essential for optimal current production by *Geobacter sulfurreducens*," *Applied and Environmental Microbiology*, vol. 76, no. 12, pp. 3999–4007, 2010.

[5] J. E. Butler, N. D. Young, and D. R. Lovley, "Evolution of electron transfer out of the cell: comparative genomics of six *Geobacter* genomes," *BMC Genomics*, vol. 11, no. 1, article 40, 2010.

[6] B. C. Kim, C. Leang, Y. H. R. Ding, R. H. Glaven, M. V. Coppi, and D. R. Lovley, "OmcF, a putative *c*-type monoheme outer membrane cytochrome required for the expression of other outer membrane cytochromes in *Geobacter sulfurreducens*," *Journal of Bacteriology*, vol. 187, no. 13, pp. 4505–4513, 2005.

[7] X. Qian, T. Mester, L. Morgado et al., "Biochemical characterization of purified OmcS, a *c*-type cytochrome required for insoluble Fe(III) reduction in *Geobacter sulfurreducens*," *Biochimica et Biophysica Acta*, vol. 1807, no. 4, pp. 404–412, 2011.

[8] X. Qian, G. Reguera, T. Mester, and D. R. Lovley, "Evidence that OmcB and OmpB of *Geobacter sulfurreducens* are outer membrane surface proteins," *FEMS Microbiology Letters*, vol. 277, no. 1, pp. 21–27, 2007.

[9] D. R. Lovley, "Bug juice: harvesting electricity with microorganisms," *Nature Reviews Microbiology*, vol. 4, no. 7, pp. 497–508, 2006.

[10] R. Mahadevan, D. R. Bond, J. E. Butler et al., "Characterization of metabolism in the Fe(III)-reducing organism *Geobacter sulfurreducens* by constraint-based modeling," *Applied and Environmental Microbiology*, vol. 72, no. 2, pp. 1558–1568, 2006.

[11] K. Srinivasan and R. Mahadevan, "Characterization of proton production and consumption associated with microbial metabolism," *BMC Biotechnology*, vol. 10, article 2, 2010.

[12] R. O. Louro, "Proton thrusters: overview of the structural and functional features of soluble tetrahaem cytochromes c_3," *Journal of Biological Inorganic Chemistry*, vol. 12, no. 1, pp. 1–10, 2007.

[13] R. O. Louro, T. Catarino, J. LeGall, and A. V. Xavier, "Redox-Bohr effect in electron/proton energy transduction: cytochrome c_3 coupled to hydrogenase works as a 'proton thruster' in Desulfovibrio vulgaris," *Journal of Biological Inorganic Chemistry*, vol. 2, no. 4, pp. 488–491, 1997.

Fine Tuning of Redox Networks on Multiheme Cytochromes from Geobacter sulfurreducens Drives Physiological Electron/Proton Energy Transduction

77

[14] R. J. P. Williams, "The history and the hypotheses concerning ATP-formation by energised protons," *FEBS Letters*, vol. 85, no. 1, pp. 9–19, 1978.

[15] P. R. Pokkuluri, Y. Y. Londer, N. E. C. Duke, W. C. Long, and M. Schiffer, "Family of cytochrome c_7-type proteins from *Geobacter sulfurreducens*: structure of one cytochrome c_7 at 1.45 Å resolution," *Biochemistry*, vol. 43, no. 4, pp. 849–859, 2004.

[16] L. Morgado, V. B. Paixão, M. Schiffer, P. R. Pokkuluri, M. Bruix, and C. A. Salgueiro, "Revealing the structural origin of the redox-Bohr effect: the first solution structure of a cytochrome from *Geobacter sulfurreducens*," *Biochemical Journal*, vol. 441, no. 1, pp. 179–187, 2012.

[17] P. R. Pokkuluri, Y. Y. Londer, X. Yang et al., "Structural characterization of a family of cytochromes c_7 involved in Fe(III) respiration by *Geobacter sulfurreducens*," *Biochimica et Biophysica Acta*, vol. 1797, no. 2, pp. 222–232, 2010.

[18] E. S. Shelobolina, M. V. Coppi, A. A. Korenevsky et al., "Importance of *c*-type cytochromes for U(VI) reduction by *Geobacter sulfurreducens*," *BMC Microbiology*, vol. 7, article 16, 2007.

[19] Y. H. R. Ding, K. K. Hixson, M. A. Aklujkar et al., "Proteome of *Geobacter sulfurreducens* grown with Fe(III) oxide or Fe(III) citrate as the electron acceptor," *Biochimica et Biophysica Acta*, vol. 1784, no. 12, pp. 1935–1941, 2008.

[20] L. Morgado, M. Brulx, M. Pessanha, Y. Y. Londer, and C. A. Salgueiro, "Thermodynamic characterization of a triheme cytochrome family From *Geobacter sulfurreducens* reveals mechanistic And functional diversity," *Biophysical Journal*, vol. 99, no. 1, pp. 293–301, 2010.

[21] M. S. Kim and Y. J. Lee, "Optimization of culture conditions and electricity generation using *Geobacter sulfurreducens* in a dual-chambered microbial fuel-cell," *International Journal of Hydrogen Energy*, vol. 35, no. 23, pp. 13028–13034, 2010.

[22] J. T. Babauta, H. D. Nguyen, T. D. Harrington, R. Renslow, and H. Beyenal, "pH, redox potential and local biofilm potential microenvironments within *Geobacter sulfurreducens* biofilms and their roles in electron transfer," *Biotechnology and Bioengineering*. In press.

[23] A. E. Franks, K. P. Nevin, H. Jia, M. Izallalen, T. L. Woodard, and D. R. Lovley, "Novel strategy for three-dimensional real-time imaging of microbial fuel cell communities: monitoring the inhibitory effects of proton accumulation within the anode biofilm," *Energy and Environmental Science*, vol. 2, no. 1, pp. 113–119, 2009.

[24] C. I. Torres, A. K. Marcus, and B. E. Rittmann, "Proton transport inside the biofilm limits electrical current generation by anode-respiring bacteria," *Biotechnology and Bioengineering*, vol. 100, no. 5, pp. 872–881, 2008.

[25] C. I. Torres, H. S. Lee, and B. E. Rittmann, "Carbonate species as OH^- carriers for decreasing the pH gradient between cathode and anode in biological fuel cells," *Environmental Science and Technology*, vol. 42, no. 23, pp. 8773–8777, 2008.

[26] Y. Liu, H. Kim, R. R. Franklin, and D. R. Bond, "Linking spectral and electrochemical analysis to monitor *c*-type cytochrome redox status in living *Geobacter sulfurreducens* biofilms," *ChemPhysChem*, vol. 12, no. 12, pp. 2235–2241, 2011.

[27] H. Richter, K. P. Nevin, H. Jia, D. A. Lowy, D. R. Lovley, and L. M. Tender, "Cyclic voltammetry of biofilms of wild type and mutant *Geobacter sulfurreducens* on fuel cell anodes indicates possible roles of OmcB, OmcZ, type IV pili, and protons in extracellular electron transfer," *Energy and Environmental Science*, vol. 2, no. 5, pp. 506–516, 2009.

[28] Y. H. R. Ding, K. K. Hixson, C. S. Giometti et al., "The proteome of dissimilatory metal-reducing microorganism *Geobacter sulfurreducens* under various growth conditions," *Biochimica et Biophysica Acta*, vol. 1764, no. 7, pp. 1198–1206, 2006.

[29] J. R. Lloyd, C. Leang, A. L. Hodges Myerson et al., "Biochemical and genetic characterization of PpcA, a periplasmic *c*-type cytochrome in *Geobacter sulfurreducens*," *Biochemical Journal*, vol. 369, no. 1, pp. 153–161, 2003.

[30] S. Srikanth, E. Marsili, M. C. Flickinger, and D. R. Bond, "Electrochemical characterization of *Geobacter sulfurreducens* cells immobilized on graphite paper electrodes," *Biotechnology and Bioengineering*, vol. 99, no. 5, pp. 1065–1073, 2008.

[31] E. Marsili, J. B. Rollefson, D. B. Baron, R. M. Hozalski, and D. R. Bond, "Microbial biofilm voltammetry: direct electrochemical characterization of catalytic electrode-attached biofilms," *Applied and Environmental Microbiology*, vol. 74, no. 23, pp. 7329–7337, 2008.

10

Melanin-Based Coatings as Lead-Binding Agents

**Karin Sono, Diane Lye, Christine A. Moore, W. Christopher Boyd,
Thomas A. Gorlin, and Jason M. Belitsky**

Department of Chemistry and Biochemistry, Oberlin College, 119 Woodland Street, Oberlin, OH 44074, USA

Correspondence should be addressed to Jason M. Belitsky, jason.belitsky@oberlin.edu

Academic Editor: Patrick Bednarski

Interactions between metal ions and different forms of melanin play significant roles in melanin biochemistry. The binding properties of natural melanin and related synthetic materials can be exploited for nonbiological applications, potentially including water purification. A method for investigating metal ion-melanin interactions on solid support is described, with lead as the initial target. 2.5 cm discs of the hydrophobic polymer PVDF were coated with synthetic eumelanin from the tyrosinase-catalyzed polymerization of L-dopa, and with melanin extracted from human hair. Lead (Pb^{2+}) binding was quantified by atomic absorption spectroscopy (flame mode), and the data was well fit by the Langmuir model. Langmuir affinities ranged from $3.4 \cdot 10^3$ to $2.2 \cdot 10^4\,M^{-1}$. At the maximum capacity observed, the synthetic eumelanin coating bound ~9% of its mass in lead. Binding of copper (Cu^{2+}), zinc (Zn^{2+}), and cadmium (Cd^{2+}) to the synthetic-eumelanin-coated discs was also investigated. Under the conditions tested, the Langmuir affinities for Zn^{2+}, Cd^{2+}, and Cu^{2+} were 35%, 53%, and 77%, respectively, of the Langmuir affinity for Pb^{2+}. The synthetic-eumelanin-coated discs have a slightly higher capacity for Cu^{2+} on a per mole basis than for Pb^{2+}, and lower capacities for Cd^{2+} and Zn^{2+}. The system described can be used to address biological questions and potentially be applied toward melanin-based water purification.

1. Introduction

While the term "melanin" is widely recognized by the general public, our understanding of the bioorganic chemistry of melanins is far less advanced than for other biopolymers. However, there have been recent advances in the understanding of melanin structure, function, and properties [1, 2], as well as in the development of well-defined synthetic mimics [3, 4] and melanin-based materials [5, 6]. Melanins display extensive molecular recognition chemistry, with the ability to bind a wide range of organic and inorganic species. The interaction of melanins with metal ions plays a particularly large role in their biochemistry [7]. Current research suggests that the black-to-brown pigment eumelanin is formed via supramolecular assembly, and metal ions may be integral components of this assembly process [7, 8]. In human melanocytes, eumelanin is produced along with pheomelanin, a chemically distinct yellow-to-red pigment, in organelles known as melanosomes. Melanosomes are thought to play a role in calcium homeostasis [9]. Interactions between metals and melanin have been proposed to play a role in the

development of melanoma [10] and may be leveraged for potential treatment [11]. In the substancia nigra region of the brain, the redox balance between iron, dopamine, neuromelanin, and sources of oxidative stress is thought to play a role in Parkinson's disease [12]. Recently, a new form of neuromelanin found widely throughout the brain, nanoparticle aggregates with a eumelanin-like exterior, was shown to be a protective reservoir for heavy metals, including lead [13]. The sequestration of metal ions by melanin has been exploited in forensics and toxicology [14], and most recently, paleontology [15]. Interactions between melanins and metal ions can also be leveraged for nonbiological applications [16, 17], potentially including water purification [18, 19].

Eumelanin biosynthesis is initiated by the enzyme tyrosinase, which carries out two successive oxidations of tyrosine, first to L-dopa and then to dopaquinone (Figure 1(a)). Intramolecular ring closure ultimately leads, with and without loss of the carboxylic acid, to the eumelanin monomers 5,6-dihydroxyindole (DHI) and 5,6-dihydroxyindole-carboxylic acid (DHICA), respectively. These are further oxidized to highly reactive indolequinones, which react with

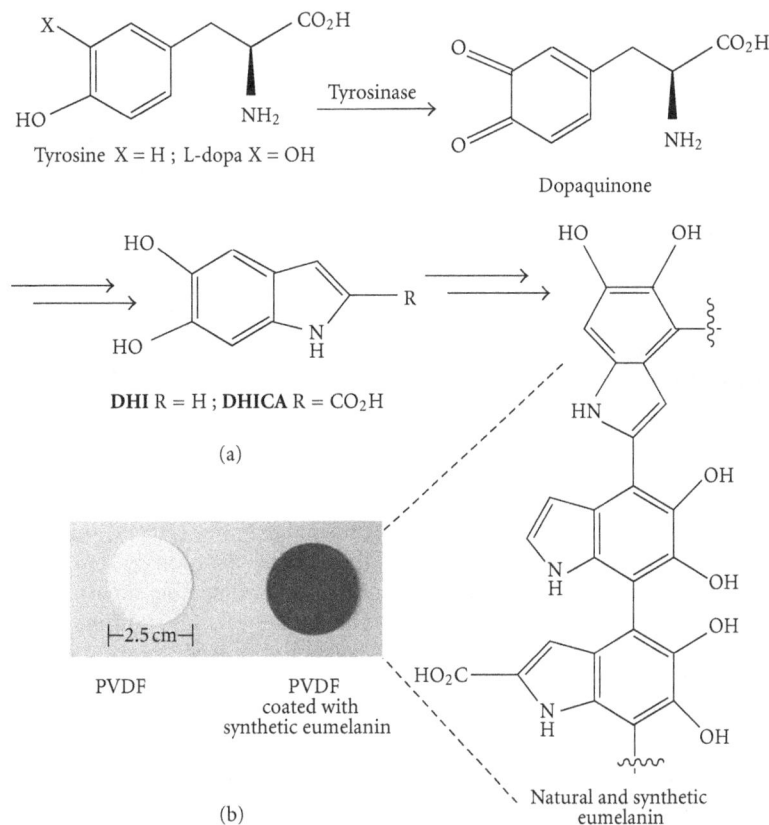

FIGURE 1: (a) Biosynthesis of natural eumelanin and biomimetic production of synthetic eumelanin. (b) 2.5 cm PVDF disc before (left) and after (right) coating with synthetic eumelanin.

each other and other components of the melanosome to form eumelanin. DHI and DHICA are now thought to form relatively short covalent oligomers that are heterogeneous with respect to both linkages between dihydroxyindole units and their oxidation states. These heterogeneous oligomers aggregate to form nanoparticles, and eumelanin superstructures are formed from these nanoparticles [1–4, 8]. Along with hydrogen bonding and aromatic stacking, metal ions (particularly Fe^{3+}, Cu^{2+}, Zn^{2+}, Mg^{2+}, and Ca^{2+}) are thought to provide key bridging interactions between both oligomers and nanoparticles and may also act as templates for different oligomeric and supramolecular structures [7, 8]. Changes in the aggregation and nanoscale morphology of preformed synthetic eumelanin have been observed upon the addition of Cu^{2+} and K^+ [8, 20]. Endogenously bound metal ions can be removed from natural samples such as *Sepia* eumelanin from cuttlefish ink and rebound with high affinity [21].

In vitro, under ambient oxygen conditions, conversion of L-dopa to a DHI/DHICA-containing synthetic eumelanin material (Figure 1) is thermodynamically spontaneous and kinetically facile from the dopaquinone oxidation state [22]. Thus, numerous methods can be used to initiate the formation of synthetic eumelanin from L-dopa, including autooxidation, chemical oxidants, and biomimetic catalysis with tyrosinase [22]. The quinone intermediates in these polymerizations are sufficiently reactive that many other compounds can potentially be incorporated covalently into the resulting materials, potentially tuning their properties.

(Nicotine is a physiologically relevant example [23].) Additionally, Messersmith, Ball, and others have found that eumelanin-related polydopamine coatings can adhere to nearly any surface [5, 24]. Thus, synthetic eumelanin can be readily prepared, modified, and presented in a variety of formats. Whereas synthetic chemistry offers the potential for greater variation, the ability to make use of melanin from renewable natural resources is attractive from a sustainability perspective. Wherever there are people there is hair, such that melanin from human hair [25–27] could be useful for water purification in areas where there are few other resources.

As part of a broader program to understand and modulate [28] eumelanin self-assembly and subsequent molecular recognition, we have developed a convenient method for studying metal ion binding to both synthetic and naturally derived melanin-based materials on solid support. Lead was chosen as initial target due to its toxicity, environmental distribution, and accumulation in neuromelanin [13]. There is great interest in the development of heavy metal sequestration agents based on synthetic systems [29, 30], as well as natural, renewable, and/or otherwise-discarded materials [31, 32]. Two recent studies have investigated melanin from squid ink for binding lead (Pb^{2+}) and cadmium (Cd^{2+}) [18] and fungal melanin secretions for binding chromium (Cr^{6+}) [19]. For water purification applications under batch conditions, an advantage of the presentation of melanin as a coating on solid support compared to a free suspension is that the solid support obviates the need for filtration or sedimentation steps to remove the suspension. As a research method, the advantages of coatings rather than free suspensions include ease of handling, minimization of the volume of heavy metal waste, and potential for combinatorial variation of synthetic coatings. Importantly, the solid support chosen, polyvinylidene di-fluoride (PVDF), is amenable to coating with both natural and synthetic melanin samples, allowing their direct comparison under identical conditions. Overall the process described here is complementary to other recent approaches for investigating melanin-metal ion interactions [7, 8, 33]. Here we introduce a metal binding assay using PVDF discs coated with synthetic eumelanin and melanin extracted from human hair, with a focus on lead binding.

2. Experimental

2.1. Synthetic-Eumelanin-Coated Discs. The production of synthetic eumelanin via the biomimetic oxidation of L-dopa with tyrosinase follows standard practice [22, 23, 34–36]. The discs are prepared in batches of twelve. Polyvinylidene di-fluoride (PVDF) discs are prepared from BioTrace PVDF membrane sheets (Pall) using a 2.5 cm diameter arch punch. 2.5 cm PVDF discs are shaken in MeOH for ~1 min, followed by two water washes (6 discs per 50 mL H_2O, 5 min each). A mushroom tyrosinase (Aldrich) stock solution of 1000 units/mL in sodium phosphate buffer (50 mM, pH 7.0) is prepared. L-dopa is dissolved in the same buffer to a concentration of 5.08 mM (1 mg/mL). 84 mL of freshly prepared L-dopa solution is added to a wide-mouth glass jar followed by 850 μL of the tyrosinase stock solution. Twelve PDVF discs are added to this solution (5.02 mM L-dopa, 10 units/mL tyrosinase, 50 mM sodium phosphate, pH 7.0), and the jar is shaken at 250 rpm for ~22 hrs at room temperature. After shaking for ~22 hrs, the discs are coated synthetic eumelanin and black in appearance. They are washed twice in the sodium phosphate buffer (6 discs per 50 mL buffer, 5 min each). Synthetic-eumelanin-coated discs are stored immersed in sodium phosphate buffer (50 mM, pH 7.0), in the dark at room temperature.

To determine the amount of synthetic eumelanin per disc, ≥10 discs from different polymerizations were stored in a dessicator until dryness and compared to blank PVDF discs treated in the same manner. Approximately 1.5 mg are added per disc as a result of the biomimetic polymerization (average mass of "blank" PDVF disc 36.87 ± 0.38 mg, average mass of synthetic-eumelanin-coated disc 38.40 ± 0.40 mg).

2.2. Hair-Extract-Coated Discs. The extraction of melanin from human hair via an acid-base method [26, 27], described in detail below, follows the procedure of Haywood et al. with a wash/precipitation sequence modified from Liu et al. to include a step of removal of endogenous metal ions. The presence of extracted melanin throughout the procedure and its reduction following deposition on PVDF was verified by spectroscopic comparison with synthetic eumelanin standards [25, 26].

Five grams of human hair produce 40 coated discs by the following procedure. Black (Indian) hair (5 g) cut in ~5 mm pieces is added to 250 mL 1 M NaOH (aq.) in a 1 L beaker at

90°C, and the mixture is stirred for 10 minutes at 90°C. The beaker is cooled on ice, and the total volume is returned to 250 mL by adding water. The mixture is centrifuged at 650 g for 5 min. The supernatant is transferred to new centrifuge vessels and adjusted to pH ≤ 3 by addition of con. HCl, inducing precipitation. The precipitate is collected by centrifugation at 14680 g for 10 min. The supernatant is discarded and the precipitate is resuspended in 200 mL sodium phosphate buffer (50 mM, pH 4.50). The pH of this suspension is adjusted to between pH 4.5 and pH 5.0 by addition of 3 M NaOH. The suspension is mixed thoroughly, then centrifuged at 14680 g for 10 min. The supernatant is discarded and the precipitate is resuspended in 200 mL sodium ethylenediaminetetraacetic acid buffer (Na_2EDTA, 100 mM, pH 4.50). After thorough mixing, the suspension is centrifuged at 14680 g for 10 min, and then the supernatant is discarded. The precipitate is resuspended in 200 mL sodium phosphate buffer (50 mM, pH 4.50), mixed thoroughly, and then the suspension is centrifuged at 14680 g for 10 min. After discarding the supernatant, the precipitate is resuspended in 152 mL sodium phosphate buffer (50 mM, pH 4.50) and mixed thoroughly. Forty 2.5 cm PVDF discs are shaken in MeOH for ~1 min, followed by two water washes (up to 6 discs per 50 mL H_2O, 5 min each). The 152 mL melanin suspension in pH 4.50 phosphate buffer is adjusted to pH 2.20 (±0.05) and distributed to 4 wide-mouth glass jars, 38 mL per jar. 12.667 mL of formamide (to 25% v/v, 50.667 mL total volume) is added to each jar, followed by 10 washed PVDF discs. The jars are shaken at 250 rpm for ~23 hours at room temperature, during which time melanin deposits onto the PVDF discs from the 1 : 3 formamide : phosphate buffer (50 mM, pH 2.20) suspension, forming a macroscopically spotty coating. Hair-extract-coated discs are stored immersed in water, in the dark at room temperature.

To determine the amount of hair extract per disc, ≥10 discs from different depositions were stored in a dessicator until dryness and compared to blank PVDF discs treated in the same manner. Approximately 3 mg are added per disc as a result of the hair extract deposition (average mass of "blank" PDVF disc 36.87 ± 0.38 mg, average mass of hair-extract-coated disc 40.03 ± 0.52 mg).

2.3. Metal Ion-Binding Experiments. Each titration includes 10 concentrations of the nitrate salts of Pb^{2+} (15.1 μM to 2.265 mM), Cu^{2+} (21.5 μM to 3.221 mM), Cd^{2+} (16.2 μM to 3.240 mM), or Zn^{2+} (33.6 μM to 6.720 mM). Pb^{2+} solutions are prepared from $Pb(NO_3)_2$ in pure H_2O (pH uncorrected) or in 50 mM $NaNO_3$ (aq.), with the final pH for each concentration adjusted to pH 4.00, 4.75, or 5.50. Cu^{2+}, Cd^{2+}, and Zn^{2+} solutions are prepared from $Cu(NO_3)_2 \cdot 2.5H_2O$, $Cd(NO_3)_2 \cdot 4H_2O$, and $Zn(NO_3)_2 \cdot 6H_2O$, respectively, in 50 mM $NaNO_3$ (aq.) with the final pH for each concentration adjusted to pH 4.75.

Synthetic-eumelanin- and hair-extract-coated discs are washed twice in water (6 discs per 50 mL H_2O, 5 min each). Uncoated, blank PVDF discs are shaken in MeOH for 1 min, followed by two water washes (6 discs per 50 mL H_2O, 5 min each). Using six-well polystyrene plates, a single disc is added

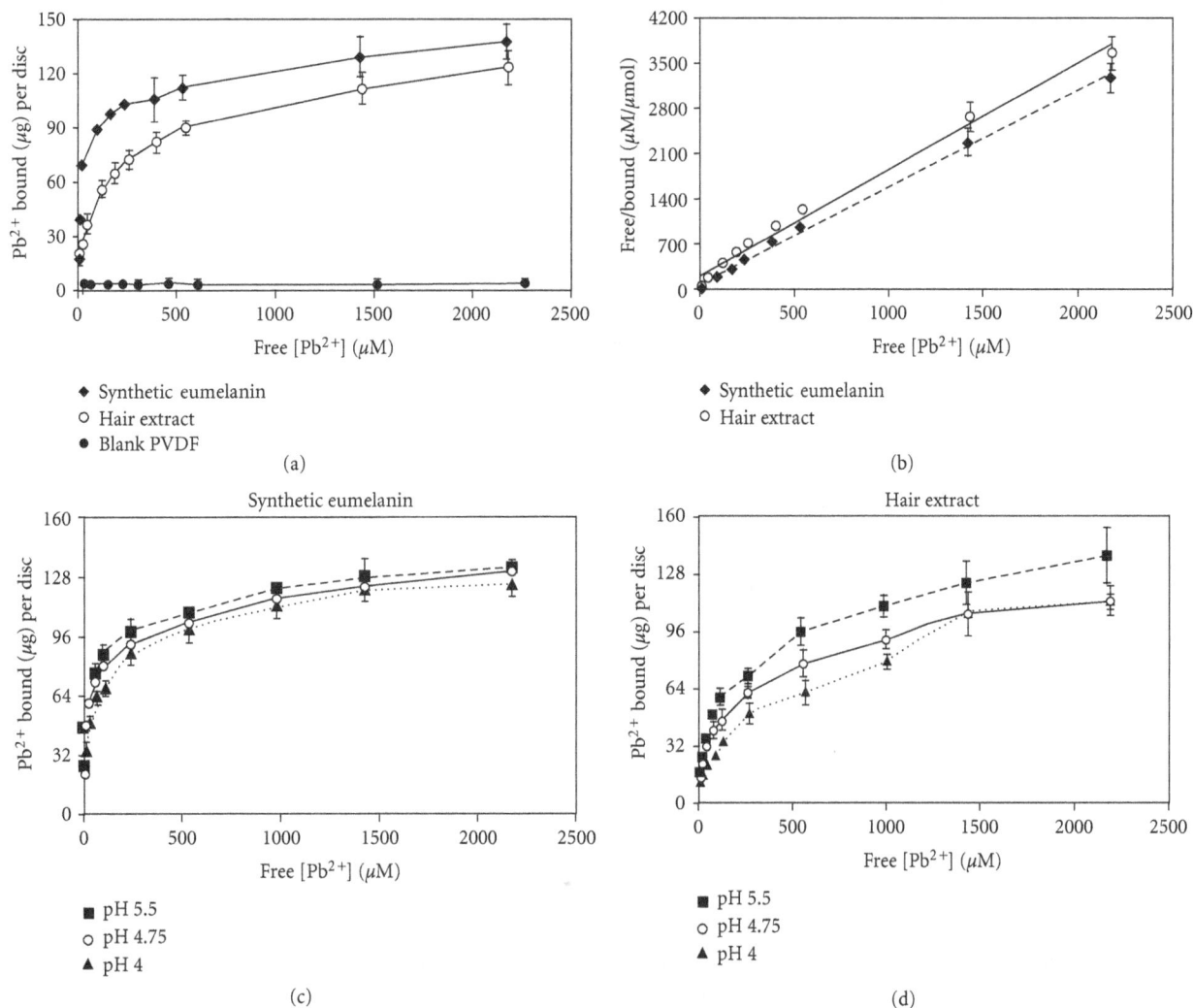

FIGURE 2: (a) Binding isotherms of Pb^{2+} ions to synthetic-eumelanin-coated discs, hair-extract-coated discs, and blank PVDF discs; pH uncorrected titrations of $Pb(NO_3)_2$ in H_2O. (b) Linearized Langmuir isotherm representation of the data for synthetic-eumelanin- and hair-extracted-coated discs in (a), $R^2 \geq 0.98$. (c) Binding isotherms to Pb^{2+} ions to synthetic-eumelanin-coated discs; titrations of $Pb(NO_3)_2$ in 50 mM $NaNO_3$ (aq.) at initial pH 5.50, 4.75, and 4.00. (d) Binding isotherms to Pb^{2+} ions to hair-extract-coated discs; titrations of $Pb(NO_3)_2$ in 50 mM $NaNO_3$ (aq.) at initial pH 5.50, 4.75, and 4.00. (a–d) Data are average values from at least three independent titrations, each with different discs; error bars represent one standard deviation.

to 7 mL per well of a metal ion-containing solution. The six-well plates are covered with tinfoil and shaken at room temperature for ~22 hrs. At the end of this equilibration, discs are removed and washed individually (25 mL H_2O, 5 min, 2x). A single disc is added to 7 mL of EDTA solution (100 mM Na_2EDTA, pH 4.50) per well in six-well polystyrene plates. The six-well plates are covered with tinfoil and shaken at room temperature for 30 min. The discs are removed and the EDTA solutions were analyzed using atomic absorption spectroscopy in flame mode. Lead samples were analyzed using a Perkin-Elmer 1100B atomic absorption spectrometer (air-acetylene flame, 0.7 nm slot width, hollow cathode lamp, 283.3 nm). A Perkin-Elmer AAnalyst 700 atomic absorption spectrometer was used to analyze copper (air-acetylene flame, 0.7 nm slot width, hollow cathode lamp, 324.8 nm), cadmium (air-acetylene flame, 0.7 nm slot width, hollow cathode lamp, 228.8 nm), and zinc (air-acetylene flame,

0.7 nm slot width, hollow cathode lamp, 213.9 nm). Metal ion concentrations (in $\mu g/mL$) were determined with respect to a standard curve generated during the same session. The total amount of metal ion per EDTA solution ($\mu g/mL \times 7$ mL volume) is taken to be the total amount (μg) of metal ion bound per disc. The data shown in Figures 2 and 3 and Tables 1, 2, and 3 represent average values from at least three independent titrations, each with different discs.

3. Results and Discussion

Our procedure for investigating melanin-metal ion interactions on solid support begins with the formation of a melanin-based coating on discs of the hydrophobic polymer polyvinylidene di-fluoride (PVDF). The coated discs are then allowed to equilibrate with varying concentrations of the target ion for 22 hours, washed, and immersed in a solution

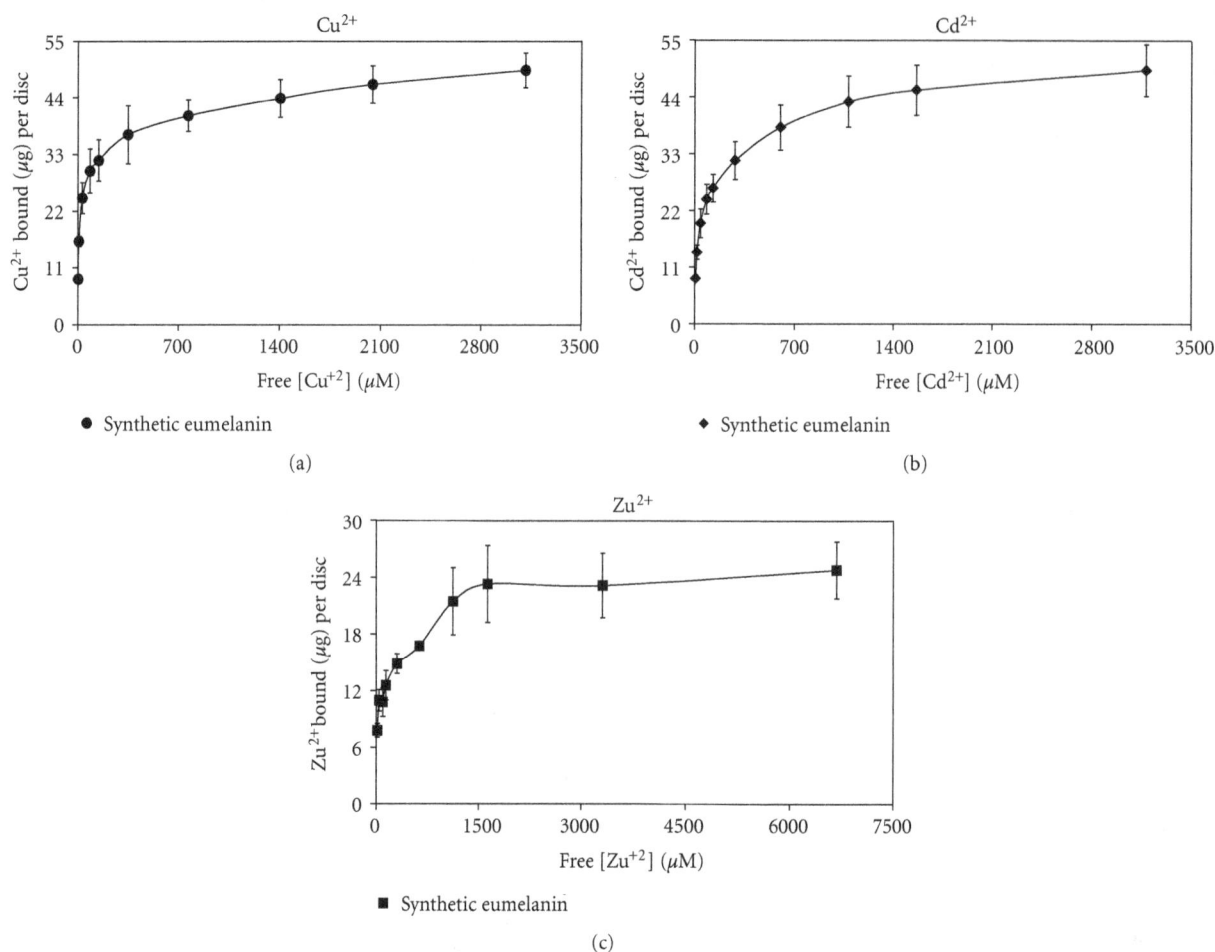

FIGURE 3: (a) Binding isotherm of Cu^{2+} ions to synthetic-eumelanin-coated discs; titration of $Cu(NO_3)_2 \cdot 2.5H_2O$ in 50 mM $NaNO_3$(aq.) at initial pH 4.75. (b) Binding isotherm of Cd^{2+} ions to synthetic-eumelanin-coated discs; titration of $Cd(NO_3)_2 \cdot 4H_2O$ in 50 mM $NaNO_3$(aq.) at initial pH 4.75. (c) Binding isotherm of Zn^{2+} ions to synthetic-eumelanin-coated discs; titration of $Zn(NO_3)_2 \cdot 6H_2O$ in 50 mM $NaNO_3$(aq.) at initial pH 4.75. (a–c) Data are average values from at least three independent titrations, each with different discs; error bars represent one standard deviation.

TABLE 1: Lead-binding parameters.

Synthetic-eumelanin-coated discs		
pH[a]	K_L (M^{-1})[b]	Capacity $\mu g/disc$[b]
Uncorrected	$1.74 (\pm 0.12) \cdot 10^4$	137.8 ± 10.1
5.50	$2.21 (\pm 0.22) \cdot 10^4$	133.0 ± 4.2
4.75	$1.83 (\pm 0.02) \cdot 10^4$	130.3 ± 5.5
4.00	$1.40 (\pm 0.15) \cdot 10^4$	119.1 ± 6.5
Hair-extract-coated discs		
pH[a]	K_L (M^{-1})[b]	Capacity $\mu g/disc$[b]
Uncorrected	$7.58 (\pm 0.39) \cdot 10^3$	126.1 ± 9.5
5.50	$7.04 (\pm 0.56) \cdot 10^3$	139.1 ± 14.7
4.75	$6.74 (\pm 1.04) \cdot 10^3$	115.6 ± 5.4
4.00[c]	$3.40 (\pm 0.20) \cdot 10^3$	110.5 ± 9.3

[a] Uncorrected titrations are $Pb(NO_3)_2$ in H_2O, other titrations are $Pb(NO_3)_2$ in 50 mM $NaNO_3$ with the initial pH adjusted as shown.
[b] Values calculated from linearized Langmuir isotherms from at least three independent titrations with different discs; standard deviations shown.
[c] Linearized Langmuir isotherm $R^2 \geq 0.96$, all others $R^2 \geq 0.98$.

TABLE 2: Percentage of Pb^{2+} bound at equilibrium.

Initial [Pb^{2+}]	Synthetic eumelanin[a]			Hair extract[a]		
	pH 5.50	pH 4.75	pH 4.00	pH 5.50	pH 4.75	pH 4.00
15.1 μM	>95%	>95%	>95%	82.2%	68.2%	42.6%
30.2 μM	>95%	>95%	73.3%	58.6%	49.7%	29.4%
60.4 μM	68.8%	68.7%	52.6%	41.2%	36.1%	22.5%
105.7 μM	49.8%	46.7%	38.7%	32.1%	26.7%	16.7%
151.0 μM	39.3%	36.8%	29.4%	26.9%	21.0%	14.9%
302.0 μM	22.4%	21.0%	18.3%	16.1%	14.1%	10.0%
604.0 μM	12.4%	11.8%	10.6%	10.8%	8.8%	6.3%
1057.0 μM	8.0%	7.6%	6.7%	7.2%	5.7%	4.9%
1510.0 μM	5.6%	5.6%	5.2%	5.6%	4.8%	4.3%
2265.0 μM	4.0%	4.0%	3.6%	4.2%	3.4%	3.2%

[a] Percentage of Pb^{2+} bound at equilibrium for each initial concentration from the titrations shown in Figures 2(c) and 2(d).

TABLE 3: Binding parameters of synthetic-eumelanin-coated discs with different metal ions.

	K_L (M^{-1})[a]	Capacity μg/disc[a]	Capacity nmol/disc[a]	Highest % bound (initial [M^{2+}])[b]
Pb^{2+c}	1.83 ($\pm$0.02) $\cdot 10^4$	130.3 ($\pm$5.5)	629 ($\pm$27)	>95% (15.1–30.2 μM)
Cu^{2+}	1.41 ($\pm$0.33) $\cdot 10^4$	49.0 ($\pm$3.2)	771 ($\pm$51)	92.3% (21.5 μM)
Cd^{2+}	9.70 ($\pm$0.74) $\cdot 10^3$	49.8 ($\pm$5.0)	443 ($\pm$45)	68.9% (16.2 μM)
Zn^{2+}	6.44 ($\pm$0.19) $\cdot 10^3$	25.0 ($\pm$3.2)	383 ($\pm$49)	47.9% (33.6 μM)

[a] Titrations in 50 mM $NaNO_3$, initial pH 4.75, binding isotherms are shown in Figures 2(c) and 3. Values calculated from linearized Langmuir isotherms ($R^2 \geq 0.99$) from at least three independent titrations with different discs; standard deviations shown.
[b] For each metal ion, the highest observed percentage bound at equilibrium is listed, with the initial concentration where this occurs shown in parentheses.
[c] Values for Pb^{2+} are repeated from Tables 1 and 2.

containing the chelating agent EDTA for 30 minutes. The metal ion content of this solution is then analyzed by atomic absorption spectroscopy (flame mode). Preliminary kinetic investigations show that 22 hours is more than sufficient for equilibration. Similarly, 30 minutes in a 100 mM Na_2EDTA solution (pH 4.50) was sufficient for metal ion extraction, in that further washes did not yield additional removal of metal ions. This does not rule out the possibility of an extremely tight-binding fraction, and weakly bound ions may be lost in the washing steps, such that the procedure may underestimate the total amount of metal ions bound. However, direct measurement of the decrease in metal ion concentration in the equilibration solutions agrees well with the amount recovered in the EDTA washes (data not shown).

Commercially available PVDF membranes (BioTrace, Pall) with 40 micron pores are typically used for bioanalytical applications such as Western blots. PVDF has been previously used as solid-support for metal ion-binding polymer membranes [37] and was chosen for use here because it displays little to no binding to lead ions on its own (Figure 2(a)). Preparation of synthetic eumelanin and deposition onto PVDF was accomplished in a single overnight step. Initiating the biomimetic tyrosinase-catalyzed oxidative polymerization of L-dopa [22] in the presence of 2.5 cm PVDF discs and allowing the reaction to proceed for ~22 hrs at room temperature while shaking at a rate fast enough (250 rpm) to avoid settling or self-aggregation of the hydrophobic discs provides a black synthetic eumelanin coating (Figure 1(b), see Section 2 for detailed procedure). The average mass

added per disc during this biomimetic polymerization is on the order of 1.5 mg. Compared to eumelanin samples from natural sources such as *Sepia* ink, synthetic eumelanin materials produced from L-dopa/tyrosinase tend to have a lower DHICA : DHI ratio (i.e., fewer carboxylic acids) and tend to be more amorphous [34].

PVDF discs can also be coated with melanin from human hair (see Section 2 for detailed procedure). Melanin in hair is a mixture of eumelanin and pheomelanin; black hair, used in this study, primarily contains eumelanin [25]. A straightforward acid/base extraction procedure [26, 27] was used, including a short, high temperature, basic heating step (1 M NaOH, 90°C, 10 min), which has been shown to reduce melanin degradation for black hair samples [26] and an EDTA wash to remove endogenously bound metal ions. Following extraction from hair, melanin deposits onto PVDF discs from an acidic 25% formamide solution forming a macroscopically spotty coating. The average mass added per disc as a result of this deposition is on the order of 3 mg, which represents a 2.4% recovery of the mass of hair used. This hair extract coating is likely not completely free of keratin—also a metal-binding agent—and the melanin may experience some degradation, which results in increased carboxylic acid content [27]. Both the hair extract and synthetic eumelanin coatings are stable to shaking and storage in aqueous solutions, without loss of material from the support over the course of shaking for 22 hrs during binding experiments or at least two weeks of storage in water (hair-extract-coated discs) or 50 mM sodium phosphate buffer, pH 7.0 (synthetic-eumelanin-coated discs). If the coated discs are stored at room temperature in aqueous solution on the time-scale of months, some leaching into the solution is observed.

The binding assay outlined above and described in detail in Section 2 was developed with Pb^{2+} as the initial target ion (Figure 2 and Tables 1 and 2). Coated and blank PVDF discs were immersed in aqueous lead nitrate solutions with concentrations ranging from 15.1 μM to 2.265 mM. Following a 22 hr equilibration, the discs were washed in pure water and then exposed to an EDTA solution (100 mM Na_2EDTA, pH 4.50) for 30 minutes. Lead ions that bound to the discs were extracted into the EDTA solution, which is then quantified by atomic absorption spectroscopy (flame mode, see Section 2). As shown in Figure 2(a), a plot of μg Pb^{2+} bound *per disc* versus free [Pb^{2+}], the coated discs show significant binding compared to the negligible results for the blank discs. In our first titrations (Figures 2(a) and 2(b), designated "pH uncorrected," the equilibration solutions contained only $Pb(NO_3)_2$ in pure water, such that the ionic strength and initial pH of these solutions varies with the different Pb^{2+} concentrations tested (4.75 < pH < 5.50). Titrations were also carried out in 50 mM $NaNO_3$ with varying $Pb(NO_3)_2$ concentrations with all solutions in a given titration adjusted to the same initial pH value (Figures 2(c) and 2(d), Tables 1 and 2). Solutions were set to an initial pH of 5.50, 4.75, or 4.00, values which span the initial pH range of the uncorrected titrations and extend to pH 4.00 on the acidic side, a value that is commonly tested in lead-binding experiments [30, 31]. Seven of the eight titrations are extremely good fits ($R^2 \geq 0.98$) for the Langmuir surface-binding model [38] (see linearized Langmuir isotherm, Figure 2(b), and Table 1). In the pH uncorrected titrations (Figures 2(a) and 2(b) and Table 1), the synthetic-eumelanin-coated discs display a Langmuir affinity (K_L) of $1.74 \cdot 10^4 \, M^{-1}$ and a capacity of 138 μg Pb^{2+} bound per disc. Amberlite IR-120, a commercial ion-exchange resin, has a similar affinity for lead ions ($K_L = 1.2 \cdot 10^4$) [30]. The hair-extract-coated discs display a ~2-fold-lower Langmuir affinity of $7.58 \cdot 10^3 \, M^{-1}$ and a slightly lower capacity of 126 μg Pb^{2+} bound per disc.

The effect of varying the initial pH of the equilibration solutions in the tested range is relatively modest (Figures 2(c) and 2(d) and Tables 1 and 2). The affinity and capacity of the synthetic-eumelanin-coated discs in the pH 4.00 titration are 63% and 86%, respectively, of these values in the pH 5.50 titration. For the hair extract coating in the pH 4.00 titration, the affinity and capacity are 45% and 80%, respectively, of these values in the pH 5.50 titration. Table 2 shows the percentage of Pb^{2+} bound at equilibrium for each point of these titrations; variation with pH is particularly apparent for the hair-extract-coated discs. At initial concentrations up to 30.2 μM for the pH 4.75 and 5.50 titrations, or 15.1 μM for the pH 4.00 titration, the synthetic eumelanin coating bound >95% of the Pb^{2+} originally in solution. Recently, Chen et al. reported that melanin extracted from squid ink is able to bind 95% of the Pb^{2+} initially present in solutions of 500 μM to 2.0 mM [18]. However, it is difficult to directly compare these results to the data presented in Table 2 because

of large differences in the amounts of material used (200 mg versus 1.5–3.0 mg coating per disc). Au and Potts compared the binding efficiency of synthetic eumelanin and melanin extracted from three different regions of the bovine eye (all as free suspensions), for a variety of metal ions at the same initial concentration (1.67 mM) [36]. Pb^{2+} was one of the metal ions with the highest percent bound, and the synthetic eumelanin sample had intermediate efficiency between the different natural samples; melanin extracted from the choroid was the strongest binder [36]. Here, melanin extracted from human hair is less efficient than the synthetic coating (Table 2). An enzymatic extraction procedure [27] and/or alternate deposition conditions may allow us to utilize the renewable resource of human hair for more efficient lead-binding melanin-based coatings than this first generation.

The Langmuir affinity constants (Table 1) for the hair-extract- and synthetic-coated discs are within the range of values of affinity constants (10^3 to $10^7 \, M^{-1}$) obtained by a variety of methods for different metal ions with synthetic and natural melanin samples [7, 38]. Earlier reports have suggested that natural and synthetic eumelanin samples have a high affinity for Pb^{2+} compared to other metal ions [35, 36], but affinity constants were not established in these studies, and until recently [18], current studies have generally not included lead. Given the known heterogeneity of both the natural and synthetic samples, the Langmuir affinity values reported here presumably represent an aggregate affinity rather than a microscopic per site affinity. On a molecular level, the catechol-like oxygens, carboxylic acids, the indole nitrogens, and the indole π electron density all potentially play some role in binding. In a recent study using melanin from squid ink, Chen et al. saw evidence of involvement of both the catechol and carboxylate functionalities in lead binding [18]. Affinity constants were not determined in that study but it was found that lead binding was more resistant than cadmium binding to competition with salts although this could be due differing binding modes as well as differing affinities [18].

Although the PVDF support shows little to no affinity for lead ions on its own, the structure of the support likely influences both the number and distribution of available binding sites compared to other presentations. Nevertheless, for comparison with other systems we considered binding as a function of the mass of the coating. At the maximum capacities, the synthetic eumelanin coating binds ~9% of its mass in Pb^{2+} (138 μg/1.5 mg), while the hair extract coating binds ~4.5% of its mass in Pb^{2+} (139 μg/3 mg). These values, which equate to 0.44 mmol (Pb^{2+})/g (coating) and 0.22 mmol/g for the synthetic eumelanin and hair extract, respectively, should be considered approximate, because of uncertainty in the average mass of the coating. Nevertheless, they are within the range of reported capacities for a variety of natural and synthetic melanin samples with different metal ions [7] and are at the higher end of heavy metal capacities for a variety of biosorbents [32]. Melanin extracted from squid ink was able to bind 13.5% of its mass in Pb^{2+} at capacity (0.65 mmol/g) [18].

To begin to assess the selectivity of melanin-based coatings for Pb^{2+} and probe the generality of the binding assay, titrations of the nitrate salts of copper (Cu^{2+}), cadmium (Cd^{2+}), and zinc (Zn^{2+}) were performed with synthetic-eumelanin-coated discs (Figure 3 and Table 3). Ten concentrations of each metal ion in 50 mM $NaNO_3$ (aq.), initial pH adjusted to pH 4.75, were used, for comparison with the Pb^{2+} results. All three titrations were extremely good fits ($R^2 \geq 0.99$) for the Langmuir model yielding K_L values of $1.41 \cdot 10^4 \, M^{-1}$ for Cu^{2+}, $9.70 \cdot 10^3 \, M^{-1}$ for Cd^{2+}, and $6.44 \cdot 10^3 \, M^{-1}$ for Zn^{2+}. Thus, under the conditions tested, the Langmuir affinities for Zn^{2+}, Cd^{2+}, and Cu^{2+} were 35%, 53%, and 77%, respectively, of the Langmuir affinity for Pb^{2+}. In terms of capacity, on the basis of mass bound, the discs have significantly higher capacity for lead than the other ions tested, but this is partially the result of the mass of lead itself; on per mole basis, the capacities for Zn^{2+} (383 nmol/disc, ~0.26 mmol/g coating) and Cd^{2+} (443 nmol/disc, ~0.30 mmol/g coating) are below the capacity for Pb^{2+} (629 nmol/disc, ~0.42 mmol/g coating), while the capacity for Cu^{2+} (771 nmol/disc, ~0.51 mmol/g coating) is higher. Chen et al. found that melanin from squid ink had a higher molar capacity for Cd^{2+} (0.93 mmol/g) than for Pb^{2+} (0.65 mmol/g) [18]. Copper and zinc are among the most studied metal ions for interactions with natural and synthetic melanin samples [7, 20, 33, 39]. In general Cu^{2+} is found to be the stronger binder, while the capacity for Zn^{2+} can be taken as a reflection of the DHICA content of the melanin sample [7]. For example, *Sepia* eumelanin has a higher capacity for Zn^{2+} (~1.5 mmol/g) than for Cu^{2+} (1.1 mmol/g) [7], whereas the lower capacity for zinc ions here likely reflects the lower DHICA : DHI ratio in L-dopa/tyrosinase synthetic eumelanin [34]. The generally lower capacities displayed by the synthetic-eumelanin-coating compared to natural melanin samples for the four metal ions tested likely reflect differences in both the composition of the synthetic eumelanin and its presentation on PVDF.

A study by Buszman et al. comparing the binding of metal ions including Cu^{2+}, Cd^{2+}, and Zn^{2+} (but Pb^{2+}) to a suszpension of synthetic eumelanin generated by the autoxidation of L-dopa (i.e., a different method than used here) was recently reanalyzed [38, 40]. Fitting the data to several different models Bridelli and Crippa found the affinity order $Zn^{2+} < Cd^{2+} < Cu^{2+}$ [38], which is the same order as observed here. In their study with bovine eye melanin and synthetic eumelanin (produced by the L-dopa/tyrosinase method) Potts and Au found that Zn^{2+} had lower binding efficiency than Pb^{2+} and Cu^{2+}, which had had similar binding efficiency (Cd^{2+} was not tested) [36]. Using a competitive binding assay against the nitrogen di-cation paraquat, Larsson and Tjalve found that Pb^{2+} was slightly more competitive than Cu^{2+} and that Pb^{2+}, Cu^{2+}, La^{3+}, and Gd^{3+} were the most competitive of eighteen metal ions tested for bovine eye melanin (however, Zn^{2+} and Cd^{2+} were not tested) [35]. These authors tested a more limited subset of metal ions with synthetic eumelanin produced by the L-dopa/tyrosinase method; Pb^{2+} was again slightly less competitive than La^{3+} and more competitive than other metal ions tested (Cu^{2+} was not tested with the synthetic sample) [35]. Overall, the results for the synthetic eumelanin coated discs with the four metal ions tested here (Table 2) are consistent with affinity trends in the literature

and add to the earlier findings that eumelanin binds Pb^{2+} strongly compared to other metal ions, with affinity similar to (or, under the conditions tested here, slightly higher than) its affinity for Cu^{2+}. In this context, it is interesting to note that, compared to the presence of these metals in the surrounding tissue, neuromelanin concentrates lead to a much greater extent than copper [13].

4. Conclusions

Melanins have many fascinating chemical properties in addition to their roles as biological pigments, including molecular recognition properties that are potentially advantageous for water purification applications. In this study we have developed a procedure for investigating melanin-metal ion interactions on solid support, with lead as the initial target. Melanin-based coatings derived by extraction of melanin from human hair and the biomimetic tyrosinase-catalyzed oxidation of L-dopa were shown to bind Pb^{2+} ions with reasonable affinity and high capacity, with the synthetic eumelanin coating binding up to ~9% of its mass in lead. Considered by mass bound, the synthetic-eumelanin-coated discs have a much higher capacity for Pb^{2+} than for Cu^{2+}, Cd^{2+}, or Zn^{2+}; however, on a per mole basis the capacity for Cu^{2+} is slightly higher than for Pb^{2+}. The affinity of the synthetic eumelanin coating for Pb^{2+} is slightly higher than Cu^{2+} and higher than for Zn^{2+} and Cd^{2+}. Various additives can be used to alter the composition of synthetic eumelanin polymerizations, such that the binding properties of the resulting coatings can be varied in a combinatorial fashion, potentially yielding materials with greater selectivity for lead. We are investigating other methods of isolation/deposition of natural melanin from human hair (including different color samples for varied pheomelanin content), as well as second generation synthetic coatings. Results bearing on melanin biochemistry, heavy metal toxicology, and melanin-based water purification efforts will be disclosed in due course.

Acknowledgments

Acknowledgment is made to the donors of the American Chemical Society Petroleum Research Fund for partial support of this research. The authors thank Research Corporation (Cottrell College Science Award) and Oberlin College for additional support.

References

[1] A. Huijser, A. Pezzella, and V. Sundström, "Functionality of epidermal melanin pigments: current knowledge on UV-dissipative mechanisms and research perspectives," *Physical Chemistry Chemical Physics*, vol. 13, no. 20, pp. 9119–9127, 2011.

[2] J. D. Simon and D. N. Peles, "The red and the black," *Accounts of Chemical Research*, vol. 43, no. 11, pp. 1452–1460, 2010.

[3] K.-Y. Ju, Y. Lee, S. Lee, S. B. Park, and J.-K. Lee, "Bioinspired polymerization of dopamine to generate melanin-like nanoparticles having an excellent free-radical-scavenging property," *Biomacromolecules*, vol. 12, no. 3, pp. 625–632, 2011.

[4] M. Arzillo, A. Pezzella, O. Crescenzi et al., "Cyclic structural motifs in 5,6-dihydroxyindole polymerization uncovered: biomimetic modular buildup of a unique five-membered macrocycle," *Organic Letters*, vol. 12, no. 14, pp. 3250–3253, 2010.

[5] F. Bernsmann, V. Ball, F. Addiego et al., "Dopamine-melanin film deposition depends on the used oxidant and buffer solution," *Langmuir*, vol. 27, no. 6, pp. 2819–2825, 2011.

[6] M. D'Ischia, A. Napolitano, A. Pezzella, P. Meredith, and T. Sarna, "Chemical and structural diversity in eumelanins: unexplored bio-optoelectronic materials," *Angewandte Chemie - International Edition*, vol. 48, no. 22, pp. 3914–3921, 2009.

[7] L. Hong and J. D. Simon, "Current understanding of the binding sites, capacity, affinity, and biological significance of metals in melanin," *Journal of Physical Chemistry B*, vol. 111, no. 28, pp. 7938–7947, 2007.

[8] P. Borghetti, A. Goldoni, C. Castellarin-Cudia et al., "Effects of potassium on the supramolecular structure and electronic properties of eumelanin thin films," *Langmuir*, vol. 26, no. 24, pp. 19007–19013, 2010.

[9] M. J. Hoogduijn, N. P. Smit, A. Van Der Laarse, A. F. Van Nieuwpoort, J. M. Wood, and A. J. Thody, "Melanin has a role in Ca^{2+} homeostasis in human melanocytes," *Pigment Cell Research*, vol. 16, no. 2, pp. 127–132, 2003.

[10] F. L. Meyskens and S. Yang, "Thinking about the role (Largely Ignored) of heavy metals in cancer prevention: hexavalent chromium and melanoma as a case in point," in *Clinical Cancer Prevention*, H.-J. Senn and F. Otto, Eds., chapter 5, pp. 65–74, Springer, Berlin, Germany, 2011.

[11] P. J. Farmer, S. Gidanian, B. Shahandeh, A. J. Di Bilio, N. Tohidian, and F. L. Meyskens, "Melanin as a target for melanoma chemotherapy: pro-oxidant effect of oxygen and metals on melanoma viability," *Pigment Cell Research*, vol. 16, no. 3, pp. 273–279, 2003.

[12] A. M. Snyder and J. R. Connor, "Iron, the substantia nigra and related neurological disorders," *Biochimica et Biophysica Acta*, vol. 1790, no. 7, pp. 606–614, 2009.

[13] L. Zecca, C. Bellei, P. Costi et al., "New melanic pigments in the human brain that accumulate in aging and block environmental toxic metals," *Proceedings of the National Academy of Sciences of the United States of America*, vol. 105, no. 45, pp. 17567–17572, 2008.

[14] D. J. Tobin, Ed., *Hair in Toxicology: An Important Bio-Monitor*, Royal Society of Chemistry, Cambridge, UK, 2005.

[15] R. A. Wogelius, P. L. Manning, H. E. Barden et al., "Trace metals as biomarkers for eumelanin pigment in the fossil record," *Science*, vol. 333, no. 6049, pp. 1622–1626, 2011.

[16] A. G. Orive, P. Dip, Y. Gimeno et al., "Electrocatalytic and magnetic properties of ultrathin nanostructured iron-melanin films on Au(111)," *Chemistry—A European Journal*, vol. 13, no. 2, pp. 473–482, 2007.

[17] G. S. Huang, M. T. Wang, C. W. Su, Y. S. Chen, and M. Y. Hong, "Picogram detection of metal ions by melanin-sensitized piezoelectric sensor," *Biosensors and Bioelectronics*, vol. 23, no. 3, pp. 319–325, 2007.

[18] S. Chen, C. Xue, J. Wang et al., "Adsorption of Pb(II) and Cd(II) by squid *Ommastrephes bartrami* melanin," *Bioinorganic Chemistry and Applications*, vol. 2009, Article ID 901563, 7 pages, 2009.

[19] X. Yu, Z. Gu, R. Shao, H. Chen, X. Wu, and W. Xu, "Study on adsorbing chromium(VI) ions in wastewater by aureobacidium pullulans secretion of melanin," *Advanced Materials Research*, vol. 156-157, pp. 1378–1384, 2011.

[20] J. M. Gallas, K. C. Littrell, S. Seifert, G. W. Zajac, and P. Thiyagarajan, "Solution structure of copper ion-induced molecular aggregates of tyrosine melanin," *Biophysical Journal*, vol. 77, no. 2, pp. 1135–1142, 1999.

[21] Y. Liu and J. D. Simon, "Metal-ion interactions and the structural organization of *Sepia* eumelanin," *Pigment Cell Research*, vol. 18, no. 1, pp. 42–48, 2005.

[22] G. Prota, *Melanins and Melanogenesis*, Academic Press, San Diego, Calif, USA, 1992.

[23] D. L. Dehn, D. J. Claffey, M. W. Duncan, and J. A. Ruth, "Nicotine and cotinine adducts of a melanin intermediate demonstrated by matrix-assisted laser desorption/ionization time-of-flight mass spectrometry," *Chemical Research in Toxicology*, vol. 14, no. 3, pp. 275–279, 2001.

[24] H. Lee, S. M. Dellatore, W. M. Miller, and P. B. Messersmith, "Mussel-inspired surface chemistry for multifunctional coatings," *Science*, vol. 318, no. 5849, pp. 426–430, 2007.

[25] S. Ito and K. Wakamatsu, "Human hair melanins: what we have learned and have not learned from mouse coat color pigmentation," *Pigment Cell and Melanoma Research*, vol. 24, no. 1, pp. 63–74, 2011.

[26] R. M. Haywood, M. Lee, and C. Andrady, "Comparable photoreactivity of hair melanosomes, eu- and pheomelanins at low concentrations: low melanin a risk factor for UVA damage and melanoma?" *Photochemistry and Photobiology*, vol. 84, no. 3, pp. 572–581, 2008.

[27] Y. Liu, V. R. Kempf, J. B. Nofsinger et al., "Comparison of the structural and physical properties of human hair eumelanin following enzymatic or acid/base extraction," *Pigment Cell Research*, vol. 16, no. 4, pp. 355–365, 2003.

[28] J. M. Belitsky, "Aryl boronic acid inhibition of synthetic melanin polymerization," *Bioorganic and Medicinal Chemistry Letters*, vol. 20, no. 15, pp. 4475–4478, 2010.

[29] Q. F. Lü, M. R. Huang, and X. G. Li, "Synthesis and heavy-metal-ion sorption of pure sulfophenylenediamine copolymer nanoparticles with intrinsic conductivity and stability," *Chemistry—A European Journal*, vol. 13, no. 21, pp. 6009–6018, 2007.

[30] A. Demirbas, E. Pehlivan, F. Gode, T. Altun, and G. Arslan, "Adsorption of Cu(II), Zn(II), Ni(II), Pb(II), and Cd(II) from aqueous solution on Amberlite IR-120 synthetic resin," *Journal of Colloid and Interface Science*, vol. 282, no. 1, pp. 20–25, 2005.

[31] D. M. Vieira, A. C. A. Da Costa, C. A. Henriques, V. L. Cardoso, and F. P. De França, "Biosorption of lead by the brown seaweed *Sargassum filipendula*—batch and continuous pilot studies," *Electronic Journal of Biotechnology*, vol. 10, no. 3, pp. 368–375, 2007.

[32] A. V. Jamode, M. Rao, B. S. Chandak, V. S. Jamode, and A. V. Parwate, "Applications of the inexpensive adsorbents for the removal of heavy metals from industrial wastewater: a brief review," *Journal of Industrial Pollution Control*, vol. 19, no. 1, pp. 114–134, 2003.

[33] L. Najder-Kozdrowska, B. Pilawa, A. B. Wieckowski, E. Buszman, and D. Wrześniok, "Influence of copper(II) ions on radicals in DOPA-melanin," *Applied Magnetic Resonance*, vol. 36, no. 1, pp. 81–88, 2009.

[34] J. B. Nofsinger, S. E. Forest, L. M. Eibest, K. A. Gold, and J. D. Simon, "Probing the building blocks of eumelanins using scanning electron microscopy," *Pigment Cell Research*, vol. 13, no. 3, pp. 179–184, 2000.

[35] B. Larsson and H. Tjalve, "Studies on the melanin-affinity of metal ions," *Acta Physiologica Scandinavica*, vol. 104, no. 4, pp. 479–484, 1978.

[36] A. M. Potts and P. C. Au, "The affinity of melanin for inorganic ions," *Experimental Eye Research*, vol. 22, no. 5, pp. 487–491, 1976.

[37] Y. Zhai, Y. Liu, X. Chang, X. Ruan, and J. Liu, "Metal ion-small molecule complex imprinted polymer membranes: preparation and separation characteristics," *Reactive and Functional Polymers*, vol. 68, no. 1, pp. 284–291, 2008.

[38] M. G. Bridelli and P. R. Crippa, "Theoretical analysis of the adsorption of metal ions to the surface of melanin particles," *Adsorption*, vol. 14, no. 1, pp. 101–109, 2008.

[39] B. Szpoganicz, S. Gidanian, P. Kong, and P. Farmer, "Metal binding by melanins: studies of colloidal dihydroxyindole-melanin, and its complexation by Cu(II) and Zn(II) ions," *Journal of Inorganic Biochemistry*, vol. 89, no. 1-2, pp. 45–53, 2002.

[40] E. Buszman, B. Kwasniak, and A. Bogacz, "Binding capacity of metal ions to synthetic Dopa-Melanins," *Studia Biophysica*, vol. 125, no. 2, pp. 143–153, 1988.

11

Nanocrystalline Hydroxyapatite/Si Coating by Mechanical Alloying Technique

Ahmed E. Hannora,[1] Alexander S. Mukasyan,[2] and Zulkhair A. Mansurov[3]

[1] *Faculty of Petroleum and Mining Engineering, Suez Canal University, Suez 43721, Egypt*
[2] *Department of Chemical and Biomolecular Engineering, University of Notre Dame, Notre Dame, IN 46556, USA*
[3] *Institute of Combustion Problems, Almaty 050012, Kazakhstan*

Correspondence should be addressed to Ahmed E. Hannora, ahannora@yahoo.com

Academic Editor: Nick Katsaros

A novel approach for depositing hydroxyapatite (HA) films on titanium substrates by using mechanical alloying (MA) technique has been developed. However, it was shown that one-hour heat treatment at 800°C of such mechanically coated HA layer leads to partial transformation of desired HA phase to beta-tri-calcium phosphate (β-TCP) phase. It appears that the grain boundary and interface defects formed during MA promote this transformation. It was discovered that doping HA by silicon results in hindering this phase transformation process. The Si-doped HA does not show phase transition to β-TCP or decomposition after heat treatment even at 900°C.

1. Introduction

Natural bone is a nanocomposite that consists of mineral fraction, including small apatite crystals and nonstoichiometric calcium phosphate, and organic fraction, which together confer mechanical resistance. In order to simulate the nature bone structure, the synthesis of hydroxyapatite (HA: $Ca_{10}(PO_4)_6(OH)_2$) has received attention in recent years. However, the HA coatings of metallic implants often flake off as a result of poor ceramic/metal interface bonding, which may cause surgery to fail [1]. This problem may be solved by fabrication of metal/HA composites. Some work has been reported on the preparation of Ti-based alloy/HA composite materials or composite coatings for biomedical applications [2–4]. But most synthetic apatites are formed via high temperature and complicate processes, resulting in a well-crystallized structure and a larger particle size, which have little or no activity in bioresorption. Since kinetic solubility is dependent on particle size, there has been great interest in nanosized HA-based cement as bone substitute materials [5, 6]. Our group also has recently described the effect of high energy ball milling treatment on HA and Ti substrate [7, 8].

It was shown that, during this type of mechanical treatment, a significant decrease of the HAs' particle and crystalline size takes place which leads to formation of nanoscaled coating on the Ti substrate. For example, it was demonstrated that one hour of such mechanical treatment with optimum ball to powder ratio equals to 40 : 1 leads to the reduction of surface particle size from 1.0 μm to 80 nm.

Recently [9, 10], experimental work on HA and its substituted counterparts such as silicon-doped HA (SiHA) has focused on finding a measurable indicator of biocompatibility. It was found that dissolution or mineralization of HA may start from dislocations and defects at the grain boundaries in dental HA. Within synthetically produced hydroxyapatite, these boundaries are "clean" with few defects or voids. More close investigation of the grain boundary structures in hydroxyapatite and silicon-substituted hydroxyapatite revealed that while there was no significant difference in dislocation density between HA and SiHA there was certainly an increase in triple junctions within creased silicon doping. Also it is shown that the dissolution does occur preferentially from grain boundaries and triple junctions. Additionally, at triple junctions, it was the smallest grains that showed the greatest dissolution, suggesting that a decrease in grain size would lead to increased solubility and hence greater biocompatibility. It is of particular interest to note that after sintering, SiHA has been shown to have a much finer grain structure than pure HA-phase, suggesting that silicon also inhibits grain growth. Thus SiHA, can be considered as an important candidate as an enhanced bioactive material.

The objective of the present work is to develop high-energy ball milling approach for one-step production of nanosized SiHA coating on the surface of titanium substrate. Special attention is paid on the influence of silicon doping on the microstructure of the produced coatings.

2. Materials and Methods

The following procedure was used to prepare hydroxyapatite/silicon coatings on the titanium substrate. The mixtures of the hydroxyapatite (HA) (Wako, 99.5%, 250 μm) and silicon (Si) powders, as well as Ti substrate (plate with dimensions $10 \times 8 \times 2$ mm) are placed into the vibration chamber and undergo the high-energy ball milling (HEBM) with different mixture to ball weight ratios ($W_p : W_b$). The milling process was carried out in static air without any process control agent. During HEBM, two main processes contribute for the formation of coated layer on the Ti substrate (i) fracturing of the HA and Si particles, (ii) cold welding of the mixture and its adhesion to the metal (Ti) surface. As-prepared (coated) samples were then annealed in vacuum of 10^{-5} Pa, for 1 hour, at temperatures ranging between 600°C and 1000°C.

The microstructure of the coatings was analyzed by using different material science techniques. XRD analysis was performed using DRON-6 system, by using Cu K$_\alpha$ radiation (wavelength $\lambda = 0.15406$ nm) with a nickel filter at 25 kV and 25 mA. The diffractometer was operated within range of $20 < 2\theta < 60$ with steptime: 3 seconds and stepsize: 0.02 degree. The microstructural features of the surface layer and coating distribution on substrate surface have been systematically investigated using a Solver PRO scanning probe microscope JSPM-5200 (JEOL) and scanning electron microscopy Quanta 3D 200i. The sample composition was analyzed by energy dispersive x-ray spectroscopy (EDS) using JEOL JSM-6490 LA analytical SEM. A Perkin-Elmer Optima 2000 Dual View inductively coupled plasma (ICP) optical emission spectrometer (USA) was used to carry out the chemical analysis of the used powders. Transmission electron microscopy was performed with JEOL JEM CX at an accelerating voltage of 100 kV.

3. Results and Discussion

3.1. Mechanical Milling of Hydroxyapatite Powder. The function of HA in all of its applications is largely determined by its morphology, composition, crystal structure, and crystal size distribution. Thus, to control the mechanical properties of hydroxyapatite, the influence of synthesis conditions on such characteristics as particles' morphology and size distribution, as well as agglomeration have to be studied [11]. It is known

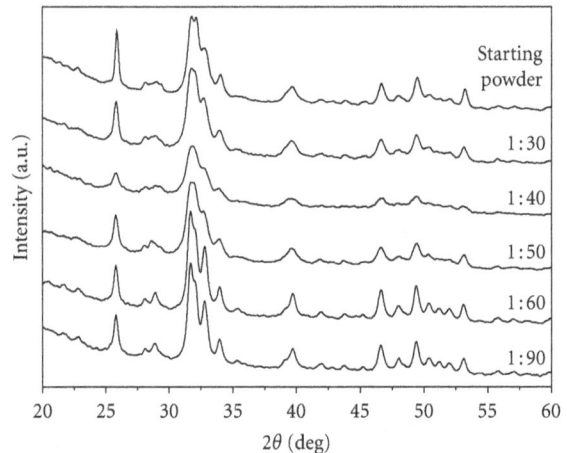

FIGURE 1: XRD patterns of hydroxyapatite powder milled one hour for various $W_p : W_b$ ratio.

that, for high-energy ball milling process, the weight ratio between powder mixture and milling media (balls) is one of the most important parameter, which affects the final powder microstructure. Figure 1 shows the XRD of the HEBM-HA powders processed for one hour with different powder/ball weight ratios ($W_p : W_b$). It can be seen that powders after milling under conditions with $W_p : W_b$ ratio 1 : 30 and 1 : 40 are characterized by XRD patterns with notable broader peaks with lower intensity, which indicates the decreasing of phase crystalline size and powder amorphization. This result can be explained by corresponding increase in the kinetic energy of interaction between mill media and the powders [12, 13]. However, further increase of the weight of the mill media (e.g., $W_p : W_b = 1 : 60$; $1 : 90$) leads to the recrystallization of HA powder. The latter effect may be related to the extremely high kinetic energy of the ball mill charge which transforms into heat, thus, promoting the recrystallization process.

Using Scherrer equation, the unit cell dimensions were calculated for the 1 : 60 and 1 : 90 samples (Table 1). The slight variation of lattice parameters could be due to strain accumulated through the HEBM process. Indeed, since large plastic deformation is induced into the powder particles during mechanical milling, the crystals are strained, and the deformation occurs in an inhomogeneous manner. From the XRD data, it can also be concluded that, decomposition of HA phase or formation of secondary phases, such as tricalcium phosphate, tetracalcium phosphate, and calcium oxide do not take place throughout the milling process.

3.2. Hydroxyapatite Powder Morphology. Scanning electron microscopy (SEM) and transmission electron microscopy (TEM) were used to analyze the morphology of the HA powders after high-energy ball milling. As mentioned in the above section, cold welding and fracturing are the two essential processes involved in the HEBM process. Fracture tends to break individual particles into smaller pieces and deagglomerates particles that have been cold welded. The morphologies of the HA powder particles before ball milling

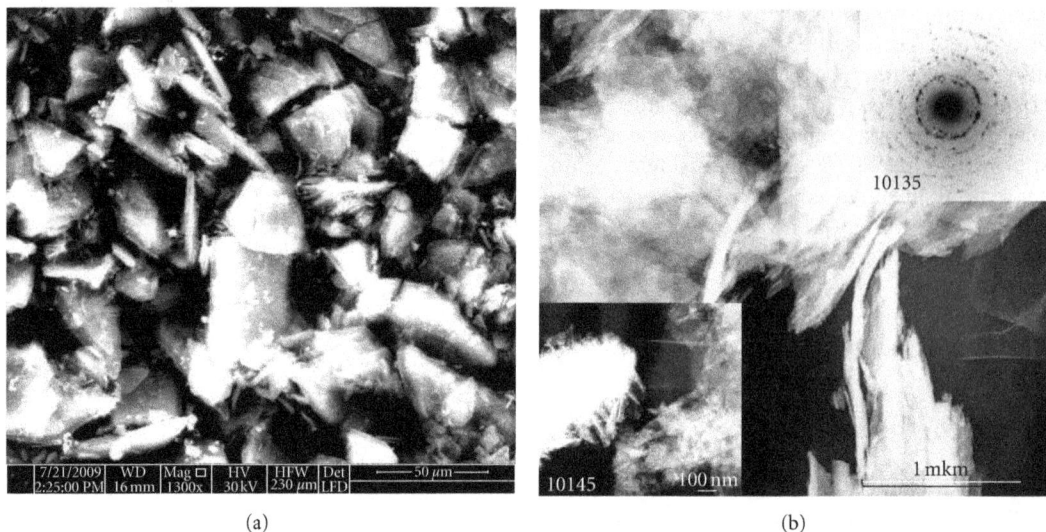

FIGURE 2: Needle-like crystallite of HA sample before milling (a) SEM and (b) TEM, respectively.

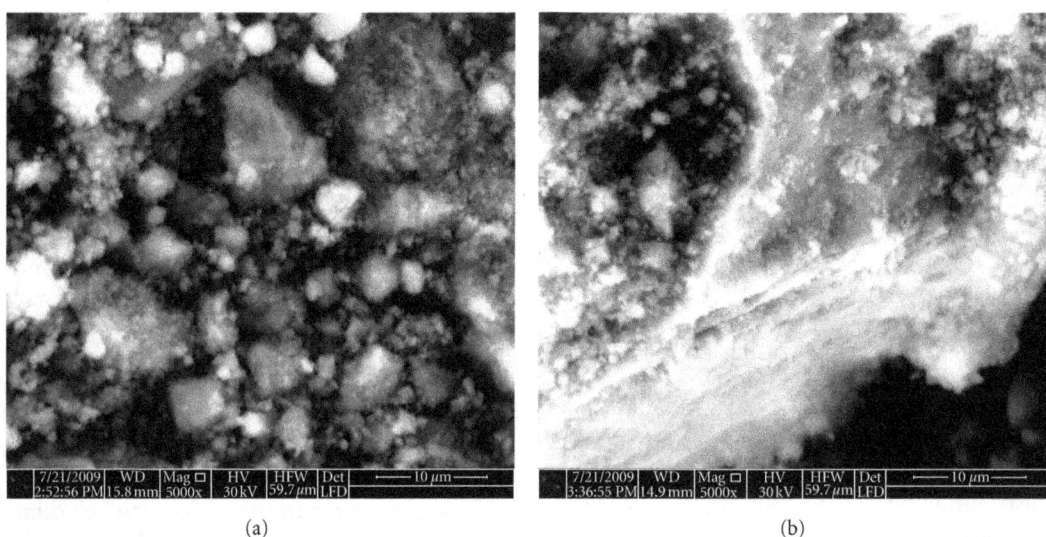

FIGURE 3: (a) and (b) Morphologies of HA powder particles after one hour of milling with $W_p : W_b = 1 : 40$.

TABLE 1: Unit cell parameters of the mechanical milled HA.

Sample	Lattice constant (Å)	Standard deviation	R_{wp} (%)
Standard card (24–0033)	$a = b = 9.432$ $c = 6.881$	—	
1 : 60 ratio	$a = b = 9.443$ $c = 6.898$	0.0014	7.6
1 : 90 ratio	$a = b = 9.430$ $c = 6.898$	0.0018	8.9

R_{wp} (weighted residual error).

are shown in Figure 2. It can be seen that HA powder consists of crystallites which have needles and platelets-like morphologies. Higher magnification shows that large HA particles composed of very smaller ones with the size of less than 100 nm. After 1 hour of HEBM ($W_p : W_b = 1 : 40$), the morphology of the initial large particles significantly changes (Figure 3) due to microforging, fracture, agglomeration, and deagglomeration processes. Thin layered (laminated) microstructure dominates in this sample.

However, more close inspection by TEM (Figure 4) shows two types of morphologies, that is, needle-like (a) and round shaped (b). It is also can be seen that these small particles formed the agglomerates as a result of cold welding phenomenon during ball milling.

Mechanical, physical, and chemical properties of powders may be altered if they are contaminated. The most common contaminations in mechanical milling are Fe and Cr

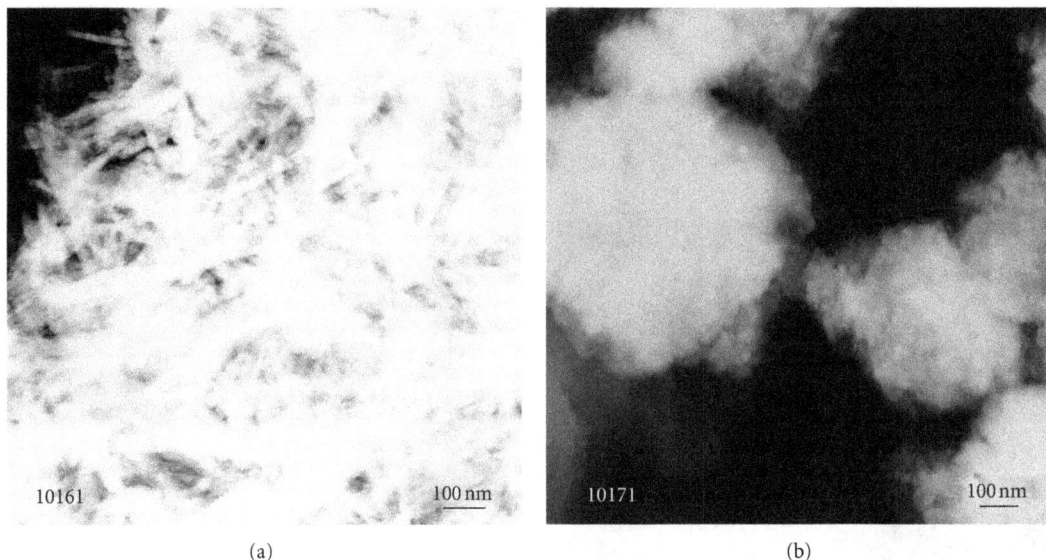

(a)   (b)

FIGURE 4: Hydroxyapatite powder after one hour of milling with $W_p : W_b = 1 : 40$.

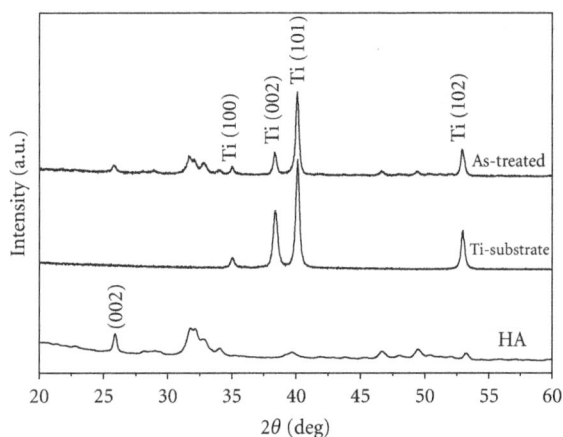

FIGURE 5: XRD patterns of ball-milled hydroxyapatite-coated Ti substrate after one hour of milling.

elements that come from the milling tools (vial and balls) since most of them are made from those types of elements. The Fe and Cr elements content after one hour of HEBM is shown in Table 2. As the $W_b : W_p$ ratio increases the amount of contamination increases. It was shown that, for the lower ratio (e.g., 10 : 1), a thin coating of the milling balls by the HA powder is formed which reduces Fe and Cr contamination. For higher $W_b : W_p$ ratio (60 : 1), the number of collisions and friction increases, which lead to a significant increase in Fe and Cr contamination.

3.3. X-Ray Diffraction of HA Coating on Ti-Substrate.

The XRD patterns shown in Figure 5 illustrate the effect of mechanical treatment on Ti substrate milled with HA powder. After 1 hour of HEBM, the (002) peak which is the most distinct reflection in the XRD pattern for HA possesses a notable intensity reduction. The (101) peak of the Ti substrate only slightly shifted to higher values, $\Delta d =$

-0.0024 and also (002) and (102) peaks are shifted with $\Delta d = -0.0015$ and $\Delta d = -0.0016$, respectively.

As reported [6], the process of covering metallic surfaces with HA at elevated temperatures (e.g., plasma spraying) has a tendency to eliminate the functional group OH in the HA matrix (dehydration) and results in the decomposition of HA into α-tricalcium phosphate, β-tricalcium phosphate, and tetracalcium phosphate. Also the condition of high substrate temperature promoted the oxidation of the substrate surface prior to the growth of the HA layer. The oxidation layer degraded the adhesion of the coating to the substrate. As can be concluded from XRD patterns in Figure 5, the repeated ball collision with the Ti substrate resulted in the deposition of HA powder on its surface without any trace of HA decomposition or Ti oxidation.

3.4. Morphology of HA Coating on Ti Substrate.

Typical morphology of the coated substrate surface is shown in

(a)

(b)

(c)

(d)

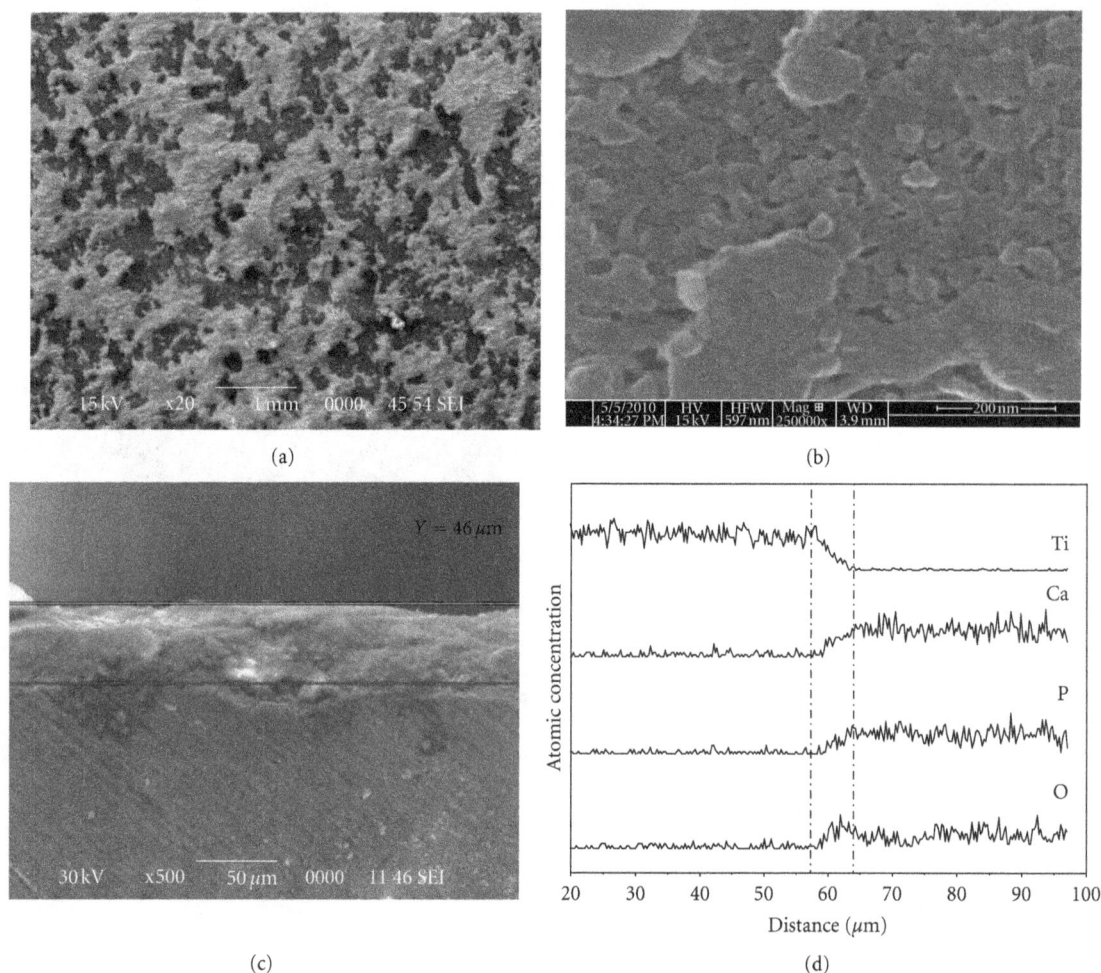

FIGURE 6: (a) and (b) Surface morphologies of the as-synthesized HA coating after one hour of milling (c) cross-section microstructure and (d) concentration profile of as-synthesized HA coating.

TABLE 2: Fe and Cr contamination in one-hour milled HA powder.

	Elements			
	Fe		Cr	
$W_b : W_p$ ratios	ppmw	%	ppmw	%
Before milling	32.3 ± 1.9	0.0032	<1	$<10^{-4}$
10 : 1	302 ± 9	0.03	3.3 ± 0.2	3.3×10^{-4}
30 : 1	480 ± 14	0.048	6.2 ± 0.3	6.2×10^{-4}
60 : 1	4787 ± 144	0.479	60 ± 2	6.0×10^{-3}

Figure 6. It can be seen that after one hour of HEBM, the substrate was covered with HA powder. The inhomogeneous distribution over the entire coated sample could be due to powder particles repeatedly fractured and cold welded on the substrate. The similar result, that is, inhomogeneous distributed precipitates was observed for calcium and phosphorus ion implanted in a dose of 10^{17} ions/cm^2 [14]. The broad face morphology of the as-coated HA (Figure 6(b)) implies that the agglomeration of particles occurs due to cold welding process. The dome-shaped morphology was also

formed in physical-vapor-deposited and laser-deposited HA films [15].

Typical cross-sections of as-treated HA coating is shown in Figure 6(c) where the coating thickness was about 50 μm. The composition of the HA-coated sample was analyzed by energy dispersive X-ray spectroscopy (EDS). Figure 6(d) shows the cross-section microstructure and concentration profile of HA coating produced by mechanical treatment method. Cold welding between particles and substrate under repeated ball collisions led to the formation of a composite

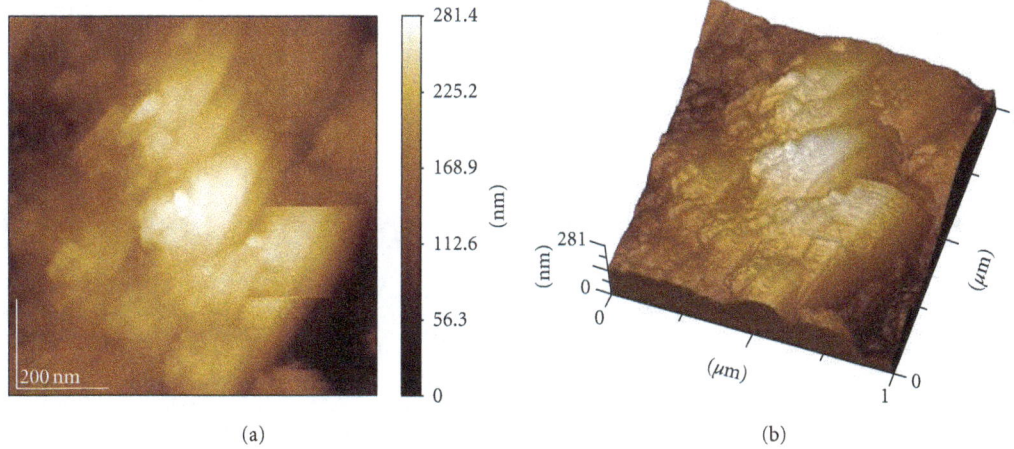

(a) (b)

FIGURE 7: 2-dimension and 3D AFM image of the as-coated HA.

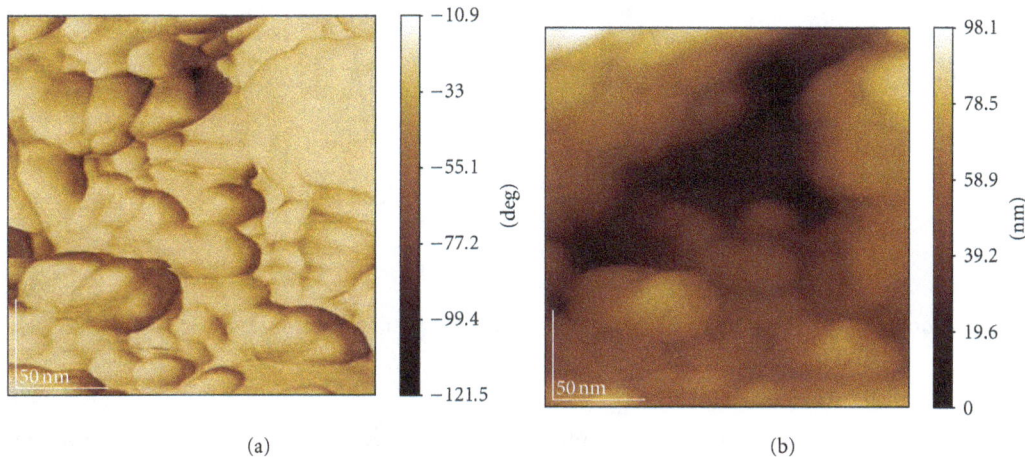

(a) (b)

FIGURE 8: Phase and topography of as-coated HA.

coating. HA (Ca, P, O-elements) flowed into the pores between Ti particles under the impact of balls and vice versa. The EDS analysis of the as-treated sample indicated that the average value of the Ca/P ratio of coated HA was (1.803 ± 0.18) while the stochiometric molar ratio is 1.67 [16].

Atomic force microscopy (AFM) image of the as-coated HA, Figure 7 showed that the HA was composed of numerous spherical-shaped aggregates of different sizes. The higher magnification micrograph, Figure 8, revealed that each spheroid-type aggregate involves a large number of much smaller grains.

3.5. X-Ray Diffraction of the Annealed HA Coating on Ti-Substrate.
XRD patterns of deposited HA heated at different temperatures are presented in Figure 9. X-ray peaks of the formed phase were matched to the ICDD (JCPDS) standard, α-tricalcium phosphate (09–0348), β-tricalcium phosphate (03–0690), and titanium oxide (29–1361). The changes in the XRD patterns give an indication of the influence of temperature on the structure stability of the samples. According to XRD patterns of the coated samples, it is clear that up to

700°C HA phase is stable; that is, no any phase transition or decompositions were observed. With increasing heat treatment, temperature the intensity of HA peaks increases as compared to that of as-treated sample. As the heat treatment temperature raised to 800°C, the slight decomposition of the HA to $Ca_3(PO_4)_2$ (TCP) can be observed. The amount of TCP increases after 900°C.

The relatively low (800°C) decomposition temperature could be due to diffusivity enhancement in HA because of Ti presence and/or effect of MA. It is reported [17] that the HA thermal decomposition occurs in two steps: dehydroxylation and decomposition. Dehydroxylation to oxyhydroxyapatite proceeds at temperatures in the range 850–900°C. The decomposition to TCP and TTCP occurs at temperatures greater than 900°C.

Also, since both the dehydroxylation and decomposition reactions include water vapor as a product, the rates at which these reactions proceed depend on the partial pressure of H_2O in the furnace. Therefore, the secondary phase formation during sintering could be suppressed by controlling the moisture content in the sintering atmosphere. The high

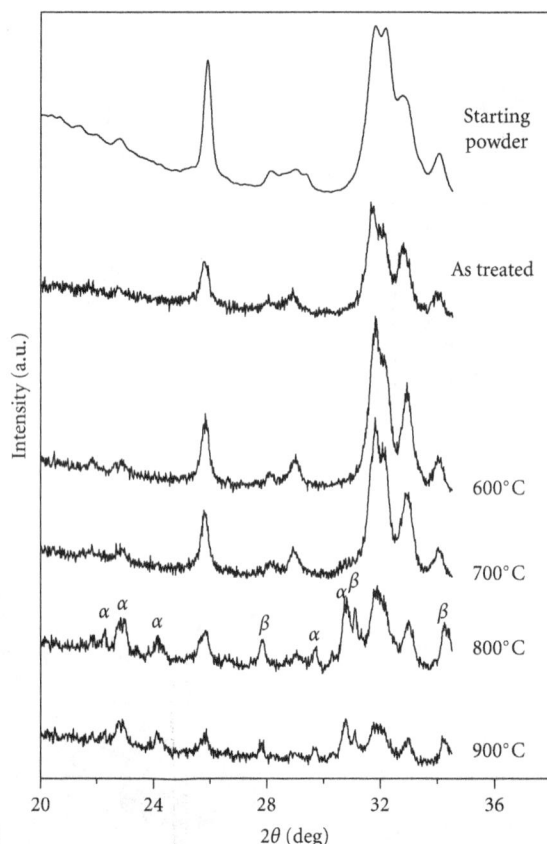

FIGURE 9: XRD patterns of HA deposited on Ti substrate after heated at various temperatures.

moisture content has the tendency to slow down the decomposition rate by preventing the dehydration of the OH group from the HA matrix. The difference in result in the present work with others could in part be also attributed to the difference in humidity in the sintering atmosphere and also to the nature of the deposited material (mechanical treated samples). Finally, the transformation from HAP to β-TCP phase may occurr at relatively low temperature because of the nanosize nature of HEBM-HAP coating, since fine particles are less stable for heat treatment.

According to [18], heat treatment of HA at 800°C for 5 h was sufficient for HA to β-TCP transformation. Also, it was shown [19] that, for Ti-doped samples (100 and 200 ppm of Ti), an endothermic peak, which corresponds to the HA → β-TCP transformation, occurs at 794°C. Note that no reactions were observed in the pure and 50 ppm Ti-doped HA samples. The formation of β-TCP is also a sensitive indicator for Ca/P ratio in HA composition.

Thus, the most likely explanation for the existence of β-TCP phase in the heated HEBM-synthesized HA-Ti sample is that the presence of Ti in the incubation media leads to the formation of a calcium-deficient apatite that has a Ca/P ratio less than 1.67. Calcium-deficient HA is less thermally stable than stoichiometric HA. The enhanced diffusivity is believed to be due to the introduction of structural defects such as grain boundaries and interfaces. The correlation between the enhanced diffusivity and the composite microstructure

is often reflected by the decrease of reaction temperature with refinement of the composite microstructure. It appears that the interface plays a major role in reducing the reaction temperature. With the reduction of layer thickness and the increase of interface area, the reaction becomes easier. When the reaction becomes easy enough, it can occur during milling, leading to mechanical alloying [20].

3.6. Morphology of the Annealed HA Coating on Ti Substrate. A set of backscattered electron images is shown in Figure 10, which illustrate the microstructure evolution of the HA surface during samples heat treatment. It can be seen that cracks are formed, while no any other microstructural changes can be detected. The presence of the cracks could be due to the difference in coefficients of thermal expansion of HA and Ti-alloy substrate. According to linear elastic fracture mechanics, a constrained film, subjected to stress, would crack when the strain energy released in the process exceeds the energy required to form the crack [21]. Also, cracks could form during cooling of annealed ceramic films deposited on a substrate due to differences between the thermal expansion coefficients of the substrate and the coated film. Finally, MA is an effective way of creating localized plastic deformation, so the cracks formation after heat treated also could be due to mechanical energy released.

Figure 11 illustrates a typical SEM view of cross-sections for samples heat treated under different temperatures.

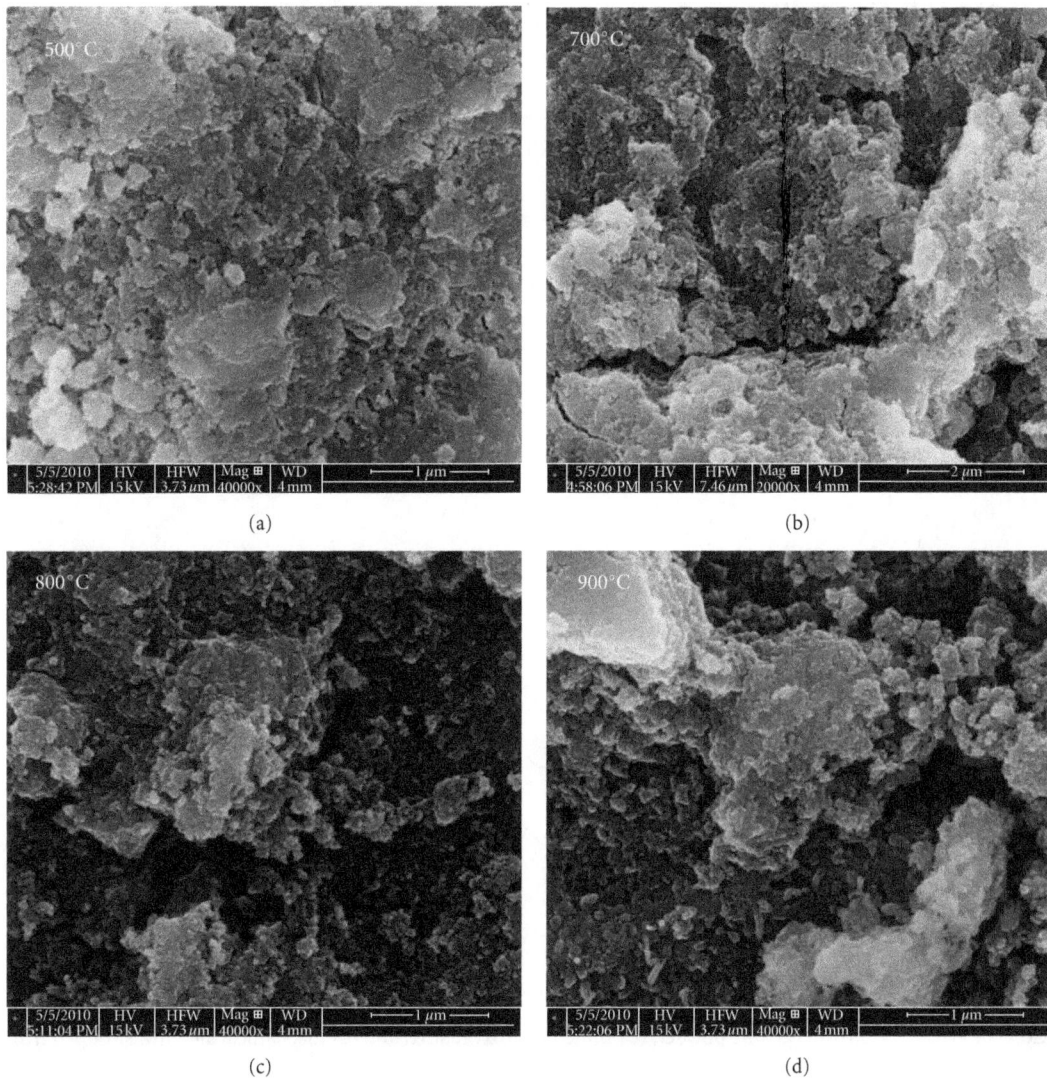

Figure 10: SEM of heat-treated hydroxyapatite coatings.

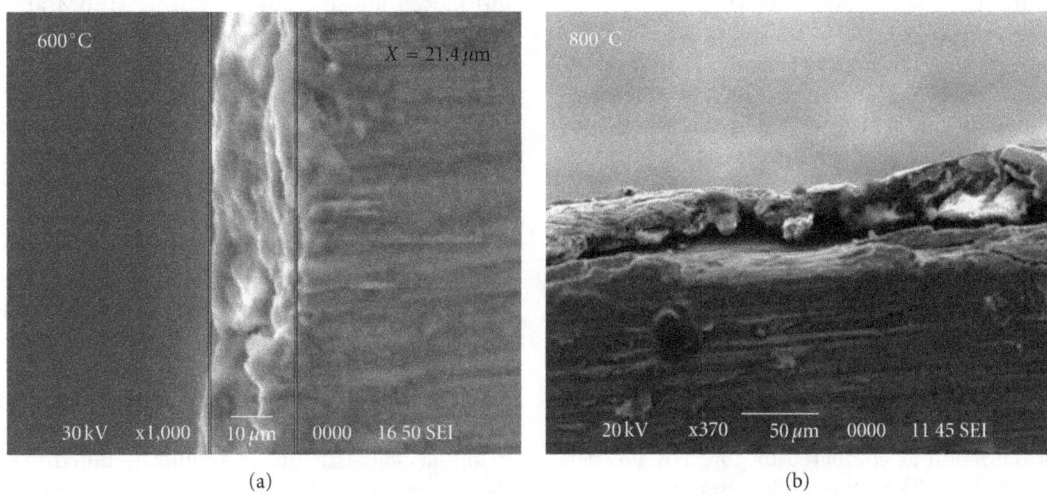

Figure 11: Cross-section microstructure of heat-treated HAe layer.

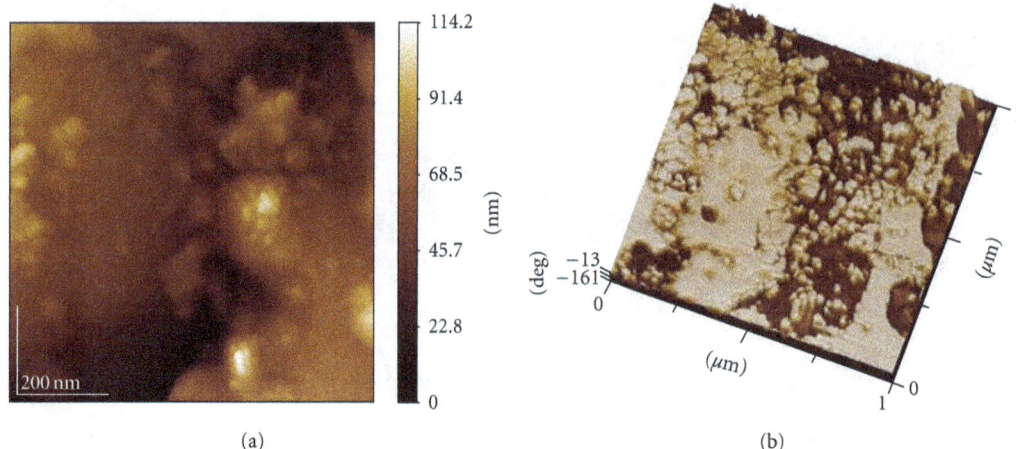

FIGURE 12: Surface structure analyzed by means of AFM of heat-treated hydroxyapatite at 700°C. (a) Topography and (b) Phase.

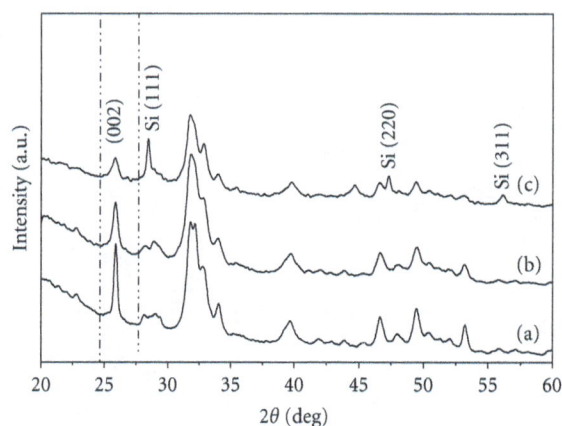

FIGURE 13: XRD patterns of (a) hydroxyapatite powder, (b) milled hydroxyapatite, and (c) mechanical alloying Si-doped hydroxyapatite.

The multilayer coatings can be readily seen after heat treatment at 600°C (Figure 11(a)), which could be due to repeated fracture. The coating layer was partially dissociated after 800°C (Figure 11(b)).

AFM observation of the coating (Figure 12) showed the presence of nanosized HA. This feature could not be resolved by SEM. The high specific area generated by this topography may be associated to the high bioactivity of this coating. According to the AFM measurements, the HA range of particle sizes for the annealed sample at 700°C is 40–70 nm.

3.7. X-Ray Diffraction Silicon-Doped Hydroxyapatite. Figures 13(a) and 13(b) shows the XRD patterns of the pure HA powder before and after milling for 1 h, while Figure 13(c) show such patterns for Si-doped HA powder. Significant changes were detected in the XRD patterns when after HEBM. The Si-doped sample shows a notable broadening and intensity reduction as compared to pure HA powders before and after milling. Thus, the phase crystallinity strongly decreases when silicon enters into the HA structure. The

same behavior was observed in [22] with Si content less than 2.41 weight percent. Due to the mechanical deformation introduced into the powder during HEBM and because Si enhances solubility, the particle and crystallite for HA phase could be refined.

Also, the (002) peak for SiHA sample (Figure 13) shows a notable broadening and intensity reduction comparing with pure HA. The wide peaks of the deposited materials should be due to very small crystallite size and microstrain. Issues about crystal structure changes in the HA with incorporation of silicon have been addressed by several experimental groups. Importantly it has been found that the incorporation of silicon into the crystal has little effect on the crystal structure. X-ray diffraction studies revealed that the intensity, width, and position of peaks for 0.4, 0.8, and 1.5 wt% silicon HA were very similar to those of phase-pure hydroxyapatite. Although there are no dramatic changes in the crystal structure, it should be noted that increasing the silicon content does produce a change in HA lattice parameters. Recent experimental work has shown that there

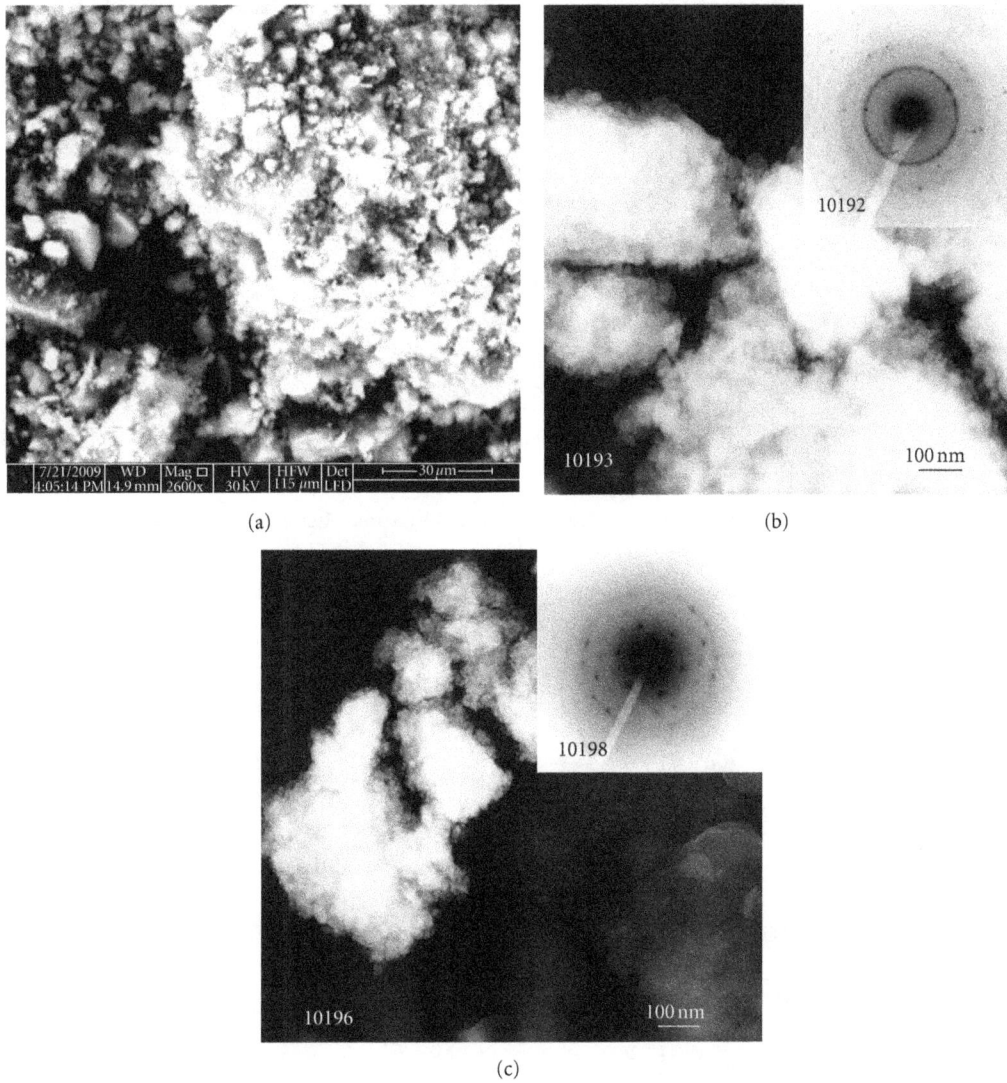

(a)

(b)

(c)

FIGURE 14: SEM (a) and TEM ((b), (c)) images Si-doped HA after 1 h of HEBM.

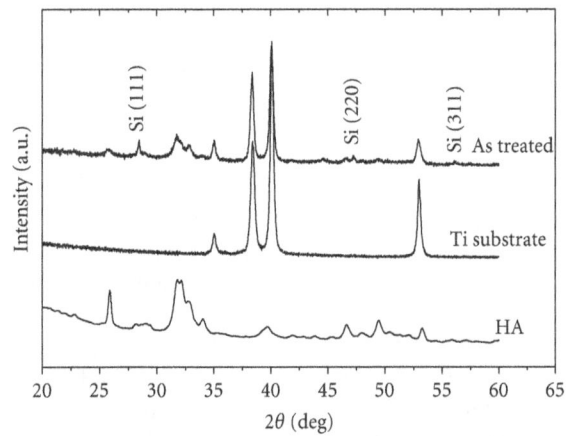

FIGURE 15: XRD patterns of the SiHA-coated Ti substrate.

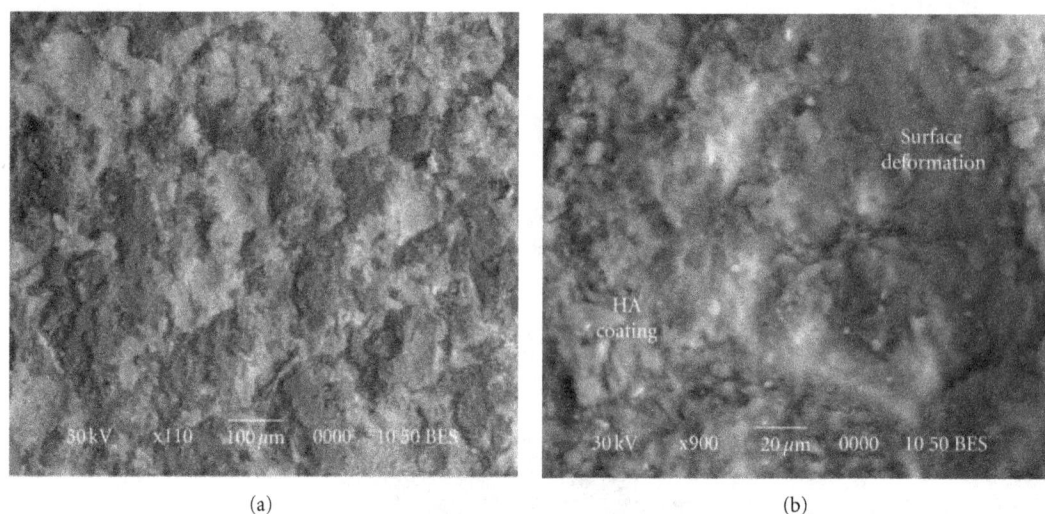

FIGURE 16: Backscattered electron images of the as-synthesized SiHA coating after one hour of mechanical treatment.

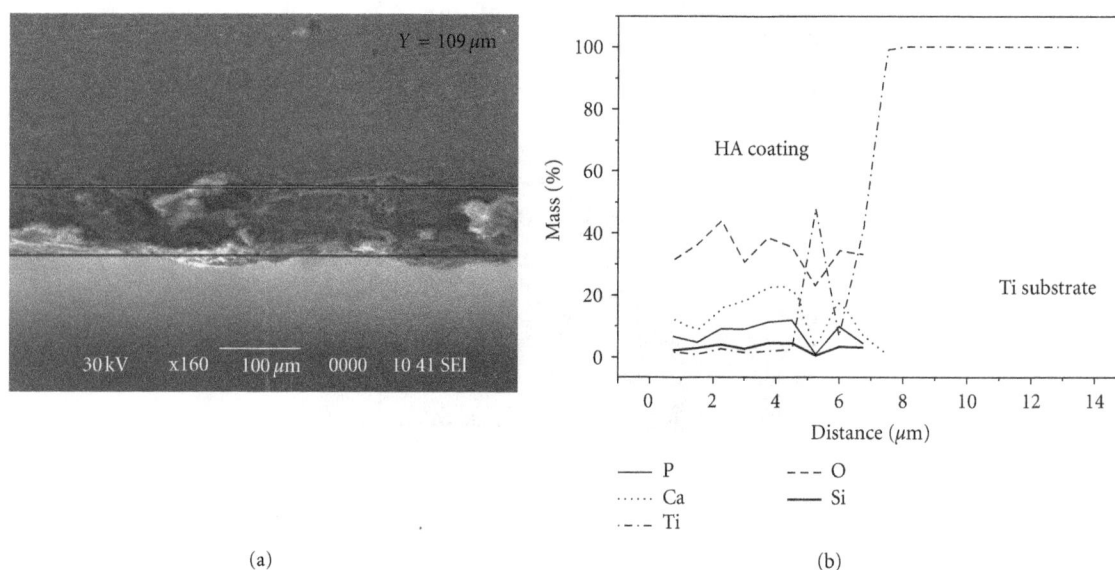

FIGURE 17: Structure of as-synthesized SiHA coating produced by MA method: (a) cross-section microstructure, (b) concentration profile.

is a systematic increase in the c-lattice parameter and a concomitant increase in cell volume with increasing silicon content. It is important to recognize that the silicate has a formal charge of −4, compared to −3 for the phosphate group. It was experimentally found that, for compensating of this excess of a negative charge, one hydroxyl group leaves the crystal for every silicon substitution [23].

3.8. Morphology of Silicon-Doped Hydroxyapatite. Morphology of the Si-dope HA powder was examined using SEM and TEM techniques. Cold welding and fracturing are the two essential processes involved in the mechanical milling/alloying process. After HEBM ($W_p : W_b = 1 : 40$) the parti-

cles morphology changes due to microforging, fracture, agglomeration, and deagglomeration. Thus the particle becomes smaller in size due to fracturing and agglomerated by cold welding through the milling process (Figure 14(a)). The TEM pictures of mechanically milled Si-dope HA sample are presented in Figures 14(b) and 14(c). It can be seen that a small round-shaped crystals with size of ∼25 nm are formed during milling process.

3.9. X-Ray Diffraction of Si-Doped HA Coating. XRD patterns of deposited HA and SiHA are presented in Figure 15. Coated SiHA on Ti substrate demonstrates the foundation of HA, as well as Si peaks. Comparison of these patterns with those obtained in [7, 8] shows the influence of high-energy

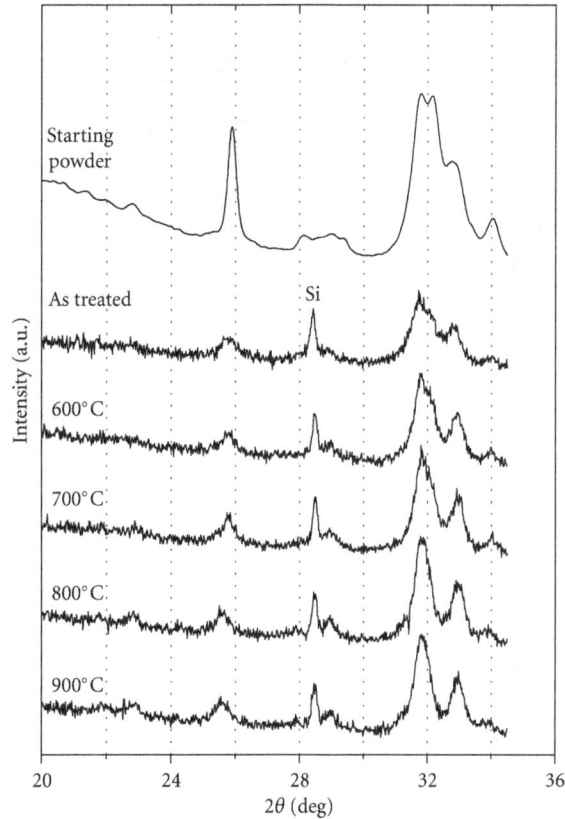

FIGURE 18: XRD patterns of SiHA deposited on Ti substrate after being heated at various temperatures (all peaks, except of shown Si peak, belong to HA phase).

ball milling process on the structure stability of the formed phases.

3.10. Morphology of Si-Doped HA Coating.

Backscattered electron images of the Si-doped HA surfaces "deposited" on Ti-alloy substrate are shown in Figure 16. After 1 hour of HEBM, the substrate was covered by HA layer in much more extent as compared to pure HA powder. This could be due to Si addition which improves the adhesion of HA to substrate. At higher magnification (Figure 16(b)), one can see the Ti-surface deformation and HA adhesion to the substrate. Figure 17(a) shows the microstructure of the cross-section of the Si-doped HA coating. The composition along the coated sample was analyzed by energy dispersive X-ray spectroscopy (EDS). Figure 17(b) shows the corresponding concentration profile at the boundary between the intermixing region and the coating. It can be concluded that cold welding between particles and substrate under repeated ball collisions led to the formation of a composite coating where the HA (Ca, P, and O elements) and Si phases flowed into the pores in Ti substrate. The homogeneous distribution of Si particles in the HA formed by HEBM improves the adhesion between coating and substrate.

3.11. X-Ray Diffraction of Heat Treatment Si-Doped HA Coating.

XRD patterns of Si-doped HA heat treated at different temperatures are presented in Figure 18. The changes in the XRD patterns give an indication of the influence of temperature on the structure stability of the samples. X-ray peaks of the formed phase were matched to the ICDD (JCPDS) standard. XRD pattern of the as-treated sample shows wide peaks owing to very small size of formed crystallites. After heat treatment, peaks shifted slightly to higher diffraction angles due to strain relaxation. It is important that in the whole range of heat treatment conditions (up to 900°C), Si-doped HA samples have not shown phase transition or decomposition. With increasing the heat treatment temperature, only the intensity of HA peaks increases due to recrystallization process.

3.12. Morphology of Heat Treatment Si-Doped HA Coating.

Figure 19 shows set of backscattered electron images, indicating the microstructures of coating surfaces obtained under various heat treatment conditions. It is worth noting that no any visible changes in microstructure were observed after treatment in the range of 600–800°C. The cracks are detected only after treatment at high temperature of ~900°C. However, the interface between HA and a metallic implant has been another matter of concern in terms of the mechanical properties and biocompatibility of the implant. In order for the HA coating to be effective and reliable, it must be strongly bonded to the metallic surface. The analysis of cross-section of heat-treated Si-doped HA samples (thickness $130\,\mu m$ at 700°C and $55\,\mu m$ at 900°C)

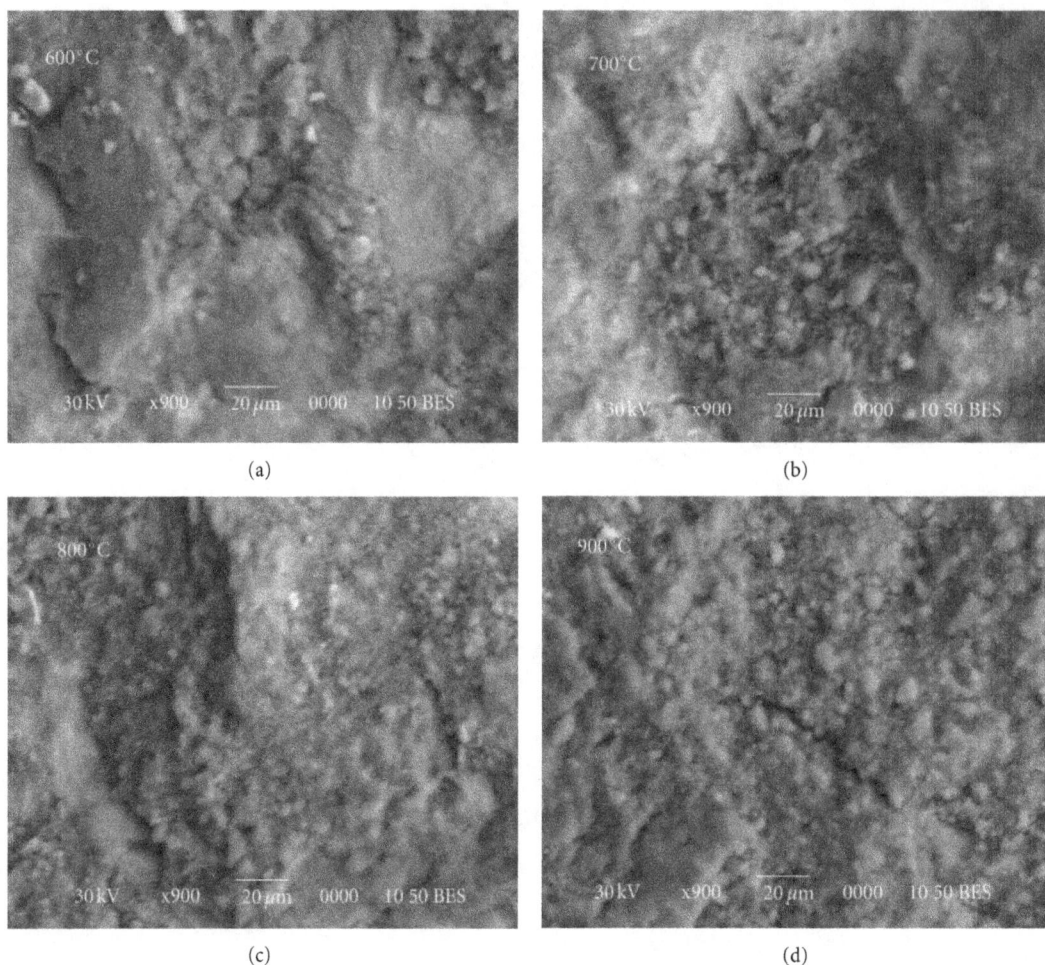

(a)

(b)

(c)

(d)

FIGURE 19: Backscattered electron images of the heat-treated SiHA coating by MA.

shows that the coatings possess good adhesion to the Ti substrates.

4. Conclusion

Thus, based on the detailed studies of using the high-energy ball milling technique for depositing ceramic coating on the metal substrate, a novel approach for production of hydroxyapatite (HA) films on titanium substrates has been developed. During HEBM process, the impacts of the milling balls activate the metal surface and lead to the robust cold welding of HA particles to the metal surface. It was shown that HEBM also results in significant decrease of the HA particle and crystalline size, forming the nanoscale structure. It was demonstrated that a heat treatment of the mechanically coated HA at 800°C for one hour leads to partial transformation of HA phase to β-TCP. It appears that the grain boundary and interface defects formed during HEBM reduce the transformation temperature. Also, it was shown that Ti incorporation into the HA structure not only causes the lattice shrinkage and reduction of its grain size as compared to pure HA, but also promotes the phase transformation of HA to TCP during heat treatment. It is important that doping

HA by silicon, while significantly decreases the crystal size of HA layer, it results in hindering of the phase transformation process. The Si-doped HA does not show phase transition or decomposition after heat treatment even at 900°C.

References

[1] S. Johnson, M. Haluska, R. J. Narayan, and R. L. Snyder, "In situ annealing of hydroxyapatite thin films," *Materials Science and Engineering C*, vol. 26, no. 8, pp. 1312–1316, 2006.

[2] O. Albayrak, O. El-Atwani, and S. Altintas, "Hydroxyapatite coating on titanium substrate by electrophoretic deposition method: effects of titanium dioxide inner layer on adhesion strength and hydroxyapatite decomposition," *Surface and Coatings Technology*, vol. 202, no. 11, pp. 2482–2487, 2008.

[3] K. A. Khor, C. S. Yip, and P. Cheang, "Ti-6Al-4V/hydroxyapatite composite coatings prepared by thermal spray techniques," *Journal of Thermal Spray Technology*, vol. 6, no. 1, pp. 109–115, 1997.

[4] C. K. Wang, J. H. Chern Lin, C. P. Ju, H. C. Ong, and R. P. H. Chang, "Structural characterization of pulsed laser-deposited hydroxyapatite film on titanium substrate," *Biomaterials*, vol. 18, no. 20, pp. 1331–1338, 1997.

[5] Y. Xiao, D. Li, H. Fan, X. Li, Z. Gu, and X. Zhang, "Preparation of nano-HA/PLA composite by modified-PLA for controlling

the growth of HA crystals," *Materials Letters*, vol. 61, no. 1, pp. 59–62, 2007.

[6] W. J. Shih, Y. F. Chen, M. C. Wang, and M. H. Hon, "Crystal growth and morphology of the nano-sized hydroxyapatite powders synthesized from $CaHPO_4 \cdot 2H_2O$ and $CaCO_3$ by hydrolysis method," *Journal of Crystal Growth*, vol. 270, no. 1-2, pp. 211–218, 2004.

[7] A. Hannora, A. Mamaeva, and Z. Mansurov, "X-ray investigation of Ti-doped hydroxyapatite coating by mechanical alloying," *Surface Review and Letters*, vol. 16, no. 5, pp. 781–786, 2009.

[8] A. Hannora, A. Mamaeva, N. Mofa, S. Aknazarov, and Z. Mansurov, "Formation of hydroxyapatite coating by mechanical alloying method," *Eurasian Chemico-Technological Journal*, vol. 11, no. 1, pp. 37–43, 2009.

[9] M. Lopez-A lvarez, E. L. Solla, P. Gonzalez et al., "Silicon-hydroxyapatite bioactive coatings (Si-HA) from diatomaceous earth and silica. Study of adhesion and proliferation of osteoblast-like cells," *Journal of Materials Science: Materials in Medicine*, vol. 20, no. 5, pp. 1131–1136, 2009.

[10] S. Zou, J. Huang, S. Best, and W. Bonfield, "Crystal imperfection studies of pure and silicon substituted hydroxyapatite using Raman and XRD," *Journal of Materials Science: Materials in Medicine*, vol. 16, no. 12, pp. 1143–1148, 2005.

[11] W. J. Shih, J. W. Wang, M. C. Wang, and M. H. Hon, "A study on the phase transformation of the nanosized hydroxyapatite synthesized by hydrolysis using in situ high temperature X-ray diffraction," *Materials Science and Engineering C*, vol. 26, no. 8, pp. 1434–1438, 2006.

[12] C. Suryanarayana, "Mechanical alloying and milling," *Progress in Materials Science*, vol. 46, no. 1-2, pp. 1–184, 2001.

[13] M. S. El-Eskandarany, K. Aoki, H. Itoh, and K. Suzuki, "Effect of ball-to-powder weight ratio on the amorphization reaction of Al50Ta50 by ball milling," *Journal of The Less-Common Metals*, vol. 169, no. 2, pp. 235–244, 1991.

[14] D. Krupa, J. Baszkiewicz, J. A. Kozubowski et al., "Effect of calcium and phosphorus ion implantation on the corrosion resistance and biocompatibility of titanium," *Bio-Medical Materials and Engineering*, vol. 14, no. 4, pp. 525–536, 2004.

[15] J. R. Díaz-Estrada, E. Camps, L. Escobar-Alarcón, and J. A. Ascencio, "Mechanical improvement of hydroxyapatite by TiOx nanoparticles deposition," *Journal of Materials Science*, vol. 42, no. 4, pp. 1360–1368, 2007.

[16] T. White, C. Ferraris, J. Kim, and S. Madhavi, "Apatite—an adaptive framework structure," *Reviews in Mineralogy and Geochemistry*, vol. 57, pp. 307–401, 2005.

[17] S. Ramesh, C. Y. Tan, I. Sopyan, M. Hamdi, and W. D. Teng, "Consolidation of nanocrystalline hydroxyapatite powder," *Science and Technology of Advanced Materials*, vol. 8, no. 1-2, pp. 124–130, 2007.

[18] I. Mayer, F. J. G. Cuisinier, S. Gdalya, and I. Popov, "TEM study of the morphology of Mn^{2+}-doped calcium hydroxyapatite and β-tricalcium phosphate," *Journal of Inorganic Biochemistry*, vol. 102, no. 2, pp. 311–317, 2008.

[19] C. C. Ribeiro, I. Gibson, and M. A. Barbosa, "The uptake of titanium ions by hydroxyapatite particles—structural changes and possible mechanisms," *Biomaterials*, vol. 27, no. 9, pp. 1749–1761, 2006.

[20] D. L. Zhang, "Processing of advanced materials using high-energy mechanical milling," *Progress in Materials Science*, vol. 49, no. 3-4, pp. 537–560, 2004.

[21] M. Mahé, J. M. Heintz, J. Rödel, and P. Reynders, "Cracking of titania nanocrystalline coatings," *Journal of the European Ceramic Society*, vol. 28, no. 10, pp. 2003–2010, 2008.

[22] S. Sprio, A. Tampieri, E. Landi et al., "Physico-chemical properties and solubility behaviour of multi-substituted hydroxyapatite powders containing silicon," *Materials Science and Engineering C*, vol. 28, no. 1, pp. 179–187, 2008.

[23] H. F. Chappell and P. D. Bristowe, "Density functional calculations of the properties of silicon-substituted hydroxyapatite," *Journal of Materials Science: Materials in Medicine*, vol. 18, no. 5, pp. 829–837, 2007.

12

Synthesis and *In Vitro* Antitumor Activity of Two Mixed-Ligand Oxovanadium(IV) Complexes of Schiff Base and Phenanthroline

Yongli Zhang,[1] **Xiangsheng Wang,**[1] **Wei Fang,**[1] **Xiaoyan Cai,**[1] **Fujiang Chu,**[2] **Xiangwen Liao,**[3] **and Jiazheng Lu**[3]

[1] *Department of Biology, School of Basic Courses, Guangdong Pharmaceutical University, Guangzhou, Guangdong 510006, China*
[2] *Guangdong Provincial Key Laboratory of Pharmaceutical Bioactive Substances, Guangzhou, Guangdong 510006, China*
[3] *Chemistry Department, School of Pharmacy, Guangdong Pharmaceutical University, Guangzhou, Guangdong 510006, China*

Correspondence should be addressed to Jiazheng Lu; lujia6812@163.com

Academic Editor: Takao Yagi

Two oxovanadium(IV) complexes of [VO(msatsc)(phen)], (1) (msatsc = methoxylsalicylaldehyde thiosemicarbazone, phen = phenanthroline) and its novel derivative [VO (4-chlorosatsc)(phen)], (2) (4-chlorosatsc = 4-chlorosalicylaldehyde thiosemicarbazone), have been synthesized and characterized by elemental analysis, IR, ES-MS, ^{1}H NMR, and magnetic susceptibility measurements. Their antitumor effects on BEL-7402, HUH-7, and HepG2 cells were studied by MTT assay. The antitumor biological mechanism of these two complexes was studied in BEL-7402 cells by cell cycle analysis, Hoechst 33342 staining, Annexin V-FITC/PI assay, and detection of mitochondrial membrane potential ($\Delta\Psi$m). The results showed that the growth of cancer cells was inhibited significantly, and complexes **1** and **2** mainly caused in BEL-7402 cells G0/G1 cell cycle arrest and induced apoptosis. Both **1** and **2** decreased significantly the $\Delta\Psi$m, causing the depolarization of the mitochondrial membrane. Complex **2** showed greater antitumor efficiency than that of complex **1**.

1. Introduction

Schiff bases are an important class of ligands because such ligands and their transition metal complexes have a variety of applications including biological, clinical, and analytical applications [1]. The development of the field of bioinorganic chemistry has increased the interest in Schiff base complexes, because it has been recognized that N and S atoms play a key role in the coordination of transition metals at the active sites of many metallobiomolecules [2, 3]. The importance of metal ions in biological systems is well established. One of the most interesting features of metal-coordinated systems is the concerted spatial arrangement of the ligands around the metal ion [3–5].

Among the various transition metal ions used in pharmacological studies, Vanadium and its derivatives have been reported to display different biological effects including antitumor, antimicrobial, antihyperlipidemia, antihypertension, antiobesity, enhancement of oxygen affinity of hemoglobin and myoglobin, insulin-enhancing effects, and so on [6–8]. Vanadium complexes have also been explored for lowering of glucose levels [9–12], diuretic and natriuretic effects, antitumor activity against chemical carcinogenesis in animals and malignant cell lines (*in vitro*). Much effort has been done for vanadyl species coordinated to organic ligands on the research of their mimetic effects in hopes of developing vanadodrugs [13–15].

Because V(IV) complexes have no charge, they are perceived to be candidates for easy bioabsorption. On the other hand, vanadyl(IV) complexes incorporating thiosemicarbazones have been studied extensively for their insulin-like effects which result in the inhibition of glycerol release and enhancement of glucose uptake by rat adipocytes and have been used in the treatment of tuberculosis [16–20]. In view of inquisitive response of oxovanadium(IV) in biology, we have reported that four oxidovanadium(IV) complexes present highly cytotoxic activities against Myeloma cell (Ag8.653) and Gliomas cell (U251) lines [21]. To continue our

Synthesis and In Vitro Antitumor Activity of Two Mixed-Ligand Oxovanadium(IV) Complexes of Schiff Base and Phenanthroline

103

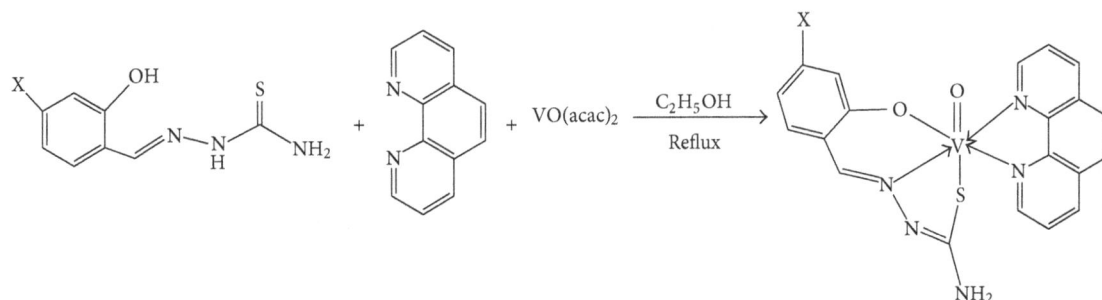

SCHEME 1: Synthesis of [VO(msatsc)(phen)] 1, X = –OCH$_3$ and [VO(4-chlorosatsc)] 2, X = Cl.

research in this project, in the present paper, an oxovanadium complex [VO(msatsc)(phen)] **1** (msatsc = methoxylsalicylaldehyde thiosemicarbazone, phen = phenanthroline) and its novel derivative [VO (4-chlorosatsc)(phen)] and (**2**) (4-chlorosatsc = 4-chlorosalicylaldehyde thiosemicarbazone) (Scheme 1) **2** have been synthesized and characterized. Their antitumor effects on BEL-7402 human liver (Bel7402), HUH-7, HepG2 cells were studied by MTT assay. The reported compounds may be an addition of new class of compounds as the metal-based drugs.

2. Experimental

2.1. Materials and Physical Measurements.
VO(acac)$_2$ (acac = acetylacetonate) and 1,10-phenanthroline (phen) were commercially available and used as received. DMSO, CHCl$_3$ were purchased from Aldrich (USA). Other chemicals and reagents of analytical grade were obtained commercially without further purification unless specifically mentioned.

BEL-7402, HUH-7, and HepG2 cell lines were purchased from The Cell Bank of Type Culture Collection of Chinese Academy of Sciences (Shanghai, China). RPMI 1640 medium was purchased from Hyclone (Logan, USA), Trypsin, fetal calf serum and Annexin V-FITC/PI apoptosis detection Kit were purchased from GIBCO Company (USA). Hoechst 33342 staining solution was purchased from Beyotime Institute of Biotechnology (China). MTT and Rhodamine123 were purchased from Sigma Company (USA).

Microanalysis (C, H, and N) was carried out with a Perkin-Elmer 240Q elemental analyzer. Electrospray mass spectra (ES-MS) were recorded on a LCQ system (Finnigan MAT, USA) using methanol as mobile phase. ^{1}H NMR spectra were recorded on a Varian-500 spectrometer. All chemical shifts are given relative to tetramethylsilane (TMS). Infrared spectra were recorded on a Bomen FTIR model MB102 instrument using KBr pellets. UV-Vis spectra were recorded on a Schimadzu UV-3101 PC spectrophotometer at room temperature. Emission spectra were recorded on a Perkin-Elmer Lambda 55 spectrofluorophotometer. Magnetic susceptibility measurements were recorded on a MPMSXL-7 (Quantum Design, USA), at room temperature. Solutions of compounds were freshly prepared 2 h prior to biochemical evaluation.

Cell viability assay was carried out with a microplate reader (Model 680 Microplat Reader, BIO-RAD, USA). Cell cycle analysis, Annexin V-FITC/PI assay of apoptotic cells and detection of mitochondrial membrane potential were recorded on a FACScan flow cytometer (BD FACSCaliburTM, USA). Fluorescence microscopy of Apoptosis assays was carried out with the fluorescence microscope (Olympus OX31, Olympus Corporation, Japan).

2.2. Synthesis and Characterization.
Methoxylsalicylaldehyde thiosemicarbazone (msatsc) and 4-chlorosalicylaldehyde thiosemicarbazone (4-chlorosatsc) were prepared with a method similar to that described earlier [19–21]. An equimolar methanolic solution of desired thiosemicarbazide (0.0182 g, 10 mmol) and corresponding methoxylsalicylaldehyde (0.0248 g, 10 mmol) or 4-chlorosalicylaldehyde (0.0560 g, 10 mmol) was refluxed for 3 h and then the precipitates were filtered off, washed with methanol and dried under vacuum. The products were recrystallized in ethanol. Msatsc: White solid. Yield: 80%. Anal. Calcd. for C$_9$H$_{11}$N$_3$O$_2$S: C, 48.01; H, 4.89; N, 18.67; S, 14.22%; Found: C, 47.89; H, 4.69; N, 18.53; S, 14.11%. ^{1}H NMR (500 MHz; DMSO-d$_6$; δ, ppm; s, singlet; d, doublet; t, triplet; m, multiplet): 11.34 (s, 2H, –NH), 9.88 (s, 1H, H–C=N), 7.21 (t, 1H, J = 7.1 Hz, –ph), 6.86 (d, 1H, J = 7.7 Hz, –ph), 6.81 (d, 1H, J = 7.6 Hz, –ph), 6.79 (t, 1H, J = 7.6 Hz, –ph), 3.53 (d, 3H, J = 7.5 Hz, –OCH$_3$). ES-MS (CH$_3$OH): m/z 226.0 ([M+H]$^+$). UV λ_{max}, nm (ε, M^{-1} cm^{-1}) in DMSO: 340 (26310). 4-chlorosatsc: White solid. Yield: 83%. Anal. Calcd. for C$_8$H$_8$N$_3$OSCl: C, 42.01; H, 3.49; N, 18.38; S, 14.00%; Found: C, 41.85; H, 3.24; N, 18.13; S, 13.92%. ^{1}H NMR (500 MHz; DMSO-d$_6$; δ, ppm): 11.54 (s, 2H, –NH), 9.97 (s, 1H, H–C=N), 8.16 (s, 1H, –ph), 7.75 (s, 1H, –ph) 6.86 (d, 1H, J = 7.7 Hz, –ph),. ES-MS (CH$_3$OH): m/z 230.5 (M$^+$). UV λ_{max}, nm (ε, M^{-1} cm^{-1}) in DMSO: 346 (18655).

2.2.1. [VO(msatsc)(phen)].
A mixture of methoxylsalicylaldehyde thiosemicarbazone (msatsc) (0.0844 g, 0.375 mmol) and 1,10-phennanthroline·H$_2$O (phen·H$_2$O) (0.0743 g, 0.375 mmol) in absolute alcohol (20 cm^3) were heated at 72°C under argon for 30 min. After dissolution, VO(acac)$_2$ (0.1000 g, 0.375 mmol) was added and the brown red precipitate was formed immediately. This suspension was kept stirring under reflux for about 3.5 h and then the coloured solid formed was filtered off from the hot solution, washed with mixed solvents of ethanol and ether for three times and dried under vacuum. Yield: 90%. Anal. Calcd for

TABLE 1: Selected IR data for complexes and their corresponding ligands $\nu(cm^{-1})$.

Compound	$\nu(C=S)$	$\nu(C-S)$	$\nu(O-H)$	$\nu(C=N)$	$\nu(V=O)$	$\nu(V-O)$	$\nu(N-H)$
VO(msatsc)(phen)	—	769	—	1625	965	624	3289
Msatsc	859	—	3467	1614	—	—	3356
VO(4-cholrobrsatsc) (phen)	—	767	—	1624	940	598	3284
4-Cholrobrsatsc	829	—	3443	1603	—	—	3319

$C_{21}H_{17}N_5O_3SV$: C, 53.63; H, 3.62; N, 14.89; S, 6.81%. Found: C, 53.54; H, 3.56; N, 14.63; S, 6.64%. ^1H NMR (500 MHz; DMSO-d_6; δ, ppm): 9.11 (d, 2H, J = 8.9 Hz, –NH), 8.51 (t, 1H, J = 9.5 Hz, H–C=N), 7.97 (d, 6H, J = 7.5 Hz, –phen), 7.78 (d, 2H, J = 7.8 Hz, –ph), 7.22 (t, 1H, J = 3.9 Hz, –ph), 6.86 (d, 1H, J = 7.7 Hz, –ph), 6.82 (d, 1H, J = 7.7 Hz, –ph), 6.77 (t, 1H, J = 7.6 Hz, –ph), 3.65 (d, 3H, J = 7.5 Hz, –OCH$_3$). ES-MS (CH$_3$OH): m/z 470.9 ([M+H]$^+$). UV λ_{max}, nm (ε, M^{-1} cm^{-1}) in DMSO: 265 (41770), 340 (15255), 383 (6525), 760 (20), 797 (20). Magnetic moment: μ_{eff}: 1.70 BM. Molar conductance Ω_M (Ω^{-1} cm^2 mol^{-1}): 9.65.

2.2.2. [VO (4-chlorosatsc)(phen)].

This complex was synthesized with the same method described for **1**. Yield: 89%. Anal. Calcd for $C_{20}H_{14}N_5O_2ClSV$: C, 50.59; H, 2.95; N, 14.76; S, 6.75% Found: C, 50.45; H, 2.68; N, 14.53 S, 6.44%. ^1H NMR (500 MHz; DMSO-d_6; δ, ppm): 9.11 (d, 2H, J = 8.8 Hz, –NH), 8.51 (d, 1H, J = 8.1 Hz, H–C=N), 8.01 (s, 6H, –phen), 7.79 (d, 2H, J = 7.8 Hz, –ph), 7.78 (d, 3H, J = 7.9 Hz, –ph). ES-MS (CH$_3$OH): m/z 475.5 (M$^+$). UV λ_{max}, nm (ε, M^{-1} cm^{-1}) in DMSO: 264 (41015), 347 (10975), 400 (7815), 786 (20). Magnetic moment: μ_{eff}: 1.73 BM. Molar conductance Ω_M (Ω^{-1} cm^2 mol^{-1}): 10.56.

2.3. In Vitro Antitumor Activity

2.3.1. Cell Culture.

The cells were routinely cultured in RPMI-1640 medium, supplemented with 10% fetal calf serum. The culture was maintained at 37°C with a gas mixture of 5% CO$_2$/95% air. The medium was changed every two days and the cells were subcultured every three days.

2.3.2. Cell Viability Assay.

Cell viability was determined using the MTT assay. Briefly, the cells were collected and resuspended in RPMI1640 medium at 4 × 10^4 cells/mL, 100 μL aliquots were added to each well of 96-well flat-bottomed microtiter plates, followed by addition of 100 μL of the complexes **1** and **2**. Three replicate wells were used for each data point in the experiments. After incubation for the indicated intervals, 20 μL of MTT (5 mg/mL in PBS) solution was added to each well and plates were then incubated for 4 h at 37°C. The medium with MTT was removed from the wells. Intracellular formazan crystals were dissolved by adding 150 μL of DMSO to each well, and the plates were shaken for 10 min. The absorbance was read at 490 nm with a microplate reader. Percentage of survival was calculated as a fraction of the negative control (medium only). The half-maximal inhibitory concentration (IC$_{50}$) was obtained from the dose-response curve with an original 6.0 software.

2.3.3. Cell Cycle Analysis.

Analysis of the cell cycle of control and treated cancer cells was determined. Using standard methods, the DNA of cells was stained with PI, and the proportion of non-apoptotic cells in different phases of the cell cycle was recorded. The cancer cells were treated with complexes **1** and **2**, harvested by centrifugation at 1000 ×g for 5 min, and then washed with ice-cold PBS. The collected cells were fixed overnight with cold 70% ethanol, and then stained with PI solution consisting of 50 μg/mL PI, 10 μg/mL RNase. After 10 min incubation at room temperature in the dark, fluorescence-activated cells were sorted in a FACScan flow cytometer using CellQuest 3.0.1 software.

2.3.4. Fluorescence Microscopy of Apoptosis Assays.

This method was modified from a previous report [18, 22]. Briefly, after exposed to complexes **1** and **2** for 48 h, BEL-7402 cells were washed twice with PBS, then stained with 10 μg/mL Hoechst 33342 staining solution at 37°C for 30 min according to the manufacturer's instructions. Finally, the cells were observed under the fluorescence microscope.

2.3.5. Annexin V-FITC/PI Assay of Apoptotic Cells.

BEL-7402 cells treated with complexes **1** and **2** for 48 h were determined by flow cytometry using a commercially available Annexin V-FITC/PI apoptosis detection Kit. After treatment, cells were harvested and washed twice in ice-cold PBS and resuspended in 500 μL of binding buffer at 1–5 × 10^5 cells/mL. The samples were incubated with 5 μL of Annexin V-FITC and 5 μL propidium iodide in the dark for 15 min at room temperature. Finally, samples were analyzed by flow cytometry and evaluated based on the percentage of cells for Annexin V positive.

2.3.6. Detection of Mitochondrial Membrane Potential ($\Delta\Psi m$).

In this study, $\Delta\Psi m$ was measured by using Rhodamine123, and 5-Fluorouracil (5-FU, 50 μM) is the positive control. Treated with complexes **1** and **2** for 48 h, BEL-7402 cells were incubated with Rhodamine123 (2 μM/mL) at 37°C for 30 min, and washed with PBS. The cell pellets were collected by centrifugation (1000 ×g, 5 min) and resuspended in 500 μL PBS. Green fluorescence intensities of Rhodamine 123 in cells were analyzed by FL-1 channel of the flow cytometer.

2.4. Statistical Analysis.

Data were expressed as the mean ± SD from these independent experiments. Statistic analysis was performed using the SPSS 13.0 for Windows. Comparisons between two groups were performed by unpaired t-test. Multiple comparisons between more than two groups were performed by one-way analysis of variance (ANOVA). Significance was accepted at P value lower than 0.05.

Synthesis and In Vitro Antitumor Activity of Two Mixed-Ligand Oxovanadium(IV) Complexes of Schiff Base and Phenanthroline

105

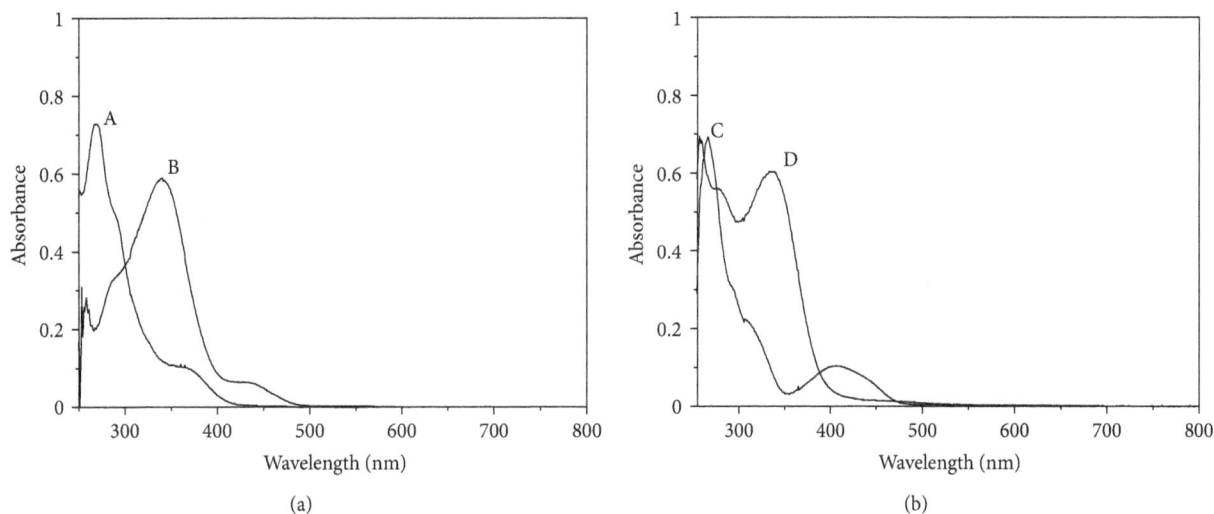

FIGURE 1: Absorption spectra of VO(msatsc)(phen) (A) and its ligand salsem (B) in (a) and VO(4-cholrobrsatsc)(*phen*) (C) and its ligand 4-cholrobrsatsc (D) in (b), respectively.

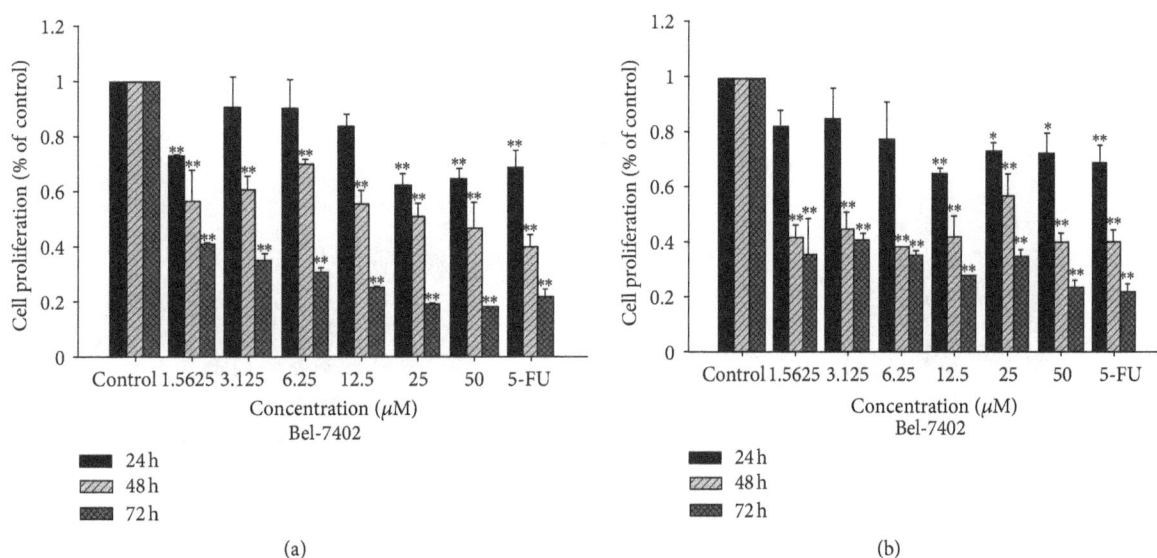

FIGURE 2: Antiproliferative activity of complexes 1 (a) and 2 (b) detected by MTT assay after 24, 48, and 72 h of treatment on BEL-7402 cells.

3. Results and Discussion

3.1. Complex Characterization. Oxovanadium(IV) complexes of formulation [VO(msatsc)(phen)] 1 (msatsc = methoxylsalicylaldehyde thiosemicarbazone, phen = phenanthroline) and its novel derivative [VO(4-chlorosatsc)(phen)] (4-chlorosatsc = 4-chlorosalicylaldehyde thiosemicarbazone) 2 are prepared in high yield from a general synthetic procedure in which vanadylacetylacetonate is reacted with the ligands in ethanol (Scheme 1).

To study the binding mode of the ligand to vanadium in the new complexes, IR spectra of the free ligands were compared with the spectra of the vanadium complexes. Selected IR data for complexes and their metal free ligands were given in Table 1. For the ligands, characteristic stretching

vibration bands appear at 859 cm^{-1} and/or 829 cm^{-1} corresponding to the C=S vibration. A strong band is observed in the free ligands at 1614 cm^{-1} or 1603 cm^{-1}, characteristic of the imine (C=N) [19, 23–25]. In the spectra of new complexes, the band due to v(C=N) showed a red shift to 1625 or 1624 cm^{-1}, indicating coordination of the nitrogen to vanadium (Table 1) [26, 27]. Medium intensity band, at 3467 and 3443 cm^{-1} in the free ligands due to $v_{(OH)}$, was absent in the complexes, indicating deprotonation of the Schiff base prior to coordination [25, 27]. In addition, the complexes exhibit the characteristic v(V=O) bands at 965 and 940 cm^{-1}, and v(V–O) bands at 624 and 598 cm^{-1} for complex 1 and 2, respectively.

The ^{1}H NMR spectra of complexes 1 and 2 are in excellent agreement with the proposed structures. In the ^{1}H

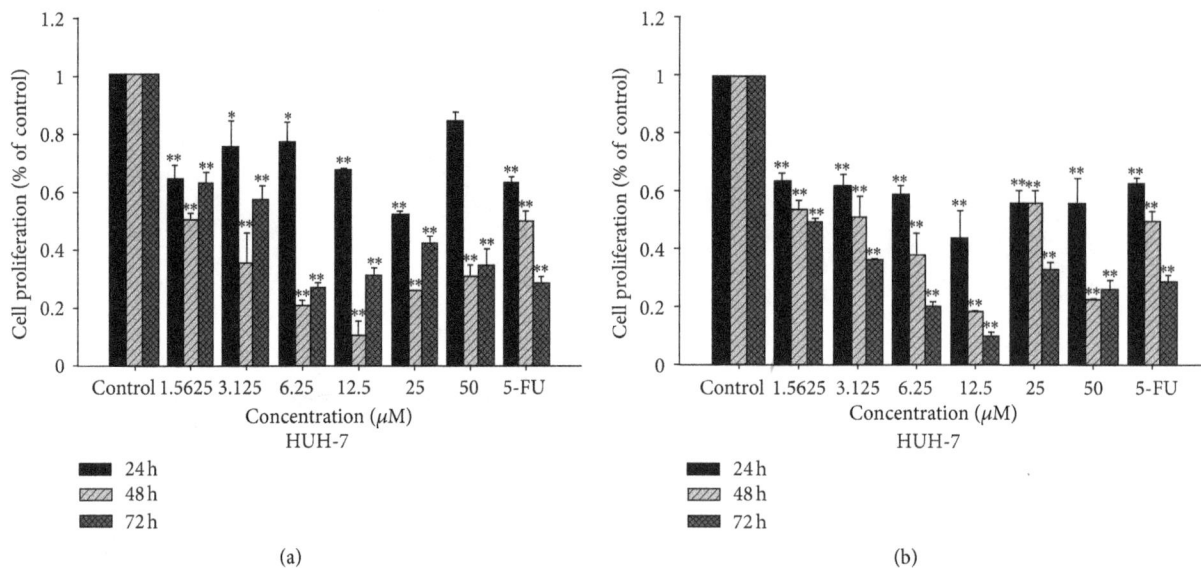

FIGURE 3: Antiproliferative activity of complexes **1** (a) and **2** (b) detected by MTT assay after 24, 48, and 72 h of treatment on HUH-7 cells.

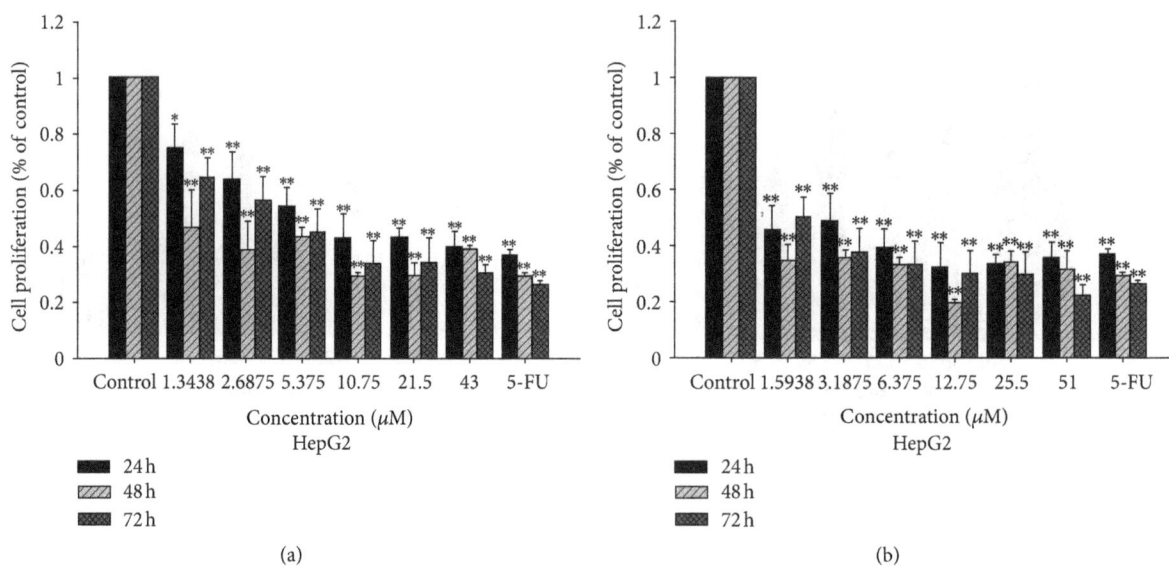

FIGURE 4: Antiproliferative activity of complexes **1** (a) and **2** (b) detected by MTT assay after 24, 48, and 72 h of treatment on HepG2 cells.

NMR spectra of the two complexes, the chemical shifts of the phenolic hydroxy protons of 11.34 ppm and 11.54 ppm, and amino protons linking directly to the imine groups of 9.88 ppm and 9.97 ppm for complexes **1** and **2**, respectively, in comparison with the metal-free ligands are unobserved. These facts also affirm that the free ligands are coordinated to metal.

The electronic spectra of the two complexes and their ligands were shown in Figure 1. The complexes show an intense band at ca. 265 nm assignable to π-π^* transitions of aromatic rings of phenanthroline [18–20, 28]. A medium band is observed near 400 nm, which is attributed to a ligand-to-metal charge-transfer transition (LMCT) as a charge transfer from a p-orbital on the lone-pair of ligands oxygen atoms

to the empty d-orbital of the vanadium atom [27, 29]. The remaining bands appearing in the UV-region (320–350 nm) are assignable to the intraligand transitions of the Schiff base [18, 26–28]. Complexes of oxovanadium(IV) with coordination numbers 5 and 6 are usually square pyramidal/trigonal bipyramidal and distorted octahedral, respectively [19, 21, 29]. From the above obtained spectral data, it indicates that the Schiff bases bonded through the phenolate oxygen, imine nitrogen, and thiolate sulfur atoms leaving the thiomethyl as the pendant group. This implies that complexes **1** and **2** bear the central V (IV) atom in a square-pyramidal geometry [18, 29, 30].

The complexes are one-electron paramagnetic giving a magnetic moment value of ~1.70 BM at room temperature.

Synthesis and In Vitro Antitumor Activity of Two Mixed-Ligand Oxovanadium(IV) Complexes of Schiff Base and Phenanthroline

107

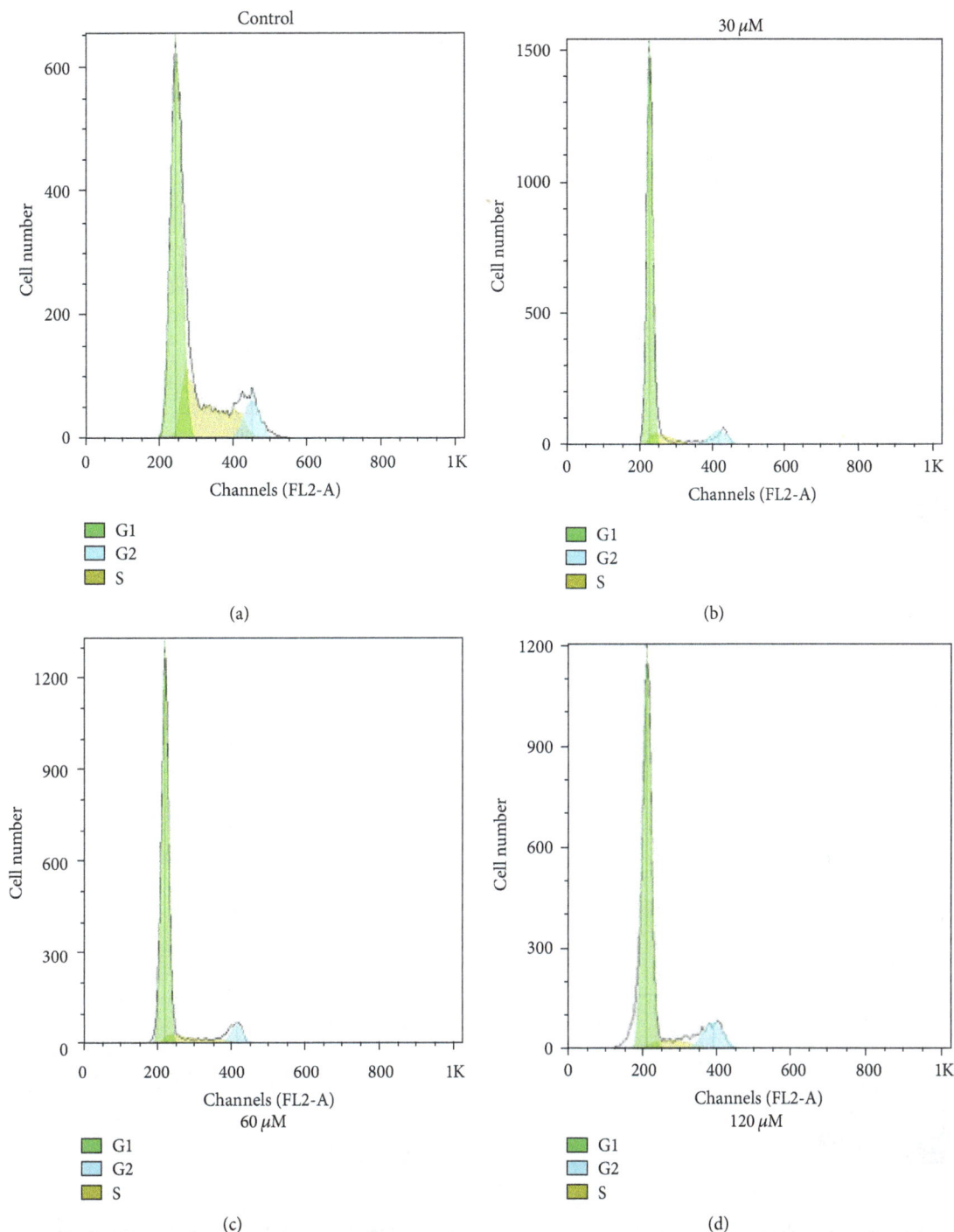

FIGURE 5: DNA content and cell cycle analysis of BEL-7402 cells after complex **1** treatment. BEL-7402 cells were cultured with either 0.1% DMSO (control), 30 μM, 60 Mm, and 120 μM of complex **1** for 48 h. The percentage of nonapoptotic cells within each cell cycle was determined by flow cytometry.

These values of magnetic susceptibility also confirm that the vanadium complexes are in the V (IV) state, with d^1 configuration [18, 19, 30, 31]. The molar conductance values of the two complexes in DMF indicated that these two oxidovanadium complexes show nonelectrolytic nature.

In addition, the assignments of the two complexes were made on the basis of elemental analyses and mass spectral data, confirming the proposed structures. The molecular ion peaks of complexes at m/z 441.1 and 475.6, for complexes **1** and **2**, respectively, were obtained by ESI-MS.

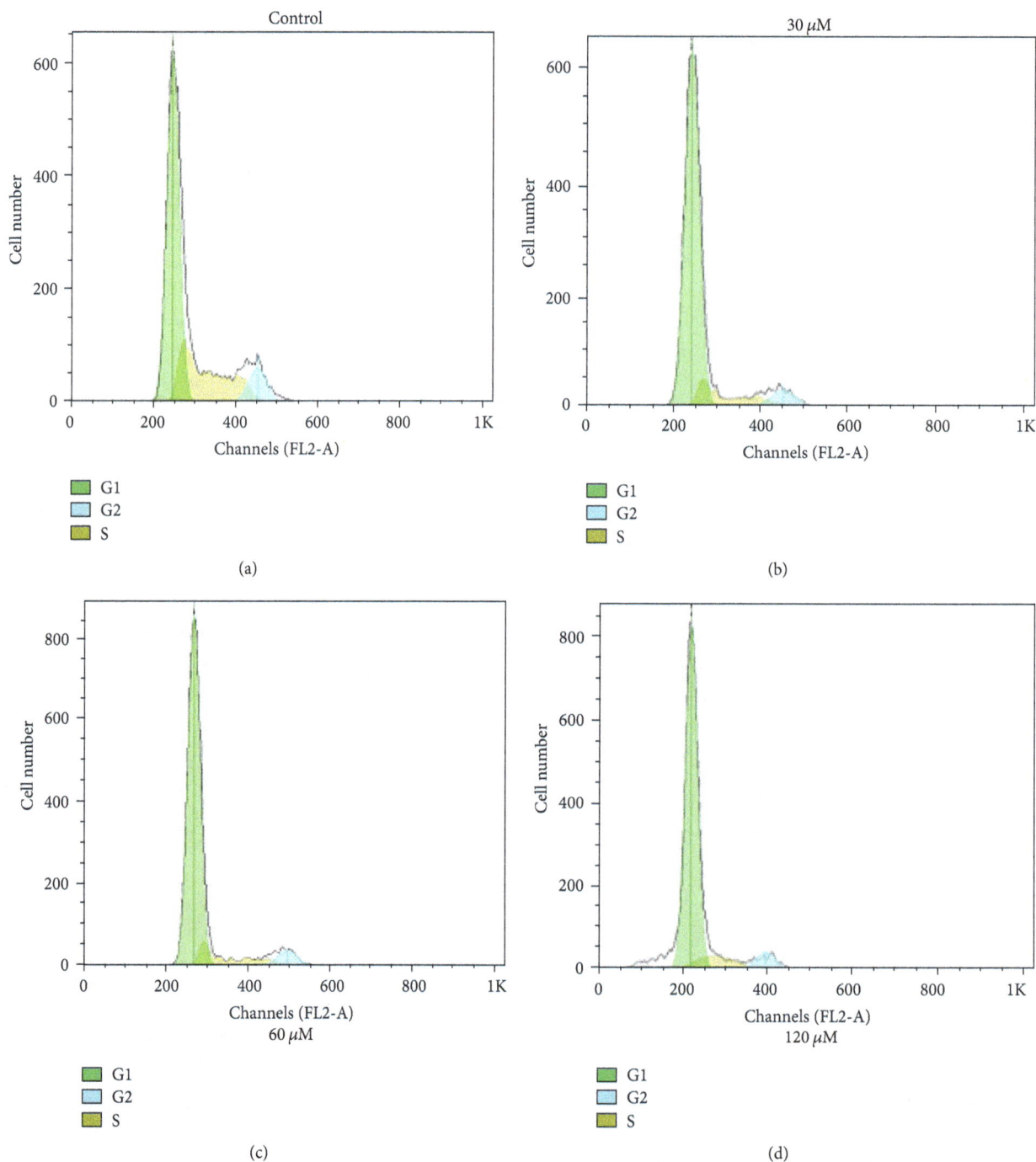

FIGURE 6: DNA content and cell cycle analysis of BEL-7402 cells after complex **2** treatment. BEL-7402 cells were cultured with either 0.1% DMSO (control), 30 μM, 60 μM, and 120 μM of complex **2** for 48 h. The percentage of nonapoptotic cells within each cell cycle was determined by flow cytometry.

3.2. In Vitro Antitumor Studies

3.2.1. Antiproliferative Effect of the Complexes 1 and 2 on BEL-7402, HUH-7, and HepG2 Hepatoma Cells.
Complexes **1** and **2** were evaluated for their ability to inhibit the growth of BEL-7402, HUH-7, and HepG2 human hepatoma cell lines using MTT assay. The inhibition was expressed as cell viability

relative to control without **1** and **2** treatments. In the present study, BEL-7402, HUH-7, and HepG2 human hepatoma cells were used which have been recently characterized as a suitable model for *in vitro* assessment of hepatoma toxicity [32, 33]. And 5-Fluorouracil (5-FU, 30 μM) was used as a positive control, which has been used extensively as an efficient anticancer drug in clinical trials [32–34]. After

Synthesis and In Vitro Antitumor Activity of Two Mixed-Ligand Oxovanadium(IV) Complexes of Schiff Base and Phenanthroline

109

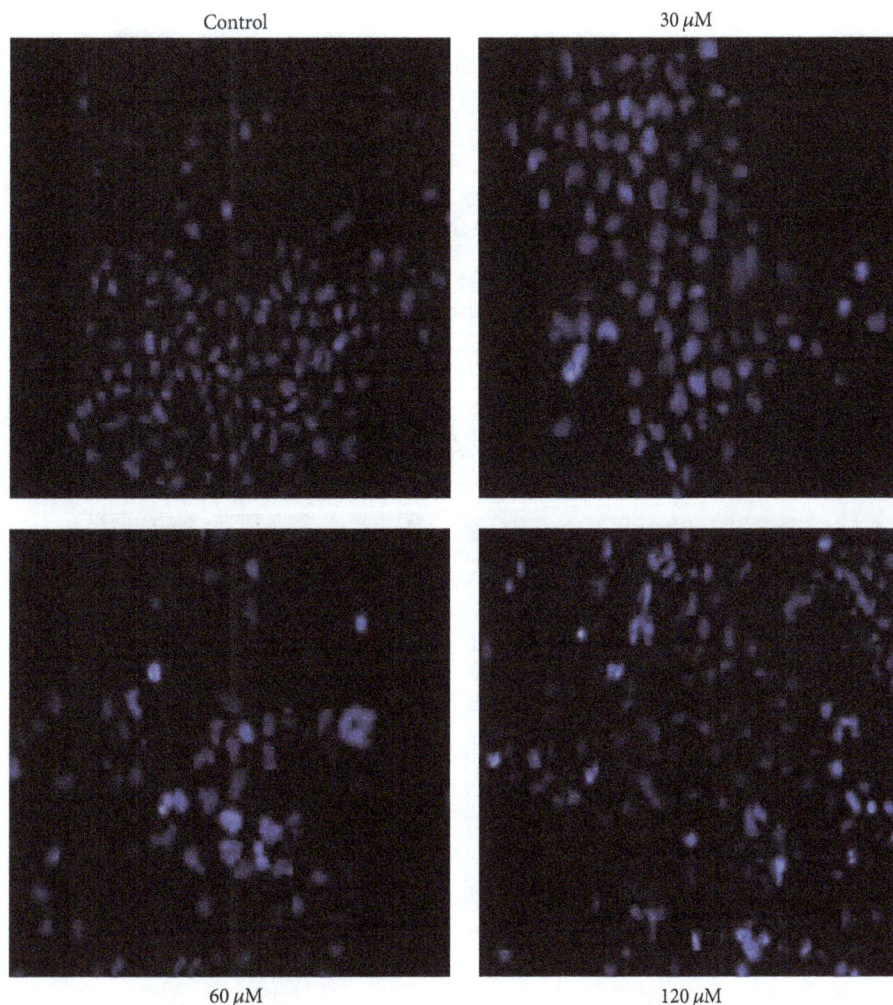

FIGURE 7: Effects of complex **1** on the morphology of Bel-7402 cells were assayed by Hoechst 33342 staining. After treatment with complex **1** for 48 h, apoptotic cells were detected by Hoechst 33342 staining and examined by fluorescence microscopy (original magnification 400x).

TABLE 2: Comparison of IC_{50} values obtained from the MTT assay on BEL-7402, HUH-7, and HepG2 hepatoma cell lines after treated for 24 h, 48 h, and 72 h using the complexes **1** and **2**.

Cell line	Subject	IC_{50} values obtained from the MTT assay (μM)		
		24 h	48 h	72 h
BEL-7402	1	294.30 ± 95.74	30.80 ± 13.05	0.51 ± 0.21
	2	166.32 ± 20.48	17.02 ± 3.69	0.90 ± 0.13
HUH-7	1	69.91 ± 11.08	2.87 ± 0.23	3.12 ± 0.39
	2	79.16 ± 25.18	1.98 ± 0.72	0.28 ± 0.09
HepG2	1	12.21 ± 1.09	1.81 ± 0.38	5.02 ± 0.14
	2	4.32 ± 0.98	1.33 ± 0.37	1.71 ± 0.10

Data are the mean ± SD of at least three independent experiments. $^*P < 0.05$; $^{**}P < 0.01$. $^*P < 0.05$ versus the control, the difference was significant. $^{**}P < 0.01$ versus the control, the difference was markedly significant.

treated for 24 h, 48 h, and 72 h on the selected three cell lines, the cells viability is showed in Figures 2, 3, and 4 and the IC_{50} values is summarized in Table 2.

As shown in Table 2 and Figures 2–4, the oxovanadium complexes exhibit broad inhibition on the three texted human cancer cell lines with the IC_{50} values ranging from 1.68 to 55.40 μM, respectively. The results indicate that both of the oxidovanadium complexes **1** and **2** exhibit antiproliferative effect to human hepatoma cells BEL-7402, HUH-7, and HepG2 in a time and does-dependent manner with increasing the concentrations of **1** and **2**. The IC_{50} values of complex **2** on BEL-7402, HUH-7, and HepG2 cells after treated for 24 h, 48 h, and 72 h were less than that of complex **1** (Figures 2–4, Table 2). It suggests that complex **2** possessed more potent inhibitory effect against the cancer cells. This difference may be attributed to the introduction of one chlorine on the 4 positions of the aromatic chromophore of salicylaldehyde thiosemicarbazone [18, 34, 35]. It implies that the electronic effect of salicylaldehyde thiosemicarbazone may be one of the factors in determining the anti-cancer activities. The results of the MTT-dye reduction assay unambiguously indicate that complexes **1** and **2** exert potent cytotoxic/antiproliferative effect which in some cases is comparable to that of the referent

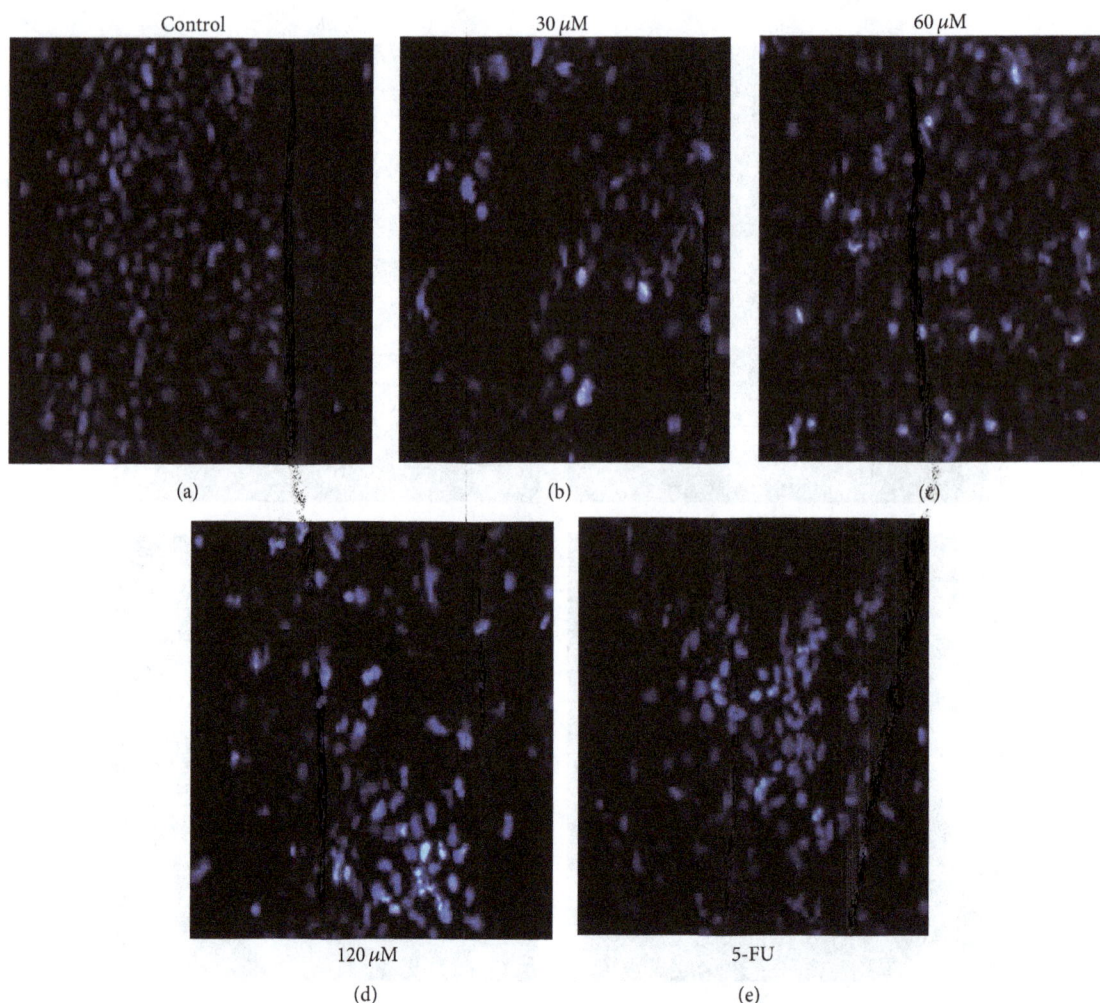

FIGURE 8: Effects of complex **2** on the morphology of Bel-7402 cells were assayed by Hoechst 33342 staining. After treatment with complex **2** for 48 h, apoptotic cells were detected by Hoechst 33342 staining and examined by fluorescence microscopy (original magnification 400x).

TABLE 3: The cell cycle analysis of the BEL-7402 cells induced by complexes **1** and **2** for 48 h.

Concentration (μM)		The relative proportion of different phase in the cell cycle (%)		
		G0/G1	S	G2/M
Control		71.65 ± 1.81	14.28 ± 0.37	12.34 ± 1.78
1	30	$81.05 \pm 0.23^{**}$	$7.94 \pm 0.81^{**}$	$9.11 \pm 0.57^{*}$
	60	$81.78 \pm 0.42^{**}$	$8.54 \pm 0.40^{**}$	$8.66 \pm 1.03^{*}$
	120	$81.96 \pm 1.88^{**}$	$8.18 \pm 0.97^{**}$	$8.58 \pm 0.60^{*}$
2	30	$85.46 \pm 2.58^{**}$	$5.53 \pm 0.23^{**}$	$7.74 \pm 1.82^{*}$
	60	$86.03 \pm 1.23^{**}$	$5.33 \pm 0.20^{**}$	$7.53 \pm 1.21^{*}$
	120	$86.00 \pm 0.99^{**}$	$5.25 \pm 0.51^{**}$	$7.89 \pm 0.63^{*}$

Data are the mean $\pm$ SD of at least three independent experiments. $^{*}P < 0.05$; $^{**}P < 0.01$. $^{*}P < 0.05$ versus the control, the difference was significant. $^{**}P < 0.01$ versus the control, the difference was markedly significant.

cytotoxic drug 5-Fluorouracil. Among the cell lines under evaluation, the HUH-7 and HepG2 human hepatoma cells proved to be more sensitive to V(IV)-complexes treatment, actually the IC50 values in these cells were somewhat higher, but more or less comparable to that of 5-Fluorouracil.

3.2.2. Cell Cycle Analysis. In order to elucidate the mechanisms underlying the observed antiproliferative effect of complexes **1** and **2** on cancer cells, the cells cycle distribution was analyzed by flow cytometry with PI staining [17, 33–39]. The BEL-7402 cells were treated with 30 μM, 60 μM, and

Synthesis and In Vitro Antitumor Activity of Two Mixed-Ligand Oxovanadium(IV) Complexes of Schiff Base and Phenanthroline

111

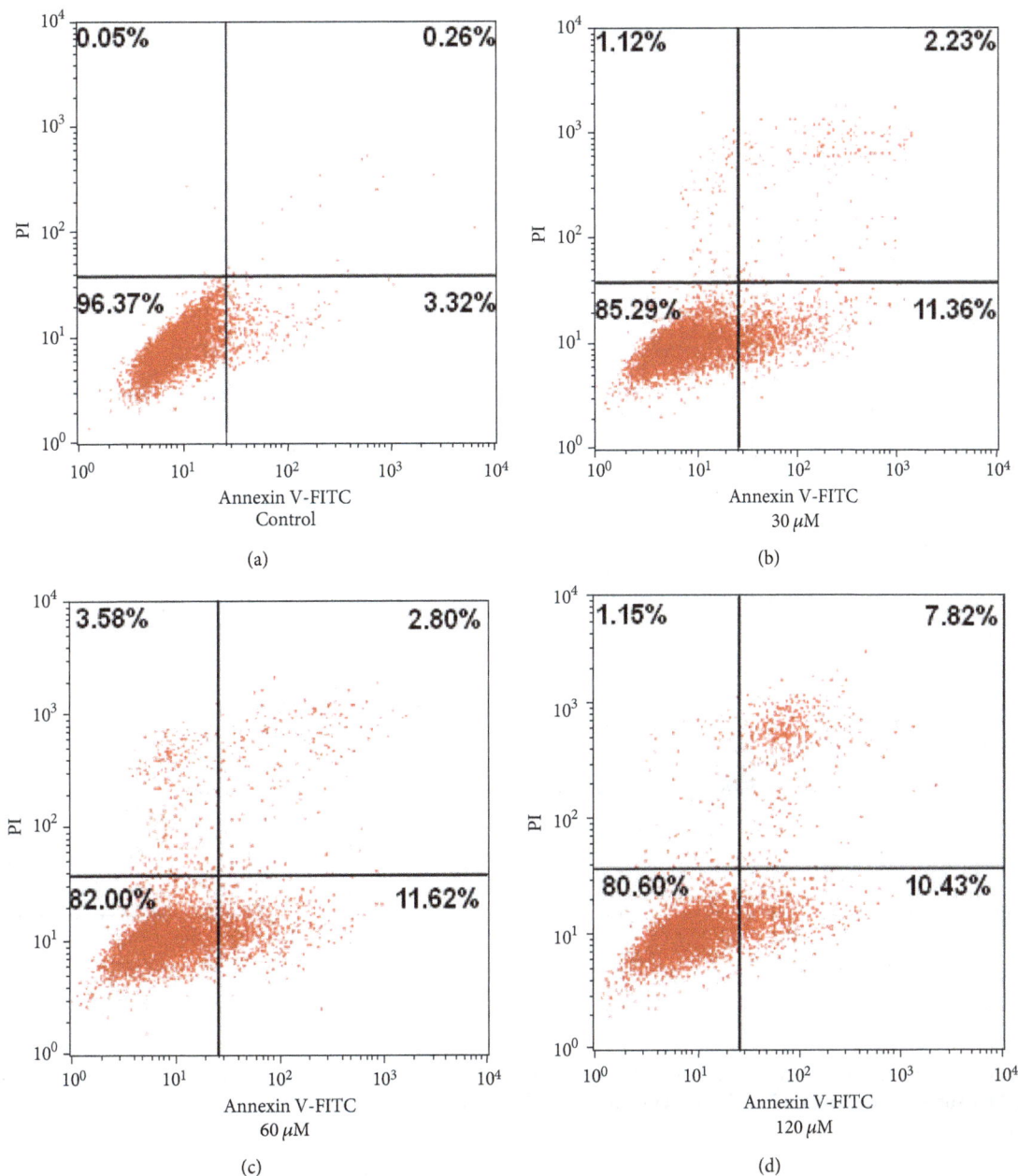

FIGURE 9: Distribution map of cell apoptosis. BEL-7402 cells were incubated with different concentrations of complex 1 (30, 60, and 120 μM) for 48 h, subjected to Annexin V-FITC/PI staining, analyzed by flow cytometry.

120 μM of 1 and 2 for 48 h, respectively. The results were shown in Figures 5 and 6. According to the results of Figures 5 and 6, it indicated that there were significantly increased rates of cells at G0/G1 phase and decreased the rates of cells at S and G2/M phase after BEL-7402 cells were exposed to 1 and 2 compared with untreated cells.

As shown in Table 3, the oxovanadium complexes 1 and 2 inhibited growth of BEL-7402 cells by inducing a block in the G0/G1 phase of the cell cycle. The results in this work showed that there were significantly increased G0/G1 phase distribution and decreased G2/M phase distribution in a dose-dependent manner, indicating the induction of G0/G1-phase arrest by complexes 1 and 2, and the arrested effect

of 2 is stronger than that of 1. The results also suggested that the two oxovanadium complexes induced proliferative suppression of BEL-7402 cells may be via the induction of apoptosis [34–38].

3.2.3. Induction of Apoptosis as Evidenced by Hoechst 33342 Staining.
Apoptosis is an important continuous process of destruction of undesirable cells during development or homeostasis in multicellular organisms. It is widely accepted that this process is characterized by distinct morphological changes including membrane blebbing, cell shrinkage, dissipation mitochondrial membrane potential ($\Delta\Psi$m), chromatin condensation, and DNA fragmentationer [17, 18, 40,

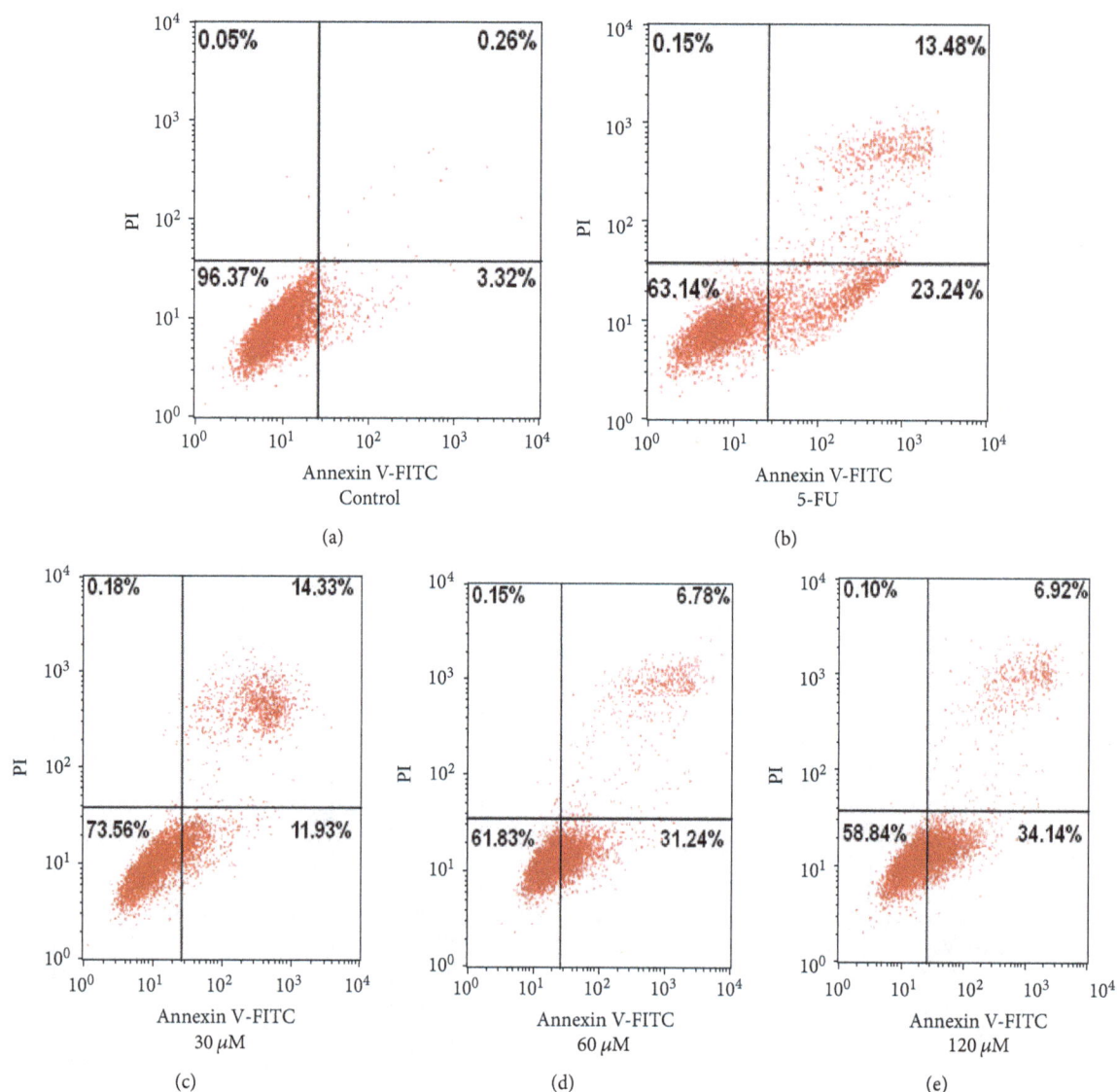

FIGURE 10: Distribution map of cell apoptosis. BEL-7402 cells were incubated with different concentrations of the complex **2** (30, 60, and 120 μM) for 48 h, subjected to Annexin V-FITC/PI staining, and analyzed by flow cytometry.

41]. It has also been known that cancer was caused by the disruption of cellular homeostasis between cell death and cell proliferation [39, 42–44] and compounds which can induce apoptosis are considered to have potential as anticancer drugs [17, 18, 45].

In an attempt to elucidate whether the G0/G1 phase arrest in the BEL-7402 cells induced by these complexes was associated with apoptosis, the occurrence of apoptosis by complexes **1** and **2** was identified with Hoechst 33342 staining. Figures 6 and 7 show representative Hoechst 33342 staining fluorescence photomicrographs of cultured BEL-7402 cells treated with or without **1** and **2**, respectively.

As shown in Figures 7 and 8, compared with the control, the BEL-7402 cells exhibited typical apoptotic features after treated with complexes **1** and **2** for 48 h including cellular morphological change, the apoptotic bodies, condensation of chromatin (brightly stained), and the apoptotic cells

significantly increased in a dose-dependent manner with increasing the concentrations (30, 60, and 120 μM, resp.) of **1** and **2**.

3.2.4. Apoptosis Assessment by Annexin V-FITC/PI Assay. To further investigate the induced apoptosis effects of complexes **1** and **2**, Annexin V-FITC/PI staining was performed to determine early, late apoptotic and necrotic cells, and 5-Fluorouracil (5-FU, 30 μM) was used as the positive control.

It is known that Phosphatidylserine (PS) externalization is an early feature of apoptosis and can be detected by the binding of Annexin V to PS on the cell surface [40–45], and the apoptosis assessment method by Annexin V-FITC/PI assay is well recognized and accurate. As shown in Figures 9 and 10, with increasing amounts of complexes (30 μM, 60 μM, 120 μM, resp.), the percentage of Annexin V-FITC/PI stained cells both the early and late apoptotic cells increases

Synthesis and In Vitro Antitumor Activity of Two Mixed-Ligand Oxovanadium(IV) Complexes of Schiff Base and Phenanthroline

113

TABLE 4: Depolarization of mitochondrial membrane potential of the BEL-7402 cells treated with complexes **1** and **2** for 48 h.

Concentration (μM)	Complex **1**		Complex **2**	
	Mean value of green fluorescence intensity ($\overline{X} \pm SD$)	% of control	Mean value of green fluorescence intensity ($\overline{X} \pm SD$)	% of control
Control	59.84 ± 0.21	100.00 ± 0.35	16.01 ± 0.09	99.99 ± 0.58
60	44.86 ± 3.99	74.96 ± 6.67	19.33 ± 2.66	120.78 ± 16.60
90	$76.99 \pm 3.06^{**}$	$128.65 \pm 5.11^{**}$	$38.85 \pm 1.35^{**}$	$242.73 \pm 8.41^{**}$
135	$85.02 \pm 1.37^{**}$	$142.07 \pm 2.28^{**}$	$58.23 \pm 0.50^{**}$	$363.78 \pm 3.13^{**}$

Data are the mean $\pm$ SD of at least three independent experiments. $^{*}P < 0.05$; $^{**}P < 0.01$. $^{*}P < 0.05$ versus the control, the difference was significant. $^{**}P < 0.01$ versus the control, the difference was markedly significant.

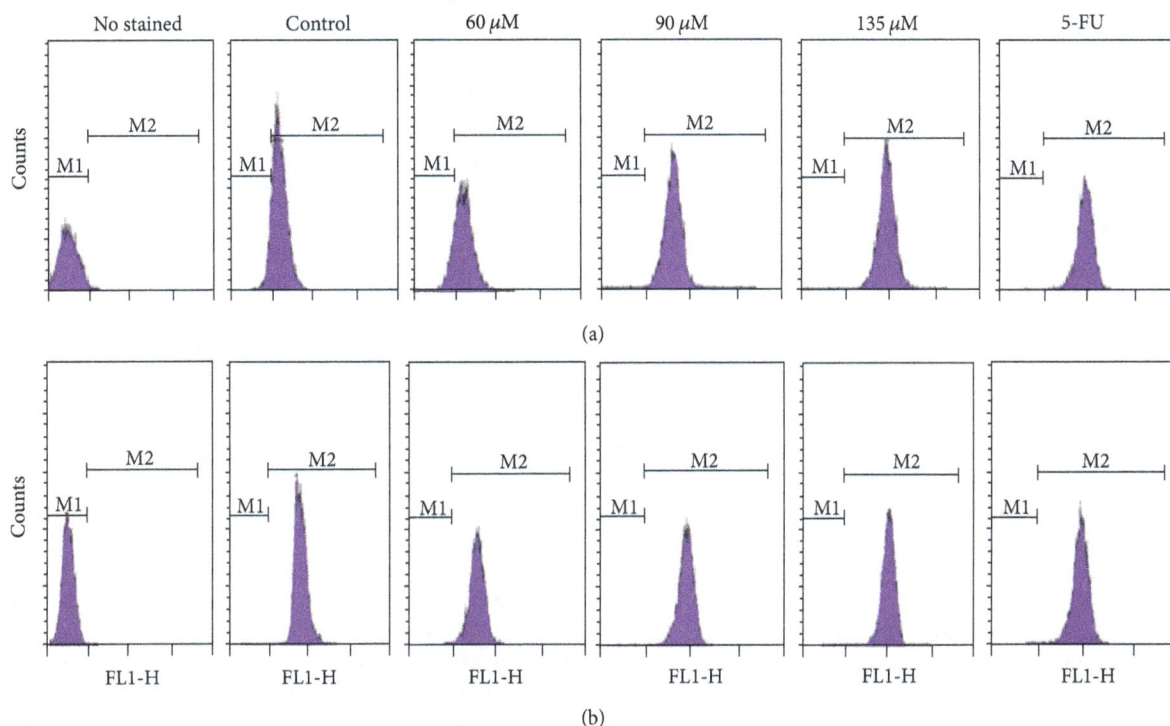

(a)

(b)

FIGURE 11: Distribution map of mitochondrial membrane potential with complex **1** (bottom) and complex **2** (top) for 48 h.

significantly. The results indicate that both complexes **1** and **2** induced proliferative suppression of BEL-7402 cells were via the induction of apoptosis, and the induced apoptosis effect of **2** is stronger than that of **1**.

3.2.5. Loss of Mitochondrial Membrane Potential ($\Delta\Psi m$). Disruption of the $\Delta\Psi m$ is one of the earliest intracellular events that occur following the induction of apoptosis [38–42], and it is also a hallmark for apoptosis. To evaluate the situation of mitochondria in **1** and **2** induced apoptosis, the ability of the two compounds to induce alterations in the mitochondrial membrane potential was investigated.

As shown in Figure 11 and Table 4 upon increasing the concentrations of complexes **1** and **2** (60, 90, and 135 μM), the mean value of green fluorescence intensity obtained from the flow cytometry on BEL-7402 cells increases steadily after treated for 48 h. It is indicated that a lot of Rhodamine 123

from mitochondria matrix was released to the cytoplasm in a dose dependent manner, complexes **1** and **2** can affect the mitochondrial function and causing the depolarization of the mitochondrial membrane, and lead to the $\Delta\Psi m$ value significantly decreased. The percentage relative to control of the complex **2** is stronger than that of the complex **1** (Table 4). These results imply that the induction of apoptosis by the complexes may be associated with the mitochondrial pathway [17, 22, 43–47]. However, further mechanism researches involved in the induction of apoptosis remain to be elucidated.

In summary, two oxovanadium(IV) complexes of [VO (satsc)(phen)] (**1**) (satsc = salicylaldehyde thiosemicarbazone, phen = phenanthroline) and [VO (4-chlorossatsc)(phen)] and (**2**) (4-chlorosatsc = 4-chlorosalicylaldehyde thiosemicarbazone) have been synthesized and characterized. Their antitumor effects on BEL-7402, HUH-7, and

HepG2 cells were studied by MTT assay. *In vitro* experimental results show that both **1** and **2** possess significant antiproliferative effects. The results also showed that they can cause G0/G1 phase arrested of the cell cycle and decreased significantly the $\Delta\Psi m$, causing the depolarization of the mitochondrial membrane. Both complexes exhibit significant induced apoptosis in BEL-7402 cells and displayed typical morphological apoptotic characteristics, and complex **2** showed greater antitumor efficiency than that of complex **1**. These results may provide further evidence to exploit the potential medicine compounds from the metal complexes.

Abbreviations

MTT:	3-(4, 5-Dimethylthiazoyl-2-yl) 2, 5-diphenyltetrazoliumbromide
IC_{50}:	50% inhibition concentrations
Matsc:	Methoxylsalicylaldehyde thiosemicarbazone
Phen:	1, 10-Phenanthroline
4-chlorosatsc:	4-Chlorosalicylaldehyde thiosemicarbazone.

Conflict of Interests

The authors have declared that there is no conflict of interests.

Acknowledgments

This work was supported by the National Natural Science Foundation of China (no. 81102753), the Science and technology Research Project of Guangdong Province (no. 2012B031800431 and no. 83046), China, and the healthy industry base special project of Zhongshan City (no. 2009H021), China.

References

[1] X. G. Ran, L. Y. Wang, D. R. Cao, Y. C. Lin, and J. Hao, "Synthesis, characterization and *in vitro* biological activity of cobalt(II), copper(II) and zinc(II) Schiff base complexes derived from salicylaldehyde and D,L-selenomethionine," *Applied Organometallic Chemistry*, vol. 25, no. 1, pp. 9–15, 2011.

[2] P. K. Sasmal, A. K. Patra, M. Nethaji, and A. R. Chakravarty, "DNA cleavage by new oxovanadium(IV) complexes of *N*-salicylidene α-amino acids and phenanthroline bases in the photodynamic therapy window," *Inorganic Chemistry*, vol. 46, no. 26, pp. 11112–11121, 2007.

[3] J. Vanco, O. Svajlenova, E. Racianska, J. Muselik, and J. Valentova, "Antiradical activity of different copper(II) Schiff base complexes and their effect on alloxan-induced diabetes," *Journal of Trace Elements in Medicine and Biology*, vol. 18, no. 2, pp. 155–162, 2004.

[4] J. A. Obaleye, J. F. Adediji, and M. A. Adebayo, "Synthesis and biological activities on metal complexes of 2,5-diamino-1,3,4-thiadiazole derived from semicarbazide hydrochloride," *Molecules*, vol. 16, no. 7, pp. 5861–5874, 2011.

[5] X. Tai, X. Yin, Q. Chen, and M. Tan, "Synthesis of some transition metal complexes of a novel Schiff base ligand derived from 2,2′-bis(p-methoxyphenylamine) and salicylicaldehyde," *Molecules*, vol. 8, no. 5, pp. 439–443, 2003.

[6] P. Noblía, E. J. Baran, L. Otero et al., "New vanadium(V) complexes with salicylaldehyde semicarbazone derivatives: synthesis, characterization, and *in vitro* insulin-mimetic activity—crystal structure of [V^VO_2(salicylaldehyde semicarbazone)]," *European Journal of Inorganic Chemistry*, vol. 2004, no. 2, pp. 322–328, 2004.

[7] P. Pattanayak, J. L. Pratihar, D. Patra et al., "Synthesis, structure and reactivity of azosalophen complexes of vanadium(IV): studies on cytotoxic properties," *Dalton Transactions*, no. 31, pp. 6220–6230, 2009.

[8] A. Messerschmidt and R. Wever, "X-ray structure of a vanadium-containing enzyme: chloroperoxidase from the fungus *Curvularia inaequalis*," *Proceedings of the National Academy of Sciences of the United States of America*, vol. 93, no. 1, pp. 392–399, 1996.

[9] D. C. Crans and A. S. Tracey, "The chemistry of vanadium in aqueous and nonaqueous solution," *ACS Symposium Series*, vol. 711, pp. 2–29, 1998.

[10] R. Liasko, T. A. Kabanos, S. Karkabounas et al., "Beneficial effects of a vanadium complex with cysteine, administered at low doses on benzo(α) pyrene-induced leiomyosarcomas in wistar rats," *Anticancer Research*, vol. 18, no. 5A, pp. 3609–3613, 1998.

[11] P. Prasad, P. K. Sasmal, R. Majumdar, R. R. Dighe, and A. R. Chakravarty, "Photocytotoxicity and near-IR light DNA cleavage activity of oxovanadium(IV) Schiff base complexes having phenanthroline bases," *Inorganica Chimica Acta*, vol. 363, no. 12, pp. 2743–2751, 2010.

[12] P. K. Sasmal, S. Saha, R. Majumdar, R. R. Dighe, and A. R. Chakravarty, "Photocytotoxic oxovanadium(IV) complexes showing light-induced DNA and protein cleavage activity," *Inorganic Chemistry*, vol. 49, no. 3, pp. 849–859, 2010.

[13] A. A. Holder, "Inorganic pharmaceuticals," *Annual Reports on the Progress of Chemistry A*, vol. 107, pp. 359–378, 2011.

[14] Y. Sun, L. E. Joyce, N. M. Dickson, and C. Turro, "DNA photocleavage by an osmium(ii) complex in the PDT window," *Chemical Communications*, vol. 46, no. 36, pp. 6759–6761, 2010.

[15] S. P. Dash, S. Pasayat, S. H. R. Dash, S. Das, R. J. Butcher, and R. Dinda, "Oxovanadium(V) complexes incorporating tridentate aroylhydrazoneoximes: synthesis, characterizations and antibacterial activity," *Polyhedron*, vol. 31, no. 1, pp. 524–529, 2012.

[16] S. A. Patil, V. H. Naik, A. D. Kulkarni, U. Kamble, G. B. Bagihalli, and P. S. Badami, "DNA cleavage, *in vitro* antimicrobial and electrochemical studies of Co(II), Ni(II), and Cu(II) complexes with m-substituted thiosemicarbazide Schiff bases," *Journal of Coordination Chemistry*, vol. 63, no. 4, pp. 688–699, 2010.

[17] O. S. Frankfurt and A. Krishan, "Apoptosis-based drug screening and detection of selective toxicity to cancer cells," *Anti-Cancer Drugs*, vol. 14, no. 7, pp. 555–561, 2003.

[18] H. L. Xu, X. F. Yu, S. C. Qu et al., "Anti-proliferative effect of Juglone from *Juglans mandshurica Maxim* on human leukemia cell HL-60 by inducing apoptosis through the mitochondria-dependent pathway," *European Journal of Pharmacology*, vol. 645, no. 1–3, pp. 14–22, 2010.

[19] J.-Z. Lu, Y.-F. Du, H.-W. Guo, J. Jiang, X.-D. Zeng, and L.-Q. Zang, "Two oxovanadium complexes incorporating thiosemicarbazones: synthesis, characterization, and DNA-binding studies," *Journal of Coordination Chemistry*, vol. 64, no. 7, pp. 1229–1239, 2011.

Synthesis and In Vitro Antitumor Activity of Two Mixed-Ligand Oxovanadium(IV) Complexes of Schiff Base and Phenanthroline

115

[20] Y. F. Du, J. Z. Lu, H. W. Guo, and J. Jiang, "DNA binding and photocleavage properties of two mixed-ligand oxovanadium complexes," *Transition Metal Chemistry*, vol. 35, pp. 859–864, 2010.

[21] J. Z. Lu, X.D. Zeng, H. W. Guo et al., "Synthesis and characterization of unsymmetrical oxidovanadium complexes: DNA-binding, cleavage studies and antitumor activities," *Journal of Inorganic Biochemistry*, vol. 112, pp. 39–48, 2012.

[22] I. Kostova, "Titanium and vanadium complexes as anticancer agents," *Anti-Cancer Agents in Medicinal Chemistry*, vol. 9, no. 8, pp. 827–842, 2009.

[23] S. K. S. Hazari, J. Kopf, D. Palit, S. Rakshit, and D. Rehder, "Oxidovanadium(IV) complexes containing ligands derived from dithiocarbazates—models for the interaction of VO^{2+} with thiofunctional ligands," *Inorganica Chimica Acta*, vol. 362, no. 4, pp. 1343–1347, 2009.

[24] B. Khera, A. K. Sharma, and N. K. Kaushik, "Bis(indenyl) titanium(IV) and zirconium(IV) complexes of monofunctional bidentate salicylidimines," *Polyhedron*, vol. 2, no. 11, pp. 1177–1180, 1983.

[25] M. Tümer, H. Köksal, S. Serin, and M. DöğÆrak, "Antimicrobial activity studies of mononuclear and binuclear mixed-ligand copper(II) complexes derived from Schiff base ligands and 1,10-phenanthroline," *Transition Metal Chemistry*, vol. 24, no. 1, pp. 13–17, 1998.

[26] M. Tümer, C. Çelik, H. S. K. ksal, and S. Serin, "Transition metal complexes of bidentate Schiff base ligands," *Transition Metal Chemistry*, vol. 24, no. 5, pp. 525–532, 1999.

[27] S. K. Dutta, S. B. Kumar, S. Bhattacharyya, E. R. T. Tiekink, and M. Chaudhury, "Intramolecular electron transfer in $(BzImH)[(LOV)_2O]$ $(H_2L = S$-methyl 3-((2-Hydroxyphenyl)methyl)dithiocarbazate): a novel μ-Oxo dinuclear oxovanadium(IV/V) compound with a trapped-valence $(V_2O_3)^{3+}$ core," *Inorganic Chemistry*, vol. 36, no. 22, p. 4954, 1997.

[28] A. A. Nejo, G. A. Kolawole, A. R. Opoku, J. Wolowska, and P. O'Brien, "Synthesis, characterization and preliminary insulin-enhancing studies of symmetrical tetradentate Schiff base complexes of oxovanadium(IV)," *Inorganica Chimica Acta*, vol. 362, no. 11, pp. 3993–4001, 2009.

[29] R. L. Farmer and F. L. Urbach, "Stereochemistry and electronic structure of oxovanadium(IV) chelates with tetradentate Schiff base ligands derived from 1,3-diamines," *Inorganic Chemistry*, vol. 13, no. 3, pp. 587–592, 1974.

[30] V. B. Arion, V. C. Kravtsov, R. Goddard et al., "Oxovanadium(IV) and oxovanadium(IV)-barium(II) complexes with heterotopic macrocyclic ligands based on isothiosemicarbazide," *Inorganica Chimica Acta*, vol. 317, no. 1-2, pp. 33–44, 2001.

[31] X. G. Ran, L. Y. Wang, D. R. Cao, Y. C. Lin, and J. Hao, "Synthesis, characterization and *in vitro* biological activity of cobalt(II), copper(II) and zinc(II) Schiff base complexes derived from salicylaldehyde and D,L-selenomethionine," *Applied Organometallic Chemistry*, vol. 25, no. 1, pp. 9–13, 2011.

[32] B. Banik, P. K. Sasmal, S. Roy, R. Majumdar, R. R. Dighe, and A. R. Chakravarty, "Terpyridine oxovanadium(IV) complexes of phenanthroline bases for cellular imaging and photocytotoxicity in HeLa cells," *European Journal of Inorganic Chemistry*, no. 9, pp. 1425–1435, 2011.

[33] R. W. Johnstone, A. A. Ruefli, and S. W. Lowe, "Apoptosis: a link between cancer genetics and chemotherapy," *Cell*, vol. 108, no. 2, pp. 153–164, 2002.

[34] W. Hu and J. J. Kavanagh, "Anticancer therapy targeting the apoptotic pathway," *The Lancet Oncology*, vol. 4, no. 12, pp. 721–729, 2003.

[35] J. D. Ly, D. R. Grubb, and A. Lawen, "The mitochondrial membrane potential ($\Delta\Psi$ m) in apoptosis; an update," *Apoptosis*, vol. 8, no. 2, pp. 115–128, 2003.

[36] S. H. Kaufmann and M. O. Hengartner, "Programmed cell death: alive and well in the new millennium," *Trends in Cell Biology*, vol. 11, no. 12, pp. 526–534, 2001.

[37] C. B. Thompson, "Apoptosis in the pathogenesis and treatment of disease," *Science*, vol. 267, no. 5203, pp. 1456–1462, 1995.

[38] S. J. Martin and D. R. Green, "Apoptosis and cancer: the failure of controls on cell death and cell survival," *Critical Reviews in Oncology/Hematology*, vol. 18, no. 2, pp. 137–153, 1995.

[39] I. Vermes, C. Haanen, H. Steffens-Nakken, and C. Reutelingsperger, "A novel assay for apoptosis. Flow cytometric detection of phosphatidylserine expression on early apoptotic cells using fluorescein labelled Annexin V," *Journal of Immunological Methods*, vol. 184, no. 1, pp. 39–51, 1995.

[40] Y. J. Liu and C. H. Zeng, "Synthesis of ruthenium(II) complexes and characterization of their cytotoxicity *in vitro*, apoptosis, DNA-binding and antioxidant activity," *European Journal of Medicinal Chemistry*, vol. 45, no. 7, pp. 3087–3095, 2010.

[41] G. Klopman, "Chemical reactivity and the concept of charge- and frontier-controlled reactions," *Journal of the American Chemical Society*, vol. 90, no. 2, pp. 223–229, 1968.

[42] A. Klein, P. Holko, J. Ligeza, and A. M. Kordowiak, "Sodium orthovanadate affects growth of some human epithelial cancer cells (A549, HTB44, DU145)," *Folia Biologica*, vol. 56, no. 3-4, pp. 115–121, 2008.

[43] M. O'Connor, A. Kellett, M. McCann et al., "Copper(II) complexes of salicylic acid combining superoxide dismutase mimetic properties with DNA binding and cleaving capabilities display promising chemotherapeutic potential with fast acting in vitro cytotoxicity against cisplatin sensitive and resistant cancer cell lines," *Journal of Medicinal Chemistry*, vol. 55, no. 5, pp. 1957–1968, 2012.

[44] R. Kim, "Recent advances in understanding the cell death pathways activated by anticancer therapy," *Cancer*, vol. 103, no. 8, pp. 1551–1558, 2005.

[45] R. S. Ray, B. Rana, B. Swami, V. Venu, and M. Chatterjee, "Vanadium mediated apoptosis and cell cycle arrest in MCF7 cell line," *Chemico-Biological Interactions*, vol. 163, no. 3, pp. 239–247, 2006.

[46] J. Benítez, L. Becco, I. Correia et al., "Vanadium polypyridyl compounds as potential antiparasitic and antitumoral agents: new achievements," *Journal of Inorganic Biochemistry*, vol. 105, no. 2, pp. 303–312, 2011.

[47] A. Bishayee, A. Waghray, M. A. Patel, and M. Chatterjee, "Vanadium in the detection, prevention and treatment of cancer: the *in vivo* evidence," *Cancer Letters*, vol. 294, no. 1, pp. 1–12, 2010.

13

Fast Disinfection of *Escherichia coli* Bacteria Using Carbon Nanotubes Interaction with Microwave Radiation

Samer M. Al-Hakami,[1] Amjad B. Khalil,[2] Tahar Laoui,[3] and Muataz Ali Atieh[1]

[1] *Department of Chemical Engineering, KFUPM, Dhahran 31261, Saudi Arabia*
[2] *Department of Biology, KFUPM, Dhahran 31261, Saudi Arabia*
[3] *Department of Mechanical Engineering, KFUPM, Dhahran 31261, Saudi Arabia*

Correspondence should be addressed to Muataz Ali Atieh; motazali@hotmail.com

Academic Editor: Mehmet Emin Argun

Water disinfection has attracted the attention of scientists worldwide due to water scarcity. The most significant challenges are determining how to achieve proper disinfection without producing harmful byproducts obtained usually using conventional chemical disinfectants and developing new point-of-use methods for the removal and inactivation of waterborne pathogens. The removal of contaminants and reuse of the treated water would provide significant reductions in cost, time, liabilities, and labour to the industry and result in improved environmental stewardship. The present study demonstrates a new approach for the removal of *Escherichia coli* (*E. coli*) from water using as-produced and modified/functionalized carbon nanotubes (CNTs) with 1-octadecanol groups (C_{18}) under the effect of microwave irradiation. Scanning/transmission electron microscopy, thermogravimetric analysis, and FTIR spectroscopy were used to characterise the morphological/structural and thermal properties of CNTs. The 1-octadecanol (C_{18}) functional group was attached to the surface of CNTs via Fischer esterification. The produced CNTs were tested for their efficiency in destroying the pathogenic bacteria (*E. coli*) in water with and without the effect of microwave radiation. A low removal rate (3–5%) of (*E. coli*) bacteria was obtained when CNTs alone were used, indicating that CNTs did not cause bacterial cellular death. When combined with microwave radiation, the unmodified CNTs were able to remove up to 98% of bacteria from water, while a higher removal of bacteria (up to 100%) was achieved when CNTs-C_{18} was used under the same conditions.

1. Introduction

Safe drinking water is one of mankind's most basic needs. Safe drinking water is generally defined as water that does not pose any health risk to humans. The World Health Organization (WHO) defines safe drinking water as water that has chemical, microbial, and physical characteristics that comply with both WHO guidelines for drinking water quality and the respective country's drinking water standard. Good-quality water (i.e., water free of contaminants) is essential to human health and is a critical feedstock in a variety of key industries, including the oil and gas, petrochemical, pharmaceutical, and food industries. The available supplies of water are decreasing due to (i) low precipitation, (ii) increased population growth, (iii) more stringent health-based regulations, and (iv) competing demands from a variety of users, for example, industrial, agricultural, and urban development.

Consequently, water scientists and engineers are seeking alternative sources of water. These alternative sources include seawater, storm water, wastewater (e.g., treated sewage effluent), and industrial wastewater. Water recovery/recycle/reuse has proven to be effective and successful in creating a new and reliable water supply while not compromising public health [1]. Waterborne pathogens are a primary public health concern in developing countries and lead to millions of deaths per year [1]. Worldwide, waterborne diseases remain the leading cause of death in many developing nations. According to the 2004 WHO report, at least one-sixth of the world's population (1.1 billion people) lack access to safe water. The consequences are daunting: diarrhoea kills approximately 2.2 million people every year, mostly children under the age of 5 [2].

Most of the remediation technologies available today, while effective, very often are costly and time-consuming methods. *E. coli* is a bacterium of enteric origin whose occurrence and abundance allow for its use in defining the sanitary quality of water and wastewater. WHO has established a maximum level of 1000 faecal coliform units (FCU)/100 mL for Category A water quality [3]. Chlorination is the most widely used wastewater disinfection method, even though it has drawbacks due to the formation of trihalomethanes and organochlorinated compounds, which are carcinogens. An alternative disinfection method is the use of some metals, either alone or combined, such as Fe, Cu, or Ag in the solid state [4, 5], in ionic form [6–11], in combination with UV light [12], or as formulations in which metal ions of Al, Cu, or Ag are added to a solid matrix, such as zeolites [13, 14], ceramic material [15], silicates [16], colloids, and metal nanoparticles [17–19]. These metal particles cannot be feasibly used directly in water treatment because their toxicity to humans is not yet known, although their elemental forms are toxic to humans at high levels of exposure. However, our experiences using metals for disinfecting wastewater are limited, and they mainly involve using metal ions in combination with other chemical disinfectants, such as chlorine, hydrogen peroxide, or peracetic acid (PAA). These combinations of disinfectants have been applied to influents from advanced primary treatment (APT), biological effluents, or raw water [20–23]. Advances in nanoscale science and engineering suggest that many of the current problems involving water quality could be resolved or greatly ameliorated using nanosorbents, nanocatalysts, bioactive nanoparticles, nanostructured catalytic membranes, and nanoparticle-enhanced filtration, among other products and processes resulting from the development of nanotechnology [24]. Over the last twenty years, carbon nanotubes (CNTs) have received considerable attention from many researchers due to their interesting properties and wide range of applications. In addition to their outstanding mechanical characteristics, CNTs exhibit excellent electrical and thermal properties. These superior properties provide exciting opportunities to produce advanced materials for new applications [25–35]. Because nanoparticles (NPs), CNTs in particular, have the ability to slip past the immune system or directly into the brain or blood cells, some highly qualified research centres around the world are looking into connecting machines to individual cells to provide treatment, inject drugs, and perform many tasks related to health issues [36]. The effect of CNTs on bacteria and viruses has not received much attention, most likely due to the difficulty of dispersing CNTs in water. The antimicrobial activity of CNTs requires direct contact between CNTs and the target microorganisms [37]. Because CNTs are highly hydrophobic materials, this finding suggests that the suspension of nonfunctionalised CNTs in water is very difficult and does not provide enough CNT-microbe contact for disinfection. Few studies available have credited SWNTs with antimicrobial activity towards Gram-positive and Gram-negative bacteria, and the damages inflicted were attributed to either a physical interaction or oxidative stress that compromises the cell membrane integrity. CNTs may therefore be useful for inhibiting microbial attachment and biofouling formation on surfaces [38–41]. For example, Kang [37] immobilised SWNTs on a membrane filter surface and

observed 87% *E. coli* die-off in 2 hours. Srivastava et al. [42] showed that CNTs could be incorporated into hollow fibres and achieve effective inactivation of *E. coli* and poliovirus. Brady-Estévez and Flimelech [43] achieved complete retention and effective inactivation of *E. coli* and up to 5–7 log removal of MS2 bacteriophages using a PVDF microporous membrane coated with a thin layer of SWNTs. In most cases, to achieve total inactivation of test microorganisms, the contact time tends to be large, that is, up to 2 h [36].

The interaction of microwaves with CNTs is an interesting topic for a variety of potential applications. Microwaves with CNTs have been used for the inactivation of microorganisms, providing a technique for simple, green, and large-scale water purification. As an innovative application, the combination of microwaves with well-aligned CNTs produced a new technology for water treatment. Moreover, the microwave-absorbing properties of CNTs and their different behaviour from typical organic compounds may open the door to the preparation of a wide range of new materials useful in many fields. The present study provides a new approach for the removal of *E. coli* from water using modified and nonmodified CNTs with and without the heating effect of microwave radiation.

2. Experimental Methods and Materials

2.1. Carboxylation Treatment of Carbon Nanotubes. Multiwall carbon nanotubes (CNTs) were purchased from Nanostructured & Amorphous Materials, Inc., USA. The purity of CNTs is >95% and the outside and inside diameters are 10–20 nm and 5–10 nm, respectively. The length of these CNTs is 10–30 μm. Three hundred millilitres of a concentrated nitric acid of AnalaR (69%) was added to 2 g of as-received CNTs. The mixture was refluxed for 48 h at 120°C and then cooled to room temperature. The mixture was diluted with 500 mL of deionised water and then vacuum-filtered through a filter paper (3 μm porosity). This washing operation was repeated until the pH became the same as the deionised water pH and was followed by drying in a vacuum oven at 100°C. Such conditions lead to the removal of the catalysts from the CNTs, opening of the tube caps, and the formation of holes in the sidewalls, followed by an oxidative etching along the walls with the concomitant release of carbon dioxide. This less vigorous condition minimised the shortening of the tubes, and the chemical modification was then limited mostly to the opening of the tube caps and the formation of functional groups at defect sites along the sidewalls. The final products were nanotube fragments whose ends and sidewalls were decorated with various oxygen-containing groups (mainly carboxyl groups) (Figure 1) [44]. Moreover, the percentage of carboxylic functions on the oxidised CNT surface does not exceed 4%, which corresponds to the percentage of CNT structural defects [15, 16].

2.2. Esterification of Carbon Nanotubes. The Fischer esterification is an equilibrium reaction, whereas other esterification routes do not involve equilibrium. To shift the equilibrium to favour the production of esters, it is customary to use an

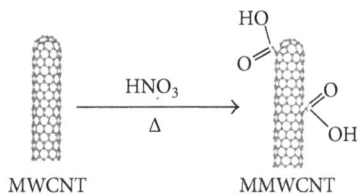

FIGURE 1: Chemical modification of carbon nanotubes through thermal oxidation [44].

excess of one of the reactants, either the alcohol or the acid. In the present reactions, an excess of 1-octadecanol (Merck, 97% purity) was used because it is cheaper and easier to remove than the CNTs. Additionally, water formed in this reaction was removed by evaporation during the reaction.

The oxidatively introduced carboxyl groups represent useful sites for further modifications [17], as they enable the covalent coupling of molecules through the creation of ester (Figure 2) [44]. In a 250 mL beaker, 10 g of the 1-octadecanol was melted on a hotplate at 90°C, and 1 g of oxidatively modified carbon nanotubes (M-CNTs) was added. The mixture was stirred for 10 minutes before a few drops of sulphuric acid were added as a catalyst. After the addition of catalyst, the reaction was kept on a hotplate and stirred for 2 hours. After completion of the reaction, the mixture was poured into 250 mL of benzene and vacuum-filtered through a filter paper (3 μm porosity). This washing operation was repeated five times and was followed by washing with petroleum ether three times and THF three times. The product was washed with deionised water and acetone a few times, and then the functionalised M-CNT material that was produced was dried in a vacuum oven at 90°C.

The mechanism for this reaction involves the nucleophilic addition of the alcohol or amine to the carbonyl group of the protonated acid in carbon nanotubes, followed by elimination of a proton. The tetrahedral intermediate is unstable under the acidic conditions of the reaction and undergoes dehydration to form the ester or amide. The key steps of this mechanism involve activation of the carbonyl group by protonation of the carbonyl oxygen, nucleophilic addition to the protonated carbonyl to form a tetrahedral intermediate, and elimination of water from the tetrahedral intermediate to restore the carbonyl group.

2.3. FTIR Measurements. Fourier transform infrared (FTIR) spectroscopy has shown a limited ability to probe the structure of CNTs. A factor that has hindered the advancement of FTIR spectroscopy as a tool for CNT analysis is the poor infrared transmittance of CNTs. To overcome this problem, KBr preparations of nanotube samples were utilized. Because of their black character, the CNTs exhibit strong absorbance and are often unable to be distinguished from background noise, for that a very low concentration of CNTs in a KBr powder was used. However, the greater vibrational freedom of attached polymeric species produces much more pronounced peaks, and so the attached species are typically the focus of attention in FTIR results. Despite this, with very

careful sample preparation, some researchers have managed to elucidate peaks corresponding to surface-bound moieties, such as carboxylic acid groups at wavenumbers of 1791, 1203, and 1080 cm^{-1}. The spectra of samples were recorded by a PERKIN ELMER 16F PC FTIR instrument. FTIR samples were prepared by grinding ~0.03 wt% dry material into potassium bromide.

2.4. Thermal Analysis. The thermogravimetric analysis (TGA) technique measures the changes in the weight of a sample with increasing temperature. The moisture content and presence of volatile species can be determined with this technique. Computer-controlled graphics can calculate the weight percent losses. The dynamic thermogravimetric experiments were performed using a Netzsch model STA 449 F3 Jupiter simultaneous thermal analyser, which allows measurement of the mass change and the associated phase transformation energetics. The system was equipped with a PtRh furnace capable of operation from 25 to 1500°C. The temperature was measured with a type S thermocouple. The system is vacuum tight, allowing measurements to be conducted under a controlled atmosphere. Differential scanning calorimetry (DSC) measurements were also taken to study the phase transitions and exothermic/endothermic decompositions taking place in the investigated samples. The TGA-DSC analysis was performed on small samples (approximately 6 mg) mounted on platinum crucibles with Al$_2$O$_3$ liners and pierced lids in a flow of inert atmosphere (flow rate of argon gas, 70 mL/min) for the degradation study and under air for the same conditions (flow rate of air gas, 70 mL/min) for the oxidation study. The temperature range was varied from room temperature to 1400°C, and the typical heating rate was 20°C/min, while the digital resolution of the balance is 1 μg/digit.

2.5. Surface Characterisation of CNTs. The morphological and structural study of CNTs was conducted using field-emission scanning electron microscopy (JEOL JSM-6700F) and transmission electron microscopy (Philips CM200-FEG). To prepare TEM samples, some alcohol was dropped on the nanotube film, which was transferred with a pair of tweezers to a carbon-coated copper grid.

2.6. Microorganisms and Culture Conditions. The *E. coli* strain ATCC number 8739 (supplied by King Fahd University, Petroleum and Minerals Clinic) was used throughout this work. *E. coli* was grown overnight in nutrient broth at 37°C on a rotary shaker (160 rpm). Aliquots of the pre-culture were inoculated into fresh medium and incubated in the same conditions to an absorbance at 600 nm of 0.50. The cells were harvested by centrifugation at 4000 rpm for 10 min at 4°C, washed twice with a sterile 0.9% NaCl solution at 4°C, and dispersed in the solution containing CNTs and CNTs-C$_{18}$ to a concentration of 3.5×10^7 CFU/mL.

The bactericidal rate K can be calculated by the following formula:

$$K = \frac{(A - B)}{A} \times 100\%, \tag{1}$$

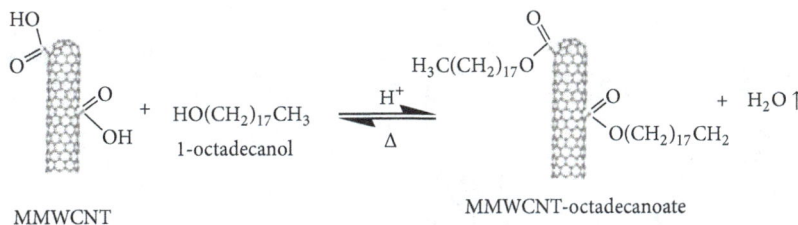

FIGURE 2: Chemical esterification of modified carbon nanotubes (M-CNTs) [44].

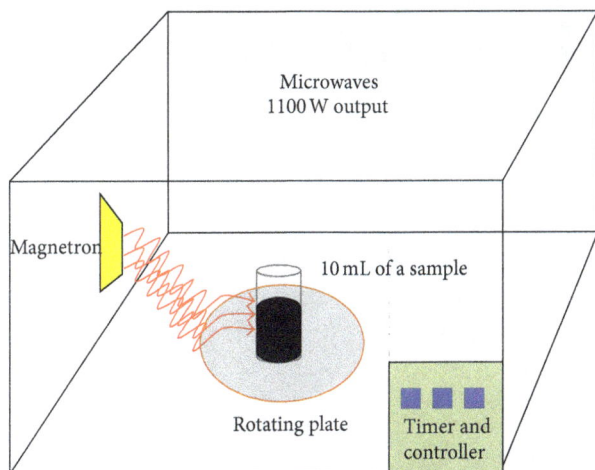

FIGURE 3: Illustration of experimental setup: the microwaves with 1100W output power was provided with a rotating plate, so that the radiation circumference covered an entire sample under study.

where K is The bactericidal rate; A is colony forming units (CFU)/mL of the control sample; B is colony forming units (CFU)/mL of the tested sample. The optimum bactericidal rate can be accounted in each Petri dish should be in the range of that 30–300 CFU/mL.

2.7. Microwave Application. The above solution of nanomaterial was sonicated before being mixed with the bacterial solution. For each type of carbon nanomaterial used, the mixture of carbon nanomaterial and bacteria was tested with and without exposure to microwave radiation for 0, 5, and 10 seconds (Figure 3). Cultured bacteria (tested bacteria with different carbon nanomaterials with and without microwave radiation) were analysed by plating on nutrient agar plates after serial dilution in 0.9% saline. The colonies were counted after a 48 h incubation at 37°C. *E. coli* control experiments were performed in parallel with each CNTs-C_{18} material tested.

3. Results and Discussion

The present study reports the results obtained using CNTs and CNTs functionalised with 1-octadecanol groups (C_{18}) for the removal of *E. coli* bacteria dispersed in water as well as the additional effect of using microwave radiation.

3.1. Characterisation of Carbon Nanotubes. The CNTs were observed by FE-SEM and TEM. The diameter of the CNTs

was found to vary from 20 to 40 nm with an average diameter at 24 nm, while the length of the CNTs reached up to a few microns. Figure 4(a) shows SEM image of CNTs at low magnification, while Figure 4(b) shows a higher-magnification image. From the SEM observation, the product represents relatively high-quality CNTs. TEM was also performed to characterise the structure of the nanotubes (Figure 5). To prepare TEM samples, some alcohol was dropped onto the nanotube film, which was transferred with a pair of tweezers to a carbon-coated copper grid. It is obvious from the images that all the nanotubes are hollow and tubular in shape. In some of the images, catalyst particles can be observed inside the nanotubes. Figure 5(b) shows the high-resolution transmission electron microscopy (HRTEM) of the CNTs and that a highly ordered crystalline CNT structure is present.

3.2. Surface Modification of Carbon Nanotubes. Figure 6 shows the IR spectra of as-received multiwall CNTs and CNTs functionalised with 1-octadecanol (CNT-C_{18}). The IR spectrum of MWCNTs shows an absorption band at 2920 cm^{-1}, which is attributed to asymmetric and symmetric CH$_2$ stretching. The band 1698 cm^{-1} is assigned to carboxylic C=O stretching, and 1097 cm^{-1} corresponds to C–O stretching in alcohols. The presence of these functional groups on the surface of pure MWCNTs indicates their introduction during the removal of metal catalysts in nanotube purification processes. The carboxylic C=O stretching peak observed at 1693 cm^{-1} can be attributed to an acid treatment of MWCNTs, which is reported elsewhere. Substitution of the MWCNT-COOH group with 1-octadecanol gives an indicative peak at 2920 cm^{-1} corresponding to CH$_2$ stretching of the long alkyl molecule of octadecanol. The peaks 1473, 1458, and 1068 cm^{-1} correspond to the ether formation of octadecanol with the –COOH groups in CNTs.

3.3. Thermal Oxidation and Degradation of Modified and Unmodified CNTs. The study of the thermal degradation of materials is of major importance because it can, in many cases, determine the upper temperature limit of use for a material. In addition, considerable attention has been directed towards the exploitation of thermogravimetric data for the determination of functional groups. For this purpose, thermogravimetric analysis (TGA) is a technique widely used because of its simplicity and the information afforded by a simple thermogram. Figure 7 depicts the TGA-DSC results for the CNTs functionalised with amine groups (CNT-C_{18}). In these figures, TG% refers to the temperature-dependent

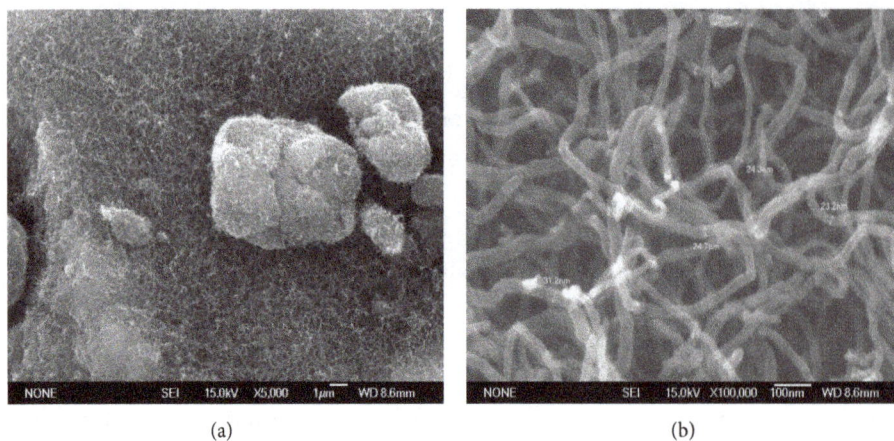

FIGURE 4: SEM Images of carbon nanotubes at (a) low resolution (b) high resolution.

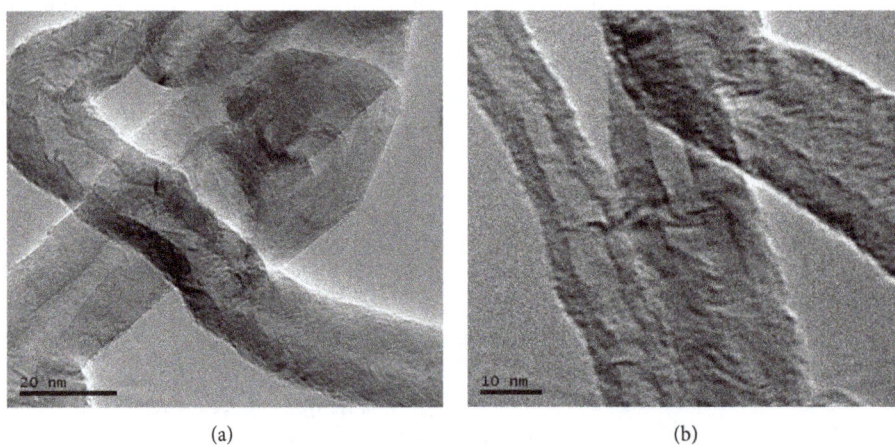

FIGURE 5: TEM images of CNTs at (a) low resolution and (b) high resolution.

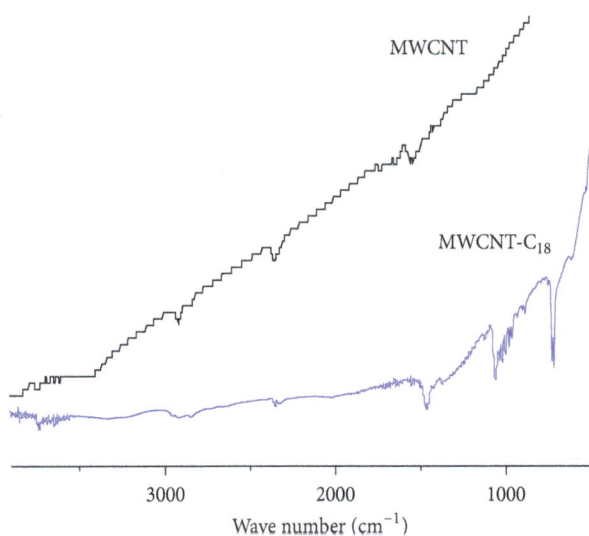

FIGURE 6: FTIR spectra of MWCNT and 1-octadecanol (C_{18}) modified MWCNT.

mass change in percent, DTG (%/min) to the rate of mass change (derivative of TG curve), and DSC (mW/mg) to the heat flow rate of the considered sample. Several mass loss steps were observed, which are due to the release of moisture (below ~150°C) and the decomposition of the associated organic groups. The mass loss steps were accompanied by endothermic effects visible in the DSC signal except for the sample. Figure 7 displays the degradation of CNT-C_{18}. Two peaks appear at 265°C and approximately 400°C, corresponding, respectively, to the maximum degradation of C_{18} and the carboxylic acid group. From Figure 8, it seems that a small amount of phenol has been attached to the carboxylic group, therefore yielding a small DTG peak appearing at approximately 250°C corresponding to the maximum degradation of phenol followed by the peak at 376°C showing the maximum degradation of the carboxylic group. It is well known that carbon will decompose in atmospheric environment. For that, the samples were burned under air atmosphere to reveal the purity of raw carbon nanotubes. The TG thermograms were carried out in air, and it was noted that there was some residual remains of the samples when they was heated to approximately 900°C, as shown in Figure 8, this residual

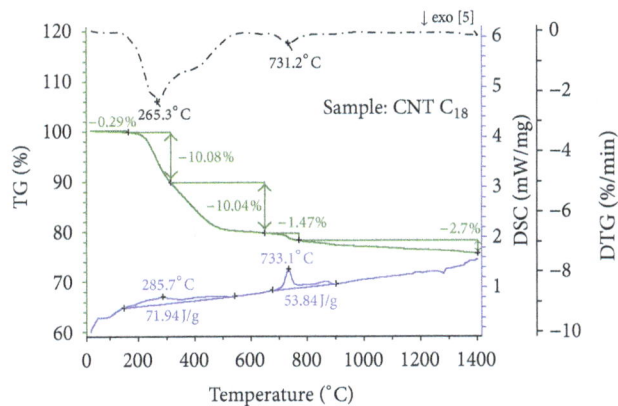

FIGURE 7: Temperature-dependent mass change (TG), rate of mass change (DTG), and heat flow rate (DSC) of the sample CNT-C$_{18}$.

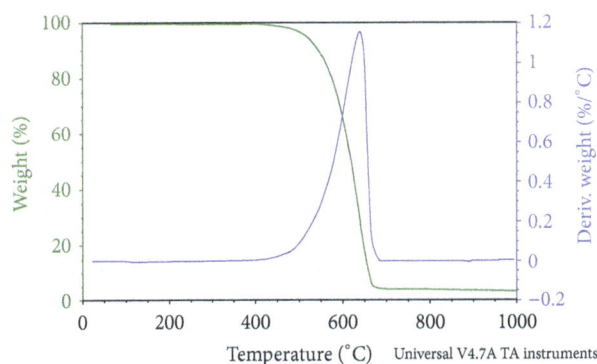

FIGURE 8: Temperature-dependent mass change (TG), rate of mass change (DTG) of raw CNTs.

TABLE 1: The percentage of E. coli removal under the effect of CNTs and CNTs-C$_{18}$.

Type of carbon nanotubes	Number of E. coli cells Control sample	% of E. coli removal (K)
Raw CNTs	$3.50E + 07$	5
CNTs-C$_{18}$	$3.50E + 07$	3

being the catalyst. It can be observed that this decomposition process is a single-stage decomposition reaction in which the procedural decomposition temperatures are well defined.

3.4. Effect of CNTs and Functionalised CNTs on the Removal of E. coli Bacteria.

This novel nanomaterial consisting of CNTs functionalised with 1-octadecanol (C$_{18}$H$_{38}$O) or simply C$_{18}$ group was investigated thoroughly in this study. The dosing amount of the CNTs-C$_{18}$ during all experiments was fixed at 0.2 g of CNTs/100 mL of NaCl autoclaved solution. Table 1 presents the percentage of E. coli bacteria removed by adding CNTS and CNTs-C$_{18}$ nanomaterial.

As shown in Table 1, the results indicate that the percentage of E. coli bacteria removal by CNTs and CNTs-C$_{18}$ is relatively small (3–5%), which could be due to experimental

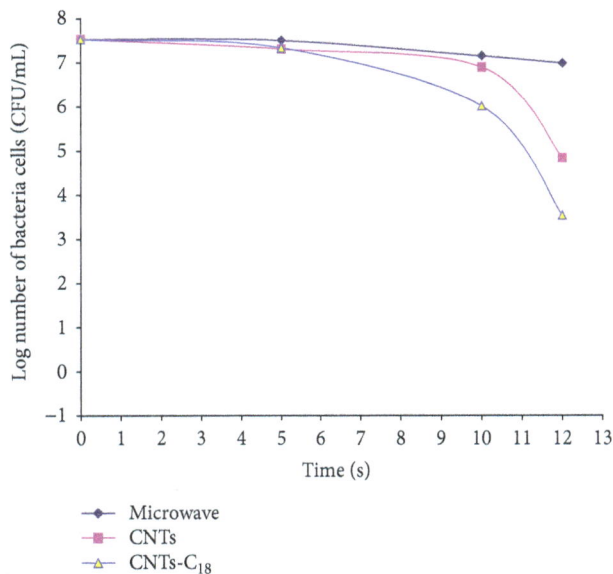

FIGURE 9: Effect of CNTs and functionalised CNTs-C$_{18}$ with and without microwave radiation on removal of bacteria from drinking water.

error, meaning no significant removal occurred. The antimicrobial activity of CNTs requires direct contact between CNTs and target microorganisms [41]. Our experiments provide direct evidence that no major removal of E. coli bacteria would take place if the as-received CNTs or CNTs modified by carboxylic or 1-octadecanol (C$_{18}$H$_{38}$O) functional groups were used.

3.5. Removal of E. coli Bacteria by Microwave Radiation Interaction with CNTs.

In this investigation, microwaves were utilised as a source of heat for the removal of E. coli bacteria. The disinfection of E. coli bacteria from drinking water before and after treating it with CNTs and functionalised CNTs with and without microwave radiation is shown in Figure 9. Under microwave radiation alone, the die-off was very limited, approximately 0.04 to 0.4 log reduction of the number of bacterial cells after 5 and 10 sec, respectively, while using microwaves in the presence of CNTs gave a reduction of approximately 0.2 to 0.6 log. A higher reduction was obtained with CNTs-C$_{18}$, yielding approximately 0.2 to 1.5 log after 5 and 10 sec, respectively. Increasing the radiation time to 12 sec, a sharp reduction in the number of bacterial cells was obtained using CNTs-C$_{18}$, yielding a reduction of 4 log (99.99% removal), 2.7 log (98.9%) with CNTs, and 0.6 log (72%) with microwave radiation alone. The bacterial die-off, which was enhanced after using microwave heating, might be due to the effect of polarisation on the cell wall. This polarisation is expected to exert acute and excessive potentials and heat, leading to bursting of the cell wall.

Figure 10 shows the schematic diagram of the removal of the E. coli bacteria under the effect of microwave radiation using as-received and functionalised CNTs. The functionalisation of CNTs with C$_{18}$ (C–C bonds) groups provides further enhancement of the thermal properties of CNTs due to the

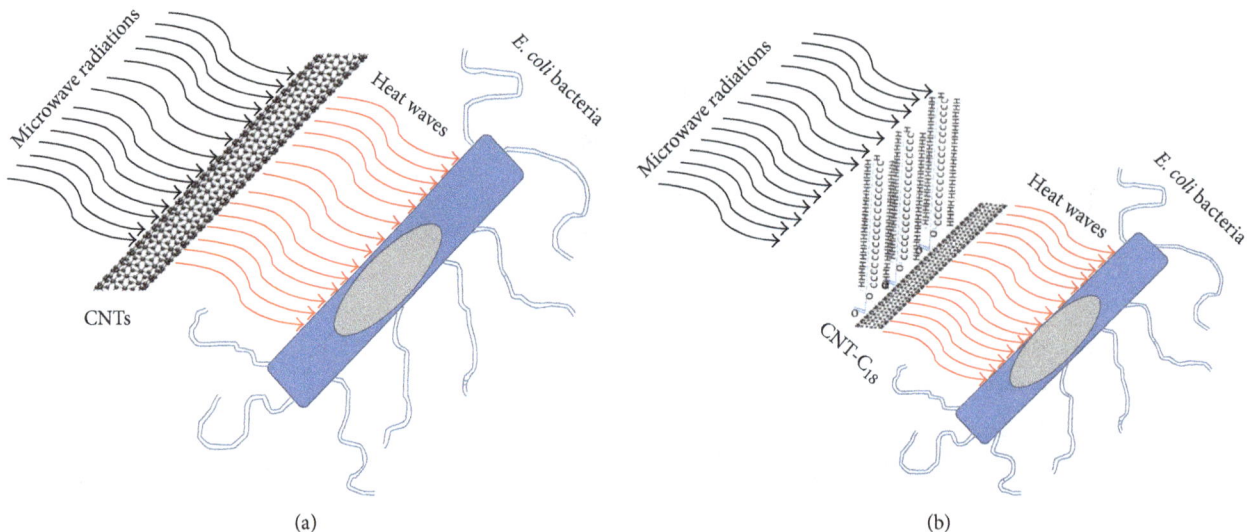

FIGURE 10: Schematic diagram of the thermal effect CNTs and CNT-C_{18} under microwave radiation on *E. coli*.

long carbon chains, which increased the absorption rate of the microwave heat attached to the surface of CNTs, leading to higher heat absorption and thus an overall increased temperature, particularly after a prolonged time period (12 sec).

4. Conclusion

The interaction of microwaves with carbon nanotubes (CNTs) is in fact an interesting topic and has a wide range of recent applications. The present study highlights a new approach for the removal of *E. coli* from water using as-received CNTs and CNTs modified with 1-octadecanol groups (C_{18}) under the effect of microwave irradiation. The morphology of the as-received CNTs was characterised using field-emission scanning electron microscopy (FE-SEM) and transmission electron microscopy (TEM) to determine the diameter and the length of the CNTs. The diameter of the CNTs varied from 20 to 40 nm with an average diameter of 24 nm; the length was 10 micrometres. The surface of the CNTs was functionalised with 1-octadecanol (C_{18}) functional groups via the Fischer esterification, which was confirmed by fourier transform Infrared (FTIR) spectroscopy, thermogravimetric analysis (TGA), and differential scanning calorimetric (DSC). The unmodified and modified CNTs were tested for their efficiency in destroying the pathogenic bacteria from water, with and without the effect of microwave radiation. A low removal (3–5%) of *E. coli* bacteria was observed when CNTs alone were used, indicating that the CNTs alone do not cause bacterial death. A high removal of *E. coli* bacteria was obtained when microwave radiation was used. Almost complete removal of *E. coli* bacteria (100%) was obtained using CNTsC_{18} followed by CNTs (98. %) and then a microwave source (71%) for 12 sec. CNTs functionalised with carbon-18 functional groups with microwave radiation generally showed the highest antibacterial activity when compared with non-functionalised carbon nanotubes interacting with microwave radiation and microwave radiation without

a carbon source. These significant results were obtained due to multiple chains of C_{18} (C–C bonds), which increased the absorption rate of the microwave heat. As an innovative application, the combination of microwaves with modified and unmodified CNTs appears to be promising and can complement the currently employed disinfection methods. Moreover, the microwave absorbing properties of CNTs and their unique behaviour compared with typical organic compounds may open the door to the preparation of a wide range of new materials useful in many fields. Therefore, extensive and deeper scientific research studies are needed in the field of "Bionanotechnology", not only for human purposes but also for the environment around us (flora and fauna). Unknown illnesses, new viruses, and thousands of dangerous bacteria are filling the world, and these crucial issues are pushing scientists to deploy new techniques for illness treatment, water purification applications, and many other environmental issues (ir 23, 8670–8673).

References

[1] U. S. Environmental Protection Agency, 2007, http://www.epa.gov/region09/water/recycling/index.html.

[2] http://www.who.int/whr/2004/annex/en/.

[3] http://www.who.int/whr/1998/en/.

[4] R. L. Davies and S. F. Etris, "The development and functions of silver in water purification and disease control," *Catalysis Today*, vol. 36, no. 1, pp. 107–114, 1997.

[5] Y. You, J. Han, P. C. Chiu, and Y. Jin, "Removal and inactivation of waterborne viruses using zerovalent iron," *Environmental Science and Technology*, vol. 39, no. 23, pp. 9263–9269, 2005.

[6] M. W. Craig, "Coping with resistance to Copper/Silver disinfection," *Water-Engineering & Management*, vol. 148, no. 11, p. 27, 2001.

[7] J. Q. Jiang, S. Wang, and A. Panagoulopoulos, "The exploration of potassium ferrate(VI) as a disinfectant/coagulant in water and wastewater treatment," *Chemosphere*, vol. 63, no. 2, pp. 212–219, 2006.

[8] J. Q. Jiang, A. Panagoulopoulos, M. Bauer, and P. Pearce, "The application of potassium ferrate for sewage treatment," *Journal of Environmental Management*, vol. 79, no. 2, pp. 215–220, 2006.

[9] J. Q. Jiang, S. Wang, and A. Panagoulopoulos, "The role of potassium ferrate(VI) in the inactivation of Escherichia coli and in the reduction of COD for water remediation," *Desalination*, vol. 210, no. 1-3, pp. 266–273, 2007.

[10] S. Silva-Martínez, A. Alvarez-Gallegos, and E. Martínez, "Electrolytically generated silver and copper ions to treat cooling water: an environmentally friendly novel alternative," *International Journal of Hydrogen Energy*, vol. 29, no. 9, pp. 921–932, 2004.

[11] N. Silvestry-Rodriguez, E. Sicairos-Ruelas, P. C. Gerba, and K. Bright, "Silver as a disinfectant," *Reviews of Environmental Contamination and Toxicology*, vol. 191, pp. 23–45, 2007.

[12] J. Y. Kim, C. Lee, M. Cho, and J. Yoon, "Enhanced inactivation of E. coli and MS-2 phage by silver ions combined with UV-A and visible light irradiation," *Water Research*, vol. 42, no. 1-2, pp. 356–362, 2008.

[13] M. Rivera-Garza, M. Olguín, I. García-Sosa, D. Alcántara, and G. Rodríguez-Fuentes, "Silver supported on natural Mexican zeolite as an antibacterial material," *Microporous and Mesoporous Materials*, vol. 39, pp. 431–444, 2000.

[14] Gómez I. De La Rosa, M. Olguín, and T. Alcántara, "Bactericides of coliform microorganisms from wastewater using silver-clinoptilolite rich tuffs," *Applied Clay Science*, vol. 40, no. 1-4, pp. 45–53, 2008.

[15] J. Kim, M. Cho, B. Oh, S. Choi, and J. Yoon, "Control of bacterial growth in water using synthesized inorganic disinfectant," *Chemosphere*, vol. 55, no. 5, pp. 775–780, 2004.

[16] M. Kawashita, S. Toda, H. M. Kim, T. Kokubo, and N. Masuda, "Preparation of antibacterial silver-doped silica glass microspheres," *Journal of Biomedical Materials Research A*, vol. 66, no. 2, pp. 266–274, 2003.

[17] K. Chaloupka, Y. Malam, and A. M. Seifalian, "Nanosilver as a new generation of nanoproduct in biomedical applications," *Trends in Biotechnology*, vol. 28, no. 11, pp. 580–588, 2010.

[18] K. H. Cho, J. E. Park, T. Osaka, and S. G. Park, "The study of antimicrobial activity and preservative effects of nanosilver ingredient," *Electrochimica Acta*, vol. 51, no. 5, pp. 956–960, 2005.

[19] O. Choi, K. K. Deng, N. J. Kim, L. Ross, R. Y. Surampalli, and Z. Hu, "The inhibitory effects of silver nanoparticles, silver ions, and silver chloride colloids on microbial growth," *Water Research*, vol. 42, no. 12, pp. 3066–3074, 2008.

[20] R. Pedahzur, O. Lev, B. Fattal, and H. I. Shuval, "The interaction of silver ions and hydrogen peroxide in the inactivation of E. coli: a preliminary evaluation of a new long acting residual drinking water disinfectant," *Water Science and Technology*, vol. 31, no. 5-6, pp. 123–129, 1995.

[21] M. T. Orta De Velásquez, I. Yáñez-Noguez, B. Jiménezcisneros, and V. Luna-Pabello, "Adding silver and Copper to hydrogen peroxide and peracetic acid in the disinfection of an advanced primary treatment effluent," *Environmental Technology*, vol. 29, no. 11, pp. 1209–1217, 2008.

[22] V. M. Luna-Pabello, M. M. Ríos, B. Jiménez, and M. T. Orta De Velasquez, "Effectiveness of the use of Ag, Cu and PAA to disinfect municipal wastewater," *Environmental Technology*, vol. 30, no. 2, pp. 129–139, 2009.

[23] M. Miranda-Ríos, V. M. Luna-Pabello, M. T. Orta de Velásquez, and J. A. Barrera-Godínez, "Removal of Escherichia coli from biological effluents using natural and artificial mineral aggregates," *Water SA*, vol. 37, no. 2, pp. 45–53, 2011.

[24] N. Savage and M. Diallo, "Nanomaterials and water purification: opportunities and challenges," *Journal of Nanoparticle Research*, vol. 7, pp. 331–342, 2005.

[25] S. Iijima, "Helical microtubules of graphitic carbon," *Nature*, vol. 354, no. 6348, pp. 56–58, 1991.

[26] C. H. Kiang, W. A. Goddard, R. Beyers, and D. S. Bethune, "Carbon nanotubes with single-layer walls," *Carbon*, vol. 33, no. 7, pp. 903–914, 1995.

[27] N. G. Chopra, L. X. Benedict, V. H. Crespi, M. L. Cohen, S. G. Louie, and A. Zettl, "Fully collapsed carbon nanotubes," *Nature*, vol. 377, no. 6545, pp. 135–138, 1995.

[28] M. G. Dresselhaus and S. Riichiro, *Physical Properties of Carbon Nanotubes*, Imperial College Press, London, UK, 1998.

[29] J. Fan, M. Wan, D. Zhu, B. Chang, Z. Pan, and S. Xie, "Synthesis, characterizations, and physical properties of carbon nanotubes coated by conducting polypyrrole," *Journal of Applied Polymer Science*, vol. 74, no. 11, pp. 2605–2610, 1999.

[30] X. Gong, J. Liu, S. Baskaran, R. D. Voise, and J. S. Young, "Surfactant-assisted processing of carbon nanotube/polymer composites," *Chemistry of Materials*, vol. 12, no. 4, pp. 1049–1052, 2000.

[31] P. G. Collins and P. Avouris, "Nanotubes for electronics," *Scientific American*, vol. 283, no. 6, pp. 62–69, 2000.

[32] M. Dresselhaus, G. Dresselhaus, and P. Avouris, *Carbon Nanotubes: Synthesis, Structure, Properties, and Applications*, Springer, Berlin, Germany, 2001.

[33] P. Harris, *Carbon Nanotubes and Related Structures: New Materials For the Twenty First Century*, Cambridge University Press, Cambridge, UK, 2001.

[34] D. Qian, G. J. Wagner, W. K. Liu, M. F. Yu, and R. S. Ruoff, "Mechanics of carbon nanotubes," *Applied Mechanics Reviews*, vol. 55, no. 6, pp. 495–532, 2002.

[35] D. Srivastava, C. Wei, and K. Cho, "Nanomechanics of carbon nanotubes and composites," *Applied Mechanics Reviews*, vol. 56, no. 2, pp. 215–229, 2003.

[36] Q. Li, S. Mahendra, D. Y. Lyon et al., "Antimicrobial nanomaterials for water disinfection and microbial control: potential applications and implications," *Water Research*, vol. 42, no. 18, pp. 4591–4602, 2008.

[37] S. Kang, M. Herzberg, D. F. Rodrigues, and M. Elimelech, "Antibacterial effects of carbon nanotubes: size does matter!," *Langmuir*, vol. 24, no. 13, pp. 6409–6413, 2008.

[38] L. Qi, Z. Xu, X. Jiang, C. Hu, and X. Zou, "Preparation and antibacterial activity of chitosan nanoparticles," *Carbohydrate Research*, vol. 339, no. 16, pp. 2693–2700, 2004.

[39] J. R. Morones, J. L. Elechiguerra, A. Camacho et al., "The bactericidal effect of silver nanoparticles," *Nanotechnology*, vol. 16, no. 10, pp. 2346–2353, 2005.

[40] D. Y. Lyon, L. K. Adams, J. C. Falkner, and P. J. J. Alvarez, "Antibacterial activity of fullerene water suspensions: effects of preparation method and particle size," *Environmental Science and Technology*, vol. 40, no. 14, pp. 4360–4366, 2006.

[41] S. Kang, M. Pinault, L. D. Pfefferle, and M. Elimelech, "Single-walled carbon nanotubes exhibit strong antimicrobial activity," *Langmuir*, vol. 23, no. 17, pp. 8670–8673, 2007.

[42] A. Srivastava, O. N. Srivastava, S. Talapatra, R. Vajtai, and P. M. Ajayan, "Carbon nanotube filters," *Nature Materials*, vol. 3, no. 9, pp. 610–614, 2004.

[43] A. S. Brady-Estévez and S. M. Elimelech, "A single-walled-carbon-nanotube filters for removal of viral and bacterial pathogens," *Small*, vol. 4, no. 4, pp. 481–484, 2008.

[44] F. A. Abuilaiwi, T. Laoui, M. Al-Harthi, and M. A. Atieh, "Modification and functionalization of multiwalled carbon nanotube (MWCNT) via fischer esterification," *Arabian Journal for Science & Engineering*, vol. 35, no. 1, pp. 37–48, 2010.

14

Synthesis, CP-MAS NMR Characterization, and Antibacterial Activities of Glycine and Histidine Complexes of Cd(SeCN)$_2$ and Hg(SeCN)$_2$

Bassem A. Al-Maythalony,[1] **M. Monim-ul-Mehboob,**[1] **Mohammed I. M. Wazeer,**[1] **Anvarhusein A. Isab,**[1] **M. Nasiruzzaman Shaikh,**[2] **and Saleh Altuwaijri**[3]

[1] *Department of Chemistry, King Fahd University of Petroleum and Minerals, Dhahran 31261, Saudi Arabia*
[2] *Center of Research Excellence in Nanotechnology (CENT), King Fahd University of Petroleum and Minerals, Dhahran 31261, Saudi Arabia*
[3] *Clinical Research Laboratory, Saad Research & Development Center, Saad Specialist Hospital, Al-Khobar 31952, Saudi Arabia*

Correspondence should be addressed to Anvarhusein A. Isab; aisab@kfupm.edu.sa

Academic Editor: Imre Sovago

The synthesis and characterization of cadmium and mercury complexes of selenocyanate of the type [(L)M(SeCN)$_2$] are described, where L is L-Histidine (His) or L-Glycine (Gly) and M is Cd^{2+} or Hg^{2+}. These complexes are obtained by the reaction of 1 equivalent of respective amino acids with metal diselenocyanate precursor in a mixture of solvents (methanol : water = 1 : 1). These synthesized compounds are characterized by analytical and various spectroscopic techniques such as elemental analysis (EA), IR, ^{1}H, and ^{13}C NMR in solution and in the solid state for ^{13}C and ^{15}N. The *in vitro* antibacterial activities of these complexes have been investigated with standard type cultures of *Escherichia coli* (MTCC 443), *Klebsiella pneumoniae* (MTCC 109), *Pseudomonas aeruginosa* (MTCC 1688), *Salmonella typhi* (MTCC 733), and *Staphylococcus aureus* (MTCC 737).

1. Introduction

Since the metal ions play vital roles in a number of biological processes such as biomolecules stabilizations, enzyme regulations, transportation of fluids through transmembrane channels, and so forth [1–3], numerous metal ions amino acids complexes also act as potent antifungal, antibacterial, and anticancer drugs [4–6]. Therefore, the extensive studies of these metallic species have been dedicated to understand the impact on living systems. It is well known that metal-binding proteins cover a large fraction of the total protein, and they are actively participating in several essential life processes [2, 3]. Therefore, understanding of the physicochemical and biochemical properties of metals with amino acids becomes indispensible and a broad area of research for several years [7–11].

Among all amino acids found in nature, Histidine is often found as a ligand in various types of metalloenzymes because it is the key amino acids residue in many enzymatic reactions [12]. This is may be due to its stereochemical location of the coordinating atom in Histidine. The Histidine skeleton contains the imidazole side group having two nitrogen atoms capable of participating in metal-ligand coordination sphere thereby it can take on various metal-bound forms in proteins. Thus, it is important to know the coordination modes of the Histidine (His) and Glycine (Gly) ligands to understand the reaction mechanism of metalloenzymes [13].

The coordination modes of various metal ions with amino acids have been the topic of discussion for a long period of time, and the ideas to get the binding modes are not easy to predict for amino acids with large side chain such as Histidine [16], because of different types of donor atoms present in

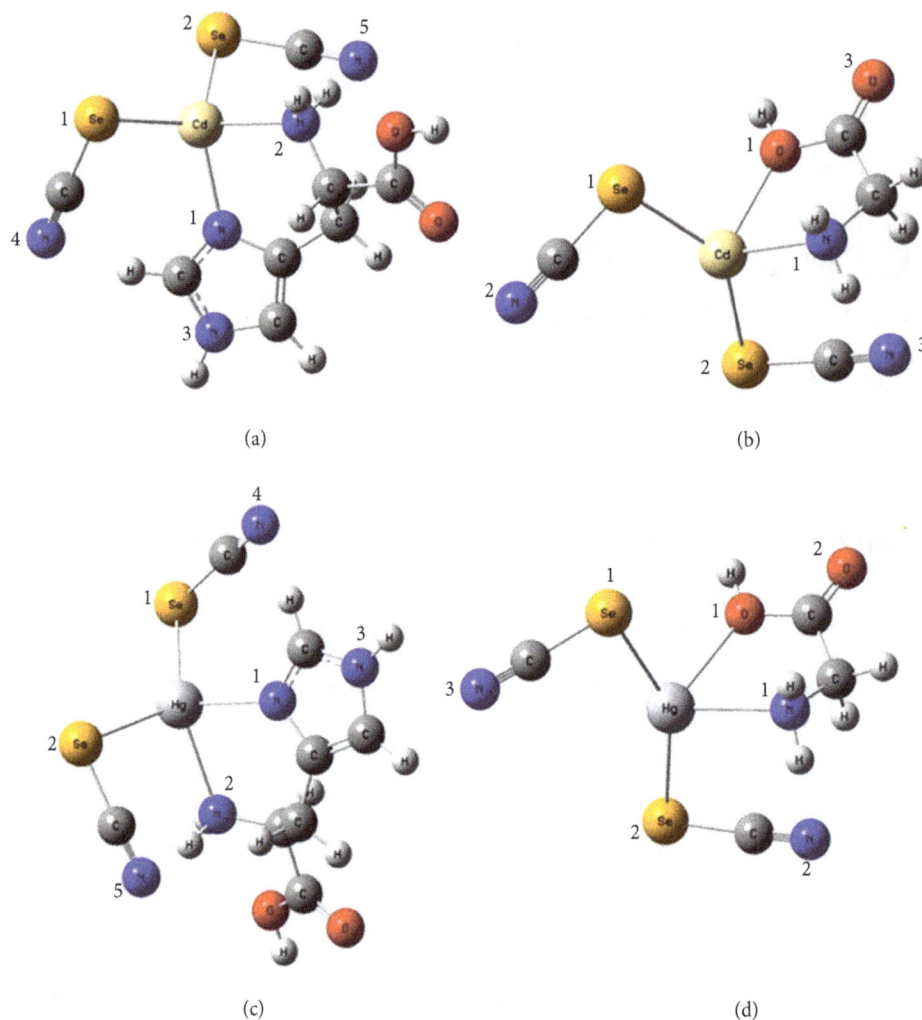

(a)

(b)

(c)

(d)

SCHEME 1: Optimized geometries of [LM(SeCN)$_2$] (a), (b), (c), and (d), obtained at the B3LYP/LanL2DZ level of theory using Gaussian 09, Revision A. 1. L refers to Histidine or Glycine, while M refers to Hg or Cd.

TABLE 1: Elemental analysis of the prepared complexes.

Complex	M. Pt. (C)	Found (Calcd.)%		
		H	C	N
(Gly)Cd(SeCN)$_2$	Decomp. at 184	1.32 (1.27)	12.15 (12.09)	10.78 (10.57)
(Gly)Hg(SeCN)$_2$	>300	1.10 (1.04)	10.04 (9.89)	9.00 (8.65)
(His)Cd(SeCN)$_2$	Decomp. > 205	2.00 (1.90)	20.44 (20.12)	14.88 (14.67)
(His)Hg(SeCN)$_2$	Decomp. > 140	1.70 (1.60)	17.11 (16.99)	12.64 (12.38)

amino acid backbone. In this point it has become necessary to study its active sites and binding affinity to transition metals at both theoretical and experimental levels. In line with efforts made by theoretical studies it has appeared that the stereochemical suitability of the metal ions play a critical role in determining the location of bond formations [17].

Selenocyanate ligand can have versatile binding modes [18, 19]; nevertheless soft Se center is expected to coordinate more preferably to soft metals leaving the harder N uncoordinated [20, 21].

In order to gain better understanding of the interaction of the metal ions with macromolecules involving amino acids, knowledge of the structure and the energetic of the metal ions coordination to amino acids are required. In an effort to obtain a more complete picture, we have synthesized a number of hitherto unknown cadmium and mercury

Synthesis, CP-MAS NMR Characterization, and Antibacterial Activities of Glycine and Histidine Complexes of Cd(SeCN)₂ and Hg(SeCN)₂

127

TABLE 2: IR frequencies, $\nu(\text{cm}^{-1})$ Hg(SeCN)$_2$ and Cd(SeCN)$_2$ complexes theoretical versus experimental.

Species	ν(C=O) Exp.	ν(C=O) Theo.	ν(SeCN) Exp.	ν(SeCN) Theo.	ν(NH$_2$) Exp.	ν(NH$_2$) Theo.
KSeCN	—	—	2070[a]	—	—	—
L-Gly	1606 s	—	—	—	3424	—
L-Hist	1634 s	—	—	—	3127	—
Cd(SeCN)$_2$	—	—	2107	—	—	—
(L-Gly)Cd(SeCN)$_2$	1611 s	1759	2107	2127	3450	3455
(L-Hist)Cd(SeCN)$_2$	1631 s	1718	2109	2112	3460	3402
Hg(SeCN)$_2$	—	—	2127	—	—	—
(L-Gly)Hg(SeCN)$_2$	1611 s	1742	2130	2137	3447	3476
(L-Hist)Hg(SeCN)$_2$	1636 s	1716	2111	2118	3422	3423

[a][14].

TABLE 3: ^{13}C NMR chemical shifts of Hg(SeCN)$_2$ and Cd(SeCN)$_2$ complexes in DMSO-d_6.

Species	SeCN	C=O	C-1	C-2	C-3	C-4	C-5
His	—	174.7	136.2	135.0	117.9	55.1	29.0
Gly	—	173.1	42.5				
Cd(SeCN)$_2$	116.9						
(His)Cd(SeCN)$_2$	115.4	173.0	136.6	134.5	117.0	53.5	28.1
(Gly)Cd(SeCN)$_2$	119.0	194.9					
Hg(SeCN)$_2$	103.3						
(His)Hg(SeCN)$_2$	109.8	170.3	135.0	132.3	116.4	53.3	27.4
(Gly)Hg(SeCN)$_2$	116.5	189.2					

TABLE 4: ^{77}Se NMR chemical shifts of Hg(SeCN)$_2$ and Cd(SeCN)$_2$ complexes in DMSO-d_6.

Species	^{77}Se (in ppm)
Cd(SeCN)$_2$	−272.94
Hg(SeCN)$_2$	−109.18
(His)Hg(SeCN)$_2$	−169.71

TABLE 5: Solid-state ^{13}C Isotropic Chemical Shifts (δ_{iso}) and Principle Shielding Tensors(σ_{xx})[a] of complexes Cd(II)-Selenocyanate complexes with Glycine and Histidine ligands.

Complex	Nucleus	δ_{iso}	σ_{11}	σ_{22}	σ_{33}	$\Delta\sigma$	η
	^{113}Cd	211.9	322	283	30	291	0.73
Cd(SeCN)$_2$	^{77}Se	−119.6	53	41	−452	505	0.96
	^{13}C	117.0	222	205	−76	298	0.89
(Gly)Cd(SeCN)$_2$	^{13}C	170.8	242	171	98	−109	0.98
	^{13}C	119.9	212	124	23	−146	0.90
	^{13}C	169.3	236	169	102	−101	0.99
	^{13}C	108.4	181	103	41	−101	0.86
(His)Cd(SeCN)$_2$	^{13}C	132.0	202	136	58	−111	0.90
	^{13}C	129.2	196	130	61	−102	0.96
	^{13}C	119.5	213	120	23	−142	0.97

[a]Isotropic shielding, $\sigma_i = (\sigma_{11} + \sigma_{22} + \sigma_{33})/3$; $\Delta\sigma = \sigma_{33} - 0.5(\sigma_{11} + \sigma_{22})$; $\eta = 3(\sigma_{22} - \sigma_{11})/2\Delta\sigma$.

selenocyanate complexes and their characterization using various important spectroscopic techniques.

2. Experimental

2.1. General Remarks

2.2. Preparation of Cd(II) and Hg(II) Complexes. A solution of CdCl$_2$ in 10 mL dist. water was mixed with a stoichiometrically equivalent amount of ligand (Histidine or Glycine) in 10 mL solvent mixture (methanol : water = 1 : 1 in volume), produced solution stirred for 30 min, then two equivalents KSeCN water solution was added, the resulting mixture fluxed with nitrogen gas with stirring for 15 min then heat it for ~1.5 hour at 70°C. The product was filtered and dried. The same procedure was applied for mercury complexes using HgCl$_2$ instead of CdCl$_2$.

2.3. Spectroscopic Measurements. The measurements of solid-state IR and solution NMR were recorded as described in the literature [22, 23]. The solution NMR chemical shifts of ligands along with corresponding complexes are given in Tables 3 and 4.

2.4. Solid-State NMR Studies. Natural abundance ^{13}C solid-state NMR spectra were obtained on a JEOL LAMBDA 500 spectrometer operating at 110.85 MHz, corresponding to a magnetic field of 11.74 T, at ambient temperature of 25°C. Samples were packed into 6 mm zirconia rotors. Cross-polarization and high power decoupling were employed. Pulse delay of 7.0 s and a contact time of 5.0 ms were used for carbon observations in the CPMAS experiments, whereas the pulse delay of 10 s and a contact time of 6.0 ms were

TABLE 6: Solid-state ^{15}N isotropic chemical shifts (δ_{iso}) and principle shielding tensors (δ_{xx})[a] of complexes, Hg(II)-selenocyanate complexes.

Complex	Nucleus	δ_{iso}	δ_{11}	δ_{22}	δ_{33}	$\Delta\sigma$	η
His	^{15}N	−202.55	−97.76	−181	−328.85	−189.45	0.66
	^{15}N	−331.02	—	—	—	—	—
(His)Hg(SeCN)$_2$	^{15}N	−156.73	−27.66	—	−272.80	−174.11	0.80
	^{15}N	−146.5	—	169.72	—	—	—
Gly	^{15}N	−345.56	—	—	—	—	—
(Gly)Hg(SeCN)$_2$	^{15}N	−311.01	—	—	—	—	—

[a]Isotropic shielding, σ_i: $(\sigma_{11} + \sigma_{22} + \sigma_{33})/3$; $\Delta\sigma$: $\sigma_{33} - 0.5(\sigma_{11} + \sigma_{22})$; η: $3(\sigma_{22} - \sigma_{11})/2\Delta\sigma$.

TABLE 7: Selected bond lengths (Å) for [LM(SeCN)$_2$] for optimized structure using B3LYP/LanL2DZ; L refers to Histidine and Glycine, while M refers to Hg or Cd.

Hg(SeCN)$_2$ + His		Cd(SeCN)$_2$ + His		Hg(SeCN)$_2$ + Gly		Cd(SeCN)$_2$ + Gly	
Hg-Se1	2.765	Cd-Se1	2.718	Hg-Se1	2.699	Cd-Se1	2.656
Hg-Se2	2.795	Cd-Se2	2.745	Hg-Se2	2.731	Cd-Se2	2.691
Hg-N1	2.434	Cd-N1	2.302	Hg-N1	2.576	Cd-N1	2.415
Hg-N2	2.537	Cd-N2	2.399	Hg-O1	2.656	Cd-O1	2.423
C=O	1.236	C=O	1.236	C=O	1.232	C=O	1.228
C–O	1.386	C–O	1.386	C–O	1.402	C–O	1.412

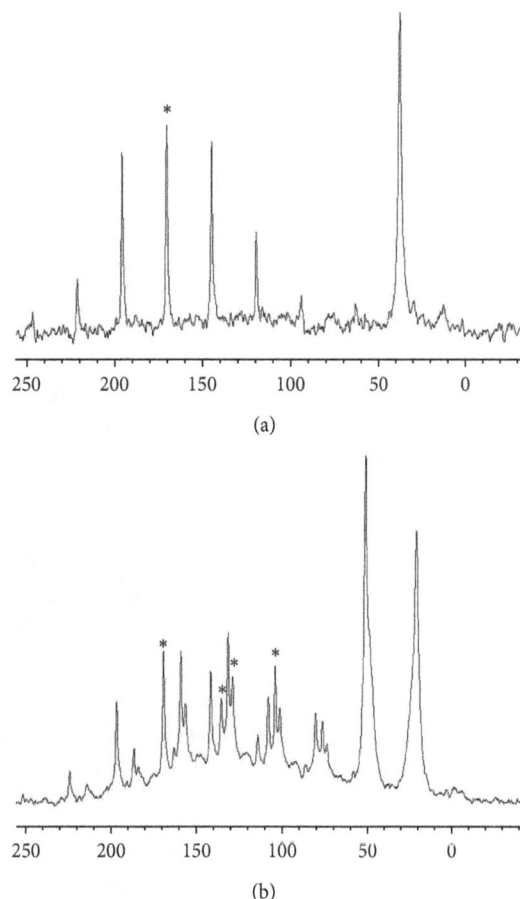

FIGURE 1: ^{13}C CPMAS spectra of (a) (Gly)Cd(SeCN)$_2$, (b) (His)Cd(SeCN)$_2$. The center peak is denoted by "*."

used in the selenium observation. The magic angle spinning rates were from 3000 to 5000 Hz. ^{13}C chemical shifts were referenced to TMS by setting the high-frequency isotropic peak of solid adamantane to 38.48 ppm.

The ^{15}N NMR spectrum was recorded at 50.55 MHz using $^{15}NH_4 NO_3$ as external reference, which lies at −358.62 ppm relative to pure MeNO$_2$ [24]. The spectral conditions for ^{15}N were 32 K data points, 0.721 sec acquisition time, 2.50 sec delay time, 60° pulse angle, and approximately 5000 scans. The chemical shift of nitrogen was initially referenced with respect to liquid NH$_3$, by setting the ^{15}N peak in enriched solid $^{15}NH_4Cl$ to 40.73 ppm [25] and then converted to the standard nitromethane by a shift of −380.0 ppm [24] for ammonia. The ^{13}C and ^{15}N spectra containing spinning sideband manifolds were analyzed using a computer program WSOLIDS developed at Dalhousie and Tubingen universities [26].

2.5. Computational Study. Geometry optimization was done for the built structures and optimized by DFT level of theory with LanL2DZ (Los Alamos ECP plus double zeta) [27, 28] basis sets using Gaussian 09, Revision A. 1 program package [29].

2.6. Test of Bacterial Strains. Standard type cultures of *Escherichia coli* (MTCC 443), *Klebsiella pneumoniae* (MTCC 109), *Pseudomonas aeruginosa* (MTCC 1688), *Salmonella typhi* (MTCC 733), and *Staphylococcus aureus* (MTCC 737) were obtained from Microbial Type Culture Collection (MTCC) Chandigarh, India). The agar well-diffusion technique [30] was used to screen the antibacterial activity. *In*

Synthesis, CP-MAS NMR Characterization, and Antibacterial Activities of Glycine and Histidine Complexes of Cd(SeCN)$_2$ and Hg(SeCN)$_2$

129

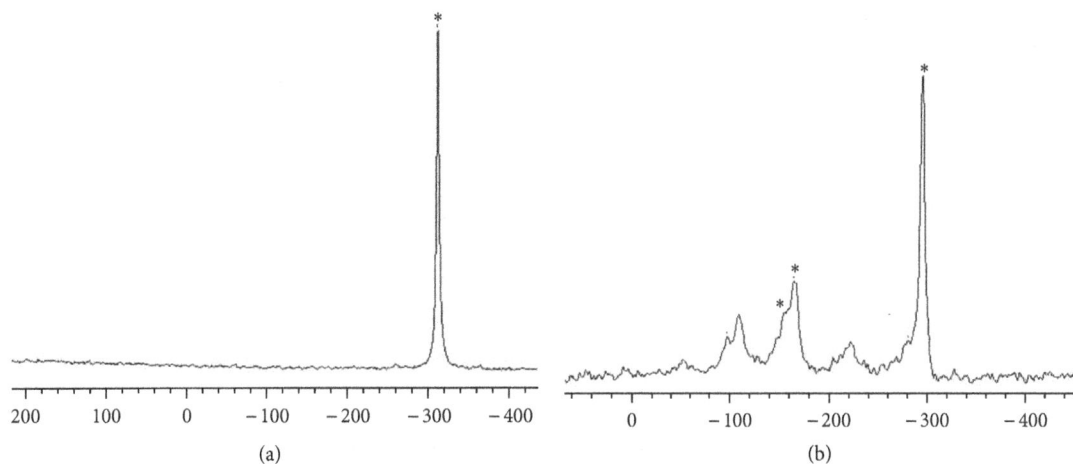

FIGURE 2: ^{15}N NMR spectra of (a) (Gly)Hg(SeCN)$_2$ and (b) (His)Hg(SeCN)$_2$.

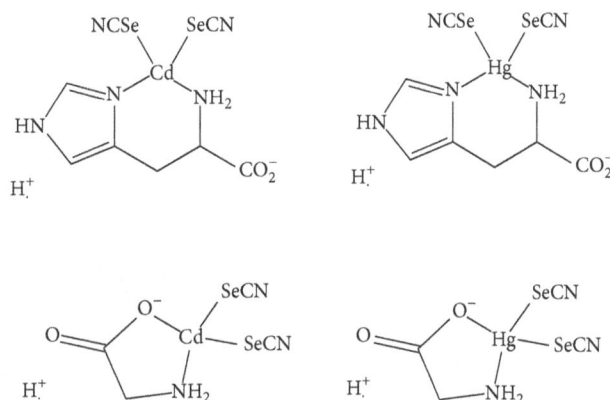

FIGURE 3: Possible binding sites for Cd^{2+} and Hg^{2+} His and Gly complexes.

3. Results and Discussions

vitro antibacterial activities were screened by using nutrient agar plates obtained from HiMedia (Mumbai, India). The plates were prepared by pouring 20 mL of molten media into a sterile Petri dish and allowed to solidify for 5 minutes (Table 1). A sterile cork borer of diameter 6.0 mm was used to make wells in the agar plates. Inoculums were swabbed uniformly on the surface of agar plates. 0.1 mg/well were loaded on 6.00 mm diameter wells. The plates were allowed to stand for 1 h for diffusion then incubated at 37°C for 24 hrs. At the end of incubation, inhibition zones were measured.

3.1. IR and NMR Studies. The ^{13}C solution NMR data of all complexes were shown in Table 3. The downfield chemical shifts were observed for the prepared complexes for (Gly)Cd(SeCN)$_2$ at 194.9 ppm and (Gly)Hg(SeCN)$_2$ at 189.24 with respect to the free ligand, Glycine at 173.1 ppm. These high downfield shifts resulted from the electron donation from Glycine carboxylate to metal thereby causing about 20 ppm shifts of carbonyl carbon, while this shift was not observed in the Histidine complexes because, in Histidine

complexes, imidazole nitrogen and α-amine are involved in coordination to the metal center, which agree with the reported binding mode of Histidine to mercury metal ion as shown in Figure 3 [31].

The C≡N infrared frequency for Hg(SeCN)$_2$ is higher than that for Cd(SeCN)$_2$ which means stronger C–N bond, and this lead to less electron density at the selenium atom that derives more back donation from the Hg to Se, which makes Hg–Se bond stronger. This less electron density observed also by downfield shift for the de-shielded Se bound to Hg (−109 ppm) compared with Se bound to Cd (−272 ppm) as shown by ^{77}Se NMR (Table 4). In case of Histidine complexes of Hg(SeCN)$_2$ selenium atom became more shielded and shifted upfield (−169.71 ppm) because of donation from Histidine to the metal center, which causes even stronger π-back donation to selenium. In Table 2, the IR data shows the highest red shift for selenocyanate frequency for Histidine complex of mercury which means the highest π-back donation from the metal to the selenocyanate rather than Glycine so greater donation to the antibonding π-orbitals of the cyanate from selenium atom indication of stronger Histidine mercury bonding than the rest of the complexes

TABLE 8: Selected torsion angle (°) for [LM(SeCN)$_2$] for optimized structure using B3LYP/LanL2DZ; L refers to Histidine and Glycine, while M refers to Hg or Cd.

Hg(SeCN)$_2$ + His		Cd(SeCN)$_2$ + His		Hg(SeCN)$_2$ + Gly		Cd(SeCN)$_2$ + Gly	
Se1-C-N4	178.79	Se1-C-N4	178.45	Se1-C-N3	178.03	Se1-C-N2	178.20
Se2-C-N5	176.36	Se2-C-N5	175.84	Se2-C-N2	177.04	Se2-C-N3	176.32
N1-Hg-N2	83.87	N1-Cd-N2	87.79	N1-Hg-O1	64.67	N1-Cd-O1	96.81
Se1-Hg-Se2	125.79	Se1-Cd-Se2	121.32	Se1-Hg-Se2	149.92	Se1-Cd-Se2	139.71
Se1-Hg-N1	108.92	Se1-Cd-N1	111.15	Se1-Hg-N1	107.50	Se1-Cd-N1	112.35
Se2-Hg-N2	99.18	Se2-Cd-N2	101.26	Se2-Hg-O1	111.58	Se2-Cd-O1	112.17
Se1-Hg-N2	118.69	Se1-Cd-N2	116.85	Se1-Hg-O1	89.99	Se1-Cd-O1	96.81
Se2-Hg-N1	112.60	Se2-Cd-N1	113.30	Se2-Hg-N1	100.81	Se2-Cd-N1	103.86

TABLE 9: Antibacterial activities of [LM(SeCN)$_2$] complexes.

Microorganisms	Zone of inhibition (mm)			
	Hg(SeCN)$_2$ [*]	Cd(SeCN)$_2$	(His)Cd(SeCN)$_2$	(Gly)Cd(SeCN)$_2$
E. coli	—	25	35	22
P. aeruginosa	10	20	18	32
S. typhi	10	28	32	29
S. aureus	22	20	22	20
K. pneumoniae	22	—	—	—

[*] [15]

series. This is also clear from ^{77}Se NMR data, which showed greater deshielding effect at the selenium via complexing to Histidine, which binds through two nitrogen atoms [32]. In general, good agreement of the experimental and theoretical IR starching bands observed for the prepared complexes with some blue shift of the calculated results because of the intermolecular interaction in the real IR experiment.

The CPMAS NMR spectral data for complexes (Gly)Cd(SeCN)$_2$ and (His)Cd(SeCN)$_2$ for ^{13}C and (Gly)Hg(SeCN)$_2$ and (His)Hg(SeCN)$_2$ for ^{15}N are shown in Tables 5 and 6, respectively. The solid-state ^{13}C and ^{15}N NMR spectra are shown in Figures 1 and 2, and the peaks are denoted by asterisk. The calculated chemical shift tensors are also compiled in Tables 5 and 6, along with the span, Ω, which describes the breadth of the chemical shift tensor and skew, κ, describing the shape of the powder pattern. From Table 5, solid-state ^{13}C NMR of Glycine and Histidine cadmium complexes shows increase in the chemical of the Se[13]CN NMR shift increased by ~2 ppm for mercury complex this is because the involvement of selenium in binding to the metal center, causing deshielding at the SeCN carbon. But in case of the ^{15}N NMR data in Table 6, Histidine and Glycine complexes show a significant downfield shift of the nitrogen atom signal of the amines group; additionally, Histidine shows downfield shift for the imidazole moiety, which improves imidazole nitrogen involvement in binding to metal.

3.2. Computation Study. Computational study shows that Glycine (C=O) and (C-O) bonds in the optimized structure (Scheme 1) are shorter in Cd complex than that in Hg

complex (Tables 7 and 8), which agree with experimental ^{13}C NMR results, while no differences in (C=O) and (C-O) bonds were observed in His complexes indicating less contribution of carboxylate in binding. Se-Cd is shorter than Se-Hg because of the size proximity in the sizes of Se and Cd atoms. It is worth mentioning here that the calculated bond lengths are comparable to reported experimental bond lengths for Se-Cd obtained by single crystals (2.723 and 2.828 Å) [20, 21]. It is also observed that Hg-N and Hg-O are longer than Cd-N and Cd-O because Cd is harder than Hg, and this results in better interaction. This may cause higher stability of Cd complexes in general. Nitrogen is a less electronegative atom than oxygen, so it can donate electron more easily to the metal and form stronger bonds with metals, which results in stronger chelation through two nitrogens than chelation through nitrogen and oxygen. L-Histidine complexes of Hg and Cd have shorter bonds than Glycine complexes indicating stronger bonds in Histidine complexes, and this agrees with the higher electron donation concluded from ^{77}Se NMR data.

3.3. Antibacterial Activity. The in vitro antibacterial activity studies were performed with Cd(II) complexes against activity of both gram-positive as well as gram-negative bacteria. Two complexes, (His)Cd(SeCN)$_2$ and (Gly)Cd(SeCN)$_2$, exhibited their antibacterial activity compared to Cd(SeCN)$_2$ except for K. pneumoniae which showed resistance to all the compounds tested, and Hg(SeCN)$_2$ inhibition was reported previously [15], while mercury complexes with Glycine and Histidine did not show significant antibacterial activity. The activities of the complexes are summarized in Table 9.

Synthesis, CP-MAS NMR Characterization, and Antibacterial Activities of Glycine and Histidine Complexes of Cd(SeCN)$_2$ and Hg(SeCN)$_2$

131

4. Conclusions

We have described the synthesis of Cd(II) and Hg(II) complexes of the type, [(L)M(SeCN)$_2$] (where L=Histidine or Glycine and M=Cd^{2+} or Hg^{2+}), for use as a potential antibacterial agents. Characterization of these compounds by EA, IR solution, and solid NMR of various nucleuses reveals that the metal complexes with Histidine are more strongly coordinated than that of the corresponding Glycine containing metal complexes. The Cd(II) complexes have shown good zone inhibition towards different microorganisms, and thier further biological evaluation is under process, while no significant antibacterial activity was observed for the mercury complexes.

Acknowledgment

This parer was supported by the KFUPM Research Committee under project no. IN100039.

References

[1] F. R. Keene, J. A. Smith, and J. G. Collins, "Metal complexes as structure-selective binding agents for nucleic acids," *Coordination Chemistry Reviews*, vol. 253, no. 15-16, pp. 2021–2035, 2009.

[2] J. A. Drewry and P. T. Gunning, "Recent advances in biosensory and medicinal therapeutic applications of zinc(II) and copper(II) coordination complexes," *Coordination Chemistry Reviews*, vol. 255, no. 3-4, pp. 459–472, 2011.

[3] T. Marino, N. Russo, and M. Toscano, "Potential energy surfaces for the gas-phase interaction between α-alanine and alkali metal ions (Li$^+$, Na$^+$, K$^+$). A density functional study," *Inorganic Chemistry*, vol. 40, no. 25, pp. 6439–6443, 2001.

[4] C. Orvig and M. J. Abrams, "Medicinal inorganic chemistry: introduction," *Chemical Reviews*, vol. 99, no. 9, pp. 2201–2204, 1999.

[5] Z. Guo and P. J. Sadler, "Medicinal inorganic chemistry," *Advances in Inorganic Chemistry*, vol. 49, pp. 183–306, 1999.

[6] P. C. Bruijnincx and P. J. Sadler, "New trends for metal complexes with anticancer activity," *Current Opinion in Chemical Biology*, vol. 12, no. 2, pp. 197–206, 2008.

[7] M. Peschke, A. T. Blades, and P. Kebarle, "Metalloion-ligand binding energies and biological function of metalloenzymes such as carbonic anhydrase. A study based on ab initio calculations and experimental ion-ligand equilibria in the gas phase," *Journal of the American Chemical Society*, vol. 122, no. 7, pp. 1492–1505, 2000.

[8] T. Marino, N. Russo, and M. Toscano, "Gas-phase metal ion (Li$^+$, Na$^+$, Cu$^+$) affinities of glycine and alanine," *Journal of Inorganic Biochemistry*, vol. 79, no. 1-4, pp. 179–185, 2000.

[9] T. Marino, M. Toscano, N. Russo, and A. Grannd, "Structural and electronic characterization of the complexes obtained by the interaction between bare and hydrated first-row transition-metal ions (Mn^{2+}, Fe^{2+}, Co^{2+}, Ni^{2+}, Cu^{2+}, Zn^{2+}) and glycine," *The Journal of Physical Chemistry B*, vol. 110, no. 48, pp. 24666–24673, 2006.

[10] H. Ai, Y. Bu, and K. Han, "Glycine-Zn$^+$/Zn^{2+} and their hydrates: on the number of water molecules necessary to stabilize the switterionic glycine-Zn$^+$/Zn^{2+} over the nonzwitterionic ones," *Journal of Chemical Physics*, vol. 118, no. 24, p. 10973, 2003.

[11] H. Pesonen, A. Sillanpaä, R. Aksela, and K. Laasonen, "Density functional complexation study of metal ions with poly(carboxylic acid) ligands—part 1: poly(acrylic acid) and poly(α-hydroxy acrylic acid)," *Polymer*, vol. 46, no. 26, pp. 12641–12652, 2005.

[12] K. Hasegawa, T. Ono, and T. Noguchi, "Ab initio density functional theory calculations and vibrational analysis of zinc-bound 4-methylimidazole as a model of a histidine ligand in metalloenzymes," *Journal of Physical Chemistry A*, vol. 106, no. 14, pp. 3377–3390, 2002.

[13] J. A. Leary, Z. Zhou, S. A. Ogden, and T. D. Williams, "Investigations of gas-phase lithium-peptide adducts: tandem. mass spectrometry and semiempirical studies," *Journal of the American Society for Mass Spectrometry*, vol. 1, no. 6, pp. 473–480, 1990.

[14] S. L. Li, J. Y. Wu, Y. P. Tian et al., "Design, crystal growth, characterization, and second-order nonlinear optical properties of two new three-dimensional coordination polymers containing selenocyanate ligands," *European Journal of Inorganic Chemistry*, no. 14, pp. 2900–2907, 2006.

[15] M. Nasiruzzaman Shaikh, B. A. Al-Maythalony, M. I. M. Wazeer, and A. A. Isab, "Complexations of 2-thiouracil and 2,4-dithiouracil with Cd(SeCN)$_2$ and Hg(SeCN)$_2$: NMR and antibacterial activity studies," *Spectroscopy*, vol. 25, no. 3-4, pp. 187–195, 2011.

[16] P. Barraud, M. Schubert, and F. H. T. Allain, "A strong ^{13}C chemical shift signature provides the coordination mode of histidines in zinc-binding proteins," *Journal of Biomolecular NMR*, vol. 53, no. 2, pp. 93–101, 2012.

[17] M. I. M. Wazeer, A. A. Isab, and M. Fettouhi, "New cadmium chloride complexes with imidazolidine-2-thione and its derivatives: X-ray structures, solid state and solution NMR and antimicrobial activity studies," *Polyhedron*, vol. 26, no. 8, pp. 1725–1730, 2007.

[18] J. Boeckmann, T. Reinert, and C. Näther, "Investigations on the thermal reactivity of Zn^{II} and Cd^{II} selenocyanato coordination compounds," *Zeitschrift für Anorganische und Allgemeine Chemie*, vol. 637, no. 7-8, pp. 940–946, 2011.

[19] S. N. Shukla, P. Gaur, S. Mathews, S. Khan, and A. Srivastava, "Synthesis, characterization, catalytic and biological activity of some bimetallic selenocyanate Lewis acid derivatives of N, N'- bis (2-chlorobenzylidene)ethylenediamine," *Journal of Coordination Chemistry*, vol. 61, no. 24, pp. 3913–3921, 2008.

[20] S. Sain, T. K. Maji, G. Mostafa, T. H. Lu, M. Y. Chiang, and N. R. Chaudhuri, "Self assembly towards the construction of molecular ladder and rectangular grid of cadmium(II)-selenocyanate," *Polyhedron*, vol. 21, no. 22, pp. 2293–2299, 2002.

[21] T. K. Maji, S. Sain, G. Mostafa, D. Das, T. H. Lu, and N. R. Chaudhuri, "Synthesis and crystal structure of selenocyanato bridged two dimensional supramolecular coordination compounds of cadmium(II)," *Journal of the Chemical Society, Dalton Transactions*, no. 21, pp. 3149–3153, 2001.

[22] A. A. Isab and M. I. M. Wazeer, "Complexation of Zn(II), Cd(II) and Hg(II) with thiourea and selenourea: a ^{1}H, ^{13}C, ^{15}N, ^{77}Se and ^{113}Cd solution and solid-state NMR study," *Journal of Coordination Chemistry*, vol. 58, no. 6, pp. 529–537, 2005.

[23] S. Hayashi and K. Hayamizu, "Chemical shift standards in high-resolution solid-state NMR (2) ^{15}N nuclei," *Bulletin of the Chemical Society of Japan*, vol. 64, no. 2, pp. 688–690, 1991.

[24] W. Kemp, *NMR in Chemistry*, Macmillan Education Ltd., London, UK, 1986.

[25] K. Eichele and R. E. Wasylischen, *W: Simulation Package, Version 1. 4. 4*, Dalhousie University, Halifax, Canada; University of Tubingen, Tübingen, Germany, 2001.

[26] M. J. Colaneri and J. Peisach, "A single crystal EPR and ESEEM analysis of Cu(II)-doped bis(L-histidinato)cadmium dihydrate," *Journal of the American Chemical Society*, vol. 117, no. 23, pp. 6308–6315, 1995.

[27] N. U. Zhanpeisov, M. Matsuoka, H. Yamashita, and M. Anpo, "Cluster quantum chemical ab initio study on the interaction of NO molecules with highly dispersed titanium oxides incorporated into silicalite and zeolites," *Journal of Physical Chemistry B*, vol. 102, no. 35, pp. 6915–6920, 1998.

[28] A. Nicklass, M. Dolg, H. Stoll, and H. Preuss, "Ab initio energy-adjusted pseudopotentials for the noble gases Ne through Xe: calculation of atomic dipole and quadrupole polarizabilities," *The Journal of Chemical Physics*, vol. 102, no. 22, pp. 8942–8952, 1995.

[29] M. Frisch, G. Trucks, H. Schlegel et al., *Gaussian 09, Rev. A. 1*, Gaussian Inc., Wallingford, Conn, USA, 2009.

[30] V. Navarro, M. L. Villarreal, G. Rojas, and X. Lozoya, "Antimicrobial evaluation of some plants used in Mexican traditional medicine for the treatment of infectious diseases," *Journal of Ethnopharmacology*, vol. 53, no. 3, pp. 143–147, 1996.

[31] P. Brooks and N. Davidson, "Mercury (II) complexes of imidazole and histidine," *Journal of the American Chemical Society*, vol. 82, no. 9, pp. 2118–2123, 1960.

[32] A. Renuka, K. Shakuntala, and P. C. Srinivasan, *Indian Journal of Chemistry*, vol. 37A, p. 5, 1998.

15

Osteoconductivity and Hydrophilicity of TiO$_2$ Coatings on Ti Substrates Prepared by Different Oxidizing Processes

Dai Yamamoto,[1] **Ikki Kawai,**[1] **Kensuke Kuroda,**[2] **Ryoichi Ichino,**[2]
Masazumi Okido,[2] **and Azusa Seki**[3]

[1] *Department of Materials Science and Engineering, Graduate School of Engineering, Nagoya University, Nagoya 464-8603, Japan*
[2] *EcoTopia Science Institute, Nagoya University, Nagoya 464-8603, Japan*
[3] *Hamri Co., Ltd., Tokyo 110-0005, Japan*

Correspondence should be addressed to Dai Yamamoto, yamamoto@f2.numse.nagoya-u.ac.jp

Academic Editor: Ian Butler

Various techniques for forming TiO$_2$ coatings on Ti have been investigated for the improvement of the osteoconductivity of Ti implants. However, it is not clear how the oxidizing process affects this osteoconductivity. In this study, TiO$_2$ coatings were prepared using the following three processes: anodizing in 0.1 M H$_3$PO$_4$ or 0.1 M NaOH aqueous solution; thermal oxidation at 673 K for 2 h in air; and a two-step process of anodizing followed by thermal oxidation. The oxide coatings were evaluated using SEM, XRD, and XPS. The water contact angle on the TiO$_2$ coatings was measured as a surface property. The osteoconductivity of these samples was evaluated by measuring the contact ratio of formed hard tissue on the implanted samples (defined as the R_{B-I} value) after 14 d implantation in rats' tibias. Anatase was formed by anodizing and rutile by thermal oxidation, but the difference in the TiO$_2$ crystal structure did not influence the osteoconductivity. Anodized TiO$_2$ coatings were hydrophilic, but thermally oxidized TiO$_2$ coatings were less hydrophilic than anodized TiO$_2$ coatings because they lacked in surface OH groups. The TiO$_2$ coating process using anodizing without thermal oxidation gave effective improvement of the osteoconductivity of Ti samples.

1. Introduction

Titanium has been widely used in dental and orthopedic implants because of its good biocompatibility and high corrosion resistance [1, 2]. However, Ti in itself does not always show good performance to form hard tissue on its surface in living bodies. Therefore, proper surface treatment to enhance bone-forming ability, as represented by hydroxyapatite (HAp) coating [3–10], has been studied for a long time. Similar to HAp, TiO$_2$ is also important as an osteoconductive substance because it has been shown to exhibit strong physicochemical fixation with living bone, even though it is not a component of natural bone [11]. There are many coating processes to create TiO$_2$ films on Ti substrates, such as thermal oxidation [12], chemical methods [13–15], physical vapor deposition [16, 17], and anodizing [18–21]. These processes are classified into hydroprocesses and pyroprocesses. Anodizing is a widely used hydroprocess; it is performed in various aqueous solutions at an arbitrary

applied voltage. On the other hand, thermal oxidation is one of pyroprocesses, which is performed in various atmospheres at high temperature. Both processes have been used to prepare bioactive TiO$_2$ coatings on Ti [20, 22–25].

Previously, various parameters of TiO$_2$ coating, such as crystal structure, surface roughness, and film thickness, were reported to influence the osteoconductivity of the coating [20, 27, 28]. Even though these parameters can be altered simultaneously when changing the oxidizing condition, they have not been well controlled to compare the surface properties of TiO$_2$ coatings prepared with different oxidizing processes in previous works, which made it unclear what kind of surface property had influence on the hard tissue formation in *in vivo* test. Therefore, in this study, we prepared TiO$_2$ coatings with controlled surface structure on Ti using anodizing and/or thermal oxidation and investigated the chemical influence of oxidizing processes on the osteoconductivity.

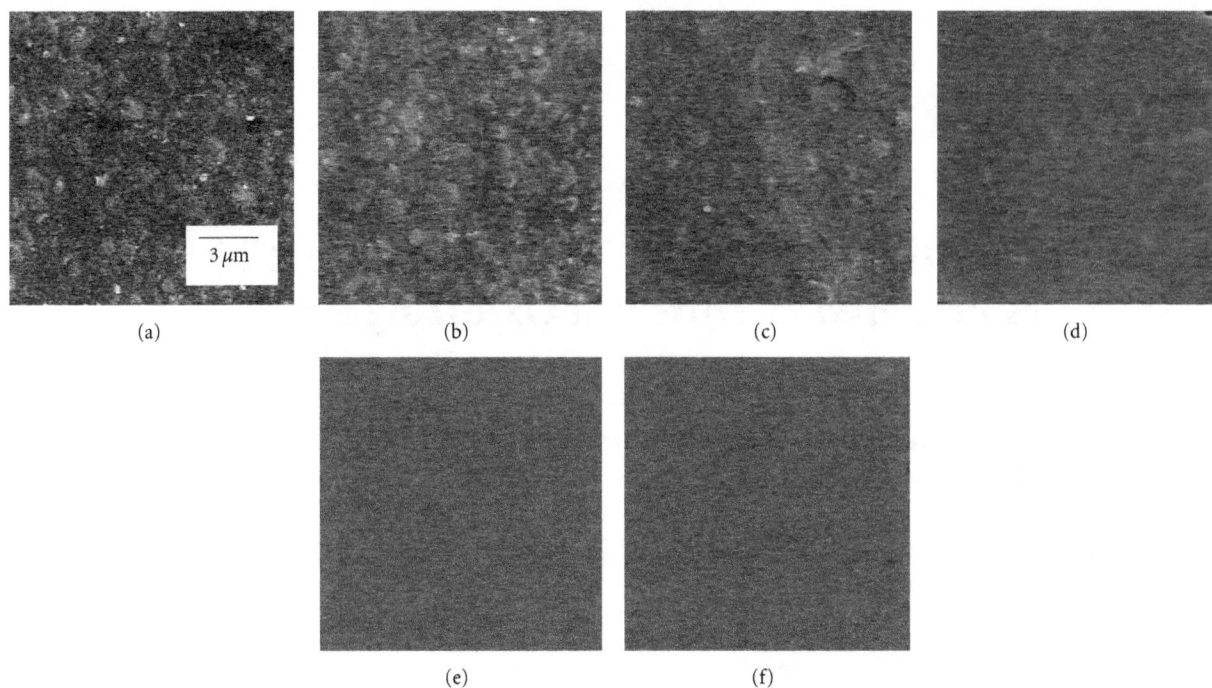

	Treatment	Oxidizing process		Crystal structure of TiO$_2$	Ra (μm)	WCA (deg.)
		Anodizing	Thermal oxidation			
(a)	AZ	H$_3$PO$_4$ 100 V	—	Anatase	0.082	28
(b)	AZ-TO	70 V	673 K, 2 h, air		0.09	65
(c)	AZ	NaOH 80 V	—		0.092	38
(d)	AZ-TO	50 V	673 K, 2 h, air		0.098	63
(e)	TO	—	673 K, 2 h, air	Rutile	0.087	66
(f)	As-polished	—	—	—	0.057	71

FIGURE 1: The surface SEM images, surface roughness Ra, and WCA of the samples prepared in this study.

2. Materials and Methods

2.1. Preparation of Ti Substrates.
Commercially pure Ti (Cp-Ti) disks (for anodizing, area = 1.13 cm^2) and plates (for thermal oxidation, area = 1 cm^2) were used as substrates to prepare coatings for surface analysis, and rods (for both anodizing and thermal oxidation, dimensions = ϕ2 $\times$ 5 mm) for *in vivo* testing. The Cp-Ti disks were covered with epoxy resin, except for the face that would be in contact with the aqueous solution. All of the substrates were polished by emery paper followed by buffing using Al$_2$O$_3$ particles (particle size = 0.05 μm). After polishing, the substrates were cleaned and then degreased with ethanol.

2.2. TiO$_2$ Coatings.
The following three methods were used to form TiO$_2$ coatings on Ti substrates. The preparation conditions were selected in each case so that all of the oxide coatings had the same film thickness to exclude the influence of that variable on the osteoconductivity of the formed coatings.

2.2.1. Anodizing in Aqueous Solutions.
A Ti substrate and a Pt coil were set as anode and cathode, respectively, and reference electrode was not used. The anodizing voltage was increased from 0 V up to 100 V in 0.1 M H$_3$PO$_4$ aqueous solution and up to 80 V in 0.1 M NaOH aqueous solution at a rate of 0.1 V s^{-1}. The aqueous solution was stirred and kept at a constant temperature (298 K) in a water bath during anodizing. This oxidation process is denoted as "AZ treatment" in the following description.

2.2.2. Thermal Oxidation.
Titanium substrates were heated to 673 K at a rate of 4 K min^{-1} in air in an electric resistance furnace, and kept at that temperature for 2 h. The substrates were then cooled in the furnace, in the same atmosphere. This oxidation process is denoted as "TO treatment."

2.2.3. Two-Step Process of Anodizing and Thermal Oxidation.
In the first step, substrates were anodized with the voltage increasing from 0 V to 70 V in 0.1 M H$_3$PO$_4$ aqueous solution or up to 50 V in 0.1 M NaOH aqueous solutions at a rate of 0.1 V s^{-1}. The anodized samples then were heated to 673 K at a rate of 4 K min^{-1} in air in an electric resistance furnace, and kept at that temperature for 2 h. The substrates were then cooled in the furnace, in the same atmosphere. This two-step oxidation process is denoted as "AZ-TO treatment."

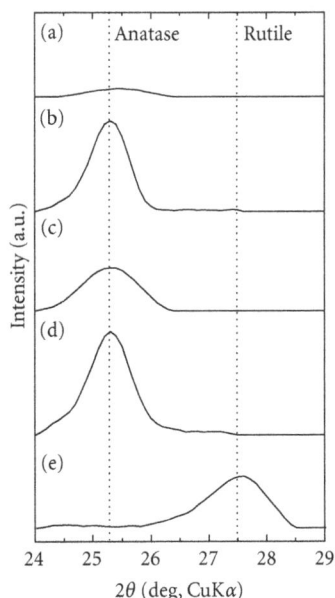

FIGURE 2: The XRD patterns of the surface coatings on samples treated by (a) AZ treatment up to 100 V in 0.1 M H$_3$PO$_4$ aq., (b) AZ-TO treatment (anodized in H$_3$PO$_4$ aq.), (c) AZ treatment up to 80 V in 0.1 M NaOH aq., (d) AZ-TO treatment (anodized in NaOH aq.), and (e) TO treatment.

2.3. Analysis of the Coatings. All of the samples were sterilized using an autoclave unit at 394 K for a period of 20 min before following. The surface morphology was observed using a scanning electron microscope (SEM). The coated films were identified using thin-film X-ray diffraction (XRD) and X-ray photoelectron spectroscopy (XPS). The surface roughness was measured using a confocal laser scanning microscope with an analysis area of 150 μm $\times$ 112 μm. The arithmetical mean of the surface roughness (Ra) was used, as this value was not distorted by any local scarring of the sample. The contact angle of distilled water (WCA) was measured at three different points for each sample using a 2 μL droplet of distilled water 24 h after autoclaving, and the average value was used as the WCA value.

2.4. In Vivo Test. In vivo tests were performed for all of the AZ-treated, TO-treated, AZ-TO-treated, and as-polished samples. Because the experimental procedure for our *in vivo* study was almost the same as described in previous reports [8], it is described only briefly here. Before surgery, all of the implants were cleaned in distilled water and immersed in a chlorhexidine gluconate solution. Ten-week-old male Sprague Dawley rats (Charles River Japan, Inc., Japan) were used in our experimental procedures. The samples were implanted in the tibial metaphysis of the rats. A slightly oversized hole, which did not pass through to the rear side of the bone, was created using a low-speed rotary drill. Subsequently, the implants were inserted into these holes, and then the subcutaneous tissue and skin were closed and sterilized.

The rats were sacrificed after a period of 14 d, and the implants with their surrounding tissue were retrieved. The samples were fixed in a 10% neutral buffered formalin

solution, dehydrated in a graded series of ethanol, and embedded in methylmethacrylate. Following polymerization, each implant block was sectioned longitudinally into 20 μm thick slices. These sections were then stained with toluidine blue.

The sum of the linear bone contact with the implant surface was measured and was expressed as a percentage over the entire implant length (the bone-implant contact ratio, R_{B-I}) in the cancellous bone and the cortical bone parts [8]. Significant differences in the bone-implant contact ratio were analyzed statistically using the Tukey-Kramer method [29]. Differences were considered statistically significant at the $P < 0.05$ level. This animal study was conducted in a laboratory accredited by AAALAC International (Association for Assessment and Accreditation of Laboratory Animal Care International).

3. Results and Discussion

3.1. Evaluation of Oxide Coatings. The surface morphology and surface roughness of these samples are shown in Figure 1. Figure 2 shows the XRD patterns of the samples after AZ treatment in (a) 0.1 M H$_3$PO$_4$ aqueous solution up to 100 V, (b) 0.1 M NaOH aqueous solution up to 80 V, (c) TO treatment at 673 K for 2 h in air, and AZ-TO treatment (anodized in (d) H$_3$PO$_4$ and (e) NaOH aqueous solution). Figure 3 shows the XPS spectra ((A) O1s, (B) P2p, and (C) Na1s) of each of the samples.

Surface was kept fine (Ra/μm < 0.1) after any oxidizing process (Figures 1(a)–1(e)). All of the formed oxide coating had the same film thickness of *c.a.* 120 nm [30]. However, the crystal structure of the oxide coating was different depending on the oxidizing process; anatase-type TiO$_2$ was formed by AZ treatment (Figures 2(a), and 2(b)), but rutile-type TiO$_2$ by TO treatment (Figure 2(e)). Each of AZ-treated samples contained PO$_4{}^{3-}$ or Na$^+$ in the oxide coating which derived from aqueous solutions (Figures 3(B)(a) and 3(B)(c)), as described in our previous report [27]. Anatase was also formed by AZ-TO treatment, which was the same crystal structure as the AZ-treated coatings (Figures 2(b), and 2(d)), and no rutile was detected in the coatings using XRD, despite the thermal oxidation.

The XPS analysis showed different O1s spectra depending on the oxidation process. When deconvoluting the O1s spectrum of the as-polished sample in the same way as in a previous work [31], the spectrum was divided into three predominant peaks (530.1, 531.5, and 532.5 eV) originating from O^{2-}, OH group, and adsorbed water (Figure 3(A)(f)), which derived from the thin natural oxide layer on the Ti substrate. These components were also present after anodizing (Figures 3(A)(a), and 3(A)(c)). However, OH groups was lost from the surface after thermal oxidation (Figures 3(A)(b), 3(A)(d), and 3(A)(e)), in the same way as a previous report by Zhao et al. [24]. Instead, large amounts of water molecule adsorbed on the surface of thermally oxidized samples. The absence of PO$_4{}^{3-}$ or Na$^+$ on the AZ-TO-treated samples (Figures 3(B)(b) and 3(B)(d)), which was detected on the anodized-only samples (Figures 3(B)(a) and 3(C)(c)), supports the idea that the oxide layer grows outward during

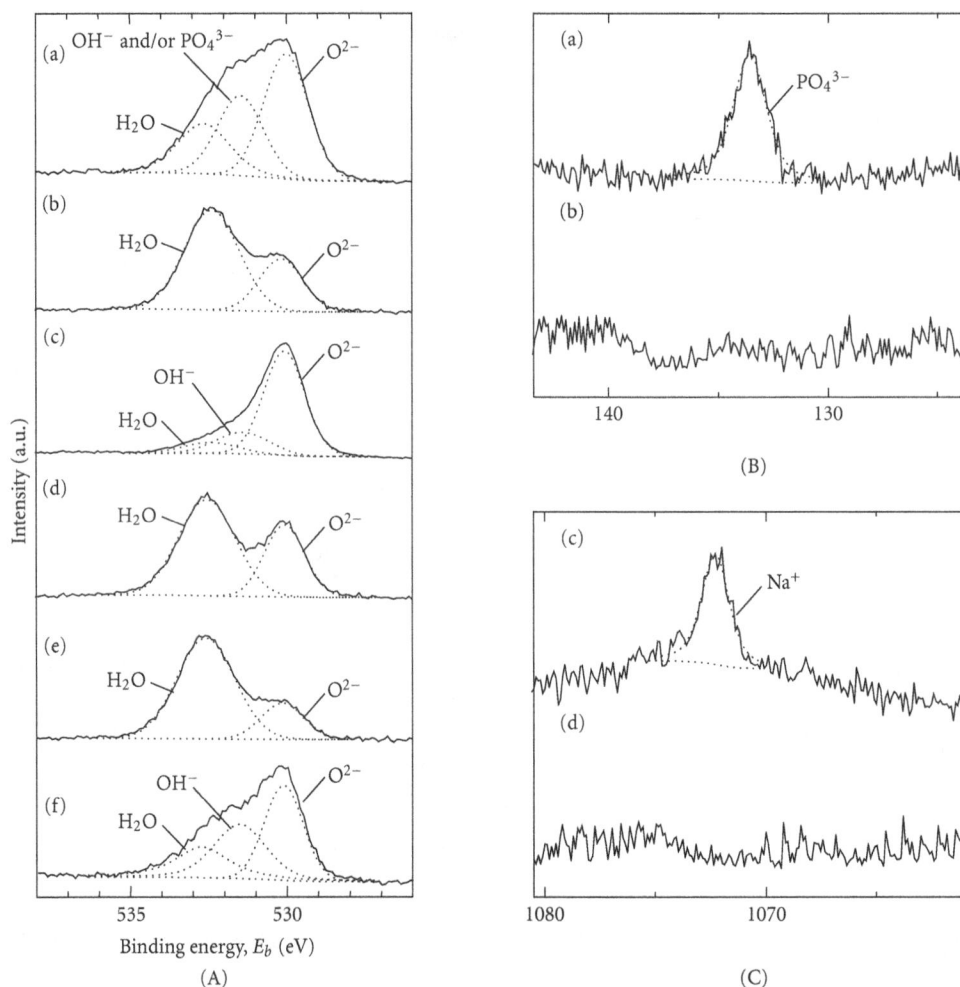

FIGURE 3: The (A) O1s, (B) P2p, and (C) Na1s XPS surface spectra of samples treated by (a) AZ treatment up 100 V in 0.1 M H_3PO_4 aq., (b) AZ-TO treatment (anodized in H_3PO_4 aq.), (c) AZ treatment up to 80 V in 0.1 M NaOH aq., (d) AZ-TO treatment (anodized in NaOH aq.), and (e) TO treatment.

thermal oxidation, as noted by Kumar et al. [32]. The sample anodized in H_3PO_4 aqueous solution is considered to have surface OH groups, but the corresponding peak could not be separated from the peak of PO_4^{3-} at the binding energy of around 531.5 eV (Figure 3(A)(a)). Although it is not shown in this figure, hydrocarbon was detected as a contaminant at the surface of all of the samples at the same adsorption level.

The surface hydrophilicity of oxidized samples was different depending on their oxidizing processes. As listed in Figure 1, WCA values decreased largely from 71 deg. (f) to less than 40 deg. ((a), (c)) after AZ treatment, regardless of the type of aqueous solutions used for anodizing. The anodizing process did not significantly change the amounts of OH group and adsorbed hydrocarbon at the surface. It was already shown that solute ions (PO_4^{3-} and Na^+) contained in anodized coatings did not contribute to the reduction of WCA [28]. On the other hand, all of the thermally oxidized samples (i.e., TO-treated sample (Figure 1(e)) and AZ-TO-treated samples (Figures 1(b), and 1(d))) had high WCA values similar to the as-polished sample. Since TO-treated sample and AZ-TO-treated samples had almost the

same WCA values in spite of different crystal structure of formed TiO_2, it is thought that the crystal structure of TiO_2, whether anatase or rutile, did not have strong influence on the hydrophilicity. In addition, the increased amount of adsorbed water molecule on thermally treated samples did not contribute to the decrease of the WCA value. Based on these results, it is reasonable to think that the absence of OH groups on the surface is responsible for the high WCA values of TO- and AZ-TO-treated samples. In other words, anodizing can improve the hydrophilicity of as-polished samples as a consequence of forming TiO_2 films having surface OH groups, but thermal oxidation does not improve the hydrophilicity because of the lack of surface OH groups on the TiO_2 films.

3.2. In Vivo Evaluation. Figure 4 shows the bone-implant contact ratios, R_{B-I}, of the samples in both the cortical bone part (A) and the cancellous bone part (B). In the cancellous bone part, the R_{B-I} values approached nearly 30% with any coating process; this value agreed well with the percentage of hard tissue in healthy cancellous bone in rats. According to

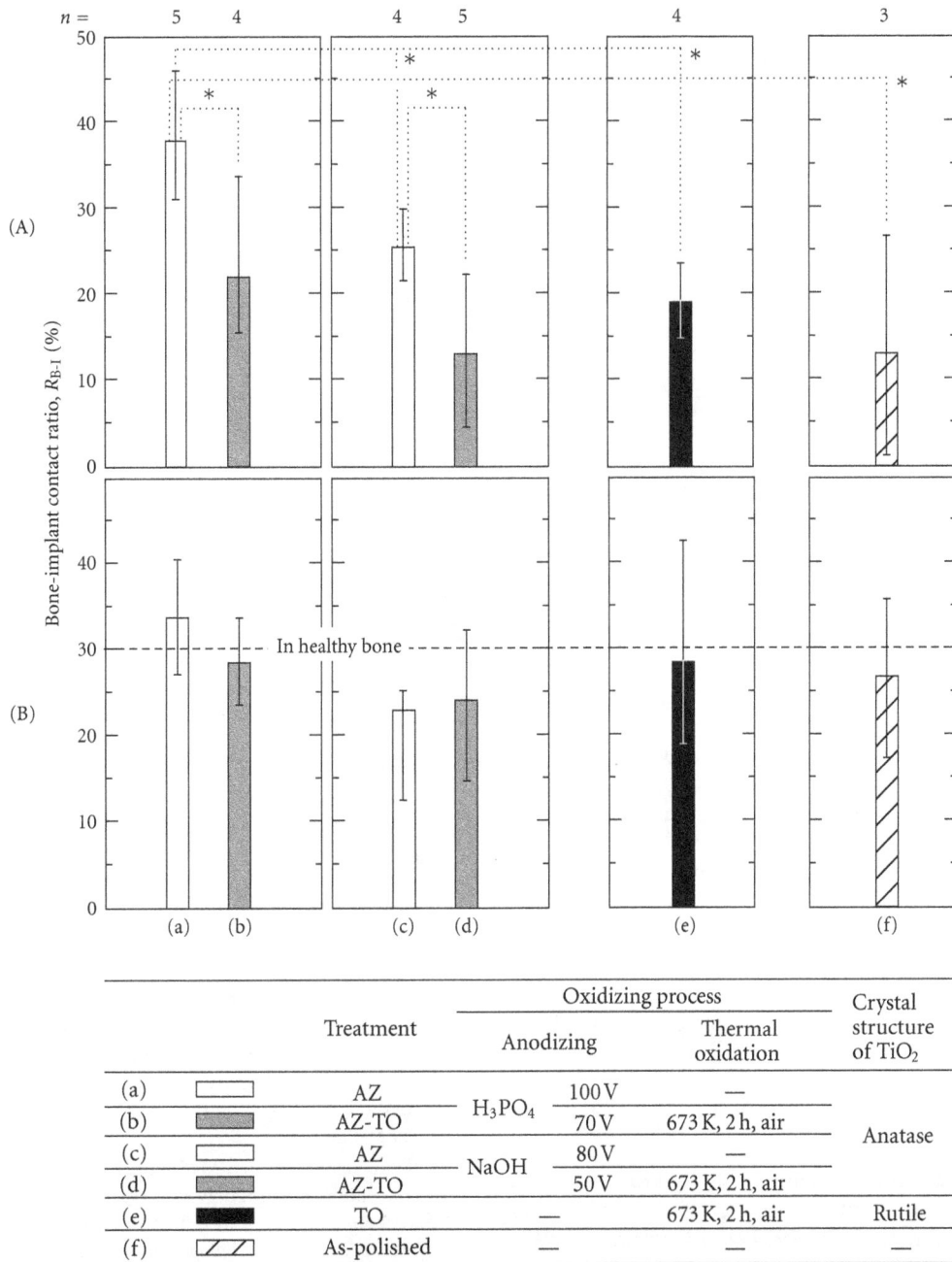

FIGURE 4: Bone-implant contact ratio, R_{B-I}, of Ti samples in both (A) the cortical bone part and (B) the cancellous bone part. $^*P < 0.05$.

this, it is difficult to judge the influence of the oxidizing process on R_{B-I} values in cancellous bone parts in the early stages after implantation. On the other hand, in the cortical bone part (A), the R_{B-I} values were obviously different depending on the oxidizing process for the slow formation rate of hard tissue. Therefore, we will focus on the cortical bone part in the following discussion.

The R_{B-I} values of samples varied significantly depending on the oxidizing process. Both AZ-treated samples (Figures 4(A)(a) and 4(A)(c)) tended to have higher R_{B-I} values than the as-polished sample (Figure 4(A)(f)). In contrast, the R_{B-I} values of thermally oxidized samples (b), (d), (e) were as low

as that of the as-polished sample. The low R_{B-I} values of the AZ-TO-treated samples mean that the R_{B-I} values of the AZ-treated samples decreased significantly after additional thermal oxidation, regardless of the type of aqueous solutions used for AZ treatment (Figures 4(A)(b), and 4(A)(d)). These results indicate that thermal oxidation attributed to decrease the osteoconductivity of TiO₂ coatings, and also that the different crystal structures of TiO₂ did not influence the osteoconductivity. Furthermore, all of the thermally oxidized samples were similar, with WCA values ranging from 60 (deg.) to 70 (deg.) and also R_{B-I} values ranging from 10% to 20%. These values followed the relation between R_{B-I} and

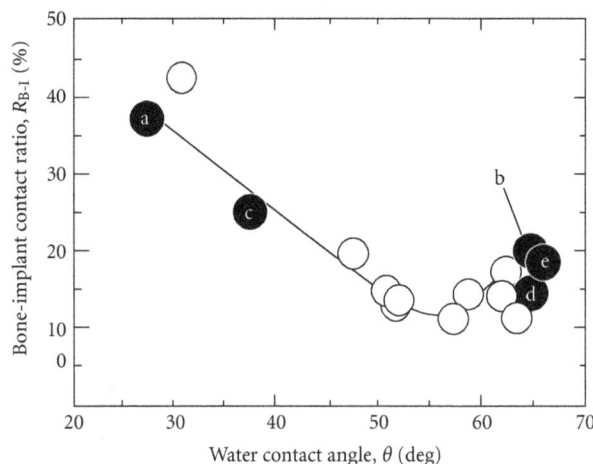

FIGURE 5: Relationship between the WCA and $R_{B\text{-}1}$ values in the cortical bone part. The symbols in closed circle correspond to the letters in Figure 4. Open circles represent the data for the sample anodized in various aqueous solutions, taken from previous work [26].

WCA reported in our previous study (Figure 5) [28]. This means that variation of surface hydrophilicity was the reason for the different osteoconductivity of TiO_2 coatings prepared with different oxidizing process. Therefore, it was thought that the surface properties, such as WCA and the amount of OH groups, strongly affected the osteoconductivity, rather than the coating substance, crystal structure, and coating process. TiO_2 coating processes using only anodizing and not thermal oxidation improved the osteoconductivity of Ti samples effectively.

4. Conclusions

In this study, we prepared TiO_2 coatings with controlled surface structure on Ti using anodizing and/or thermal oxidation and investigated the chemical influence of oxidizing processes on the osteoconductivity. The following results were obtained.

(1) Anatase was formed by anodizing, and rutile by thermal oxidation, but the difference of TiO_2 crystal structure did not influence their osteoconductivity.

(2) Anodized TiO_2 coatings were hydrophilic, but thermally oxidized TiO_2 coatings were less hydrophilic than anodized-only TiO_2 coatings because of the lack of surface OH groups.

(3) The TiO_2 coating process using only anodizing, not including thermal oxidation, gave the effective improvement of the osteoconductivity of Ti samples.

Acknowledgments

This work was partially supported by Grant-in-Aid for JSPS fellows (No. 2310401), and the Global COE program (COE for Education and Research of Micro-Nano Mechatronics) from the Japan Society for the Promotion of Science (JSPS), and a Grant-in-Aid for Scientific Research (C) (No. 21560719).

References

[1] R. Adell, B. Eriksson, U. Lekholm, P. I. Brånemark, and T. Jemt, "Long-term follow-up study of osseointegrated implants in the treatment of totally edentulous jaws," *The International Journal of Oral and Maxillofacial Implants*, vol. 5, no. 4, pp. 347–359, 1990.

[2] D. van Steenberghe, U. Lekholm, C. Bolender et al., "Applicability of osseointegrated oral implants in the rehabilitation of partial edentulism: a prospective multicenter study on 558 fixtures," *The International Journal of Oral and Maxillofacial Implants*, vol. 5, no. 3, pp. 272–281, 1990.

[3] L. L. Hench and J. Wilson, *An Introduction to Bioceramics*, chapter 1, World Scientific, Singapore, 1993.

[4] K. Kuroda, R. Ichino, M. Okido, and O. Takai, "Effects of ion concentration and pH on hydroxyapatite deposition from aqueous solution onto titanium by the thermal substrate method," *Journal of Biomedical Materials Research*, vol. 61, no. 3, pp. 354–359, 2002.

[5] K. Kuroda, R. Ichino, M. Okido, and O. Takai, "Effects of ion concentration and pH on hydroxyapatite deposition from aqueous solution onto titanium by the thermal substrate method," *Journal of Biomedical Materials Research*, vol. 61, no. 3, pp. 354–359, 2002.

[6] K. Kuroda, Y. Miyashita, R. Ichino, M. Okido, and O. Takai, "Preparation of calcium phosphate coatings on titanium using the thermal substrate method and their in vitro evaluation," *Materials Transactions*, vol. 43, no. 12, pp. 3015–3019, 2002.

[7] K. Kuroda, S. Nakamoto, R. Ichino, M. Okido, and R. M. Pilliar, "Hydroxyapatite coatings on a 3D porous surface using thermal substrate method," *Materials Transactions*, vol. 46, no. 7, pp. 1633–1635, 2005.

[8] K. Kuroda, S. Nakamoto, Y. Miyashita, R. Ichino, and M. Okido, "Osteoinductivity of HAp films with different surface morphologies coated by the thermal substrate method in aqueous solutions," *Materials Transactions*, vol. 47, no. 5, pp. 1391–1394, 2006.

[9] K. Kuroda, M. Moriyama, R. Ichino, M. Okido, and A. Seki, "Formation and in vivo evaluation of carbonate apatite and carbonate apatite/CaCO$_3$ composite films using the thermal

substrate method in aqueous solution," *Materials Transactions*, vol. 49, no. 6, pp. 1434–1440, 2008.

[10] K. Kuroda, M. Moriyama, R. Ichino, M. Okido, and A. Seki, "Formation and osteoconductivity of hydroxyapatite/collagen composite films using a thermal substrate method in aqueous solutions," *Materials Transactions*, vol. 50, no. 5, pp. 1190–1195, 2009.

[11] R. Hazan, R. Brener, and U. Oron, "Bone growth to metal implants is regulated by their surface chemical properties," *Biomaterials*, vol. 14, no. 8, pp. 570–574, 1993.

[12] S. Fujibayashi, M. Neo, H. M. Kim, T. Kokubo, and T. Nakamura, "Osteoinduction of porous bioactive titanium metal," *Biomaterials*, vol. 25, no. 3, pp. 443–450, 2004.

[13] L. Jonášová, F. A. Müller, A. Helebrant, J. Strnad, and P. Greil, "Biomimetic apatite formation on chemically treated titanium," *Biomaterials*, vol. 25, no. 7-8, pp. 1187–1194, 2004.

[14] F. Xiao, K. Tsuru, S. Hayakawa, and A. Osaka, "In vitro apatite deposition on titania film derived from chemical treatment of Ti substrates with an oxysulfate solution containing hydrogen peroxide at low temperature," *Thin Solid Films*, vol. 441, no. 1-2, pp. 271–276, 2003.

[15] J. M. Wu, S. Hayakawa, K. Tsuru, and A. Osaka, "Porous titania films prepared from interactions of titanium with hydrogen peroxide solution," *Scripta Materialia*, vol. 46, no. 1, pp. 101–106, 2002.

[16] K. R. Wu, C. H. Ting, W. C. Liu, C. H. Lin, and J. K. Wu, "Successive deposition of layered titanium oxide/indium tin oxide films on unheated substrates by twin direct current magnetron sputtering," *Thin Solid Films*, vol. 500, no. 1-2, pp. 110–116, 2006.

[17] L. S. Hsu, R. Rujkorakarn, J. R. Sites, and C. Y. She, "Thermally induced crystallization of amorphous-titania films," *Journal of Applied Physics*, vol. 59, no. 10, pp. 3475–3480, 1986.

[18] Y. T. Sul, C. B. Johansson, S. Petronis et al., "Characteristics of the surface oxides on turned and electrochemically oxidized pure titanium implants up to dielectric breakdown: the oxide thickness, micropore configurations, surface roughness, crystal structure and chemical composition," *Biomaterials*, vol. 23, no. 2, pp. 491–501, 2002.

[19] J. P. Schreckenbach, G. Marx, F. Schlottig, M. Textor, and N. D. Spencer, "Characterization of anodic spark-converted titanium surfaces for biomedical applications," *Journal of Materials Science: Materials in Medicine*, vol. 10, no. 8, pp. 453–457, 1999.

[20] B. Yang, M. Uchida, H. M. Kim, X. Zhang, and T. Kokubo, "Preparation of bioactive titanium metal via anodic oxidation treatment," *Biomaterials*, vol. 25, no. 6, pp. 1003–1010, 2004.

[21] L. A. de Sena, N. C. C. Rocha, M. C. Andrade, and G. A. Soares, "Bioactivity assessment of titanium sheets electrochemically coated with thick oxide film," *Surface and Coatings Technology*, vol. 166, no. 2-3, pp. 254–258, 2003.

[22] K. Das, S. Bose, and A. Bandyopadhyay, "Surface modifications and cell-materials interactions with anodized Ti," *Acta Biomaterialia*, vol. 3, no. 4, pp. 573–585, 2007.

[23] P. S. Vanzillotta, M. S. Sader, I. N. Bastos, and G. De Almeida Soares, "Improvement of in vitro titanium bioactivity by three different surface treatments," *Dental Materials*, vol. 22, no. 3, pp. 275–282, 2006.

[24] Y. Zhao, T. Xiong, and W. Huang, "Effect of heat treatment on bioactivity of anodic titania films," *Applied Surface Science*, vol. 256, no. 10, pp. 3073–3076, 2010.

[25] B. Feng, J. Weng, B. C. Yang, S. X. Qu, and X. D. Zhang, "Characterization of surface oxide films on titanium and adhesion of osteoblast," *Biomaterials*, vol. 24, no. 25, pp. 4663–4670, 2003.

[26] C. Larsson, P. Thomsen, J. Lausmaa, M. Rodahl, B. Kasemo, and L. E. Ericson, "Bone response to surface modified titanium implants: studies on electropolished implants with different oxide thicknesses and morphology," *Biomaterials*, vol. 15, no. 13, pp. 1062–1074, 1994.

[27] D. Yamamoto, I. Kawai, K. Kuroda, R. Ichino, M. Okido, and A. Seki, "Osteoconductivity of anodized titanium with controlled micron-level surface roughness," *Materials Transactions*, vol. 52, no. 8, pp. 1650–1654, 2011.

[28] D. Yamamoto, T. Iida, K. Arii et al., "Surface hydrophilicity and osteoconductivity of anodized Ti in aqueous solutions with various solute ions," *Materials Transactions*, vol. 53, no. 11, pp. 1956–1196, 2012.

[29] C. Y. Kramer, "Extension of multiple range tests to group means with unequal numbers of replications," *Biometrics*, vol. 12, no. 3, pp. 307–310, 1956.

[30] S. Van Gils, P. Mast, E. Stijns, and H. Terryn, "Colour properties of barrier anodic oxide films on aluminium and titanium studied with total reflectance and spectroscopic ellipsometry," *Surface and Coatings Technology*, vol. 185, no. 2-3, pp. 303–310, 2004.

[31] Y. Tanaka, M. Nakai, T. Akahori et al., "Characterization of air-formed surface oxide film on Ti-29Nb-13Ta-4.6Zr alloy surface using XPS and AES," *Corrosion Science*, vol. 50, no. 8, pp. 2111–2116, 2008.

[32] S. Kumar, T. S. N. S. Narayanan, S. G. S. Raman, and S. K. Seshadri, "Thermal oxidation of Ti6Al4V alloy: microstructural and electrochemical characterization," *Materials Chemistry and Physics*, vol. 119, no. 1-2, pp. 337–346, 2010.

16

Novel Organotin(IV) Schiff Base Complexes with Histidine Derivatives: Synthesis, Characterization, and Biological Activity

Ariadna Garza-Ortiz,[1] Carlos Camacho-Camacho,[1] Teresita Sainz-Espuñes,[1]
Irma Rojas-Oviedo,[1] Luis Raúl Gutiérrez-Lucas,[1]
Atilano Gutierrez Carrillo,[2] and Marco A. Vera Ramirez[2]

[1] *Departamento de Sistemas Biológicos, Universidad Autónoma Metropolitana-Unidad Xochimilco, Calzada de Hueso 1100, Colonia Villa Quietud, 04960 Coyoacán, D.F., Mexico*
[2] *Departamento de Resonancia Magnética Nuclear, Universidad Autónoma Metropolitana, Unidad Iztapalapa, San Rafael Atlixco, No. 186, Col. Vicentina, 09340 Iztapalapa, D.F., Mexico*

Correspondence should be addressed to Carlos Camacho-Camacho; ccamacho@correo.xoc.uam.mx

Academic Editor: Claudio Pettinari

Five novel tin Schiff base complexes with histidine analogues (derived from the condensation reaction between L-histidine and 3,5-di-*tert*-butyl-2-hydroxybenzaldehyde) have been synthesized and characterized. Characterization has been completed by IR and high-resolution mass spectroscopy, 1D and 2D solution NMR (^{1}H, ^{13}C and ^{119}Sn), as well as solid state ^{119}Sn NMR. The spectroscopic evidence shows two types of structures: a trigonal bipyramidal stereochemistry with the tin atom coordinated to five donating atoms (two oxygen atoms, one nitrogen atom, and two carbon atoms belonging to the alkyl moieties), where one molecule of ligand is coordinated in a three dentate fashion. The second structure is spectroscopically described as a tetrahedral tin complex with four donating atoms (one oxygen atom coordinated to the metal and three carbon atoms belonging to the alkyl or aryl substituents), with one molecule of ligand attached. The antimicrobial activity of the tin compounds has been tested against the growth of bacteria *in vitro* to assess their bactericidal properties. While pentacoordinated compounds **1**, **2**, and **3** are described as moderate effective to noneffective drugs against both Gram-positive and Gram-negative bacteria, tetracoordinated tin(IV) compounds **4** and **5** are considered as moderate effective and most effective compounds, respectively, against the methicillin-resistant *Staphylococcus aureus* strains (Gram-positive).

1. Introduction

The use of metal complexes as chemotherapeutic agents in the treatment of illness, which are a major public health concern, appears as a very attractive alternative. The success of cisplatin for the treatment of testicular and ovarian cancer attracted research attention to other metal-based antineoplastic agents. Metal-based compounds are of particular interest due to their physical and chemical properties. Properties such as ligand exchange rates, redox properties, oxidation states, coordination affinities, solubility, biodisponibility, and biodistribution could be modified in order to increase the therapeutic effect while reducing the side effects.

Although the design of a metal-based compound with good therapeutic index is still rather empirical, a number of potential metal-based bactericide compounds have been fully described in the literature. The evidence about specific or selective bonding of metals and organometallic species to donor sites in biological structures is very limited, so trustable mechanisms of biological activity and valid structure-activity relationships are limited as well.

One approach that could produce successful results involves the metal coordination of ligands with well-known biological activity. In this way, the designed metal-based compound combines ligands with an important biological activity and pharmacologically active metals in a single chemical moiety. This strategy could produce enhanced efficacy and reduced toxic or side effects while lowering the therapeutic doses and/or overcoming drug resistance mechanisms. Additionally the metal could act as a carrier and/or stabilizer of the drug until it is able to reach the target. At the same time, the organic ligand with well-known biological activity could transport and protect the metal, then avoiding side reactions on its route to the potential targets. The metal-ligand combined effects may result in an important improvement in the activity of the resulting coordination compounds [1, 2].

Amino acids and their derivatives are excellent ligand candidates due to their coordination properties and biological importance. In addition, their solubility properties and specific mechanisms of transport could facilitate their biological uptake. In particular histidine is a very important bioactive amino acid with multiple physiological functions.

The histidine binding to transition metal ions in biological systems has a major physicochemical role in several proteins, as the X-ray structural determination studies of metalloproteins like carbonic anhydrase, carboxypeptidase, plastocyanin, or azurin among others have demonstrated [3]. Additionally, histidine plays a major role in the zinc metabolism acting as the major zinc binding moiety in serum [4].

Schiff bases represent an important group of compounds in organic chemistry because they are starting materials in the synthesis of industrial products where carbon-nitrogen bonds are present [5]. In particular, this N–C bond is involved in several biological functions allowing the Schiff bases to behave, for instance, as antimicrobial, anti-inflamatory, antitumour, or antiviral drugs [6].

The design of organotin derivatives with biologically important ligands, like antibiotics [7, 8], anticancer drugs, [9] and some other biologically relevant substrates [10], has been explored in the past years [11]. Organotin derivatives of amino acids have been of interest as possible biocides [12]; for example, tricyclohexyltin alaninate is a powerful fungicide and bactericide [13]. This compound has been described as a substrate of peptide syntheses as well [14]. Amino acids and dipeptides which are considered potentially polydentate ligands could afford interesting structural possibilities when coordinated to organotin(IV) groups, since organotin(IV) compounds adopt higher coordination numbers whenever favorable conditions exist [15].

One of our research goals is the study of organotin(IV) compounds with important biological activity. In this order of ideas we designed, prepared, and characterized diorganotin and triorganotin derivatives of an L-histidine Schiff base analogue, with biological application as antibacterial compounds. The L-histidine derivative was obtained by condensation reaction of L-histidine and 3,5-di-tert-butyl-2-hydroxybenzaldehyde. Posterior complexation of the L-histidine analogue was achieved with tin(IV) organometallic reagents.

To the best of our knowledge there are no reports on organotin(IV) compounds of L-histidine Schiff base derivatives. We have characterized these compounds not only

spectroscopically but also with regard to their antibacterial activity against *Escherichia coli*, *Staphylococcus aureus*, *Pseudomona aeruginosa*, *Shigella dysenteriae*, *Salmonella typhimurium*, *Salmonella typhi*, *Proteus mirabilis*, *Klebsiella pneumonia*, and *Serratia marcescens*. The compounds described in this work are all new and some of them showing promising antibacterial activity for methicillin-resistant *Staphylococcus aureus* strains (Gram-positive).

2. Experimental

2.1. Materials and Methodology. Di-*n*-methyltin oxide, Di-*n*-butyltin oxide, di-*n*-phenyl oxide, bis-*n*-tributyltin oxide, triphenyltin hydroxide, L-histidine 99%, and 3,5-di-*tert*.butyl-2-hydroxybenzaldehyde 99% were purchased from Aldrich and were used without further purification. Solvents were freshly distilled before use following the standard procedures. Elemental analyses were performed in an Eager 300 analyzer. IR spectra were obtained on a Nicolet FT-55X apparatus as KBr discs of each compound. Melting points were measured on a Fisher-Johns melting point apparatus and are uncorrected. ^{1}H, ^{13}C and ^{119}Sn NMR spectra were recorded with a Varian spectrometer operating at 300 MHz using $CDCl_3$ or CD_3OD as solvent and TMS as a reference. Solid state ^{119}Sn NMR spectra were recorded with a Bruker AVANCE-II, 300 MHz NMR spectrometer, using a 4 mm CP-MAS probe. ^{119}Sn chemical shift referencing is toward tetramethyltin. High-resolution mass spectra were obtained by LC/MSD TOF on an Agilent Technologies instrument with APCI as ionization source.

2.1.1. Synthesis of Tin Derivatives

Compounds 1–5. To a solution of 3,5-di-*tert*-butyl-2-hydroxybenzaldehyde (2 mmol, 468 mg) in 16 mL of toluene at RT, L-histidine (2 mmol, 310.32 mg), 6 mL of ethanol, and the corresponding tin derivative (2 mmol) were successively added to a flask equipped with a Dean-Stark moisture trap (filled with dry toluene). After 24 h under gentle reflux, the solvents were evaporated giving yellow solids in all cases. Purification was achieved by recrystallization from dichloromethane-hexane (3 : 1) mixtures.

[($C_{21}H_{27}N_3O_3$)$Sn(CH_3)_2$](0.4CH_2Cl_2) (1). A yellow crystalline solid **1** was obtained in moderate yield (350 mg, 33.76%). Elemental analysis for $C_{23.4}H_{33.8}Cl_{0.8}N_3O_3Sn$: Calcd (%): C, 50.90; N, 7.61; H, 6.17. Found (%): C, 50.96; N, 7.27; H, 5.75. IR (KBr disk) cm^{-1}: 3445 s, 2958–2869 s, 1659 s, 1613 s, 1539 m, 1461 w, 1429 m, 1362 m, 1172 m, 784 w, 746 w, 655 w, 623 w, 530 w, 470 w. mp = 171–174°C. Electrospray mass (masses given based on ^{1}H, ^{12}C, ^{16}O, ^{14}N, and ^{120}Sn): the isotopic distributions were compared with the calculated. Only tin-containing fragments are given, *m/z*: $[M^+]$ ($[(C_{21}H_{28}N_3O_3)Sn(CH_3)_2]^+$), 520 (100%); ($3[M^+]$ + CH_2Cl_2), 542 (16%); ($[M^+]$ + M), 1037 (7%).

[($C_{21}H_{27}N_3O_3$)$Sn(CH_2CH_2CH_2CH_3)_2$](0.5CH_2Cl_2) (2). A yellow crystalline solid **2** was obtained in moderate yield (520 mg, 43.16%). Elemental analysis for $C_{29.5}H_{46}ClN_3O_3Sn$:

Calcd(%): C, 54.94; N, 6.52; H, 7.19. Found (%): C, 54.70; N, 6.08; H, 7.32. IR (KBr disk) cm^{-1}: 3447 m, 3276 m, 2958–2869 s, 1668 s, 1611 s, 1539 m, 1460–1426 m, 1363 m, 1172 m, 840 w, 746 w, 655 w, 622 w, 532 w, 480 w. mp = 110–116°C. Electrospray mass (masses given based on ^{1}H, ^{12}C, ^{16}O, ^{14}N, and ^{120}Sn): the isotopic distributions were compared with the calculated. Only tin-containing fragments are given, m/z: [M$^+$] ([($C_{21}H_{28}N_3O_3$)Sn($CH_2CH_2CH_2CH_3$)$_2$]$^+$), 604 (100%); (4[M$^+$] + CH_2Cl_2), 625 (11%); ([M$^+$] + M), 1205 (3%).

[($C_{21}H_{27}N_3O_3$)Sn(C_6H_5)$_2$] (3). A yellow solid 3 was obtained in moderate yield (660 mg, 47.66%). Elemental analysis for $C_{33}H_{37}N_3O_3Sn$: Calcd(%): C, 61.70; N, 6.54; H, 5.81. Found(%): C, 61.59; N, 6.39; H, 5.21. IR (KBr disk) cm^{-1}: 3448 m, 3306 m, 2957–2870 s, 1668 w, 1618 s, 1538 m, 1461 m, 1431 m, 1361 m, 1253 m, 1171 m, 837 w, 731 m, 654 w, 620 w, 536 w, 450 w. mp = 181–185°C. Electrospray mass (masses given based on ^{1}H, ^{12}C, ^{16}O, ^{14}N, and ^{120}Sn): the isotopic distributions were compared with the calculated. Only tin-containing fragments are given, m/z: [M$^+$] ([($C_{21}H_{28}N_3O_3$)Sn(C_5H_5)$_2$]$^+$), 643 (100%); (4[M$^+$] + CH_2Cl_2), 666 (12%); ([M$^+$] + M), 1285 (6%).

[($C_{21}H_{28}N_3O_3$)Sn($CH_2CH_2CH_2CH_3$)$_3$] (4). A yellow crystalline solid 4 was obtained in moderate yield (410 mg, 31.03%). Elemental analysis for $C_{33}H_{55}N_3O_3Sn$: Calcd(%): C, 60.01; N, 6.36; H, 8.39. Found(%): C, 60.29; N, 6.46; H, 7.99. IR (KBr disk) cm^{-1}: 3091 w br, 2955–2866 s, 1611 s, 1460–1439 m, 1366 s, 1269–1239 m, 1174 w, 818 s, 655 m, 450 w. mp = 200–203°C. Electrospray mass (masses given based on ^{1}H, ^{12}C, ^{16}O, ^{14}N, and ^{120}Sn): the isotopic distributions were compared with the calculated. Only tin-containing fragments are given, m/z: [HSn($CH_2CH_2CH_2CH_3$)$_3$$^+$], 291 (87%); [M$^+$] ([($C_{21}H_{29}N_3O_3$)Sn($CH_2CH_2CH_2CH_3$)$_3$]$^+$), 662 (100%); ([M$^+$] + Bu$_3$Sn), 951 (41%).

[($C_{21}H_{28}N_3O_3$)Sn(C_6H_5)$_3$](2.2CH_2Cl_2)(3.6$C_6H_5CH_3$) (5). A yellow solid 5 was obtained in moderate yield (380 mg, 15.52%). Elemental analysis for $C_{66.4}H_{76.2}N_3O_3SnCl_{4.4}$ Calcd(%): C, 64.37; N, 3.39; H, 6.20. Found(%): C, 64.27; N, 3.35; H, 5.56. IR (KBr disk) cm^{-1}: 3266–3060 m, 2958–2864 s, 1962–1716 w, 1631–1595 s, 1479 m, 1427 s, 1361 m, 1260 w, 1180 w, 1073 s, 729 (CH_2Cl_2) s, 697 (CH_2Cl_2) s, 657 m, 455 m, 443 m. mp = 203–205°C. Electrospray mass (masses given based on ^{1}H, ^{12}C, ^{16}O, ^{14}N, and ^{120}Sn): the isotopic distributions were compared with the calculated. Only tin-containing fragments are given, m/z: [M$^+$] ([($C_{21}H_{29}N_3O_3$)Sn(C_6H_5)$_3$]$^+$), 722 (23%); [$C_{21}H_{29}N_3O_3$]$^+$, 371 (22%); [C_6H_5]$^+$, 77 (100%).

2.1.2. Biological Activity: Antibacterial Screening

Bacterial Strains. Bacterial reference cultures and self-collection strains were used to test a broad spectrum of bacteria. The strains used in this study were Enterohemorrhagic *Escherichia coli* EDL933, *Staphylococcus aureus* MRSA, *Pseudomona aeruginosa* ATCC27853, *Shigella dysenteriae* FMUNAM98863, *Salmonella typhimurium* ATCC14028, *Salmonella typhi* FMUNAM95073, *Proteus*

mirabilis RTX339, *Klebsiella pneumoniae* ATCC8045, and *Serratia marcescens* TSUAM1.

Isolates of methicillin-resistant *S. aureus* (MRSA) were obtained from our collection [16]. Identification of *S. aureus* was performed by the coagulase test with human plasma. MRSA strains were confirmed to be methicillin resistant by testing for methicillin susceptibility and by PCR detection of the *mecA* gene.

The organotin compounds were tested *in vitro* for the antibacterial activity against clinical pathogen and reference strains using the disk diffusion test, in accordance with the procedures outlined by CLSI (formerly the NCCLS) [17]. Muller-Hinton agar was used for this method. Inocula were adjusted to 1.5×10^8 CFU mL (0.5 McFarland). BD BBLTM Sensi-Discs (Becton Dickinson and Company, Sparks, MD, USA) were used as growth inhibition controls. The antibiotics tested were Kanamycin (30 μg), Vancomycin (30 μg), Chloranphenicol (30 μg), Cloxacillin (1 μg), and Penicillin (10 IU).

The degree of effectiveness was measured by determining the diameters of the zone of inhibition caused by the compound. Compounds (1–5) were used, each one with a concentration of 6 mg per disc. The plates were incubated at 37°C for 24 h. During incubation time the complexes diffused from the disc to the medium. Effectiveness was classified into four different categories based on diameter of the zone of inhibition: (+++) most effective, (++) moderate effective, (+) slightly effective, and (−) noneffective. The results were compared against those of controls, which were screened simultaneously. The determinations were performed by triplicate and the results are reported as average values.

3. Results and Discussion

3.1. Synthesis. Compounds 1–5 are prepared by *in situ* condensation of L-histidine with 3,5-di-*tert*-butyl-2-hydroxybenzaldehyde in the presence of the corresponding tin(IV) derivatives: di-*n*-methyltin oxide, di-*n*-butyltin oxide, di-*n*-phenyltin oxide, bis-tri-*n*-butyltin oxide, or triphenyltin hydroxide. These conditions affording compounds 1, 2, 3, 4, and 5, respectively, and the proposed structures are shown in Figure 1. In all complexes, the Schiff base histidine derivative is coordinated to the alkyl- or aryl-tin(IV) ion through the oxygen atom belonging to the carboxylate group. Compounds 1, 2, and 3 are also coordinated to the Schiff derivative through the imino nitrogen and the phenolic oxygen in order to stabilize a 5-coordinated tin(IV) core. In compounds 4 and 5, tin is 4-coordinated.

All complexes are light yellow crystalline solids which are stable in air. Most of them are soluble in organic solvents. Compound 5 shows a very limited solubility in ethanol. Empirical formulae and proposed structures for all the compounds are confirmed by analytical and spectral data which are described in the following lines.

3.2. Infrared Spectroscopy. The coordination sphere on compounds 1–5 has been analyzed by means of FT-IR. The binding mode of the L-histidine Schiff base derivative to

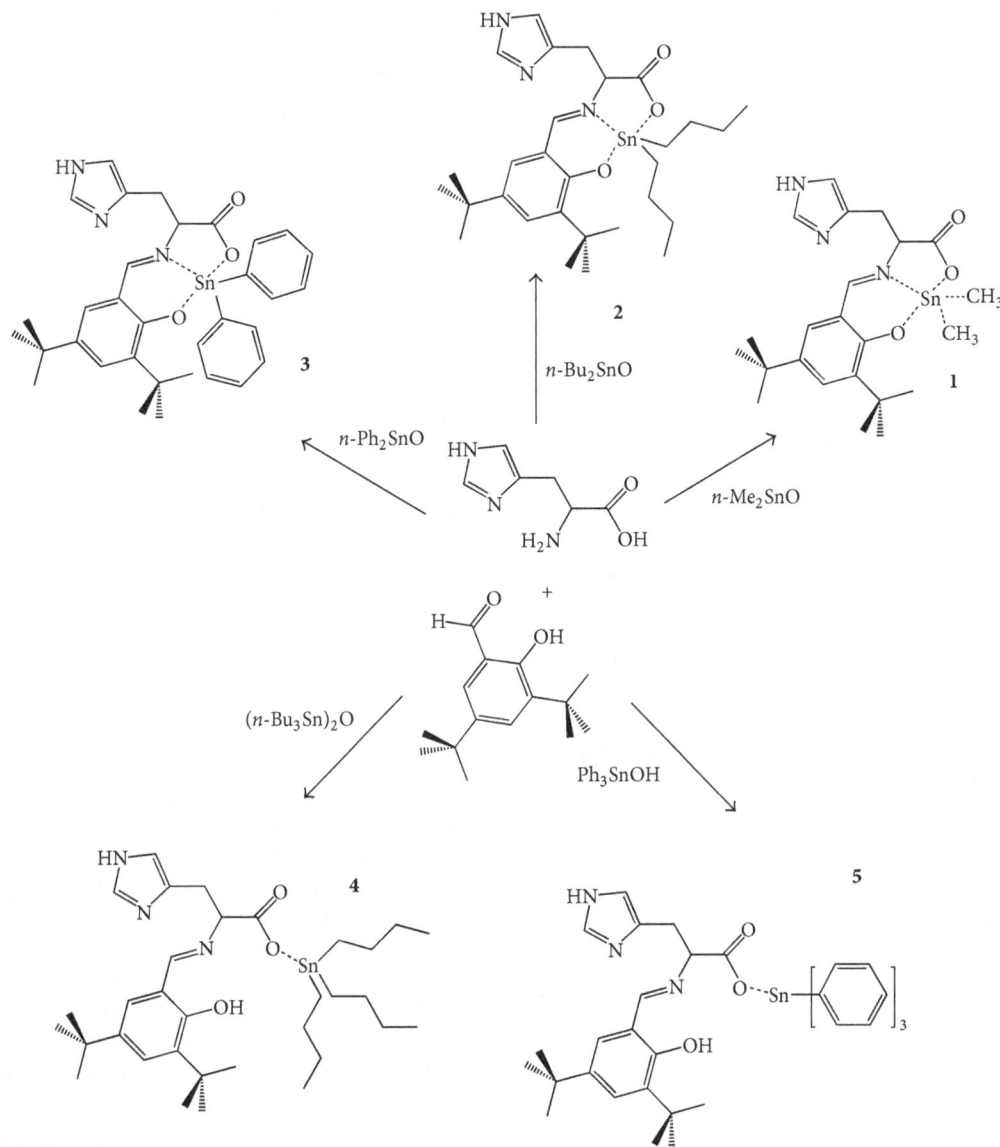

FIGURE 1: Schematic representation of compounds **1–5** and their synthetic procedure.

metal in the complexes has been studied by comparison of IR spectra of free ligand precursors and metal complexes. The absence of the broad O–H stretching frequencies in the region 3170–2870 cm^{-1} denotes that deprotonation and coordination of the carboxylate group to tin have taken part during the complex formation in all compounds.

Free L-histidine shows a strong band at 1630 cm^{-1}, which is assigned to the stretching vibration of the C=O bond. There are also medium strength bands at 1589–1571 cm^{-1} that are assigned to the C=N and C=C stretching vibration. The corresponding bands in the tin compounds are shifted to higher frequencies as expected, due to tin coordination. For pentacoordinated compounds **1**, **2**, and **3**, the observed shifts are in the range 40–30 cm^{-1}. For tetracoordinated tin compounds **4** and **5**, the shifts are smaller in the range 11–2 cm^{-1}. This observation is compatible with some other 4-coordinated structures described in the literature [9, 18].

Moreover, the azomethine band is easily observed at 1539–1538 cm^{-1} region for compounds **1**, **2**, and **3**. This red shift is considered the result of coordination of this moiety to the metal center (see Figure 1) [6]. For compounds **4** and **5**, this ν(C=N) band is not observed in such region, evidence that demonstrates that the azomethine moiety is not involved in coordination to tin.

The carboxylate moiety in compounds **1–5** has been structurally characterized by the intense and broad absorptions in the range 1660–1330 cm^{-1}, due to asymmetric and symmetric stretching modes. The difference between asymmetric and symmetric O–C=O stretching vibrations ($\Delta\nu$) has been used to determine the mode of coordination of a carboxylate ligand with metals [9, 19, 20]. Differences larger than 250 cm^{-1} are indicative of monodentate carboxylate complexes, while $\Delta\nu$ values in the range 150–250 cm^{-1} are indicative of compounds with carboxylate-bridged structures. Finally, it is presumed that a chelation is present, if this parameter is lower than 150 cm^{-1}.

The $\Delta\nu = \nu_{as}(OCO) - \nu_s(OCO)$ for the proposed pentacoordinated compounds **1**, **2**, and **3** is the following: $\Delta\nu$ = 297, 305, and 307 cm^{-1}, respectively. For the tetracoordinated compounds **4** and **5**, $\Delta\nu$ is $\Delta\nu$ = 245 and 270 cm^{-1}, respectively. All values of $\Delta\nu$ for compounds **1–5** are in the range 310–245 cm^{-1}, and this evidence strongly suggests that the carboxylate group in all compounds adopted a unidentate nature.

Some other stretching frequencies of interest are those characteristics of the ν(Sn–C), ν(Sn–O), and ν(Sn–N) bonds frequencies. All these values are consistent with those detected in a number of organotin(IV)-oxygen, organotin(IV)-carbon, and organotin(IV)-nitrogen derivatives [10, 21, 22]. Worthy mentioning is the presence of two ν(Sn–O) weak bond frequencies for compounds **1** (655, 623 cm^{-1}), **2** (655, 622 cm^{-1}), and **3** (654, 620 cm^{-1}) [23]. This evidence could be explained in views of two different Sn–O bonds. Compounds **4** and **5** only show one stretching Sn–O vibration. Then it suggests that just one type of Sn–O bond is observed, confirming the tetracoordinated sphere for tin in the compounds.

The recorded high-resolution mass spectra of **1–5** and molecular ion peaks are in agreement with the proposed structures. The MS peaks present the correct isotopomer distribution, as expected for the tin isotope distribution.

For instance, in complex **1**, the degradation pattern shows peaks at 520 (100% m/z) that indicates molecular ion peak [M$^+$, [(C$_{21}$H$_{28}$N$_3$O$_3$)Sn(CH$_3$)$_2$]$^+$], whereas peaks at 542 (16% m/z) and 1037 (7% m/z) show 3[M$^+$] + CH$_2$Cl$_2$ and [M$^+$] + M, respectively (Figure 2). Complex **2** shows the normal trends of degradation patterns (Figure 3). The various degradation peaks at 604 (100% m/z) show molecular ion; the peaks at 625 (11% m/z) show 4[M$^+$] + CH$_2$Cl$_2$; peaks at 1205 (3% m/z) show [M$^+$] + M.

Complexes **3–5** again show a similar pattern of degradation.

3.3. Multinuclear NMR Spectroscopy.

The ^{1}H, ^{13}C, and ^{119}Sn NMR spectra of compounds **1–4** have been recorded in CDCl$_3$ or MeOD and the assignments obtained through 1D and 2D experiments. When necessary, ^{119}Sn NMR spectra have also been recorded in the solid state. The very limited solubility of compound **5** prompted us to obtain the ^{13}C and ^{119}Sn spectra in the solid state only.

The ^{1}H NMR spectra assignment of compounds **1–5** is based on integration values, chemical shifts, coupling constants, and ^{1}H-^{13}C HMQC experiments. Starting and model compounds spectra were also useful for comparison reasons [18]. The expected resonances are observed for L-histidine, 3,5-di-*tert*-butyl-2-hydroxybenzaldehyde, tributyltin(IV), triphenyltin(IV), dibutyltin(IV), dimethyltin(IV), and dibutyltin(IV) moieties. Due to rapid exchange between the two nitrogen atoms of the imidazole ring and probably with deuterium from solvent, the N–H proton from the imidazole moiety is not observed in the spectra.

The results obtained are listed in Tables 1 and 2. For comparative purposes the data are tabulated using a general numbering scheme (Figure 4).

The characteristic resonance peak in the ^{1}H NMR spectra of compounds **1–5**, at 8.4–7.9 ppm, reveals the presence of the imino moiety H–C11=N, which confirms the condensation of the L-histidine derivative (Figure 1). Additionally, for compounds **1**, **2**, and **3**, coordination to Sn(IV) is confirmed by small coupling peaks associated to the H–C11=N resonance peak. This spin-spin coupling between the azomethine proton and the tin nucleus, ^3J(Sn–N=CH), presents coupling constants in the range 52–70 ppm, values that are in agreement with a magnetic coupling between the tin nuclei located in a transposition with respect to the azomethine proton [23, 24]. In complex **1**, characteristic peaks of the methyl groups attached to the tin atom are observed at 0.64 and 0.56 ppm as expected [25]. In complex **2**, the presence of the Sn–Bu$_2$ resonance peaks, Hα, Hβ, Hγ, and Hδ, confirms the chemical nature of the dibutyl moiety of **2**. In compound **4**, the butyl protons appear in the range 0.80–1.53 cm^{-1}, with multiplicity and integration values in agreement with the proposed structure. For complex **3** the signals corresponding to the phenyl groups are clearly observed.

The ^{1}H NMR spectra of all the complexes present a resonance peak assigned to H–C12 moiety. This moiety is attached to the carboxylic group and it is observed in the range 4.14–4.59 ppm. These are clearly downfield shifted when comparing with the resonance peak in the free ligand at around 3.98 ppm (D$_2$O). This evidence suggests that the downfield effect is due to coordination to tin. In this order of ideas, the ^{1}H NMR spectra of all complexes present a resonance for H–C14 with the same downfield shift effect (3.15–3.47), with a minimal shift for compounds **3** and **4**. The signals corresponding to the *tert*-butyl protons of the ligand have also been assigned and all data is presented in Table 1.

Table 2 presents the ^{13}C NMR and ^{119}Sn NMR data for the synthesized organotin compounds following the same numbering scheme presented on Figure 4. The very limited solubility of compound **5** prompted us to obtain the ^{13}C and ^{119}Sn spectra in the solid state. The ^{13}C NMR characterization was obtained through comparison with starting materials, and structural related analogues.

The ^{13}C NMR data explicitly resolved the resonances of all the distinct carbon atoms present in all the complexes. The ^{13}C NMR data for all compounds show that the resonance peak of the carboxyl carbon, C13, appears in the range from 173 to 177 ppm, in agreement with the data reported for analogous derivatives [18]. The signal of the imine carbon, C11, appears in the 166–177 ppm range, showing in some cases a marked deshielding effect with respect to the imine group. The Sn–N coordination should produce an N–C polarization. The biggest polarization effect could be observed, as expected, in compound **3** where phenyl groups are attached to tin.

For compounds **4** and **5**, the signal of the imine carbon, C11, appears in the range 165–167 ppm, which is consistent with the lack of the Sn–N coordination. The same high field shift effect could be observed in case of C2.

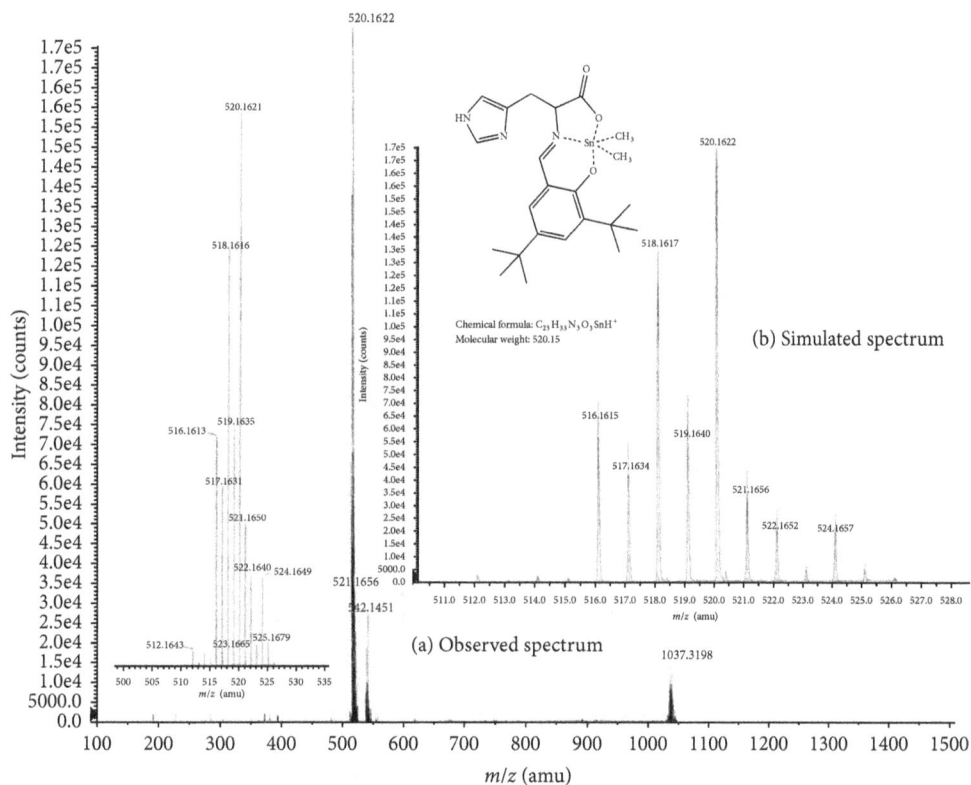

FIGURE 2: Observed mass spectrum (a) and simulated spectrum (b) of complex **1**.

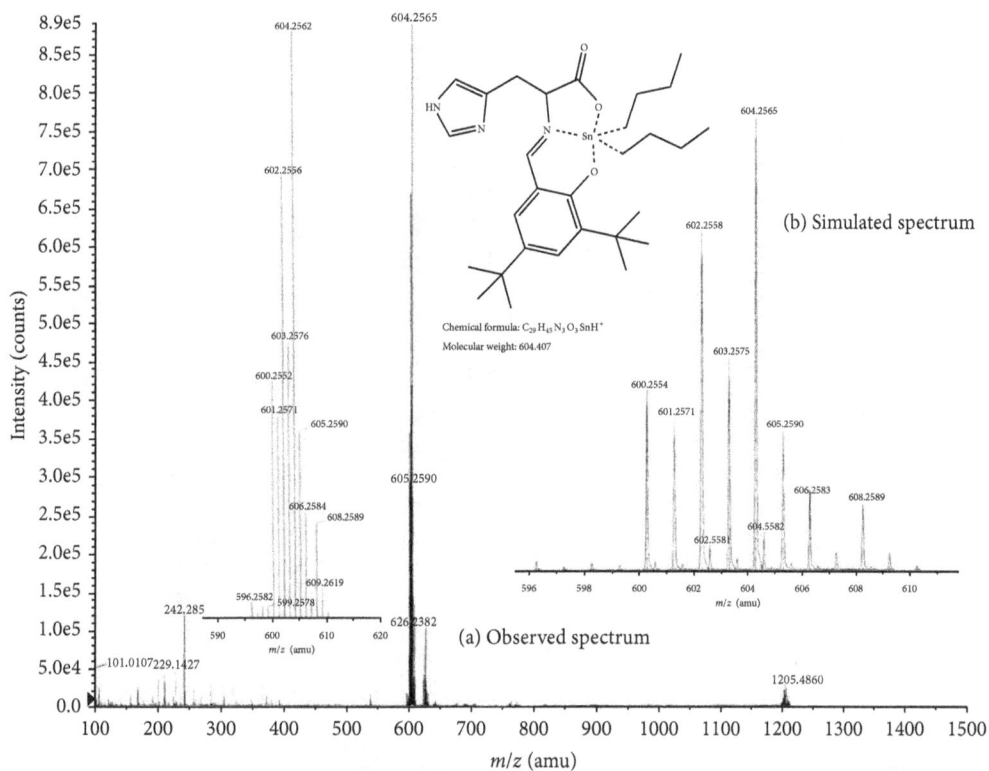

FIGURE 3: Observed mass spectrum (a) and simulated spectrum (b) of complex **2**.

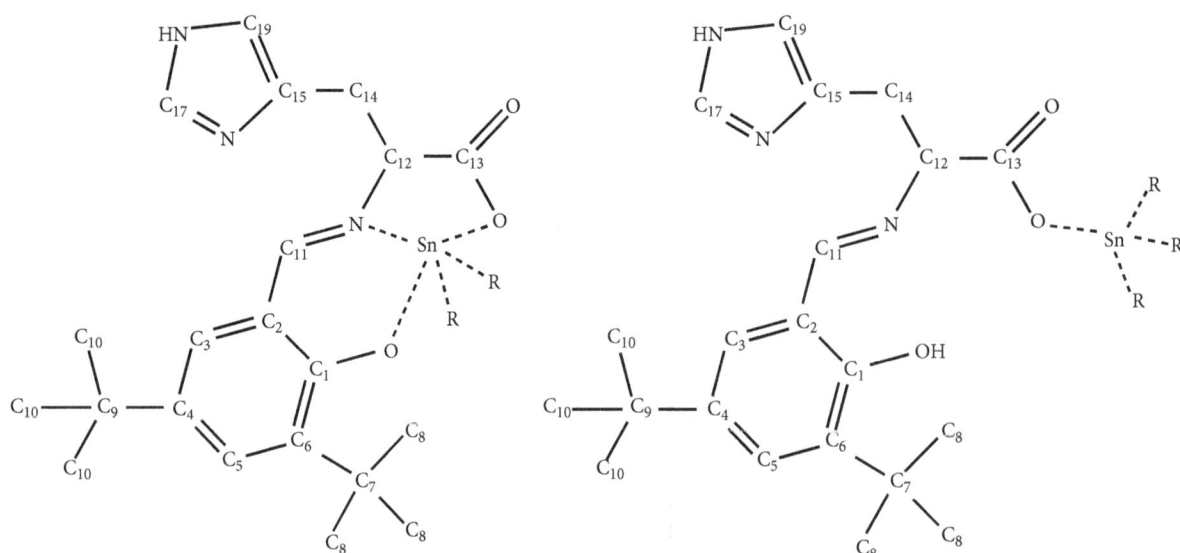

FIGURE 4: Numbering system used in pentacoordinated and tetracoordinated tin(IV) compounds for the NMR spectra assignment. R = CH$_3$–, CH$_3$CH$_2$CH$_2$CH$_2$– (CHα, CHβ, CHγ, CHδ) or C$_6$H$_5$– (CHo, CHm, CHp).

TABLE 1: ^{1}H NMR data of compounds **1–5**.

Compound numbering	**1** CDCl$_3$	**2** CDCl$_3$	**3** CD$_3$OD	**4** CDCl$_3$
H11	8.13 (s, 1H) [58]	7.91 (s, 1H) [52]	8.33 (s, 1H) [70]	8.34 (s, 1H)
H17	7.63 (s, 1H)	7.60 (s, 1H)	7.08 (s, 1H)	7.48 (s, 1H)
H5	7.54 (d, 1H) [2.4]	7.51 (d, 1H) [2.7]	7.63 (d, 1H) [2.4]	7.35 (d, 1H) [2.4]
H3	6.85 (d, 1H) [2.4]	6.76 (d, 1H) [2.7]	7.07 (d, 1H) [2.5]	7.03 (d, 1H) [2.4]
H19	6.80 (s, 1H)	6.73 (s, 1H)	6.52 (s, 1H)	6.73 (s, 1H)
H12	4.43 (m, 1H)	4.43 (m, 1H)	4.59 (m, 1H)	4.14 (m, 1H)
H14	3.39–3.25 (m, 2H)	3.47–3.44 3.11–3.07 (m, 2H)	3.27–3.16 (m, 2H)	3.31–3.15 (m, 2H)
Hγ		1.73–1.51 (m, 4H)		1.27–1.28 (m, 6H)
Hα		1.4–1.3 (m, 4H)		1.27–1.28 (m, 6H)
Hβ		1.4–1.3 (m, 4H)		1.53 (m, 6H)
Hδ		0.92 (t, 3H) [7.3] 0.79 (t, 3H) [7.3]		0.80 (t, 9H) [7.3]
Ho			7.87–7.82 (m, 4H)	
Hm			7.58–7.29 (m, 4H)	
Hp			7.58–7.29 (m, 4H)	
H8, H10 t-butyl CH$_3$	1.37 (s, 9H) 1.28 (s, 9H)	1.37 (s, 9H) 1.25 (s, 9H)	1.47 (s, 9H) 1.30 (s, 9H)	1.41 (s, 9H) 1.28 (s, 9H)
CH$_3$-Sn	0.64 (s, 6H), 0.56 (s, 6H)			

Data obtained at 500 MHz. Chemical shifts in ppm with respect to TMS; coupling constants in hertz, $^nJ(^1$H–^{1}H) and $^nJ(^1$H–$^{119/117}$Sn) between brackets. s: singlet; d: doublet; t: triplet; m: complex pattern. Assignment base in 1D and 2D NMR studies.

TABLE 2: ^{13}C and ^{119}Sn NMR data of compounds 1–5.

Compound numbering	1 CDCl$_3$	2 CDCl$_3$	3 CD$_3$OD	4 CDCl$_3$	5 Solid state
C11	174.33	174.54	177.66	167.57	166.6
C13	173.72	173.72	175.79	175.62	176.6
C2	166.59	167.0	168.61	158.13	157.55
C1	141.29	140.96	143.61	140.00	a
C4	139.04	138.64	142.00	136.67	a
C6	135.70	135.67	134.63	134.09	a
C5	133.19	133.08	132.97	127.16	a
C17	131.08 (b)	131.30 (b)	132.64 (b)	131.98 (b)	—
C3	129.49	129.47	129.95	126.24	a
C19	119.78 (b)	119.84 (b)	118.00 (b)	123.69 (b)	120.2
C15	116.61	116.51	119.66	117.98	117.5
C12	67.18	67.42	68.56	72.02	64.0
C7	35.27	35.19	35.45	35.08	34.68
C9	34.06	33.93	34.17	34.16	33.86
C14	33.97	33.90 (b)	33.98	29.50 (b)	29.5
C8	31.23	31.10	30.56	31.54	31.0
C10	29.44	29.36	29.47	29.50	29.2
CH$_3$	1.23, 0.933 [653/648]				
C$_\beta$		27.00 [38] 26.85 [n.o.]		28.07 [24]	
C$_\gamma$		26.74 [91] 26.56 [92]		27.02 [72/70]	
C$_\alpha$		22.26 [599] 22.20 [599]		17.73 [435/415]	
C$_\delta$		13.59 13.39		13.66	
C$_i$			142.67 [n.o.] 141.27 [n.o.]		a
C$_o$			137.78 [53] 137.32 [53]		a
C$_m$			130.54 [87] 130.23 [83]		a
C$_p$			132.12 [n.o.] 131.64 [16]		a
^{119}Sn	−162.16 CDCl$_3$	−197.58 CDCl$_3$	−369 CD$_3$OD	39.44 CDCl$_3$	No soluble
^{119}Sn Solid state	−169.50	−196.68	−455.87	−112.59	−122.09

Chemical shifts in ppm with respect to TMS; ^{119}Sn chemical shifts in ppm with respect to (CH$_3$)$_4$Sn; $^nJ(^{13}$C$-^{119/117}$Sn) coupling constants between square brackets. n.o.: not observed.
a: Unambiguous assignment of this resonance peak was not reached because there are multiple broad signals included in the range 143–127 ppm.

The $^1J(^{13}$C$-^{117/119}$Sn) coupling constant values are of major importance in the estimation of the coordination mode of the tin atom. The values of ^1J (in solution) can be related to the tin-coordination number. Thus these values are a qualitative evidence of the organotin(IV) compound structure [26–28]. Pentacoordinated organotin(IV) ions show ^1J values between 450–610 Hz, while tetracoordinated organotin(IV) ions show values in the range 325–440 Hz [18].

The ^1J coupling constants for compounds 1 and 2 suggest a pentacoordinated nature in solution. For compound 3 the ^1J coupling constants are not observed. For compound 4 the ^1J values are 415 and 435 Hz, suggesting a four-coordinated

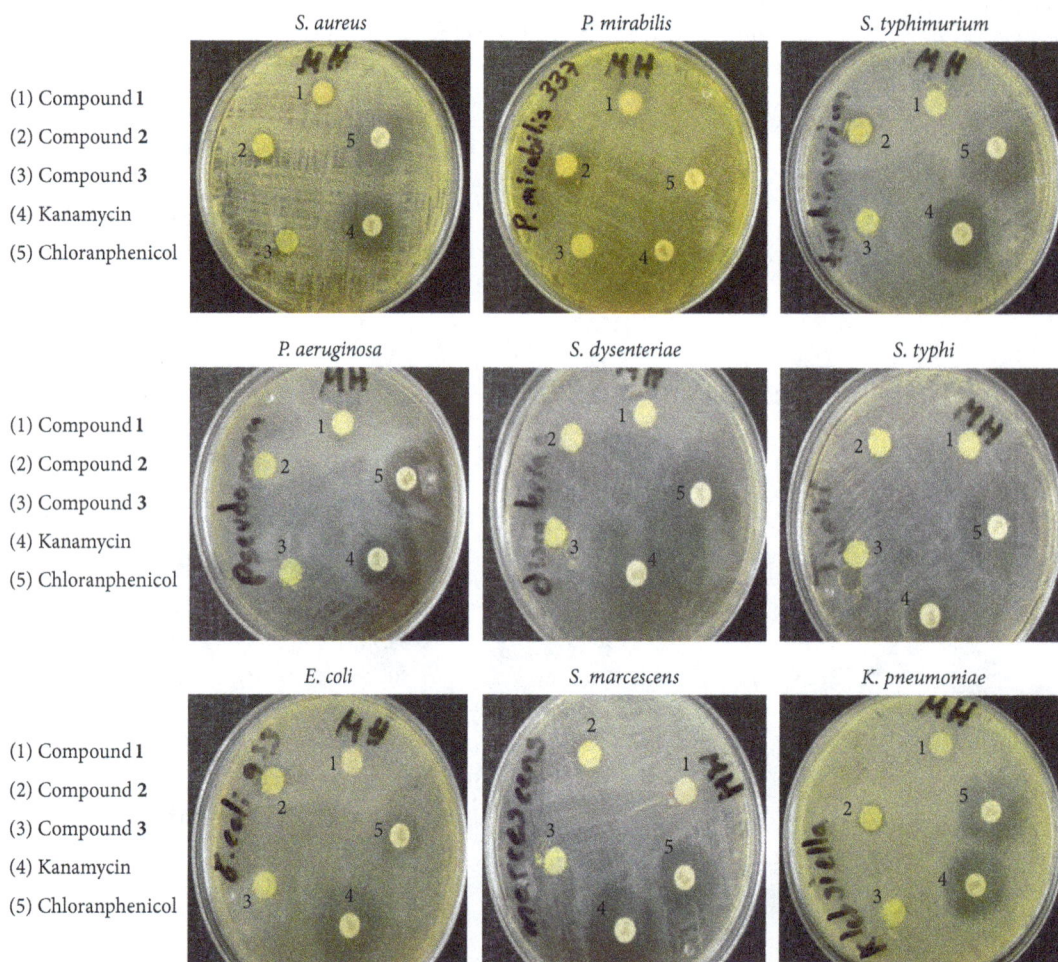

FIGURE 5: Photographs of the antimicrobial activities of compounds **1**, **2**, and **3** against *S. aureus*, *P. mirabilis*, *S. typhimurium*, *P. aeruginosa*, *S. dysenteriae*, and *S. typhi*, showing their inhibition zone.

structure in solution. The $^2J(^{13}C-^{117/119}Sn) = 72/70$ Hz also suggests a tetrahedral geometry for **4**.

The correct assignment of ^{13}C NMR resonances of the *n*-butyl group in **2** is determined by the coupling constants $^nJ(^{13}C-^{117/119}Sn)$. Compound **2** displays a resonance peak for the *n*-butyl $C\alpha$ at 21-22 ppm with two very close coupling signals. This observation provides additional evidence for the magnetic equivalence of the butyl moieties. The same criterion is followed in the assignment of ^{13}C NMR resonances of the *n*-butyl group in **4**. The magnetic equivalence of the butyl moieties is concluded from the spectra.

In order to provide further evidence for the proposed structures of the complexes in solution, ^{119}Sn NMR spectra are analyzed. The ^{119}Sn chemical shifts are extremely dependent on the coordination number for tin and the nature of the groups directly coordinated to the metal. The chemical shift values for compounds **1–3** are within the expected range for pentacoordinated tin ions with a single resonance peak in solution. There are not changes in the solid state spectra as could be appreciated by the chemical shifts observed in the corresponding solid ^{119}Sn NMR spectra. Thus, the structural

composition in the solid state is retained in solution for compounds **1**, **2**, and **3**.

^{119}Sn NMR spectrum of compound **4** exhibits a single resonance in solution, with a chemical shift of 39.44 ppm. In the solid state, the chemical shift observed is characteristic of a five-coordinated compound. This evidence suggests that in the solid state inter- or intramolecular interactions could be responsible of the increase in coordination number.

No ^{119}Sn resonance could be observed for compound **5** in solution, due to its limited solubility. In addition the chemical shift observed in the ^{119}Sn NMR spectrum in the solid state suggests the tetracoordinated nature of **5**. Despite the Lewis acidity that a phenyl group could induce in the tin atom [29], this is very low as this compound appears unable to extent its coordination sphere.

3.4. Microbial Activities. Organotin(IV) cations with biological active ligands are a major research debt for the scientific efforts in the search of new metal-based drugs, but little work is available on the bactericidal properties of organotin(IV) derivatives of amino acids in the literature. In

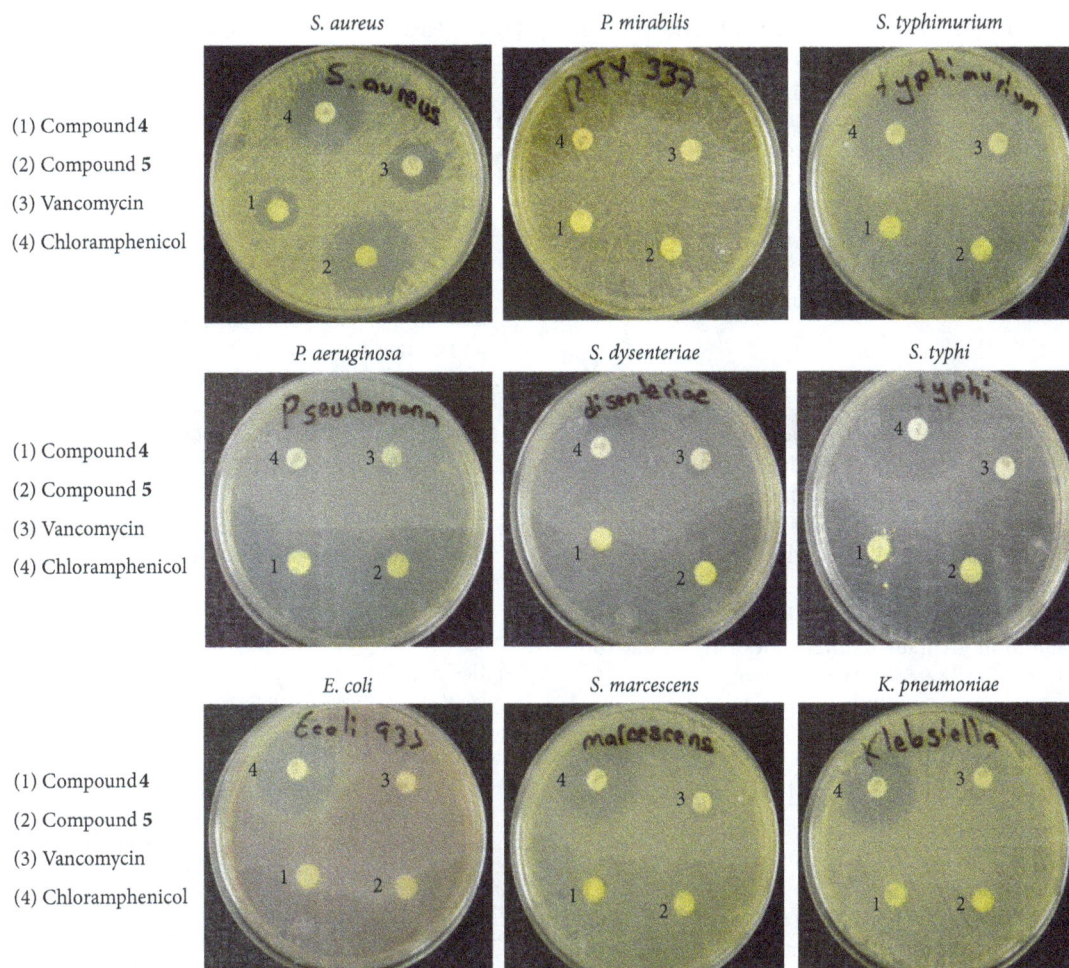

FIGURE 6: Photographs of the antimicrobial activities of compounds **4** and **5** against *S. aureus, P. mirabilis, S. typhimurium, P. aeruginosa, S. dysenteriae*, and *S. typhi*, showing their inhibition zone.

TABLE 3: Antimicrobial activity of organotin(IV) compounds **1–5**.

Strain	Compound				
	1	**2**	**3**	**4**	**5**
Staphylococcus aureus	+	−	−	++	+++
Proteus mirabilis RTX337	+	+	−	−	−
Pseudomona aeruginosa	+	−	−	−	−
Salmonella typhi	−	+	−	−	−
Salmonella typhimurium	−	+	−	−	−
Escherichia coli 933	−	+	−	−	−
Klebsiella	−	+	−	−	−
Serratia marcescens	−	−	−	−	−
Shigella dysenteriae	−	−	−	−	+

particular, organotin derivatives of amino acids containing N-heterocycles bonded to the α-carbon should be studied in more detail because of the well-known role of the histidine residue binding properties. In particular it is important to study the histidine binding properties to triorganotin compounds in order to provide evidence for important mechanisms of biological activities [30].

In order to study the potential biological properties of the synthesized compounds, the bactericidal properties are studied.

The degree of effectiveness was measured by determining the diameters of the zone of inhibition caused by the compound. The biological activity was then classified in 4 categories on the bases of their diameter of growth inhibition: (+++) most effective, (++) moderate effective, (+) slightly effective, and (−) noneffective.

Compound **1** is slightly effective against both Gram-positive and Gram-negative bacteria; **2** is moderately effective against Gram-negative bacteria; **3** does not show antibacterial activity at all. These results are comparable to related pentacoordinated tin(IV) compounds with Schiff bases derived from amino acids reported in the literature [2].

Compounds **4** and **5** are moderate effective and most effective, respectively, against *Staphylococcus aureus* strains (Gram-positive). Results are presented in Table 3 and Figures 5 and 6.

Thus the results clearly indicate that pentacoordinated tin(IV) compounds of Schiff bases derived from L-histidine have exhibited moderate to minimal bactericidal properties while tetracoordinated tin(IV) compounds have presented high activity against the Gram-positive methicillin-resistant *S. aureus*. Even more, compound **5** as observed in Figure 6 is more active when compared with standard references vancomycin and chloramphenicol against the same bacteria under identical experimental conditions. These are very promising results in particular for the triphenyltin(IV) derivative.

It has been described that triorganotin compounds are significantly more biocidally active than other classes [30], statement that is true based on the results of this work.

4. Conclusions

Five novel tin(IV) compounds have been synthesized. The compounds are coordinated to a histidine analogue. With the help of various physicochemical techniques, the structure of the newly synthesized compounds has been proposed (Figure 1). It has been spectroscopically demonstrated that the Schiff base ligand derived from L-histidine and 3,5-di-*tert*-butyl-2-hydroxybenzaldehyde coordinates to tin(IV) in a tridentate and monodentate manner producing pentacoordinated and tetracoordinated tin(IV) compounds.

Amino acids are very difficult ligands to work with due to their many donor groups and reduced solubility in nonpolar solvents. Nevertheless it was possible to isolate pure compounds. The selection of the ligand, a biologically active amino acid, has produced metal-ligand combined effects that may result in an important improvement in the activity of the resulting coordination compounds.

The triphenyltin(IV) complex **5** has been found to be most active against the Gram-positive methicillin-resistant *S. aureus*.

With these results, it is important to stress the need of more research in the synthesis, isolation, and characterization of new organotin(IV) cations with biologically active ligands, in an effort to obtain important structure-activity relationships.

Acknowledgment

Ariadna Garza-Ortiz expresses gratitude to Universidad Autonoma Metropolitana-Xochimilco for the posdoctoral fellowship.

References

[1] S. Roy, K. D. Hagen, P. U. Maheswari et al., "Phenanthroline derivatives with improved selectivity as DNA-targeting anticancer or antimicrobial drugs," *ChemMedChem*, vol. 3, no. 9, pp. 1427–1434, 2008.

[2] L. Pellerito and L. Nagy, "Organotin(IV)$^{n+}$ complexes formed with biologically active ligands: equilibrium and structural studies, and some biological aspects," *Coordination Chemistry Reviews*, vol. 224, no. 1–2, pp. 111–150, 2002.

[3] L. Casella and M. Gullotti, "Coordination modes of histidine: 4. Coordination structures in the copper(II)-L-histidine (1:2) system," *Journal of Inorganic Biochemistry*, vol. 18, no. 1, pp. 19–31, 1983.

[4] R. I. Henkin, *Metal-Albumin-Amino Acid Interactions: Chemical and Physiological Interrelationships*, Plenum Publishing, New York, NY, USA, 1974.

[5] A. Garza-Ortiz, P. U. Maheswari, M. Siegler, A. L. Spek, and J. Reedijk, "Ruthenium(III) chloride complex with a tridentate bis(arylimino)pyridine ligand: synthesis, spectra, X-ray structure, 9-ethylguanine binding pattern, and in vitro cytotoxicity," *Inorganic Chemistry*, vol. 47, no. 15, pp. 6964–6973, 2008.

[6] M. Aslam, I. Anis, N. Afza, B. Ali, and M. R. Shah, "Synthesis, characterization and biological activities of a bidentate schiff base ligand: N,N′-Bis(1-phenylethylidene)ethane-1,2-diamine and its transition metals (II) complexes," *Journal of the Chemical Society of Pakistan*, vol. 34, p. 391, 2012.

[7] F. Kayser, M. Biesemans, M. Gielen, and R. Willem, "Characterization of the dibutylstannylene derivative of pyridoxine by proton detected 2D heteronuclear correlation NMR," *Magnetic Resonance in Chemistry*, vol. 32, no. 6, pp. 358–360, 1994.

[8] J. C. Martins, R. Willem, and M. Biesemans, "A comparative investigation of the consistent valence and extensible systematic force fields. A case study on the conformation of erythromycin A in benzene," *Journal of the Chemical Society*, vol. 2, no. 7, pp. 1513–1520, 1999.

[9] C. Camacho-Camacho, I. Rojas-Oviedo, M. A. Paz-Sandoval et al., "Synthesis, structural characterization and cytotoxic activity of organotin derivatives of indomethacin," *Applied Organometallic Chemistry*, vol. 22, no. 3, pp. 171–176, 2008.

[10] M. Gielen, "Review: organotin compounds and their therapeutic potential: a report from the Organometallic Chemistry Department of the Free University of Brussels," *Applied Organometallic Chemistry*, vol. 16, no. 9, pp. 481–494, 2002.

[11] V. Valla and M. Bakola-Christianopoulou, "Chemical aspects of organotin derivatives of beta-diketones, quinonoids, steroids and some currently used drugs: a review of the literature with emphasis on the medicinal potential of organotins," *Synthesis and Reactivity in Inorganic and Metal-Organic Chemistry*, vol. 37, pp. 507–525, 2007.

[12] B. Y. K. Ho and J. J. Zuckerman, "Trialkyltin derivatives of amino acids and dipeptides," *Inorganic Chemistry*, vol. 12, pp. 1552–1561, 1973.

[13] H. Bruckner and K. Hartel, German Patent 1, 061, 561, 1959.

[14] M. Frankel, D. Gertner, D. Wagner, and A. Zilkha, "Organotin esters of amino acids and their use in peptide syntheses," *Journal of Organic Chemistry*, vol. 30, no. 5, pp. 1596–1599, 1965.

[15] B. Y. K. Ho and J. J. Zuckerman, "Structure organotin chemistry," *Journal of Organometallic Chemistry*, vol. 49, no. 1, pp. 1–84, 1973.

[16] A. Hamdan-Partida, T. Sainz-Espuñes, and J. Bustos-Martínez, "Characterization and persistence of *Staphylococcus aureus* strains isolated from the anterior nares and throats of healthy carriers in a mexican community," *Journal of Clinical Microbiology*, vol. 48, pp. 1701–1705, 2010.

[17] National Committee for Clinical Laboratory Standards, Document: M2-A9. National Committee for Clinical Laboratory Standards, Wayne, PA, 2006.

[18] C. Camacho-Camacho, I. Rojas-Oviedo, A. Garza-Ortiz, J. Cardenas, R. Alfredo Toscano, and R. Gavino, "Synthesis, structural characterization and *in vitro* cytotoxic activity of novel polymeric triorganotin(IV) complexes of urocanic acid," *Applied Organometallic Chemistry*, vol. 27, no. 1, pp. 45–51, 2013.

[19] B. S. Manhas and A. K. Trikha, "Relationships between the direction of shifts in the carbon-oxygen stretching frequencies

of carboxylate complexes and the type of carboxylate coordination," *Journal-Indian Chemical Society*, vol. 59, p. 315, 1982.

[20] K. Nakamoto, *Infrared and Raman Spectra of Inorganic and Coordination Compounds*, Wiley, New York, NY, USA, 1980.

[21] M. Gielen, "Tin-based antitumour drugs," *Coordination Chemistry Reviews*, vol. 151, pp. 41–51, 1996.

[22] M. Pellei, S. Alidori, F. Benetollo et al., "Di- and tri-organotin(IV) complexes of the new bis(1-methyl-1H-imidazol-2-ylthio)acetate ligand and the decarboxylated analogues," *Journal of Organometallic Chemistry*, vol. 693, no. 6, pp. 996–1004, 2008.

[23] F. E. Smith, R. C. Hynes, T. T. Ang, L. E. Khoo, and G. Eng, "The synthesis and structural characterization of a series of pentacoordinate diorganotin(IV) N-arylidene-α-amino acid complexes," *Canadian Journal of Chemistry*, vol. 70, no. 4, pp. 1114–1120, 1992.

[24] N. Kobakhidze, N. Farfan, M. Romero et al., "New pentaco-ordinated Schiff-base diorganotin(IV) complexes derived from nonpolar side chain α-amino acids," *Journal of Organometallic Chemistry*, vol. 695, no. 8, pp. 1189–1199, 2010.

[25] C. Camacho-Camacho, A. Esparza-Ruiz, A. Vásquez-Badillo, H. Nöth, A. Flores-Parra, and R. Contreras, "Fused hexacyclic tin compounds derived from 3-(3,5-di-t-butyl-2-hydroxy-phenylimino)-3H-phenoxazin-2-ol," *Journal of Organometallic Chemistry*, vol. 694, no. 5, pp. 726–730, 2009.

[26] M. Gielen, M. Biesemans, and R. Willem, "Organotin compounds: from kinetics to stereochemistry and antitumour activities," *Applied Organometallic Chemistry*, vol. 19, pp. 440–450, 2005.

[27] M. Gielen and E. R. T. Tiekink, *50Sn Tin Compounds and Their Therapeutic Potential*, John Wiley & Sons, Chichester, UK, 2005.

[28] R. Willem, A. Bouhdid, M. Biesemans et al., "Synthesis and characterization of triphenyl- and tri-n-butyltin pentafluo-robenzoates, -phenylacetates and -cinnamates. X-ray structure determination of tri-n-butyltin pentafluorocinnamate," *Journal of Organometallic Chemistry*, vol. 514, no. 1-2, pp. 203–212, 1996.

[29] R. Willem, I. Verbruggen, M. Gielen et al., "Correlating Mössbauer and solution- and solid-state 117Sn NMR data with x-ray diffraction structural data of triorganotin 2-[(E)-2-(2-hydroxy-5-methylphenyl)-1-diazenyl]benzoates," *Organometallics*, vol. 17, no. 26, pp. 5758–5766, 1998.

[30] K. C. Molloy, *Bioorganotin Compounds*, John Wiley and Sons, London, UK, 1989.

17

Synthesis, Characterization, and Antibacterial Studies of Pd(II) and Pt(II) Complexes of Some Diaminopyrimidine Derivatives

Peter A. Ajibade and Omoruyi G. Idemudia

Department of Chemistry, University of Fort Hare, Alice 5700, South Africa

Correspondence should be addressed to Peter A. Ajibade; pajibade@ufh.ac.za

Academic Editor: Concepción López

Pd(II) and Pt(II) complexes of trimethoprim and pyrimethamine were synthesized and characterized by elemental analysis, UV-Vis, FTIR, and NMR spectroscopy. The complexes are formulated as four coordinate square planar species containing two molecules of the drugs and two chloride or thiocyanate ions. The coordination of the metal ions to the pyrimidine nitrogen atom of the drugs was confirmed by spectroscopic analyses. The complexes were screened for their antibacterial activities against eight bacterial isolates. They showed varied activities with the active metal complexes showing more enhanced inhibition than either trimethoprim or pyrimethamine. The Pd(II) complexes of pyrimethamine showed unique inhibitory activities against *P. aeruginosa* and *B. pumilus*, and none of the other complexes or the drugs showed any activity against these bacteria isolates. The MIC and MBC determinations revealed that these Pd(II) complexes are the most active. Structure activity relationship showed that Pt(II) complexes containing chloride ions are more active, while for Pd(II) complexes containing thiocyanate ions showed more enhanced activity than those containing chloride ions.

1. Introduction

The discovery of potent group of pyrimidines with pronounced antagonistic effect on folic acid in cultures of *Lactobacilli* [1] led to the development of pyrimethamine and trimethoprim. Pyrimethamine was developed through brilliant feet of organic synthesis guided by biochemical considerations [2]. Additional modifications led to the synthesis of trimethoprim that inhibits bacterial dihydrofolate reductase like other diaminopyrimidines and its consequence selection as antibacterial agent [3–5]. Trimethoprim is a broad-spectrum antimicrobial and also exhibits antiparasitic activities [6]. Due to intensive use and misuse, resistance has emerged against trimethoprim [7].

Development of antimicrobial drugs was hailed as one of the great medical success story of the twentieth century [8]. At present, resistance against antimicrobial agents have become public health problem worldwide [9–15]. In the search for novel drugs against drug resistant diseases, the use of metal complexes has received tremendous attention [16–24] and resulted in a variety of exciting and invaluable drugs such as

cis-platin [24]. Research are being undertaken in fields such as cancer [25–27], diabetes [28–32], arthritis [33], magnetic resonance imaging [34], metal-mediated antibiotics, antibacterial, antiviral, antiparasitic and radiosensitizers [35–38]. In continuation of our efforts [39–44] to develop metal-based therapeutics agents, the synthesis, characterization, and antibacterial studies of trimethoprim and pyrimethamine are presented.

2. Experimental

2.1. Materials and Physical Measurements. All reagents and solvents were of analytical grade and used without further purification. Elemental analyses were carried out on a Perkin-Elmer elemental analyzer. Melting point determination was obtained with the Gallenkamp melting point apparatus. Molar conductivity measurement (10^{-3} M solutions in dimethylformamide) was obtained on the CON 6/TDS 6 conductivity/TDS Meter. FTIR spectra of the complexes were recorded as KBr pellets on a Perkin-Elmer paragon 2000 FT-IR spectrophotometer in the range 4000–370 cm^{-1}.

TABLE 1: Analytical data and some physical properties of the metal complexes.

| Complexes | Molecular formulae | Colour | Analytical Data (%) | | | Yield (%) | M.P (°C) | Cond. μS |
			C Found (calc.)	H Found (calc.)	N Found (calc.)			
[Pd(tmp)$_2$Cl$_2$]	C$_{28}$H$_{36}$N$_8$Cl$_2$O$_6$Pd	Orange	43.94 (44.37)	5.13 (4.79)	15.47 (14.78)	91	248–250	11.42
[Pd(tmp)$_2$(SCN)$_2$]	C$_{30}$H$_{36}$N$_{10}$O$_6$S$_2$Pd	Yellow	44.42 (44.86)	3.97 (4.52)	17.01 (17.44)	92	242–245	9.03
[Pt(tmp)$_2$Cl$_2$]	C$_{28}$H$_{26}$N$_8$Cl$_2$O$_6$Pt	Pale yellow	40.10 (39.72)	3.79 (4.29)	13.72 (13.24)	92	135–137	10.13
[Pt(tmp)$_2$(SCN)$_2$]	C$_{30}$H$_{36}$N$_{10}$O$_6$S$_2$Pt	Pale yellow	40.94 (40.40)	4.22 (4.07)	16.27 (15.70)	88	167–171	7.02
[Pd(pyrm)$_2$Cl$_2$]	C$_{24}$H$_{26}$N$_8$Cl$_4$Pd	Orange	43.65 (42.72)	4.26 (3.88)	16.69 (16.61)	90	259–261	15.62
[Pd(pyrm)$_2$(SCN)$_2$]	C$_{26}$H$_{26}$N$_{10}$Cl$_2$S$_2$Pd	Orange	43.46 (43.37)	3.76 (3.64)	19.66 (19.45)	92	80–82	40.80
[Pt(pyrm)$_2$Cl$_2$]	C$_{24}$H$_{26}$N$_8$Cl$_4$Pt	Pale yellow	37.99 (37.76)	3.88 (3.43)	14.93 (14.68)	89	200–202	39.80
[Pt(pyrm)$_2$(NCS)$_2$]	C$_{26}$H$_{26}$N$_{10}$Cl$_2$S$_2$Pt	Pale yellow	39.90 (38.62)	3.77 (3.24)	16.98 (17.32)	97	180-181	52.80

Electronic spectra of the complexes were recorded on a Perkin-Elmer Lambda 25 spectrophotometer. The ^{1}H spectra in DMSO-d$_6$ were performed and recorded on a Varian-NMR-vnmr s400 MHz spectrometer at 25°C, using high-power proton decoupling, and pulse sequence: s2pul. Proton chemical shifts in DMSO-d$_6$ were referenced to DMSO-d$_6$ (^{1}H-NMR, δ(DMSO) = 2.49 ppm). Chemical shifts for proton resonances are reported relative to tetramethylsilane. PtCl$_2$(COD)$_2$ and PdCl$_2$(CH$_3$CN)$_2$ were prepared according to literature procedures [45, 46].

2.2. Synthesis of Metal Complexes of the Type [M(L)$_2$Cl$_2$]. A solution containing 1 mmol of the respective metal salts (PtCl$_2$(COD)$_2$, 0.260 g) and (PdCl$_2$(CH$_3$CN)$_2$, 0.374 g) was added to colorless solutions of trimethoprim (2 mmol, 0.508 g) or pyrimethamine (2 mmol, 0.497 g) in 50 mL of methanol. The mixture was refluxed for 4 h and cooled to room temperature, and the solvent was removed in vacuo. The solid product was dried over CaCl$_2$.

2.3. Synthesis of Metal Complexes of the Type [M(L)$_2$(NCS)$_2$]. A solution containing 1 mmol of the respective metal salts (PtCl$_2$(COD)$_2$, 0.260 g) and (PdCl$_2$(CH$_3$CN)$_2$, 0.374 g) was added to colorless solutions of trimethoprim (2 mmol, 0.508 g) or pyrimethamine (2 mmol, 0.497 g) in 50 mL of methanol. The mixture was refluxed for 1 h, followed by the addition of a colourless solution of NH$_4$NCS (2 mmol, 0.152 g) in methanol and refluxed further for 3 h and cooled to room temperature, and the solvent was removed in vacuo. The solid product was dried over CaCl$_2$.

2.4. Antibacterial Studies. The antimicrobial activity of the synthesized compounds as well as their free ligands was studied by the zone of inhibition technique [47, 48] using *Staphylococcus aureus* (ATCC 6538), *Streptococcus faecalis* (ATCC 29212), *Bacillus cereus* (ATCC 10702), gram(−) *Escherichia coli* (ATCC 8739), *Klebsiella pneumonia* (ATCC 4352), *Proteus vulgaris* (ATCC 6830), *Pseudomonas aeruginosa* (ATCC 19582), and *Bacillus pumilus* (ATCC 14884) typed cultures as obtained from American Type Culture Collection (ATCC). The macrobroth dilution technique [49, 50] was used to determine the MIC. The MIC was taken as the lowest

concentration of the tested complexes that shows no visible bacterial growth [51]. Samples of organisms were taken from plates which were used for the MIC test that were with no visible growth and subcultured by way of streaking onto a freshly prepared Mueller Hinton agar medium. MBC was carried out with the method of Olorundare et al. [52] and was taken as the lowest concentration of antibiotic at which all bacteria are killed. These plates were incubated at 35–37°C for 24 h and results taken as the MBC.

3. Results and Discussion

Pd(II) and Pt(II) complexes, trimethoprim and pyrimethamine, have been synthesized and characterized by elemental analysis, UV-Vis, FTIR, and ^{1}H and ^{13}C-nmr spectroscopy. Conductivity measurements on the complexes showed that they are nonelectrolyte in solution. Generally all the complexes are insoluble both in polar and nonpolar solvents except polar coordinating solvents such as DMSO and DMF. The analytical data for the complexes are presented in Table 1 and proposed structures in Figure 1.

3.1. Infrared Spectra. The FTIR spectra of the ligands and metal complexes were compared and assigned on careful comparison. The N—H stretching frequencies of the pyrimidine NH$_2$ in the free trimethoprim shifted slightly in the metal complexes. It was observed in the same region, 3332–3461 cm^{-1}, as in the free ligands. The slight shift is ascribed to hydrogen bonding and other noncovalent interactions in the metal complexes. The coordination of the metal ions to trimethoprim affected the v(C=N) stretching vibrations. The v(C=N) that occur at 1635 cm^{-1} in the free trimethoprim ligand shifted to lower frequencies in all the complexes confirming that the metal ions are coordinated directly to the pyrimidine nitrogen atom. Strong vibrations at 2111 and 2120 cm^{-1} in [Pd(tmp)$_2$(NCS)$_2$] and [Pt(tmp)$_2$(NCS)$_2$], respectively, are due to v(NCS) stretching vibrations and may be attributed to the presence of the thiocyanate ion in the coordination sphere of these complexes [53]. The band observed in the complexes in the region 542–502 cm^{-1} was attributed to v(Pd—N) and v(Pt—N) [54].

Pyrimethamine possesses four potential coordination sites. A comparison of the spectra of pyrimethamine and

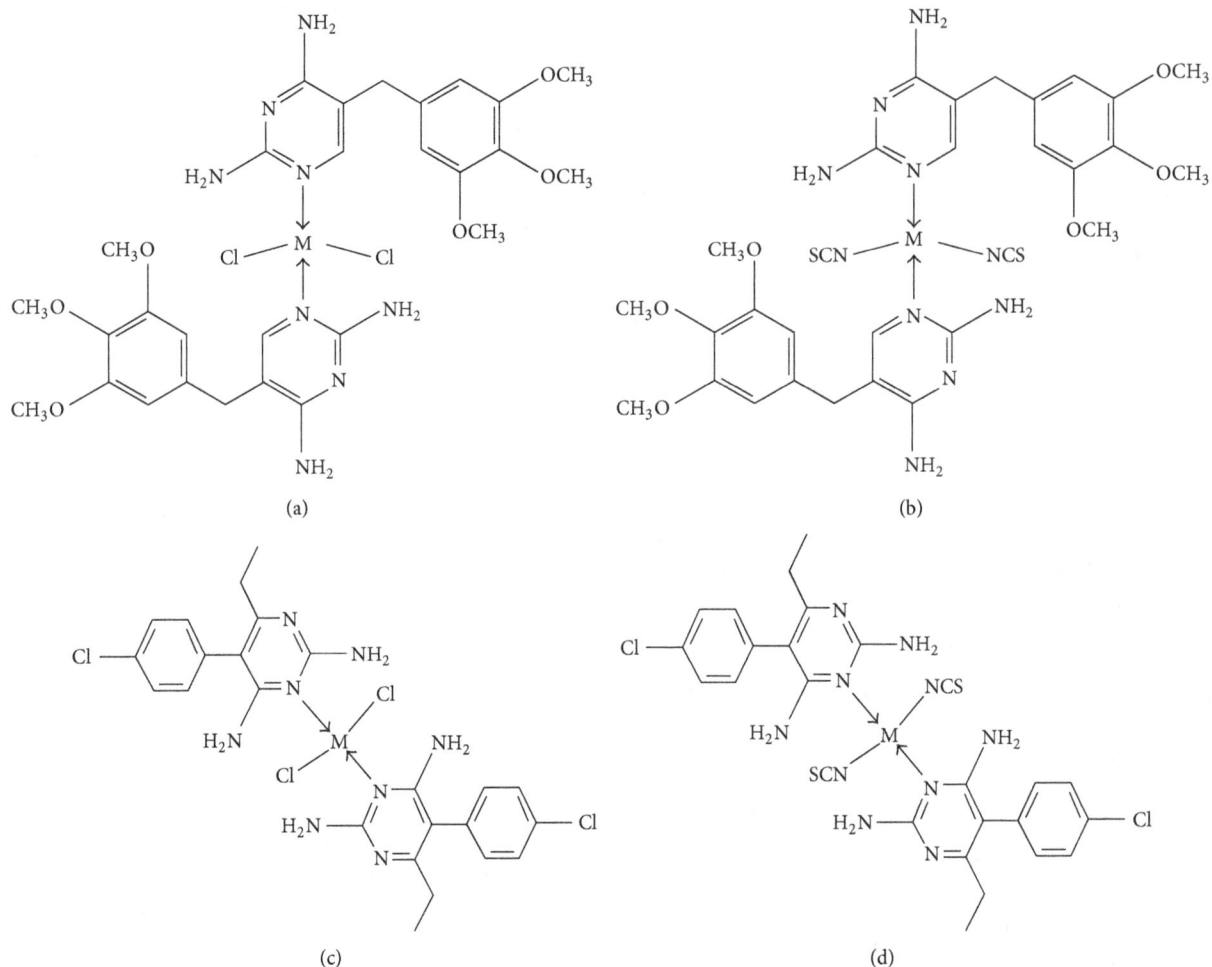

FIGURE 1: Proposed structures for the metal complexes.

the metal complexes showed that the bands due to symmetrical and asymmetrical stretching modes of NH_2 in the spectrum of pyrimethamine undergo only very slight changes in the complexes. This indicates that the metal ions bond preferentially to pyrimethamine through the nitrogen atom of pyrimidine. The absorption band at $1629 \, cm^{-1}$ in the spectrum of the pyrimethamine is attributed to the $v(C=N)$ of the pyrimidine ring. It shifted to 1612, 1619, 1639, and $1632 \, cm^{-1}$, in $[Pd(pyrm)_2Cl_2]$, $[Pd(pyrm)_2(NCS)_2]$, $[Pt(pyrm)_2Cl_2]$, and $[Pt(pyrm)_2(NCS)_2]$, respectively, which is a good indication that pyrimethamine is coordinated to Pd(II) and Pt(II) ions through the N(1) atom of the pyrimidine ring. The appearance of a prominent absorption bands observed at 2112 and $2107 \, cm^{-1}$ in the complexes $[Pd(pyrm)_2(NCS)_2]$ and $[Pt(pyrm)_2(NCS)_2]$, but absent in the free ligand one has is due to $v(NCS)$ stretching frequency of the thiocyanate ion [55].

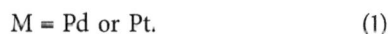

$$M = Pd \text{ or } Pt. \tag{1}$$

3.2. Electronic Spectra of the Complexes. The effect of complexation on the splitting of the d orbital is more marked for Pd(II) and Pt(II), and consequently their complexes are

diamagnetic and majority of them are square planar. The electronic spectra of Pd(II) and Pt(II) complexes like any other square planar complexes can be assigned easily. However, the situation is complicated in the Pt(II) series by the expectation that the d-p transitions will occur at comparable energies to LMCT transitions, and a clear distinction between these two types of transition may be difficult. Pt(II) complexes of trimethoprim do not show any absorption band in the visible region of their electronic spectra, but the Pd(II) complexes display weak absorption bands at around 440 nm which is assigned to $^1B_{1g} \leftarrow {}^1A_{1g}$ and $^1A_{2g} \leftarrow {}^1A_{1g}$ d-d transition of a four coordinate palladium complexes [56]. The absorption band in $[Pd(tmp)_2Cl_2]$ is stronger than that of the $[Pd(tmp)_2(NCS)_2]$, and this can be attributed to the more intense orange colour of $[Pd(tmp)_2Cl_2]$ as compared to that of yellow $[Pd(tmp)_2(NCS)_2]$. The palladium complexes of pyrimethamine show absorption bands at 553 nm, and another absorption band in the region 450–480 nm in both complexes corresponding to $^1B_{1g} \leftarrow {}^1A_{1g}$ and $^1A_{2g} \leftarrow {}^1A_{1g}$ low spin allowed d-d transition [56], respectively. The d-d transition in the platinum complexes is not seen, and this can be evident from its pale yellow colour which makes the MLCT bands dominant, as compared to the deep orange colour of palladium [57]. All the four complexes display high energy absorption band around 300 nm which can be attributed to typical charge transfer transitions in the complexes [56, 57] confirming the square planar geometries proposed for the metal complexes.

3.3. NMR Spectroscopy of the Metal Complexes.

^{1}H-nmr spectra data of the trimethoprim complexes in d^6-DMSO shows the presence of some of the proton signals as compared to that of the free trimethoprim ligand [58, 59]. The proton NMR of [Pt(tmp)$_2$(NCS)$_2$] showed three major peaks at the aromatic region integrating for only three protons at δ(ppm) 7.68, 7.46, and 6.61 ppm assigned to (s, 1H, H-4a, $J = 1.2$ Hz), (s, 1H, H-5b), and (s, 1H, H-1b), respectively. The protons of the methyl in the methoxyl group which are equivalent can be observed as a single peak at δ(ppm) 1.77 ppm integrated for three protons. In [Pd(tmp)$_2$(NCS)$_2$] four major peaks can also be observed in the aromatic region but only two of the peaks which are integrated for one proton each was successfully assigned at delta values of 8.25 and 7.96 ppm ascribed to (s, 1H, H-4a) and (s, 1H, H-5b), respectively. The ^{1}H-nmr spectrum of [Pt(tmp)$_2$Cl$_2$] could not be resolved but the ^{13}C-nmr gave useful information for the formation of the Pd(II) complex. The coordinations of the two different metals and their contributions in [Pt(tmp)$_2$(NCS)$_2$] and [Pd(tmp)$_2$(NCS)$_2$] can be seen from the slight shift of the proton nmr signals assigned in their spectra.

^{1}H and ^{13}C-nmr spectra of pyrimethamine [60] were compared to those of the complexes. There was a significant shift in the chemical shift values upon complexation. The ^{1}H-nmr spectra of [Pt(pyrm)$_2$Cl$_2$] in DMSO-d$_6$ solutions showed chemical shift at $\delta_{(ppm)}$ 7.51 (s, 1H, H-5b), 7.26 (s, 1H, H-3b), 6.75 (s, 1H, H-6b), and 6.55 (s, 1H, H-2b). The aliphatic region showed multiple signals between 2.20 and 1.00 ppm which support the presence of an ethyl group of the pyrimethamine ligand, integrating for five protons. In [Pt(pyrm)$_2$(NCS)$_2$], the chemical shift was observed at $\delta_{(ppm)}$ 7.53 (s, 2H, H-5b, 3b), 7.28 (s, 1H, H-6b), and 6.98 (s, 1H, H-2b). Once again, the aliphatic region showed multiple signals between 2.20 and 1.00 ppm which support the presence of an ethyl group integrating for five protons. The two amino groups of pyrimethamine exhibit characteristic shifts which can be seen as a broad peak at chemical shift value at around 4.00 ppm in both complexes.

The ^{13}C-nmr spectra of the metal complexes of trimethoprim were compared with that of free trimethoprim ligand [61] with a significant shift in the area that has been affected by the coordination to the metal ions. The signal at 155.91 ppm in the ligand has been shifted upfield to a value of 153.60 ppm in [Pt(tmp)$_2$Cl$_2$] assigned to C(2a, 6a), both of C=N carbon atom of the pyrimidine ring of trimethoprim. Other ^{13}C-nmr signal of trimethoprim were observed at 136.69 ppm assigned to the quaternary carbon C(5a, 5b), 106.59 ppm C(1b, 5b), 60.99 ppm assigned to C7a, the peaks at 56.64 and 32.99 ppm assigned to C(2b, 3b, 4b) and CH$_3$ bonded to the oxygen of the methoxyl. The ^{13}C-nmr of [Pt(tmp)$_2$(NCS)$_2$] and [Pd(tmp)$_2$(NCS)$_2$] both have more peaks than that of the [Pt(tmp)$_2$Cl$_2$], probably due to the presence of a thiocyanate

group NCS in the coordination sphere of these complexes. The resonance at 162.77 and 160.68 ppm in [Pt(tmp)$_2$(NCS)$_2$] and at 163.10 and 160.51 ppm in [Pd(tmp)$_2$(NCS)$_2$] corresponding to C=N (C2a, C6a) of the trimethoprim ligand shifted downfield and this can be ascribed to the coordination of the metal to the nitrogen of the pyrimidine ring. The quaternary carbon C5a, C5b chemical shift occurs at 155.94 and 153.53 ppm in [Pt(tmp)$_2$(NCS)$_2$] and at 154.83 and 153.72 ppm in [Pd(tmp)$_2$(NCS)$_2$]. The presence of a peak at 135 ppm in both complexes is an indication that the thiocyanate ions are present in the complex and coordinate through the nitrogen. The high intensity peaks at 56.67 and 56.68 ppm in [Pt(tmp)$_2$(NCS)$_2$] and [Pd(tmp)$_2$(NCS)$_2$] were assigned to the methoxyl carbon, the peaks at 60.85 ppm in [Pt(tmp)$_2$(NCS)$_2$] and 60.84 ppm in [Pd(tmp)$_2$(NCS)$_2$] were assigned to methylene C7a, and lastly the peaks at 33.19 and 33.58 ppm were assigned to the CH$_3$ of the aliphatic region.

The ^{13}C-nmr for both platinum complexes of pyrimethamine gave more information in assigning the necessary signals and affirming the probable structure of the complexes. In the spectra of [Pt(pyrm)$_2$Cl$_2$], the singlet resonance at 163.83 is assigned to carbon atom of C=N of the pyrimidine ring; there was a significant upfield shift which is probably due to the effect of coordination to the Pt(II) ion. The peak at 159.43 ppm was assigned to C5a. The phenyl ring carbon C1b–C6b can be seen in the range of 133.58–107.10 ppm. The aliphatic region consists of two peaks at 26.44 and 13.59 ppm assigned to the methylene and the methyl groups, respectively, from the ethyl of the pyrimethamine ligand. [Pt(pyrm)$_2$(NCS)$_2$] show a similar trend in the ^{13}C-nmr spectrum; the singlet peak at 164.34 ppm was assigned to C(2a) of the pyrimethamine. The thiocyanate carbon is observed at 134.02 ppm [62, 63]. The carbons of the pyrimethamine phenyl ring were found in the range of 133.34–107.65 ppm. The aliphatic region consists of complex peaks at between 29.93 and 25.62 ppm, indicative of the presence of a methylene group.

3.4. Antibacterial Screening of the Metal Complexes.

The complexes showed varied antibacterial activities against both gram-positive and gram-negative bacterial isolates (Table 2). The highest zone of inhibition of 34 mm was recorded for [Pd(tmp)$_2$Cl$_2$] B. cereus. All Pd and Pt complexes of trimethoprim are active against E. coli and P. vulgaris. Their zones of inhibition varied between 28 and 32 mm as against 16 mm shown by trimethoprim drug. Trimethoprim and all its complexes are inactive against P. aeruginosa and B. pumilus. Of all the trimethoprim complexes, [Pd(tmp)$_2$(NCS)$_2$] appear to be the least active, inhibiting only E. coli, S. faecalis, and P. vulgaris. It must also be noted that, it might be least active but it is the only complex of trimethoprim that shows a zone of inhibition of 27 mm against S. faecalis whereas the other three complexes did not show any visible inhibition. For the trimethoprim complexes, the structure activity relationship showed that both Pd(II) and Pt(II) complexes containing chloride ions are the most active. They are active against two gram positive bacteria and three gram negative bacteria isolates with the Pt complexes, [Pt(tmp)$_2$Cl$_2$], showing relatively

TABLE 2: Zone of inhibition exhibited by metal complexes at 40 mg/mL (mm).

Complexes	E. coli	P. aeruginosa	S. aureus	S. faecalis	B. cereus	B. pumilus	K. pneumonia	P. vulgaris
$[Pd(tmp)_2Cl_2]$	30.0	NI	29.0	NI	32.0	NI	26.0	32.0
$[Pd(tmp)_2(NCS)_2]$	28.0	NI	NI	27.0	NI	NI	NI	28.0
$[Pt(tmp)_2Cl_2]$	32.0	NI	32.0	NI	34.0	NI	30.0	28.0
$[Pt(tmp)_2(NCS)_2]$	30.0	NI	25.0	NI	32.0	NI	NI	30.0
$[Pd(pyrm)_2Cl_2]$	NI	30	12	NI	NI	21	14	NI
$[Pd(pyrm)_2(NCS)_2]$	NI	33	29	NI	NI	20	33	NI
$[Pt(pyrm)_2Cl_2]$	14	NI	20	26	NI	NI	NI	NI
$[Pt(pyrm)_2(NCS)_2]$	20	NI	18	NI	28	NI	NI	30
Pyrimethamine	11	NA	13	12	10	NA	15	12
Trimethoprim	15.0	NA	12.0	18.0	17.0	NA	20.0	16.0

NI: no inhibition; NA: not applicable.

TABLE 3: MIC values (mg/mL) of the metal complexes on selected bacteria.

Complexes	E. coli	P. aeruginosa	S. aureus	S. faecalis	B. cereus	B. pumilus	K. pneumonia	P. vulgaris
$[Pd(pyrm)_2Cl_2]$	NI	5.0	0.63	NI	NI	0.63	0.31	NI
$[Pd(pyrm)_2(NCS)_2]$	NI	5.0	0.31	NI	NI	0.63	0.63	NI
$[Pt(pyrm)_2Cl_2]$	20.0	NI	10.0	5.0	NI	NI	NI	NI
$[Pt(pyrm)_2(NCS)_2]$	10.0	NI	10.0	NI	5.0	NI	NI	5.0
$[Pd(tmp)_2Cl_2]$	20.0	NI	10.0	NI	5.0	NI	5.0	10.0
$[Pd(tmp)_2(NCS)_2]$	10.0	NI	NI	10.0	NI	NI	NI	NI
$[Pt(tmp)_2Cl_2]$	20.0	NI	10.0	NI	10.0	NI	5.0	10.0
$[Pt(tmp)_2(NCS)_2]$	20.0	NI	10.0	NI	10.0	NI	NI	5.0
Pyrimethamine	20.0	NA	20.0	20.0	10.0	NA	20.0	10.0
Trimethoprim	20.0	NA	10.0	20.0	10.0	NA	10.0	10.0

higher zones of inhibition against five bacteria isolates than the Pd(II) complex, $[Pd(tmp)_2Cl_2]$. It must be noted however that high zone of inhibition may not reveal for certain that a particular compound is a stronger antibacterial agent since this may be attributed to factors such as the rate of diffusion of the antibacterial agents and the amount of bacterial isolates present in a certain amount of agar solution [64].

The inhibition of the bacteria isolates by the pyrimethamine complexes are relatively fewer than those of complexes of trimethoprim (Table 2). However, the results showed that Pd(II) complexes of pyrimethamine, $[Pd(pyrm)_2Cl_2]$ and $[Pd(pyrm)_2(NCS)_2]$, have zones of inhibition of 30 and 33 mm against P. aeruginosa and 14 and 33 mm against B. pumilus. None of the complexes of trimethoprim and neither trimethoprim nor pyrimethamine showed any visible activity against these bacteria isolates. The highest zones of inhibition of 33 mm were observed for $[Pd(pyrm)_2(NCS)_2]$ against P. aeruginosa and K. pneumonia. Pyrimethamine inhibited the growth of six bacteria isolates whereas three of its metal complexes inhibited four bacteria isolates while $[Pt(pyrm)_2Cl_2]$ inhibited only three bacteria isolates. In the trimethoprim metal complexes, the complexes containing the chloride ions are generally showed better antibacterial activities than those with NCS ions. In the pyrimethamine metal complexes, the NCS ion is the anion of choice for enhanced activity. The minimum inhibition concentrations

(MICs) and minimum bactericidal concentrations (MBCs) of the compounds were evaluated and presented in Tables 3 and 4, respectively. It shows that the Pd(II) complexes of pyrimethamine are more active than the Pt(II) complexes of either trimethoprim or pyrimethamine. The lowest MIC values of 0.31 mg/mL were recorded for the Pd complexes: $[Pd(pyrm)_2Cl_2]$ against K. pneumonia and $[Pd(pyrm)_2(NCS)_2]$ against S. aureus. The complexes also have MIC values of 0.63 mg/mL against S. aureus and B. pumilus for $[Pd(pyrm)_2Cl_2]$ while $[Pd(pyrms)_2(NCS)_2]$ values are against B. pumilus and K. pneumonia. Their MBC values against these bacteria isolates are also the lowest indicating their high antibacterial activities.

4. Conclusions

We report the synthesis, characterization, and antibacterial studies of Pd(II) and Pt(II) complexes of trimethoprim and pyrimethamine. The complexes formulated as four coordinate square planar species consisting of two molecules of either trimethoprim or pyrimethamine and two chloride or thiocyanate ions. The complexes were characterized by elemental analysis, electronic, FTIR, and NMR spectroscopy. Spectroscopic analyses confirmed the coordination of the metal ions to the drug through the pyrimidine nitrogen atom. The antibacterial screening of the complexes showed

TABLE 4: MBC values (mg/mL) of the metal complexes on selected bacteria.

Complexes	E. coli	P. aeruginosa	S. aureus	S. faecalis	B. cereus	B. pumilus	K. pneumonia	P. vulgaris
[Pd(tmp)$_2$Cl$_2$]	>20.0	NI	20.0	NI	5.0	NI	10.0	20.0
[Pd(tmp)$_2$(NCS)$_2$]	10.0	NI	NI	>20.0	NI	NI	NI	NI
[Pt(tmp)$_2$Cl$_2$]	20.0	NI	>20.0	NI	10.0	NI	10.0	>20.0
[Pt(tmp)$_2$(NCS)$_2$]	>20.0	NI	10.0	NI	>20.0	NI	NI	20.0
[Pd(pyrm)$_2$Cl$_2$]	NI	10.0	2.5	NI	NI	2.5	0.31	NI
[Pd(pyrm)$_2$(NCS)$_2$]	NI	10.0	0.63	NI	NI	1.25	2.50	NI
[Pt(pyrm)$_2$Cl$_2$]	20.0	NI	10.0	10.0	NI	NI	NI	NI
[Pt(pyrm)$_2$(NCS)$_2$]	>20.0	NI	20.0	NI	10.0	NI	NI	20.0
Pyrimethamine	>20.0	NA	>20.0	>20.0	>20.0	NA	20.0	>20.0
Trimethoprim	>20.0	NA	20.0	>20.0	20.0	NA	20.0	20.0

varied activities but they are more active than the drugs. The MIC and MBC determinations revealed that the Pd(II) metal complexes are the most active.

Acknowledgment

The authors acknowledge the financial contribution of GMRDC, University of Fort Hare.

References

[1] E. A. Falco, L. G. Goodwin, G. H. Hitchings, I. M. Rollo, and P. B. Russel, "2,4-diaminopyrimidines- a new series of antimalarials," *British Journal of Pharmacology and Chemotherapy*, vol. 6, no. 2, pp. 185–200, 1951.

[2] B. Roth, E. A. Fmco, G. H. Hitchings, and S. R. M. Bushby, "5-Benzyl-2, 1-diaminopyrimidines as antibacterial agents. I. Synthesis and antibacterial activity in vitro," *Journal of Medicinal and Pharmaceutical Chemistry*, vol. 5, no. 6, pp. 1103–1123, 1962.

[3] G. H. Hitchings and S. L. Smith, "Dihydrofolate reductases as targets for inhibitors," *Advances in Enzyme Regulation*, vol. 18, pp. 347–371, 1980.

[4] E. A. Falco, L. G. Goodwin, G. A. Hitchings, I. M. Rollo, and P. B. Russel, "2:4-diaminopyrimidines- a new series of antimalarials," *British Journal of Pharmacology and Chemotherapy*, vol. 6, no. 2, pp. 185–200, 1951.

[5] G. H. Hitchings, "Species differences among dihydrofolate reductases as a basis for chemotherapy," *Postgraduate Medical Journal*, vol. 45, pp. 7–10, 1969.

[6] J. J. Burchall, J. W. Corcoran, and F. E. Hahn, Eds., *Antibiotics*, vol. 3, Springer, New York, NY, USA, 1975.

[7] R. L. Then, "History and future of antimicrobial 2, 4-diaminopyrimidines," *Journal of Chemotherapy*, vol. 5, pp. 361–368, 1993.

[8] B. M. Cook, N. Mohandas, and R. L. Copel, "Malaria and the red blood cell membrane," *Seminars in Hematology*, vol. 41, pp. 173–188, 2004.

[9] D. T. W. Chu, J. J. Plattner, and L. Katz, "New directions in antibacterial research," *Journal of Medicinal Chemistry*, vol. 39, no. 20, pp. 3853–3873, 1996.

[10] J. Verhoef and A. Fluit, "Surveillance uncovers the smoking gun for resistance emergence," *Biochemical Pharmacology*, vol. 71, no. 7, pp. 1036–1041, 2006.

[11] M. Leeb, "Antibiotics: a short in the arm," *Nature*, vol. 431, pp. 892–893, 2004.

[12] P. A. Ajibade and N. H. Zulu, "Synthesis, characterization, and antibacterial activity of metal complexes of phenylthiourea: the X-ray single crystal structure of [Zn(SC(NH$_2$)NHC$_6$H$_5$)$_2$(OOCCH$_3$)$_2$] C$_2$H$_5$OH," *Journal of Coordination Chemistry*, vol. 63, no. 18, pp. 3229–3239, 2010.

[13] N. V. Loginova, T. V. Koval'chuk, R. A. Zheldakova et al., "Synthesis and biological evaluation of copper (II) complexes of sterically hindered o-aminophenol derivatives as antimicrobial agents," *Bioorganic and Medicinal Chemistry Letters*, vol. 16, no. 20, pp. 5403–5407, 2006.

[14] World Health Organization, *The World Health Report 2004-Changing History, Annex Table 2: Deaths By Cause, Sex and Mortality Stratum in WHO Regions, Estimates For 2002*, WHO, Washington, DC, USA, 2004.

[15] K. Namba, X. Zheng, K. Motoshima et al., "Design and synthesis of benzenesulfananilides active against methicillin resistant staphylococcus aureus and vancomycin-resistant enterococcus," *Bioorganic and Medicinal Chemistry Letters*, vol. 16, no. 11, pp. 6131–6144, 2008.

[16] C. S. K. Rajapakse, A. Martinez, B. Naoulou et al., "Synthesis, characterization, and in vitro antimalarial and antitumor activity of new ruthenium(II) complexes of chloroquine," *Inorganic Chemistry*, vol. 48, no. 3, pp. 1122–1131, 2009.

[17] H. A. Wee, E. Daldini, C. Scolaro, R. Scopelliti, L. Juillerat-Jeannerat, and P. J. Dyson, "Development of organometallic ruthenium-arene anticancer drugs that resist hydrolysis," *Inorganic Chemistry*, vol. 45, no. 22, pp. 9006–9013, 2006.

[18] C. S. Allardyce, A. Dorcier, C. Scolaro, and P. J. Dyson, "Development of organometallic (organo-transition metal) pharmaceuticals," *Applied Organometallic Chemistry*, vol. 19, no. 1, pp. 1–10, 2005.

[19] M. S. Refat and S. A. El-Shazly, "Identification of a new antidiabetic agent by combining VOSO4 and vitamin E in a single molecule: studies on its spectral, thermal and pharmacological properties," *European Journal of Medicinal Chemistry*, vol. 45, no. 7, pp. 3070–3079, 2010.

[20] B. H. M. Hussein, A. A. Hassan, F. E. A. Mona, and E. F. Abdullah I, "A novel anti-tumor agent, Ln(III) 2-thioacetate benzothiazole induces anti-angiogenic effect and cell death in cancer cell lines," *European Journal of Medicinal Chemistry*, vol. 51, pp. 99–109, 2012.

[21] K. H. Thompson, J. Chiles, V. G. Yuen, J. Tse, J. H. McNeill, and C. Orvig, "Comparison of anti-hyperglycemic effect amongst vanadium, molybdenum and other metal maltol complexes," *Journal of Inorganic Biochemistry*, vol. 98, no. 5, pp. 683–690, 2004.

[22] S. M. Brichard and J.-C. Henquin, "The role of vanadium in the management of diabetes," *Trends in Pharmacological Sciences*, vol. 16, no. 8, pp. 265–270, 1995.

[23] B. S. Sekhon and L. Gandhi, "Medicinal uses of inorganic compounds-1," *Resonanceno*, vol. 11, no. 4, pp. 75–89, 2006.

[24] D. Lebwohl and R. Canetta, "Clinical development of platinum complexes in cancer therapy: an historical perspective and an update," *European Journal of Cancer*, vol. 34, no. 10, pp. 1522–1534, 1998.

[25] C. G. Hartinger, M. A. Jakupec, S. Zorbas-Seifried et al., "KP1019, a new redox-active anticancer agent-preclinical development and results of a clinical phase I study in tumor patients," *Chemistry and Biodiversity*, vol. 5, no. 10, pp. 2140–2155, 2008.

[26] G. Sava, A. Bergamoa, and P. J. Dyson, "Metal-based antitumour drugs in the post-genomic era: what comes next?" *Dalton Transactions*, vol. 40, pp. 9069–9075, 2011.

[27] R. W. Y. Sun, D. L. Ma, E. L. M. Wong, and C. M. Che, "Some uses of transition metal complexes as anti-cancer and anti-HIV agents," *Dalton Transactions*, vol. 10, pp. 4884–4892, 2007.

[28] H. Sakurai, Y. Kojima, Y. Yoshikawa, K. Kawabe, and H. Yasui, "Antidiabetic vanadium(IV) and zinc(II) complexes," *Coordination Chemistry Reviews*, vol. 226, no. 1-2, pp. 187–198, 2002.

[29] T. Kiss, T. Jakusch, D. Hollender et al., "Biospeciation of antidiabetic VO, (IV) complexes," *Coordination Chemistry Review*, vol. 252, no. 10-11, pp. 1153–1162, 2008.

[30] N. Sekar, J. Li, and Y. Shechter, "Vanadium salts as insulin substitutes: mechanisms of action, a scientific and therapeutic tool in diabetes mellitus research," *Critical Reviews in Biochemistry and Molecular Biology*, vol. 31, no. 5, pp. 339–359, 1996.

[31] K. H. Thompson and C. Orvig, "Vanadium in diabetes: 100 years from Phase 0 to Phase I," *Journal of Inorganic Biochemistry*, vol. 100, no. 12, pp. 1925–1935, 2006.

[32] K. H. Thompson, J. H. McNeill, and C. Orvig, "Vanadium compounds as insulin mimics," *Chemical Reviews*, vol. 99, no. 9, pp. 2561–2571, 1999.

[33] L. Messori and G. Marcon, "Gold complexes in the treatment of rheumatoid arthritis," *Metal Ions Biological Systems*, vol. 41, pp. 279–304, 2004.

[34] P. Caravan, J. J. Ellison, T. J. McMurry, and R. B. Lauffer, "Gadolinium(III) chelates as MRI contrast agents: structure, dynamics, and applications," *Chemical Reviews*, vol. 99, no. 9, pp. 2293–2352, 1999.

[35] A. Garoufis, S. K. Hadjikakou, and N. Hadjiliadis, "Palladium coordination compounds as anti-viral, anti-fungal, antimicrobial and anti-tumor agents," *Coordination Chemistry Reviews*, vol. 253, no. 9-10, pp. 1384–1397, 2009.

[36] C. Orvig and M. J. Abrams, "Medicinal inorganic chemistry," *Chemical Review*, vol. 99, no. 9, pp. 2201–2842, 1999.

[37] Z. Guo and P. J. Sadler, "Metals in medicine," *Angew Chemie International Edition*, vol. 38, no. 11, pp. 1512–1531, 1999.

[38] M. J. Abrams and B. A. Murrer, "Metal compounds in therapy and diagnosis," *Science*, vol. 261, no. 5122, pp. 725–730, 1993.

[39] P. A. Ajibade, G. A. Kolawole, and P. O'Brien, "Metal complexes of 4-amino-N-(2-pyrimidinyl)benzene sulfonamide: synthesis, characterization and antiprotozoal studies," *Synthesis and Reactivity in Inorganic, Metal-Organic and Nano-Metal Chemistry*, vol. 37, no. 8, pp. 653–659, 2007.

[40] P. A. Ajibade and G. A. Kolawole, "Synthesis, characterization, antiplasmodial and antitrypanosomal activity of some metal(III) complexes of sulfadiazine," *Bulletin of the Chemical Society of Ethiopia*, vol. 22, no. 2, pp. 261–268, 2008.

[41] P. A. Ajibade and G. A. Kolawole, "Synthesis, characterization and antiprotozoal studies of some metal complexes of antimalarial drugs," *Transition Metal Chemistry*, vol. 33, no. 4, pp. 493–497, 2008.

[42] P. A. Ajibade and G. A. Kolawole, "Synthesis, characterization and in vitro antiprotozoal studies of iron(III) complexes of some antimalarial drugs," *Journal of Coordination Chemistry*, vol. 61, no. 21, pp. 3367–3374, 2008.

[43] P. A. Ajibade, "Metal complexes in the management of parasitic diseases: in vitro antiprotozoal studies of metal complexes of some antimalarial drugs," *Current Science*, vol. 95, no. 12, p. 1673, 2008.

[44] P. A. Ajibade and G. A. Kolawole, "Cobalt(III) complexes of some antimalarial drugs: synthesis, characterization, and in vitro antiprotozoal studies," *Synthesis and Reactivity in Inorganic, Metal-Organic and Nano-Metal Chemistry*, vol. 40, no. 4, pp. 273–278, 2010.

[45] G. K. Anderson and M. Lehn, "Bis(benzonitrile)dichloro complexes of palladium and platinum," *Inorganic Synthesis*, vol. 28, pp. 60–62, 1990.

[46] N. Manav, A. K. Mishra, and N. K. Kaushik, "Triphenyl phosphine adducts of platinum(IV) and palladium(II) dithiocarbamates complexes: a spectral and in vitro study," *Spectrochimica Acta A*, vol. 60, no. 13, pp. 3087–3092, 2004.

[47] D. S. Grierson and A. J. Afolayan, "Antibacterial activity of some indigenous plants used for the treatment of wounds in the Eastern Cape, South Africa," *Journal of Ethnopharmacology*, vol. 66, no. 1, pp. 103–106, 1999.

[48] A. D. Russell and J. R. Furr, "The antibacterial activity of a new chloroxylenol preparation containing ethylenediamine tetraacetic acid," *Journal of Applied Bacteriology*, vol. 43, no. 2, pp. 253–260, 1977.

[49] M. B. Ibrahim, M. O. Owonubi, and J. A. Onaolapo, "Antimicrobial effects of extracts of leaf, stem, and root-bark of Anogiessus leicarpus on Staphylococcus aureus NCTC, 8190, Escherichia coli NCTC, 10418 and Proteus vulgaris NCTC, 4636," *Journal of Pharmaceutical Research and Development*, vol. 2, pp. 20–26, 1997.

[50] D. A. Akinpelu and D. O. Kolawole, "Phytochemistry and antimicrobial activity of leaf extractof Piliostigma thonningii (Schum)," *Science Focus*, vol. 7, pp. 64–70, 2004.

[51] K. Nishizawa, M. Hirano, A. Kimura et al., "Evaluation of the antimicrobial activity of carbapenem and cephem antibiotics against Pseudomonas aeruginosa isolated from hospitalized patients," *Journal of Infection and Chemotherapy*, vol. 4, no. 4, pp. 174–176, 1998.

[52] E. E. Olorundare, T. S. Emudianughe, G. S. Khasar, S. A. Koteyi, and N. Irobi, "Antibacterial properties of leave extract of Cassia alata," *Biological Research*, vol. 4, pp. 113–117, 1992.

[53] S. A. Shaker, Y. Farina, S. Mahmmod, and M. Eskender, "Preparation and study of mixed ligand complexes of caffeine and cyanate with some metal ions," *Australian Journal of Basic and Applied Sciences*, vol. 3, no. 4, pp. 3337–3340, 2009.

[54] J. Liu, B. Zhang, B. Wu, K. Zhang, and S. Hu, "The direct electrochemical synthesis of Ti(II), Fe(II), Cd(II), Sn(II), and Pb(II) cwith N, N-bis(Salicylidene)-o-phenylenediamine," *Turkish Journal of Chemistry*, vol. 31, pp. 623–629, 2007.

[55] S. A. Shaker and Y. Farina, "Preparing and characterization of some mixed ligand Complexes of 1, 3, 7-Trimethylxanthin, γ-Picoline and thiocyanate with some metal ions," *American Journal of Scientific Research*, vol. 5, pp. 20–26, 2009.

[56] A. J. Charlson, N. T. McArdle, and E. C. Watton, "The induction of filamentous growth in Escherichia Coli by a palladium(II) complex of L-serine," *Inorganica Chimica Acta C*, vol. 56, pp. L35–L36, 1981.

[57] G. J. Grant, A. M. Goforth, D. G. van Derver, and W. T. Pennigton, "Homoleptic platinum(II) and palladium(II) complexes of 1,5,9-trithiacyclododecane: the crystal structures of $[Pt(12S3)_2](PF_6)_2 \cdot 2CH_3NO_2$ and $[Pd(12S3)_2](BF_4)_2 \cdot 0.5H_2O$," *Inorganic Chimica Acta*, vol. 357, no. 7, pp. 2107–2114, 2004.

[58] T. F. Koetzle and G. J. B. Williams, "The crystal and molecular structure of the antifolate drug trimethoprim (2,4-diamino-5-(3,4,5-trimethoxybenzyl)pyrimidine). A neutron diffraction study," *Journal of American Chemical Society*, vol. 98, no. 8, pp. 2074–2078, 1976.

[59] J. J. Bergh, J. C. Breytenbach, and P. L. Wessels, "Degradation of trimethoprim," *Journal of Pharmaceutical Sciences*, vol. 78, no. 4, pp. 348–350, 1989.

[60] M. V. G. de Araújo, E. K. B. Vieira, G. S. Lázaro et al., "Inclusion complexes of pyrimethamine in 2-hydroxypropyl-β-cyclodextrin: characterization, phase solubility and molecular modelling," *Bioorganic and Medicinal Chemistry*, vol. 15, no. 17, pp. 5752–5759, 2007.

[61] B. Simo, L. Porello, R. Ortiz, A. Castineiras, J. Latorre, and E. Canton, "Interactions of metal ions with a 2, 4-diaminopyrimidine derivative (trimethoprim): antimicrobial studies," *Journal of Inorganic Biochemistry*, vol. 81, no. 4, pp. 275–283, 2000.

[62] M. K. Nazeeruddin, S. M. Zakeeruddin, R. Humphry-Baker, S. I. Gorelsky, A. B. P. Lever, and M. Grätzel, "Synthesis, spectroscopic and a ZINDO study of cis- and trans-(X2)bis(4,4′-dicarboxylic acid-2,2′-bipyridine)ruthenium(II) complexes (X = Cl-, H2O, NCS-)," *Coordination Chemistry Reviews*, vol. 208, no. 1, pp. 213–225, 2000.

[63] X. Li, J. Gui, H. Yang, W. Wu, F. Li, and H. C. Huang, "A new carbozole-based phenanthrenyl ruthenium complex as sensitizer for a dye-sensitized solar cell," *Inorganic Chimica Acta*, vol. 361, no. 9-10, pp. 2835–2840, 2008.

[64] J. L. Rios, M. C. Recio, and A. Villar, "Screening methods for natural products with antimicrobial activity: a review of the literature," *Journal of Ethnopharmacology*, vol. 23, no. 2-3, pp. 127–149, 1988.

18

Synthesis, Characterization, and Interaction with Biomolecules of Platinum(II) Complexes with Shikimic Acid-Based Ligands

Yan Peng, Min-Min Zhang, Zhen-Feng Chen, Kun Hu, Yan-Cheng Liu, Xia Chen, and Hong Liang

State Key Laboratory Cultivation Base for the Chemistry and Molecular Engineering of Medicinal Resources, School of Chemistry and Chemical Engineering of Guangxi Normal University, Guilin 541004, China

Correspondence should be addressed to Zhen-Feng Chen; chenzfgxnu@yahoo.com

Academic Editor: Concepción López

Starting from the active ingredient shikimic acid (SA) of traditional Chinese medicine and $NH_2(CH_2)_nOH$, ($n = 2$–6), we have synthesized a series of new water-soluble Pt(II) complexes $PtL^{a-e}Cl_2$, where L^{a-e} are chelating diamine ligands with carbon chain covalently attached to SA (L^{a-e} = SA-NH$(CH_2)_n$NHCH$_2$CH$_2$NH$_2$; L^a, $n = 2$; L^b, $n = 3$; L^c, $n = 4$; L^d, $n = 5$; L^e, $n = 6$). The results of the elemental analysis, LC-MS, capillary electrophoresis, and ^{1}H, ^{13}C NMR indicated that there was only one product (isomer) formed under the present experimental conditions, in which the coordinate mode of $PtL^{a-e}Cl_2$ was two-amine bidentate. Their *in vitro* cytotoxic activities were evaluated by MTT method, where these compounds only exhibited low cytotoxicity towards BEL7404, which should correlate their low lipophilicity. The interactions of the five Pt(II) complexes with DNA were investigated by agarose gel electrophoresis, which suggests that the Pt(II) complexes could induce DNA alteration. We also studied the interactions of the Pt(II) complexes with 5′-GMP with ESI-MS and ^{1}H NMR and found that PtL^bCl_2, PtL^cCl_2, and PtL^dCl_2 could react with 5′-GMP to form mono-GMP and bis-GMP adducts. Furthermore, the cell-cycle analysis revealed that PtL^bCl_2, PtL^cCl_2 cause cell G_2-phase arrest after incubation for 72 h. Overall, these water-soluble Pt(II) complexes interact with DNA mainly through covalent binding, which blocks the DNA synthesis and replication and thus induces cytotoxicity that weakens as the length of carbon chain increases.

1. Introduction

As Pt(II) complexes have demonstrated successful clinical application of cisplatin for its anticancer effects, new platinum-based anticancer drugs are highly desired [1–12]. Up to now, there have been five platinum-based anticancer drugs used in clinical applications, including three FDA-approved platinum compounds: cisplatin, carboplatin, oxaliplatin, nedaplatin used in Japan, and lobaplatin approved for use in China. In addition, new and nontraditional compounds picoplatin (AMD473) [13] and ProLindac [14] as well as platinum(IV) complexes, such as satraplatin [15], tetraplatin, tetrachloro-trans-R, R-cyclohexane-1,2-diamine platinum (IV), and Ipropltin (dichlorodihydroxobis(iso-propylamine)platinum (IV) [16, 17], are being evaluated for clinical trials. Nevertheless, their effectiveness is still hindered by clinical problems, such as acquired or intrinsic resistance that limits the spectrum of cancers that can be treated, and

high toxicity leading to side effects and limiting the dose that can be registered [18]. In the past three decades, substantial efforts have been directed to the tactics that can improve cellular accumulation, oral bioavailability, lifetime in blood, and tumor targeting [13].

During the past two decades, water-soluble platinum(II) complexes have been synthesized, which cannot only retain antitumor activity but also be effectively absorbed after oral administration [19]. The most common approach to prepare these compounds is a structural approach, in which the chloride ligands are replaced by chelating carboxylates, oxalate, and glycolate [20–22]. Besides, there are many other methods, such as platinum terpyridine complexes with glycosylated acetylide and arylacetylide ligands [23], water-soluble macromolecular platinum conjugates [24, 25], water-soluble porphyrin-Pt(II) conjugates [26, 27], formation of water-soluble organometallic analogues of oxaliplatin [28], or supramolecular nanoencapsulation technique [29]. However, the water-soluble platinum complexes archived by means of amine ligands instead of ammonia are most intriguing because the carrier ligands provide broad spectrum of antitumor activity [30]. For instance, water-soluble platinum(II) complexes of diamine chelating ligands bearing amino-acid type substituents [31], the carbohydrate-metal complexes, have been proved a potential effective method [32–36]. But the synthesis of this complex is extremely laborious, which also lacks clinical data. By contrast, shikimic acid (3R,4S,5R-trihydroxy-1-cyclohexane-1-carboxylic acid, SA), an active ingredient isolated from traditional Chinese medicine *Illicium verum Hook. f.* grown in Guangxi province of China, has exhibited good water solubility. It is well used as DNA vaccine carrier [37] and [^{99m}Tc(CO)$_3$]-labeled bombesin to reduce both hepatic uptake and renal retention [38–40]. Though Farrell and coauthors have reported shikimic acid complexes of platinum [41], diamine coupled shikimic acid complexes of platinum remain unstudied. In this work, we have synthesized a series of platinum complexes (PtL^{a-e}Cl$_2$) with diamine coupled shikimic acid ligands L^{a-e} (L^{a-e} = SA-NH(CH$_2$)$_n$NHCH$_2$CH$_2$NH$_2$; L^a, n = 2; L^b, n = 3; L^c, n = 4; L^d, n = 5; L^e, n = 6) and investigated their cytotoxicity and DNA-binding properties.

2. Materials and Methods

2.1. Materials. All chemicals were purchased from commercial sources and used as received. All solvents were of analytical grade and used without further purification unless otherwise specified. The synthesized Pt(II) complexes as well as the ligands were dissolved in H$_2$O at a concentration of 5.0 mM as stock solutions to prepare the DNA binding studies. Disodium salt of guanosine-5'-monophosphate (5'-GMP) was purchased from Sigma. pUC19 plasmid DNA was purchased from Takara Biotech Co. Ltd., Dalian of China. Cancer cell lines were obtained from Shanghai Institutes for Biological Sciences of China.

2.2. Instrumentation and Methods. ^{1}H and ^{13}C NMR spectra were recorded by a Bruker AV-500 NMR spectrometer with chemical shift (in ppm) relative to tetramethylsilane. Elemental analyses (C, H, N) were performed on a Perkin Elmer Series II CHNS/O 2400 analytical instrument. ESI mass spectra were measured on a Bruker HCT Electrospray Ionization Mass Spectrometer. The purity of the platinum(II) complexes was performed on Thermo Fisher Scientific Exactive LC-MS Spectrometer. Capillary electrophoresis was recorded on Agilent HP3D High Performance Capillary Electrophoresis. Flow-cytometric analysis was undertaken by using a FAC-Scan fluorescence-activated cell sorter (FACS).

2.3. Synthesis. The compounds 1a~e, 2a~e, and 3a~e were prepared following the methods reported by Holmes et al. [42, 43]. The data are in good agreement with the literatures. The compounds 4a~e, 5a~e, 6a~e (L^{a-e}), and 7a~e (PtL^{a-e}Cl$_2$) were synthesized via modified methods that were reported by Srinivas et al. [37–40, 44–47] (see Scheme 1).

Compound 1a. ^{1}H NMR (500 MHz, CDCl$_3$): δ = 7.36 (m, 5H, C$_6$H$_5$), 5.39 (br s, 1H, H(a)), 5.11 (s, 2H, CH$_2$), 3.69 (t, J = 4.5 Hz, 2H, H(c)), 3.35 (m, 2H, H(b)); ESI-MS m/z 217.84 [M + Na]$^+$.

Compound 1b. ^{1}H NMR (500 MHz, CDCl$_3$): δ = 7.36 (m, 5H, C$_6$H$_5$), 5.47 (br s, 1H, H(a)), 5.12 (s, 2H, CH$_2$), 3.66 (t, J = 5.5 Hz, 2H, H(c)), 3.32 (m, 2H, H(b)), 1.70 (m, 4H, 2 × CH$_2$); ESI-MS m/z 209.85 [M + H]$^+$.

Compound 1c. ^{1}H NMR (500 MHz, CDCl$_3$): δ = 7.39 (m, 5H, C$_6$H$_5$), 5.20 (br s, 1H, H(a)), 5.12 (s, 2H, CH$_2$), 3.67 (m, 2H, H(c)), 3.25 (m, 2H, H(b)), 1.91 (br s, 1H, H(d)), 1.61 (m, 4H, 2 × CH$_2$); ESI-MS m/z 223.85 [M + H]$^+$.

Compound 1d. ^{1}H NMR (500 MHz, CDCl$_3$): δ = 7.38 (m, 5H, C$_6$H$_5$), 5.12 (s, 2H, CH$_2$), 3.66 (t, J = 6.5 Hz, 2H, H(c)), 3.23 (m, 2H, H(b)), 1.57 (m, 4H, 2 × CH$_2$), 1.42 (m, 2H, CH$_2$); ESI-MS m/z 259.92 [M + Na]$^+$.

Compound 1e. ^{1}H NMR (500 MHz, CDCl$_3$): δ = 7.35 (m, 5H, C$_6$H$_5$), 5.09 (s, 2H, CH$_2$), 3.62 (t, J = 6.5 Hz, 2H, H(c)), 3.20 (m, 2H, H(b)), 1.66 (br s, 1H, H(d)), 1.54 (m, 4H, 2 × CH$_2$), 1.36 (m, 4H, 2 × CH$_2$); ESI-MS m/z 274.01 [M + Na]$^+$.

Compound 2a. ^{1}H NMR (500 MHz, CDCl$_3$): δ = 7.34 (m, 5H, C$_6$H$_5$), 5.02 (s, 2H, CH$_2$), 3.24 (m, 2H, H(a)), 2.90 (m, 4H, H(b), and H(c)), 2.69 (m, 2H, H(d)); ESI-MS m/z 238.10 [M + H]$^+$.

Compound 2b. ^{1}H NMR (500 MHz, CDCl$_3$): δ = 7.30 (m, 5H, C$_6$H$_5$), 5.04 (s, 2H, CH$_2$), 3.45 (m, 2H, H(a)), 3.30 (m, 4H, H(b), and H(c)), 2.89 (m, 2H, H(d)), 1.89 (m, 2H, CH$_2$); ESI-MS m/z 251.95 [M + H]$^+$.

Compound 2c. ^{1}H NMR (500 MHz, CDCl$_3$): δ = 7.34 (m, 5H, C$_6$H$_5$), 5.09 (s, 2H, CH$_2$), 3.27 (t, J = 6.0 Hz, 2H, H(a)), 2.96 (m, 2H, H(c)), 2.86 (m, 4H, H(b), and H(d)), 1.90 (m, 4H, 2 × CH$_2$); ESI-MS m/z 266.91 [M + H]$^+$.

Compound 2d. ^{1}H NMR (500 MHz, CDCl$_3$): δ = 7.37 (m, 5H, C$_6$H$_5$), 5.10 (s, 2H, CH$_2$), 3.20 (m, 2H, H(a)), 2.82 (t, J = 6.0 Hz, 2H, H(c)), 2.67 (m, 2H, H(b)), 2.41 (m, 2H, H(d)), 1.53 (m, 4H, 2 × CH$_2$), 1.36 (m, 2H, CH$_2$); ESI-MS m/z 280.01 [M+H]$^+$.

Compound 2e. ^{1}H NMR (500 MHz, CDCl$_3$): δ = 7.34 (m, 5H, C$_6$H$_5$), 5.09 (s, 2H, CH$_2$), 3.16 (m, 2H, H(a)), 2.40 (m, 6H, H(b), H(c), and H(d)), 1.50 (m, 4H, 2 × CH$_2$), 1.33 (m, 4H, 2 × CH$_2$); ESI-MS m/z 294.06 [M + H]$^+$.

Compound 3a. ^{1}H NMR (500 MHz, CDCl$_3$): δ = 7.36 (m, 5H, C$_6$H$_5$), 5.11 (s, 2H, CH$_2$), 3.33 (m, 8H, H(a), H(b), H(c), and H(d)), 1.47 (s, 9H, 3 × CH$_3$), 1.44 (s, 9H, 3 × CH$_3$); ESI-MS m/z 460.23 [M + Na]$^+$.

Compound 3b. ^{1}H NMR (500 MHz, CDCl$_3$): δ = 7.35 (m, 5H, C$_6$H$_5$), 5.09 (s, 2H, CH$_2$), 3.25 (m, 8H, H(a), H(b), H(c),

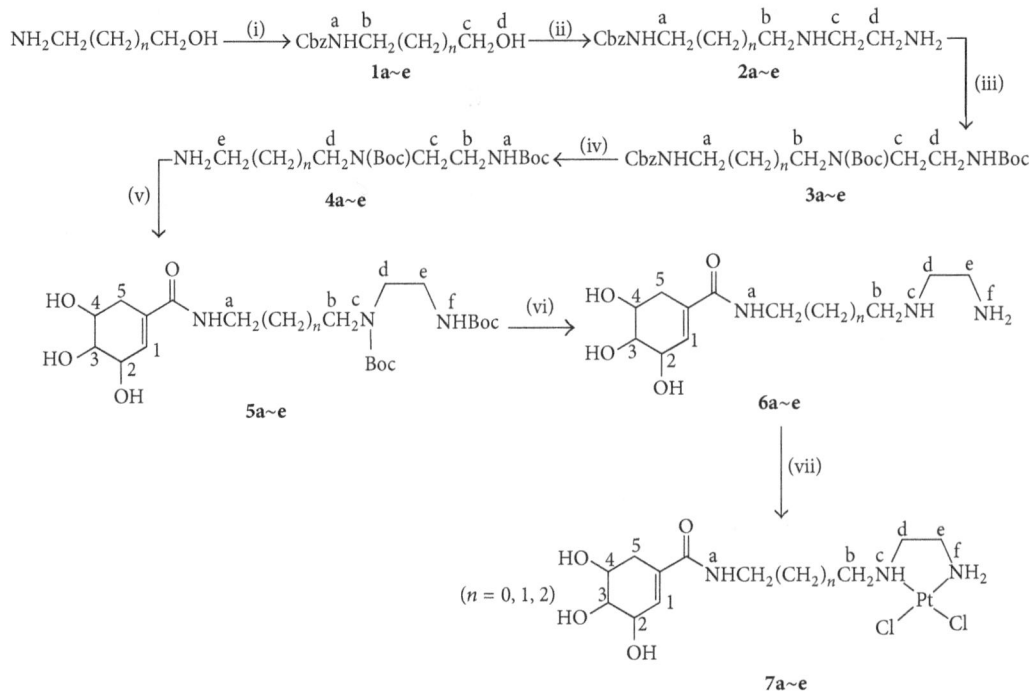

SCHEME 1: Synthesis route for the shikimic acid derivate ligands and Pt(II) complexes (n = 2, 3, 4, 5, 6): (i) CbzCl, NaHCO$_3$, H$_2$O; (ii) MsCl/py/20°C, excess NH$_2$(CH$_2$)$_2$NH$_2$; (iii) (t-BuO)$_2$CO, CH$_2$Cl$_2$; (iv) HCO$_2$NH$_4$/Pd/C; (v) Shikimic acid, DCC, HOBt; (vi) conc. HCl/MeOH/CH$_2$Cl$_2$; (vii) Na$_2$CO$_3$, K$_2$PtCl$_4$.

and H(d)), 1.46 (m, 11H, CH$_2$ and 3 × CH$_3$), 1.43 (s, 9H, 3 × CH$_3$); ESI-MS m/z 486.21, 488.16 [M + Cl]$^-$.

Compound 3c. ^{1}H NMR (500 MHz, CDCl$_3$): δ = 7.36 (m, 5H, C$_6$H$_5$), 5.11 (s, 2H, CH$_2$), 3.23 (m, 8H, H(a), H(b), H(c), and H(d)), 1.55 (m, 4H, 2 × CH$_2$), 1.47 (s, 9H, 3 × CH$_3$), 1.44 (s, 9H, 3 × CH$_3$); ESI-MS m/z 500.21, 502.18 [M + Cl]$^-$.

Compound 3d. ^{1}H NMR (500 MHz, CDCl$_3$): δ = 7.36 (m, 5H, C$_6$H$_5$), 5.08 (s, 2H, CH$_2$), 3.26 (m, 8H, H(a), H(b), H(c), and H(d)), 1.51 (m, 4H, 2 × CH$_2$), 1.44 (s, 9H, 3 × CH$_3$), 1.42 (s, 9H, 3 × CH$_3$), 1.28 (m, 2H, CH$_2$); ESI-MS m/z 514.22, 516.17 [M + Cl]$^-$.

Compound 3e. ^{1}H NMR (500 MHz, CDCl$_3$): δ = 7.34 (m, 5H, C$_6$H$_5$), 5.08 (s, 2H, CH$_2$), 3.27 (m, 8H, H(a), H(b), H(c) and H(d)), 1.49 (m, 4H, 2 × CH$_2$), 1.45 (s, 9H, 3 × CH$_3$), 1.42 (s, 9H, 3 × CH$_3$), 1.27 (m, 4H, 2 × CH$_2$); ESI-MS m/z 528.24, 530.21 [M + Cl]$^-$.

Compound 4a. The synthesis and characterization of **4a** have been well established [42], but in this work, the synthetic routine was modified [44–47]. 10% Pd/C (0.4 g) and ammonium formate (5.04 g, 80.0 mmol) suspension in 30 mL MeOH were added to a solution of **3a** (4.37 g, 10.0 mmol) in MeOH (50 mL) under nitrogen. The reaction mixture was vigorously stirred for 2.5 h at 40°C. This solution was allowed to cool and filtered through Celite

then evaporated under reduced pressure. The residual ammonium formate was removed by repeated evaporation with CH$_2$Cl$_2$. The resulting oil was obtained. Yield: 98%. ^{1}H NMR (500 MHz, CDCl$_3$): δ = 8.45 (br s, 1H, H(a)), 3.52 (t, J = 6.0 Hz, 2H, H(b)), 3.34 (m, 2H, H(c)), 3.25 (m, 2H, H(d)), 3.11 (m, 2H, H(e)), 1.46 (s, 9H, 3 × CH$_3$), 1.42 (s, 9H, 3 × CH$_3$); ^{13}C NMR (125 MHz, CDCl$_3$): δ = 167.8, 155.8 CO, 47.3, 46.1 NCH$_2$CH$_2$NH, 38.9, 38.1 NH$_2$CH$_2$CH$_2$N, 80.4, 78.8 C(CH$_3$)$_3$, 27.9 CH$_3$; ESI-MS: m/z: 304.10 [M + H]$^+$.

Compound 4b. The procedure was similar to that for **4a** except that **3b** was used. Yield: 95%. ^{1}H NMR (500 MHz, CDCl$_3$): δ = 8.37 (br s, 1H, H(a)), 3.19 (m, 4H, H(b), and H(c)), 3.12 (t, J = 6.5 Hz, 2H, H(d)), 2.82 (m, 2H, H(e)), 1.81 (m, 2H, CH$_2$), 1.34 (s, 9H, 3 × CH$_3$), 1.31 (s, 9H, 3 × CH$_3$); ^{13}C NMR (125 MHz, CDCl$_3$): δ = 168.2, 155.6 CO, 46.3, 43.6 NCH$_2$CH$_2$NH, 38.7, 36.3, 25.6 NH$_2$CH$_2$CH$_2$CH$_2$N, 79.9, 78.5 C(CH$_3$)$_3$, 27.8 6 × CH$_3$; ESI-MS: m/z: 318.16 [M + H]$^+$.

Compound 4c. The procedure was similar to that for **4a** except that **3c** was used. Yield: 98%. ^{1}H NMR (500 MHz, CDCl$_3$): δ = 8.34 (br s, 1H, H(a)), 3.22 (m, 4H, H(b), and H(c)), 3.13 (m, 2H, H(d)), 2.67 (m, 2H, H(e)), 1.41 (m, 4H, 2 × CH$_2$), 1.29 (s, 9H, 3 × CH$_3$), 1.26 (s, 9H, 3 × CH$_3$); ^{13}C NMR (125 MHz, CDCl$_3$) δ = 168.6, 155.5 CO, 46.6, 46.0 NCH$_2$CH$_2$NH, 39.5, 38.8, 27.1, 24.7 NH$_2$CH$_2$CH$_2$CH$_2$CH$_2$N, 79.9, 78.5 C(CH$_3$)$_3$, 27.7 6 × CH$_3$; ESI-MS: m/z: 332.17 [M + H]$^+$.

Compound 4d. The procedure was similar to that for **4a** except that **3d** was used. Yield: 94%. ^{1}H NMR (500 MHz, CDCl$_3$): δ = 8.37 (br s, 1H, H(a)), 3.15 (m, 4H, H(b), and H(c)), 3.09 (m, 2H, H(d)), 2.68 (m, 2H, H(e)), 1.39 (m, 4H, 2 × CH$_2$), 1.31 (s, 9H, 3 × CH$_3$), 1.28 (s, 9H, 3 × CH$_3$), 1.18 (m, 2H, CH$_2$); ^{13}C NMR (125 MHz, CDCl$_3$) δ = 168.7, 155.5 CO, 46.7, 45.9 NCH$_2$CH$_2$NH, 39.5, 38.8, 27.0, 23.1, 22.3 NH$_2$CH$_2$CH$_2$CH$_2$CH$_2$CH$_2$N, 79.0, 78.3 C(CH$_3$)$_3$, 27.7, 6 × CH$_3$; ESI-MS: *m/z*: 346.21 [M + H]$^+$.

Compound 4e. The procedure was similar to that for **4a** except that **3e** was used. Yield: 96%. ^{1}H NMR (500 MHz, CDCl$_3$): δ = 8.32 (br s, 1H, H(a)), 3.06 (m, 6H, H(b), H(c), and H(d)), 2.60 (t, J = 7.0 Hz, 2H, H(b)), 1.39 (m, 4H, 2 × CH$_2$), 1.28 (s, 9H, 3 × CH$_3$), 1.25 (s, 9H, 3 × CH$_3$), 1.19 (m, 2H, CH$_2$), 1.12 (m, 2H, CH$_2$); ^{13}C NMR (125 MHz, CDCl$_3$) δ = 168.6, 155.4 CO, 46.9, 45.8 NCH$_2$CH$_2$NH, 40.1, 38.9, 29.9, 27.3, 25.7 NH$_2$CH$_2$CH$_2$CH$_2$CH$_2$CH$_2$CH$_2$N, 78.9, 78.3 C(CH$_3$)$_3$, 27.7 6 × CH$_3$; ESI-MS: *m/z*: 360.19 [M + H]$^+$.

Compound 5a. This compound was synthesized by the modified method reported by Bowen et al. [48]. A solution of SA (1.65 g, 9.50 mmol), DCC (2.17 g, 10.5 mmol), and HOBT (1.42 g, 10.5 mmol) in DMF (20 mL) was stirred at 0°C for 1 h. Compound **4a** (0.288 g, 9.50 mmol) in CH$_2$Cl$_2$ (10 mL) was then added. After being stirred at room temperature for 24 h, the mixture was filtered and evaporated under reduced pressure to give the crude product that was purified by chromatography on silica gel using CHCl$_3$/MeOH 20 : 1 as the eluent to produce the desired product **5a**. Yield: 58%. ^{1}H NMR (500 MHz, CDCl$_3$): δ = 6.41 (br s, 1H, H(1)), 4.35 (m, 1H, H(2)), 3.96 (m, 1H, H(3)), 3.60 (m, 1H, H(4)), 3.37 (m, 4H, H(d), and H(e)), 3.30 (m, 2H, H(a)), 3.22 (m, 2H, H(b)), 2.75 (m, 1H, H(5)), 2.18 (m, 1H, H(5)), 1.43 (s, 18H, 6 × CH$_3$); ^{13}C NMR (125 MHz, CDCl$_3$): δ = 170.4, 155.9, 155.6 CO, 133.2, 130.3 C=CH, 72.2, 66.2, 65.7 CH–OH, 31.5 CH(OH)CH$_2$, 45.5 NCH$_2$CH$_2$NH, 38.9, 38.3 NHCH$_2$CH$_2$N, 80.0, 78.7 C(CH$_3$)$_3$, 27.7 CH$_3$; ESI-MS: *m/z*: 494.26, 496.20 [M + Cl]$^-$.

Compound 5b. The procedure was similar to that for **5a** except that **4b** was used. Yield: 60%. ^{1}H NMR (500 MHz, CDCl$_3$): δ = 6.34 (br s, 1H, H(1)), 4.28 (m, 1H, H(2)), 3.90 (m, 1H, H(3)), 3.54 (m, 1H, H(4)), 3.14 (m, 8H, H(a), H(b), H(d), and H(e)), 2.67 (m, 1H, H(5)), 2.14 (m, 1H, H(5)), 1.62 (m, 2H, CH$_2$), 1.35 (s, 18H, 6 × CH$_3$); ^{13}C NMR (125 MHz, CDCl$_3$) δ = 167.8, 155.6 CO, 133.4, 130.1 C=CH, 72.1, 66.2, 65.8 CH–OH, 31.5 CH(OH)CH$_2$, 46.1, 44.0 NCH$_2$CH$_2$NH, 38.9, 36.2, 31.5 NHCH$_2$CH$_2$CH$_2$N, 79.7, 78.6 C(CH$_3$)$_3$, 27.9 6 × CH$_3$; ESI-MS: *m/z*: 474.19 [M + H]$^+$, 508.21, 510.21 [M + Cl]$^-$.

Compound 5c. The procedure was similar to that for **5a** except that **4c** was used. Yield: 62%. ^{1}H NMR (500 MHz, CDCl$_3$): δ = 6.26 (br s, 1H, H(1)), 4.86 (m, 1H, H(2)), 3.86 (m, 1H, H(3)), 3.50 (m, 1H, H(4)), 3.10 (m, 8H, H(a), H(b), H(d), and H(e)), 2.63 (m, 1H, H(5)), 2.07 (m, 1H, H(5)), 1.39 (m, 4H, 2 × CH$_2$), 1.29 (s, 18H, 6 × CH$_3$); ^{13}C NMR (125 MHz, CDCl$_3$) δ = 168.0, 155.5, 155.0 CO, 133.4, 129.9 C=CH, 72.1, 66.2, 65.8 CH–OH, 31.6 CH(OH)CH$_2$, 46.4, 46.1 NCH$_2$CH$_2$NH,

38.9, 36.2, 31.5 NHCH$_2$CH$_2$CH$_2$CH$_2$N, 79.2, 78.5 C(CH$_3$)$_3$, 27.9 6 × CH$_3$; ESI-MS: *m/z*: 488.16 [M + H]$^+$, 522.21, 524.24 [M + Cl]$^-$.

Compound 5d. The procedure was similar to that for **5a** except that **4d** was used. Yield: 61%. ^{1}H NMR (500 MHz, CDCl$_3$): δ = 6.26 (br s, 1H, H(1)), 4.23 (m, 1H, H(2)), 3.85 (m, 1H, H(3)), 3.49 (m, 1H, H(4)), 3.10 (m, 8H, H(a), H(b), H(d), and H(e)), 2.63 (m, 1H, H(5)), 2.05 (m, 1H, H(5)), 1.39 (m, 4H, 2 × CH$_2$), 1.29 (s, 18H, 6 × CH$_3$) 1.14 (m, 2H, CH$_2$); ^{13}C NMR (125 MHz, CDCl$_3$) δ = 168.0, 155.5, 155.0 CO, 133.4, 129.9 C=CH, 72.1, 66.2, 65.8 CH–OH, 31.6 CH(OH)CH$_2$, 46.7, 46.0 NCH$_2$CH$_2$NH, 39.1, 29.0, 28.5, 27.2, 23.6 NHCH$_2$CH$_2$CH$_2$CH$_2$CH$_2$N, 79.1, 78.4 C(CH$_3$)$_3$, 27.8 6 × CH$_3$; ESI-MS: *m/z*: 502.21 [M + H]$^+$, 536.23, 538.20 [M + Cl]$^-$.

Compound 5e. The procedure was similar to that for **5a** except that **4e** was used. Yield: 61%. ^{1}H NMR (500 MHz, CDCl$_3$): δ = 6.29 (br s, 1H, H(1)), 4.26 (m, 1H, H(2)), 3.89 (m, 1H, H(3)), 3.52 (m, 1H, H(4)), 3.14 (m, 8H, H(a), H(b), H(d), and H(e)), 2.67 (m, 1H, H(5)), 2.10 (m, 1H, H(5)), 1.42 (m, 4H, 2 × CH$_2$), 1.33 (s, 18H, 6 × CH$_3$), 1.22 (m, 4H, 2 × CH$_2$); ^{13}C NMR (125 MHz, CDCl$_3$) δ = 167.8, 155.6, 155.0 CO, 133.7, 129.7 C=CH, 72.3, 66.2, 65.8 CH–OH, 31.7 CH(OH)CH$_2$, 46.7, 46.0 NCH$_2$CH$_2$NH, 39.2, 29.0, 28.7, 27.4, 26.1, 25.9 NHCH$_2$CH$_2$CH$_2$CH$_2$CH$_2$CH$_2$N, 79.1, 78.4 C(CH$_3$)$_3$, 27.8, 6 × CH$_3$; ESI-MS: *m/z*: 516.21 [M + H]$^+$, 550.25, 552.24 [M + Cl]$^-$.

Compound 6a (*L^a*). Compound **5a** (0.688 g, 1.50 mmol) was dissolved in CH$_2$Cl$_2$/EtOH (8 : 1, 25 mL), and 6 mol/L HCl (3 mL) was added with vigorously stirring for 20 min. Then, the solution was evaporated under reduced pressure to give a colourless oily solid. This was dissolved in distilled water (30 mL) and washed with CH$_2$Cl$_2$ (3 × 30 mL), from which the aqueous layer was concentrated to produce the desired compound **6a**. Yield: 83%. ^{1}H NMR (500 MHz, CD$_3$SOCD$_3$): δ = 9.38 (br s, 2H, H(a), and 2-OH), 8.31 (br s, 3H, H(c), H(f), and 3-OH), 8.04 (br s, 1H, 4-OH), 6.33 (br s, 1H, H(1)), 4.12 (m, 1H, H(2)), 3.76 (m, 1H, H(3)), 3.44 (m, 3H, H(4), and CH$_2$), 3.14 (m, 4H, H(d), and H(e)), 2.99 (t, J = 5.5 Hz, 2H, H(b)), 2.50 (m, 1H, H(5)), 1.98 (dd, J = 17.5 Hz, J = 4.0 Hz, 1H, H(5)); ^{13}C NMR (125 MHz, D$_2$O): δ = 170.5, CO, 132.2, 132.0, C=CH, 71.2, 66.4, 65.7 CH–OH, 30.8 CH(OH)CH$_2$, 47.8, 44.3 NHCH$_2$CH$_2$NH$_2$, 36.0, 35.5 NHCH$_2$CH$_2$NH; ESI-MS: *m/z*: 260.02 [M + H]$^+$, 294.02, 296.00 [M + Cl]$^-$.

Compound 6b (*L^b*). The procedure was similar to that for **6a** except that **5b** was used. Yield: 84%. ^{1}H NMR (500 MHz, CD$_3$SOCD$_3$): δ = 9.40 (br s, 2H, H(a), and 2-OH), 8.38 (br s, 3H, H(c), H(f), and 3-OH), 8.07 (br s, 1H, 4-OH), 6.32 (br s, 1H, H(1)), 4.17 (m, 1H, H(2)), 3.82 (m, 1H, H(3)), 3.50 (m, 1H, H(4)), 3.18 (m, 6H, CH$_2$, H(d), and H(e)), 2.91 (m, 2H, CH$_2$), 2.80 (m, 2H, H(b)), 2.46 (m, 1H, H(5)), 2.00 (dd, J = 17.5 Hz, J = 4.0 Hz, 1H, H(5)); ^{13}C NMR (125 MHz, CD$_3$SOCD$_3$): δ = 170.0 CO, 132.5, 131.3 C=CH, 71.2, 66.3, 65.6 CH–OH, 30.9 CH(OH)CH$_2$, 45.6, 44.1 NHCH$_2$CH$_2$NH$_2$, 36.1, 35.5,

25.4 $CH_2CH_2CH_2$; ESI-MS m/z: 274.01 $[M + H]^+$, 308.12, 310.00 $[M + Cl]^-$.

Compound 6c (L^c). The procedure was similar to that for **6a** except that **5c** was used. Yield: 88%. ^1H NMR (500 MHz, CD_3SOCD_3): δ = 9.19 (br s, 2H, H(a), and 2-OH), 8.23 (br s, 3H, H(c), H(f), and 3-OH), 7.85 (br s, 1H, 4-OH), 6.22 (br s, 1H, H(1)), 4.11 (m, 1H, H(2)), 3.75 (m, 1H, H(3)), 3.42 (m, 3H, H(4)), 3.09 (m, 4H, H(d), and H(e)), 3.04 (m, 2H, CH_2), 2.86 (m, 2H, H(b)), 2.39 (m, 1H, H(5)), 1.93 (dd, J = 17.5 Hz, J = 4.5 Hz, 1H, H(5)), 1.53 (m, 2H, CH_2), 1.42 (m, 2H, CH_2); ^{13}C NMR (125 MHz, CD_3SOCD_3): δ = 170.0 CO, 132.8, 130.6 C=CH, 71.0, 66.1, 65.5 CH–OH, 30.6 $CH(OH)CH_2$, 47.5, 43.7 $NHCH_2CH_2NH_2$, 38.4, 35.2, 25.1, 22.6 $CH_2CH_2CH_2CH_2$; ESI-MS m/z: 288.02 $[M + H]^+$, 322.12, 324.07 $[M + Cl]^-$.

Compound 6d (L^d). The procedure was similar to that for **6a** except that **5d** was used. Yield: 86%. ^1H NMR (500 MHz, CD_3SOCD_3): δ = 9.22 (br s, 2H, H(a), and 2-OH), 8.28 (br s, 3H, H(c), H(f), and 3-OH), 7.85 (br s, 1H, 4-OH), 6.27 (br s, 1H, H(1)), 4.17 (m, 1H, H(2)), 3.82 (m, 1H, H(3)), 3.50 (m, 1H, H(4)), 3.15 (m, 4H, H(d), and H(e)), 3.10 (m, 2H, CH_2), 2.91 (m, 2H, H(b)), 2.45 (m, 1H, H(5)), 1.96 (dd, J = 17.5 Hz, J = 5.5 Hz, 1H, H(5)), 1.62 (m, 2H, H(b)), 1.44 (m, 2H, CH_2), 1.30 (m, 2H, CH_2); ^{13}C NMR (125 MHz, CD_3SOCD_3): δ = 167.8 CO, 132.6, 132.1 C=CH, 71.3, 67.1, 66.0 CH–OH, 30.9 $CH(OH)CH_2$, 47.4, 44.5 $NHCH_2CH_2NH_2$, 39.0, 35.8, 28.9, 25.6, 23.7 $CH_2CH_2CH_2CH_2CH_2$; ESI-MS m/z: 302.05 $[M + H]^+$, 336.09, 338.06 $[M + Cl]^-$.

Compound 6e (L^e). The procedure was similar to that for **6a** except that **5e** was used. Yield: 85%. ^1H NMR (500 MHz, CD_3SOCD_3): δ = 9.54 (br s, 2H, H(a), and 2-OH), 8.50 (br s, 3H, H(c), H(f), and 3-OH), 7.80 (br s, 1H, 4-OH), 6.28 (br s, 1H, H(1)), 4.18 (m, 1H, H(2)), 3.83 (m, 1H, H(3)), 3.50 (m, 1H, H(4)), 3.20 (m, 4H, H(d), and H(e)), 3.10 (m, 2H, CH_2), 2.88 (m, 2H, H(b)), 2.48 (m, 1H, H(5)), 2.00 (dd, J = 17.5 Hz, J = 5.0 Hz, 1H, H(5)), 1.66 (m, 2H, CH_2), 1.45 (m, 2H, CH_2), 1.34 (m, 2H, CH_2), 1.27 (m, 2H, CH_2); ^{13}C NMR (125 MHz, CD_3SOCD_3) δ = 167.0 CO, 131.9, 131.7 C=CH, 71.0, 66.7, 65.6 CH–OH, 30.6 $CH(OH)CH_2$, 46.8, 44.1 $NHCH_2CH_2NH_2$, 38.6, 35.3, 28.8, 25.9, 25.6, 25.4 $CH_2CH_2CH_2CH_2CH_2CH_2$. ESI-MS m/z: 316.01 $[M + H]^+$.

Compound 7a (PtL^{a}Cl$_2$). Compound **6a** (L^a) (0.415 g, 1.25 mmol) was dissolved in EtOH (5 mL) and water (2 mL), and the pH was adjusted to 8–9 with 0.25 M aqueous Na_2CO_3. A solution of K_2PtCl_4 (0.518 g, 1.25 mmol) in water (3 mL) was added dropwise, and the resulting mixture was stirred for 12 h in the dark at room temperature. The solvent was then removed and the residue purified by silica gel chromatography using MeOH as the eluent to afford the product as a yellow solid. Yield: 78%. Elemental analysis (%) calcd. for $C_{11}H_{21}Cl_2N_3O_4Pt \cdot H_2O$: C 24.32, H 4.27, N 7.73; found C 24.11, H 4.20, N 7.78; ^1H NMR (500 MHz, CD_3SOCD_3): δ = 8.06 (m, 1H, H(a)), 7.02 (br s, 1H, H(c)), 6.32 (br s, 2H, H(f)), 6.39 (br s, 1H, H(1)), 4.82 (br s, 1H, 2-OH), 4.57 (d, J = 6.5 Hz, 1H, 3-OH), 4.74 (br s, 1H, 4-OH), 4.14 (m, 1H, H(2)), 3.77 (m, 1H, H(3)), 3.45 (m, 1H, H(4)), 3.11

(m, 2H, CH_2), 2.87 (m, 2H, H(b), Pt satellites are observed as shoulders), 2.62 (m, 4H, H(d), and H(e), Pt satellites are observed as shoulders), 1.98 (m, 1H, H(5)), 2.44 (m, 1H, H(5)); ^{13}C NMR (125 MHz, D_2O): δ = 170.3 CO, 132.9, 131.8 C=CH, 71.6, 66.6, 66.0 CH–OH, 31.1 $CH(OH)CH_2$, 55.7, 51.5 $NHCH_2CH_2NH_2$, 46.7, 36.8 $NHCH_2CH_2NH$; ESI-MS: m/z: 565.95, 566.98, 567.96, 568.97, 570.00 $[M-Cl + DMSO]^+$.

Compound 7b (PtL^{b}Cl$_2$). The procedure was similar to that for **7a** except that **6b** (L^b) was used. Yield: 82%. Elemental analysis (%) calcd. for $C_{12}H_{23}Cl_2N_3O_4Pt \cdot 2H_2O$: C 25.05, H 4.73, N 7.30; found C 25.12, H 4.64, N 7.39; ^1H NMR (500 MHz, CD_3SOCD_3): δ = 7.94 (m, 1H, H(a)), 6.96 (br s, 1H, H(c)), 6.29 (br s, 2H, H(f)), 6.24 (br s, 1H, H(1)), 4.83 (d, J = 3.0 Hz, 1H, 2-OH), 4.73 (d, J = 7.0 Hz, 1H, 3-OH), 4.57 (d, 1H, J = 4.5 Hz, 4-OH), 4.18 (m, 1H, H(2)), 3.82 (m, 1H, H(3)), 3.51 (m, 1H, H(4)), 3.15 (t, J = 6.0 Hz, 2H, CH_2), 2.84 (m, 2H, H(b), Pt satellites are observed as shoulders), 2.58 (m, 4H, H(d) and H(e), Pt satellites are observed as shoulders), 2.47 (m, 1H, H(5)), 1.98 (m, 1H, H(5)), 1.75 (m, 2H, CH_2); ^{13}C NMR (125 MHz, D_2O): δ = 170.5 CO, 133.4, 131.1 C=CH, 71.7, 66.6, 66.0 CH–OH, 31.3 $CH(OH)CH_2$, 55.7, 50.5 $NHCH_2CH_2NH_2$, 46.9, 36.9, 26.8 $CH_2CH_2CH_2$; ESI-MS: m/z: 572.08, 573.16, 574.09, 575.09, 576.07, 577.01, 578.09 $[M + Cl]^-$.

Compound 7c (PtL^{c}Cl$_2$). The procedure was similar to that for **7a** except that **6c** (L^c) was used. Yield: 85%. Elemental analysis (%) calcd. for $C_{13}H_{25}Cl_2N_3O_4Pt \cdot 2H_2O$: C 26.49, H 4.96, N 7.13; found C 26.63, H 4.87, N 7.24; ^1H NMR (500 MHz, CD_3SOCD_3): δ = 7.91 (br s, 1H, H(a)), 6.99 (br s, 1H, H(c)), 6.43 (br s, 2H, H(f)), 6.29 (br s, 1H, H(1)), 4.88 (d, J = 4.0 Hz, 1H, 2-OH), 4.79 (d, J = 6.5 Hz, 1H, 3-OH), 4.63 (d, J = 5.0 Hz, 1H, 4-OH), 4.18 (m, 1H, H(2)), 3.82 (m, 1H, H(3)), 3.51 (m, 1H, H(4)), 3.11 (t, J = 5.0 Hz, 2H, CH_2), 2.84 (m, 2H, H(b), Pt satellites are observed as shoulders), 2.59 (m, 4H, H(d), and H(e), Pt satellites are observed as shoulders), 2.47 (m, 1H, H(5)), 1.97 (dd, J = 17.5 Hz, J = 3.5 Hz, 1H, H(5)), 1.56 (m, 2H, CH_2), 1.43 (m, 2H, CH_2); ^{13}C NMR (125 MHz, CD_3SOCD_3): δ = 170.1 CO, 133.4, 131.4 C=CH, 71.7, 66.8, 66.1 CH–OH, 31.5 $CH(OH)CH_2$, 54.4, 51.6 $NHCH_2CH_2NH_2$, 45.6, 39.3, 26.3, 24.6 $CH_2CH_2CH_2CH_2$; ESI-MS: m/z: 585.99, 587.02, 588.00, 589.01, 589.98, 590.99, 592.01 $[M + Cl]^-$.

Compound 7d (PtL^{d}Cl$_2$). The procedure was similar to that for **7a** except that **6d** (L^d) was used. Yield: 84%. Elemental analysis (%) calcd. for $C_{14}H_{27}Cl_2N_3O_4Pt \cdot H_2O$: C 28.72, H 4.99, N 7.18; found C 28.65, H 4.78, N 7.25; ^1H NMR (500 MHz, CD_3SOCD_3): δ = 7.85 (br s, 1H, H(a)), 6.95 (br s, 1H, H(c)), 6.42 (br s, 2H, H(f)), 6.23 (br s, 1H, H(2)), 4.84 (d, J = 3.5 Hz, 1H, 2-OH), 4.76 (d, J = 7.0 Hz, 1H, 3-OH), 4.59 (d, J = 3.5 Hz, 1H, 4-OH), 4.13 (m, 1H, H(2)), 3.76 (m, 1H, H(3)), 3.50 (m, 1H, H(4)), 3.04 (m, 2H, CH_2), 2.82 (t, J = 5.5 Hz 2H, H(b), Pt satellites are observed as shoulders), 2.55 (m, 4H, H(d), and H(e), Pt satellites are observed as shoulders), 2.41 (m, 1H, H(5)), 1.91 (dd, J = 17.5 Hz, J = 4.5 Hz, 1H, H(5)), 1.53 (m, 2H, CH_2), 1.38 (m, 2H, CH_2), 1.20 (m, 2H, CH_2); ^{13}C NMR (125 MHz, CD_3SOCD_3):

δ = 167.6 CO, 132.4, 132.2 C=CH, 71.5, 67.2, 66.1 CH–OH, 31.1 CH(OH)CH$_2$, 54.3, 51.5 NHCH$_2$CH$_2$NH$_2$, 45.8, 39.1, 29.2, 26.7, 24.3 CH$_2$CH$_2$CH$_2$CH$_2$CH$_2$; ESI-MS: m/z: 600.08, 601.03, 602.03, 603.05, 604.01 [M + Cl]$^-$.

Compound 7e (PtL^{e}Cl$_2$). The procedure was similar to that for **7a** except that **6e** (L^e) was used. Yield: 80%. Elemental analysis (%) calcd. for C$_{15}$H$_{29}$Cl$_2$N$_3$O$_4$Pt·1.5H$_2$O: C 29.59, H 4.77, N6.90; found C 29.33, H 4.89, N 7.03; ^{1}H NMR (500 MHz, CD$_3$SOCD$_3$): δ = 7.85 (br s, 1H, H(a)), 6.86 (br s, 1H, H(c)), 6.27 (br s, 2H, H(f)), 6.21 (br s, 1H, H(1)), 4.87 (d, J = 3.5 Hz, 1H, 2-OH), 4.77 (d, J = 6.5 Hz, 1H, 3-OH), 4.59 (J = 3.5 Hz, 1H, 4-OH), 4.18 (m, 1H, H(2)), 3.80 (m, 1H, H(3)), 3.57 (m, 1H, H(4)), 3.08 (t, J = 5.0 Hz, 2H, CH$_2$), 2.85 (m, 2H, H(b), Pt satellites are observed as shoulders), 2.59 (m, 4H, H(d), and H(e), Pt satellites are observed as shoulders), 2.41 (m, 1H, H(5)), 1.96 (dd, J = 17.5 Hz, J = 4.5 Hz, 1H, H(5)), 1.56 (m, 2H, CH$_2$), 1.42 (m, 2H, CH$_2$), 1.26 (m, 4H, 2 × CH$_2$); ^{13}C NMR (125 MHz, CD$_3$SOCD$_3$) δ = 167.6 CO, 132.4, 132.2 C=CH, 71.4, 67.2, 66.0 CH–OH, 31.1 CH(OH)CH$_2$, 54.3, 51.5 NHCH$_2$CH$_2$NH$_2$, 45.8, 39.2, 29.4, 26.9, 26.6, 26.5 CH$_2$CH$_2$CH$_2$CH$_2$CH$_2$CH$_2$; ESI-MS: m/z: 614.07, 615.07, 616.08, 617.08, 618.05, 619.03, 620.05 [M + Cl]$^-$.

2.4. Cytotoxicity Assay In Vitro.

The growth-inhibitory effects of selective synthetic compounds, K$_2$PtCl$_4$ and cisplatin on the BEL7404 human cancer cell lines, in a 72 h incubation, were measured by using the MTT method. The detailed procedure has been reported in our previous work [49].

2.5. Agarose Gel Electrophoretic Assay.

DNA unwinding was determined by agarose gel electrophoretic assays through 1% (w/v) agarose gel with tris-acetate-EDTA (TAE) buffer, using pUC 19 plasmid DNA (0.5 μL, the concentration was 20 ng/μL) incubated with the ligands and compounds of various concentrations ranging from 20 to 260 μM at 37°C in the dark for 3 h. Finally, the gels were stained with ethidium bromide (0.5 μg/mL) for 30 min and visualized on a Bio-Rad gel imaging system.

2.6. ESI-MS Spectrometry.

The reaction of five Pt(II) complexes with a model compound 5′-GMP was investigated using ESI mass spectrometry. The platinum complexes and disodium salt of 5′-GMP were mixed with a molar ratio of 1 : 2 in water at 37°C for 48 h. The mass range was m/z 500–2000.

2.7. NMR Spectroscopy.

The reaction of the Pt(II) complexes with excess (1 : 3) 5′-GMP was carried out in D$_2$O within an NMR tube. The reaction mixtures were maintained at 37°C in the dark for 24 h, and then 1H NMR spectra data were recorded on a Bruker AV-500 NMR spectrometer.

2.8. Cell-Cycle Analysis.

BEL7404 cell lines were maintained in Dulbecco's modified Eagle's medium with 10% fetal calf serum in 5% CO$_2$ at 37°C. Cells were harvested by trypsinization and rinsed with PBS. After centrifugation, the pellet (105-106 cells) was suspended in 1 mL of PBS and kept on ice for 5 min. The cell suspension was then fixed by dropwise addition of 9 mL precooled (4°C) 100% ethanol under violent shaking. The mixed samples were kept at 4°C until use. For staining, cells were centrifuged, resuspended in PBS, digested with 150 mL RNAse A (250 μg/mL), and treated with 150 mL PI (100 μg/mL) then incubated at 4°C for 30 min. PI-positive cells were counted with a FACScan fluorescence-activated cell sorter (FACS). The population of cells in each cell-cycle phase was determined using Cell Modi FIT software (Becton Dickinson).

3. Results and Discussion

3.1. Synthesis and Characterization. The ligands and platinum complexes were prepared following the previous reported methods [37–40, 42–48]. Prior to synthesizing the shikimic-carboxamide ligands 6a~e, side chains shown in Scheme 1 are required. 3-aminopropanol was N-protected with benzyloxycarbonyl groups, and the alcohol group of the resulting compound (**1**) was activated with methanesulfonyl chloride and reacted with excess 1,2-diaminoethane. The newly generated amine groups of the triamine derivative **2** were then protected with Boc groups, which produced the differentially protected triamine derivative **3**. Selective deblocking the N-Cbz group gave the unstable amine **4**, which then reacted selectively with shikimic acid to produce shimiccarboxamide **5**. This was converted to shikimic carboxamide ligand **6** by deblocking the N-Boc group, the precursor ligand for the desired platinum complexes (**7**). The ligands **6a~e** (L^{a-e}) reacted easily with K$_2$PtCl$_4$ in the dark at room temperature, and gave the corresponding platinum complexes **7a~e** (PtL^{a-e}Cl$_2$). Complexes PtL^{a-e}Cl$_2$ are soluble in water, with solubilities of 40.3, 32.9, 21.7, 17.0, and 12.6 mg/mL at 298 K, respectively. These compounds were characterized by elemental analysis, ^{1}H NMR, ^{13}C NMR, and ESI-MS spectroscopies.

All ligands comprise three potential nitrogen donor sites (two amines and one amide). Hence, upon reaction with K$_2$PtCl$_4$, the ligands could bind to the metal center in a bidentate manner (two possible isomers: 2 × amine, 1 × amine, and 1 × aimde), or even in a tridentate way. Although the results of elemental analysis of the platinum(II) complexes are supportive of a bidentate binding, it does not indicate which isomer is formed, or whether both are formed. In order to confirm the isomer formed, the capillary electrophoresis, LC-MS, and detailed comparison of ^{1}H, ^{13}C NMR were carried out. As shown in Figure S1, supplementary material available online at http://dx.doi.org/10.1155/2013/565032 (see ESI†), only a single peak in their capillary electrophoresis plots was observed for PtL^{b}Cl$_2$, PtL^{c}Cl$_2$, PtL^{d}Cl$_2$, and PtL^{e}Cl$_2$ complexes, respectively, which indicated that the product (isomer) was pure and unique; furthermore, the purity of the product (isomer) was confirmed by the results of LC-MS spectra analysis (see Figure S2, ESI†). To determine the ligand coordinate mode in the product (isomer), we in detail investigated the NMR difference between the Pt(II) complexes and free ligands. As shown in Table 1, the differences of $\Delta\delta$(H^a) between L^{a-e} ligands and PtL^{a-e}Cl$_2$ complexes

TABLE 1: ^{1}H NMR shifts difference ($\Delta\delta$, ppm) between L^{a-e} ligands and $PtL^{a-e}Cl_2$ complexes.

H	$\Delta\delta$ (L^a versus PtL^aCl_2)	$\Delta\delta$ (L^b versus PtL^bCl_2)	$\Delta\delta$ (L^c versus PtL^cCl_2)	$\Delta\delta$ (L^d versus PtL^dCl_2)	$\Delta\delta$ (L^e versus PtL^eCl_2)
H^a	−0.02	−0.24	−0.06	−0.11	−0.05
H^b	+0.12	+0.17	+0.02	+0.20	+0.03
H^c	+2.36	+2.54	+2.20	+2.38	+2.68
H^d	+0.52	+0.70	+0.50	+0.72	+0.61
H^e	+0.52	+0.70	+0.50	+0.72	+0.61
H^f	+1.99	+2.20	+1.80	+1.96	+2.23

TABLE 2: ^{13}C NMR shifts difference ($\Delta\delta$, ppm) between L^{a-e} ligands and $PtL^{a-e}Cl_2$ complexes.

C	$\Delta\delta$ (L^a versus PtL^aCl_2)	$\Delta\delta$ (L^b versus PtL^bCl_2)	$\Delta\delta$ (L^c versus PtL^cCl_2)	$\Delta\delta$ (L^d versus PtL^dCl_2)	$\Delta\delta$ (L^e versus PtL^eCl_2)
C^a	+0.2	−0.5	−0.1	−0.02	−0.6
C^b	−0.8	−1.4	−4.1	−3.3	−3.9
C^c	−11.2	−11.4	−10.4	−10.0	−7.5
C^d	−7.2	−6.4	−7.9	−7.0	−7.4
C^e	−7.9	−10.1	−6.9	−6.9	−10.5

are less; however, those of $\Delta\delta(H^c)$ and $\Delta\delta(H^f)$ are very large; accompanying this trend, those of $\Delta\delta(H^b)$; $\Delta\delta(H^d)$, and $\Delta\delta(H^e)$ are moderate. The large proton chemical shift differences of two amines mainly induced by the bidentate coordinate mode of two amines of the ligands. Due to the coupling interaction, the adjacent proton chemical shifts also generated moderate changes. These observations were further confirmed by ^{13}C NMR shift differences ($\Delta\delta$) between L^{a-e} ligands and $PtL^{a-e}Cl_2$ complexes. As shown in Table 2, the differences of $\Delta\delta(C^a)$ and $\Delta\delta(C^b)$ between L^{a-e} ligands and $PtL^{a-e}Cl_2$ complexes are very less, however, those of $\Delta\delta(C^c)$; $\Delta\delta(H^d)$, and $\Delta\delta(C^e)$ are very large, which could be resulted from the bidentate coordinate mode of two amines of the ligands. It should be pointed out that platinum satellites were observed as shoulders in the ^{1}H NMR spectra of $PtL^{a-e}Cl_2$ complexes [50–52].

Based on the above mentioned, we could conclude that only one isomer of $PtL^{a-e}Cl_2$ formed through the bidentate coordinate mode of two amines under the present experimental conditions and their chemical structures were shown Scheme 1.

3.2. Cytotoxic Activity In Vitro.

In an *in vitro* assay, the platinum complexes with diamine coupled shikimic acid ligands are weakly active against BEL7404 cancer cell lines (see Table 3, the data for compound **7a** (PtL^aCl_2) is not available due to its low yield) but do not exhibit activity against SGC-7901, SK-OV-3, CNE-2, and HeLa cancer cell lines under the tested concentrations. Therefore, the platinum complexes with diamine coupled shikimic acid ligands display a certain extent selective cytotoxicity. And

TABLE 3: IC_{50} values (μM) for the water-soluble platinum(II) complexes, cisplatin in BEL7404 cancer cell lines[a].

PtL^bCl_2	PtL^cCl_2	PtL^dCl_2	PtL^cCl_2	Cisplatin
289.3 ± 18.6	298.4 ± 22.5	387.2 ± 7.8	391.7 ± 10.6	98.0 ± 17.4

[a] IC_{50} values are presented as the mean ± SD (standard error of the mean) from five separated experiments. Cisplatin was used as positive control.

their corresponding ligands do not exhibit activity. Their activities depend on the carbon linker length, in which the IC_{50} values increase at longer carbon linker. Though these platinum(II) complexes with shikimic acid-based ligands possess high water solubility, they are less active than cisplatin, Pt-shikimato complexes [41], carbohydrate-metal complexes [32], and carbohydrate-linked cisplatin analogues [20]. Such observations should be correlated their low lipophilicity. Recently, lipophilicity has been considered a crucial aspect for the cytotoxicity of platinum complexes [51–56]. The platinum antitumour agents must enter cells before reaching their main biological target, namely, DNA. Their distribution within the body and, hence, their activity are to a large extent determined by their lipophilicity [57]. It is believed that the more lipophilic a complex, the higher its cytotoxicity [58]. Since $PtL^{b-e}Cl_2$ complexes attached a high hydrophilic shikimato group which resulted in a low lipophilicity of the whole $PtL^{b-e}Cl_2$ complexes. Thus, it is expected that these platinum complexes do not effectively enter the cells and lead to a low activity [51–58].

3.3. DNA Unwinding and Cleavage Studies.

Since DNA is the primary target of Pt(II)-based antitumor complexes, the

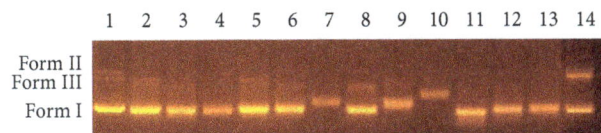

FIGURE 1: Electrophoretic mobility of pUC19 plasmid DNA. Lane 1: DNA, Lane 2~4: DNA + PtL^{a}Cl$_2$ (50, 100, 200 μM), Lane 5~7: DNA + PtL^{b}Cl$_2$ (50, 100, 200 μM), Lane 8~10: DNA + PtL^{c}Cl$_2$ (50, 100, 200 μM), and Lane 11~13: DNA + PtL^{d}Cl$_2$ (50, 100, 200 μM), and Lane 14: DNA + CDDP (40 μM).

TABLE 4: Assignments of the selected peaks in ^{1}H NMR spectra for reaction of complex PtL^{b}Cl$_2$ with 5'-GMP.

Peak	δ (ppm)	Assignment
H8	8.11	H8, free 5'-GMP
H1'	5.83	H1', ribose in free 5'-GMP
HA	6.36	HA, in free Pt(L4)Cl$_2$
H8$_A$	8.65	H8, mono-GMP adduct
H8$'_A$	8.49	H8, bis-GMP adduct
H1''	5.78	H1', ribose in GMP adduct
HA'	6.31	HA, mono-GMP adduct
HA''	6.27	HA, bis-GMP adduct

DNA binding behaviors of the PtL^{a-d}Cl$_2$ complexes have been studied via agarose gel electrophoresis.

Figure 1 and Figure S1 show the results of agarose gel electrophoresis on pUC19 plasmid DNA, in which three forms are present: supercoiled DNA (Form I) as the dominant components, nicked or open circular DNA (Form II), and linear DNA (Form III) bound with ligands and homologous Pt(II) complexes. The complexation of ligands L^{a-d} did not produce any change in the migration rate during agarose gel electrophoresis (L^c as the example, Figure S3, ESI†).

Due to these platinum(II) complexes having similar structure, herein, we just selected four of them as representatives. As shown in Figure 1, though the presence of PtL^{a}Cl$_2$ did not reduce the electrophoretic mobility of supercoiled DNA obviously, the proportion of supercoiled form decreased obviously upon increasing the concentrations of 100 and 200 μM. While the presence of PtL^{b}Cl$_2$, PtL^{c}Cl$_2$, and PtL^{d}Cl$_2$, especially PtL^{c}Cl$_2$, has induced significant reduction of the electrophoretic mobility of supercoiled DNA, all of them had concentration dependence. It seems that all platinum(II) complexes exhibit high binding affinity to DNA, and covalent binding mode of these complexes to DNA is proposed.

3.4. The Interaction with 5'-GMP. As guanine-N7 is the preferable binding site in DNA to bind with platinum-based drugs [59], we investigated the reaction of complexes PtL^{b}Cl$_2$, PtL^{c}Cl$_2$, and PtL^{d}Cl$_2$ with 5'-GMP using ESI-MS and 1H NMR spectrometry.

3.4.1. ESI-MS Analysis. Two peaks m/z 1213.4 and 1192.4 in Figure 2(a) are assigned to one negatively charged species [PtL^b(5'-GMP)^2-4H + Na]$^-$ (C$_{32}$H$_{47}$N$_{13}$NaO$_{20}$P$_2$Pt, calcd. 1213.8) and [PtL^b(5'-GMP)$_2$-3H]$^-$ (C$_{32}$H$_{48}$N$_{13}$O$_{20}$P$_2$Pt, calcd. 1191.8), respectively. Peaks m/z 1227.4 and 1205.4 in Figure 2(b) are assigned to Pt-DNA adducts [PtL^c(5'-GMP)$_2$-4H + Na]$^-$ (C$_{33}$H$_{49}$N$_{13}$NaO$_{20}$P$_2$Pt, calcd. 1227.8) and [PtL^c(5'-GMP)$_2$-3H]$^-$ (C$_{33}$H$_{50}$N$_{13}$O$_{20}$P$_2$Pt, calcd. 1205.9), respectively. Figure 2(c) shows peaks m/z 1241.4 and 1218.4, which are attributed to the species [PtL^d(5'GMP)$_2$-4H + Na]$^-$ (C$_{34}$H$_{51}$N$_{13}$NaO$_{20}$P$_2$Pt, calcd. 1241.9) and [PtL^d(5'-GMP)$_2$-3H]$^-$ (C$_{34}$H$_{52}$N$_{13}$O$_{20}$P$_2$Pt, calcd. 1219.9), respectively. These results suggested that the three complexes reacted with 5'-GMP by covalent bonding, and two chloro ligands were removed.

3.4.2. ^{1}H NMR Analysis. We took the reaction of compounds 7c (PtL^{c}Cl$_2$) with 5'-GMP as an example to confirm the coordinate site of 5'-GMP using ^{1}H NMR spectroscopy. In all cases, the typical characters H8 of 5'-GMP downshift (8.1 ppm for H8 of free 5'-GMP) and the appearance of H8 signals corresponds to platinum adducts (8.5 ppm for H8 of bis-bound and 8.6–8.8 ppm for H8 of monobound 5'-GMP) [60–62]. Dijt et al. reported that the N7 position indicated the absence of a protonation effect at low pH in the pH-dependent behavior of H8 from the end products [63].

In contrast to the individual component, the ^{1}H NMR spectroscopy of the reaction mixture of 5'-GMP and PtL^{c}Cl$_2$ showed clear changes after an incubation of 24 h at 37°C (Figure 3). The assignments for the representative peaks have been listed in Table 4. Peaks at 8.65 ppm for H8$_A$ and at 8.49 ppm for H8$'_A$ were observed at 24 h and downfield shifted from the H8 signal at 8.11 ppm, which indicates the formation of Pt-GMP mono-bound and bis-bound adducts. Meanwhile, the signal of sugar H1' at 5.83 ppm partially shifted to 5.78 ppm, and new signals at 6.31 ppm and 6.27 ppm assigned to PtL^{c}Cl$_2$ appeared after incubation with 5'-GMP. These changes suggest that both sugar H1' and alkene proton were shield due to the platination of N7 of 5'-GMP. The upfield shift of sugar H1' resonance has been observed in the ^{1}H NMR spectrum of cis-[Pt(GMP)$_2$(NH$_3$)$_2$]$^{2-}$ [64–66]. The results show that PtL^{c}Cl$_2$ can bind to 5'-GMP upon the platination of N7 of guanine, consistent with the ESI-MS results mentioned above.

In summary, these water-soluble platinum(II) complexes can bind to DNA and 5'-GMP, and the low cytotoxicity should correlate their low lipophilicity, but the cellular uptake and detailed action mechanism need further investigation in the continuing work.

3.5. S-Phase Cell-Cycle Arrest. To determine whether cellular DNA is a major target of the water-soluble Pt(II) complexes, we studied the cell-cycle profiles of PtL^{b}Cl$_2$ and PtL^{c}Cl$_2$ treated cancer cells (because compounds 7a~e have similar structures; here, only select PtL^{b}Cl$_2$ and PtL^{c}Cl$_2$ to investigate their cell-cycle profiles). Cell-cycle analysis was performed, and flow cytometry was used to assess the DNA content of cells stained with propidium iodine, which enables us to quantify the total cellular populations in different phases

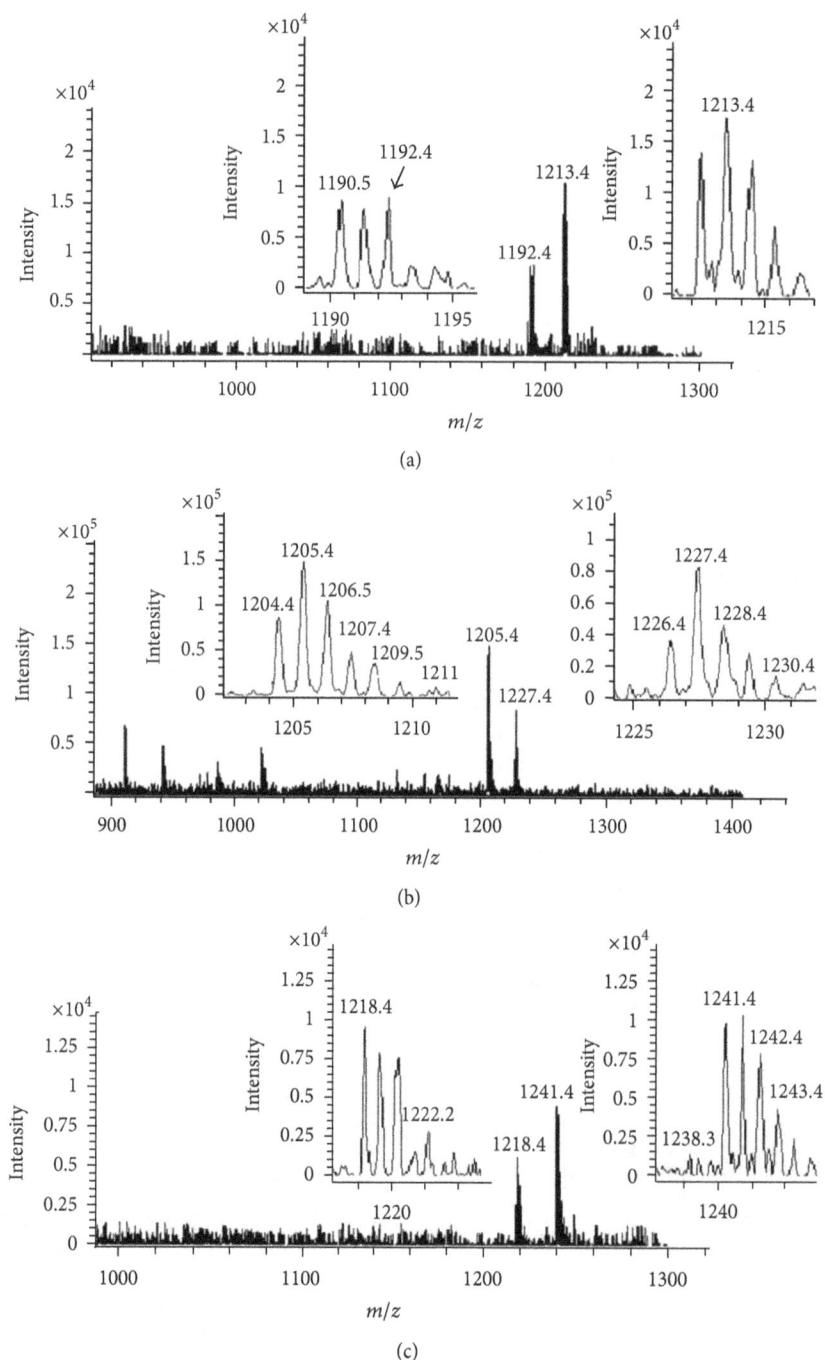

FIGURE 2: The ESI-MS spectra showing the formation of 5'-GMP adducts with complexes PtL^bCl_2 (a), PtL^cCl_2 (b), and PtL^dCl_2 (c).

of the cell cycle (G_0/G_1, S, and G_2/M). The flow-cytometric data for the BEL7404 cells treated with PtL^bCl_2 and PtL^cCl_2 were presented in Table 5. Upon treating cells with PtL^bCl_2 and PtL^cCl_2 (250 μM) for 72 h, the cell-cycle arrest has enhanced at G_2 phase, resulting in concomitant increases in the G_2 phase population and decrease in the S phase population. The direct interaction of PtL^bCl_2 and PtL^cCl_2 with DNA has been examined by agarose gel electrophoresis. It was found that PtL^bCl_2 and PtL^cCl_2 are able to alter DNA configuration. Given all these results, DNA may be a crucial cellular target of PtL^bCl_2 and PtL^cCl_2 in inducing its cytotoxicity.

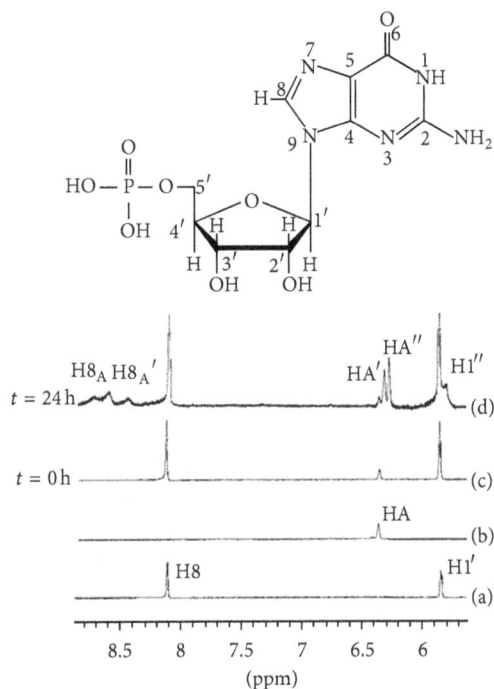

FIGURE 3: Selected ^{1}H NMR spectra for GMP (a), complex PtL^{c}Cl$_2$ (b), the reaction of complex PtL^{c}Cl$_2$ with 3 equiv. of 5′-GMP in D$_2$O after being incubated at 37°C for 0 h (c) and 24 h (d), respectively.

TABLE 5: Induction of cell-cycle arrest in BEL7404 cells after treatment with PtL^{b}Cl$_2$, PtL^{c}Cl$_2$.

	Dip G$_1$ (%)	Dip G$_2$ (%)	Dip S (%)
Pt(L^b)Cl$_2$	49.26	27.09	23.64
Pt(L^c)Cl$_2$	50.57	21.21	28.22
Untreated	49.92	14.74	35.34

4. Conclusion Remarks

We have synthesized five water-soluble platinum(II) complexes: PtL^{a-e}Cl$_2$ (L^{a-e} = SA-NH(CH$_2$)$_n$NHCH$_2$CH$_2$NH$_2$, n = 2–6) through the reactions of Pt(II) with TCM active ingredient, shikimic acid, and coupled aliphatic amine with different carbon linkers (NH$_2$(CH$_2$)$_n$NHCH$_2$CH$_2$NH$_2$, n = 2–6). These Pt(II) complexes interact with DNA by covalent binding, which blocks the DNA synthesis and replication thus inducing low cytotoxicity against cells like BEL7404. The low cytotoxicity should correlate with their low lipophilicity. In addition, PtL^{b}Cl$_2$, PtL^{c}Cl$_2$, and PtL^{d}Cl$_2$ could react with 5′-GMP to form monoGMP and bisGMP adducts via hydrolysis. The cell-cycle analysis revealed that PtL^{b}Cl$_2$ and PtL^{c}Cl$_2$ cause G$_2$-phase cell arrest.

Abbreviations

BEL7404: Liver cancer cell
BSA: Bovine serum albumin
CDDP: Cisplatin
CNE-2: Human nasopharyngeal carcinoma cells
ESI-MS: Electrospray ionization mass spectrum
5′-GMP: 5′-guanosine monophosphate
HeLa: Human cervical cancer cells
MTT: 3-[4,5-dimentylthiazole-2-yl]-2,5-diphenpyltetra-zolium bromide
PBS: Phosphate-buffered saline
SGC-7901: Human gastric adenocarcinoma cells
SK-OV-3: Human ovarian cancer cells
TBE: Tris, boric acid, EDTA.

Acknowledgments

This work was supported by the National Basic Research Program of China (no. 2012CB723501), the National Natural Science Foundation of China (no. 21271051), and Natural Science Foundation of Guangxi Province (nos. 2012GXNSFDA053005, 2012GXNSFDA385001) as well as BAGUI Scholar Program of Guangxi, China.

References

[1] Y. Jung and S. J. Lippard, "Direct cellular responses to platinum-induced DNA damage," *Chemical Reviews*, vol. 107, no. 5, pp. 1387–1407, 2007.

[2] B. Rosenberg, L. VanCamp, J. E. Trosko, and V. H. Mansour, "Platinum compounds: a new class of potent antitumour agents," *Nature*, vol. 222, no. 5191, pp. 385–386, 1969.

[3] K. S. Lovejoy and S. J. Lippard, "Non-traditional platinum compounds for improved accumulation, oral bioavailability, and tumor targeting," *Dalton Transactions*, no. 48, pp. 10651–10659, 2009.

[4] C. G. Hartinger, A. A. Nazarov, S. M. Ashraf, P. J. Dyson, and B. K. Keppler, "Carbohydrate-metal complexes and their potential as anticancer agents," *Current Medicinal Chemistry*, vol. 15, no. 25, pp. 2574–2591, 2008.

[5] K. S. Lovejoy, R. C. Todd, S. Zhang et al., "cis-Diammine(pyridine)chloroplatinum(II), a monofunctional platinum(II) antitumor agent: uptake, structure, function, and prospects," *Proceedings of the National Academy of Sciences of the United States of America*, vol. 105, no. 26, pp. 8902–8907, 2008.

[6] D. Gibson, "The mechanism of action of platinum anticancer agents—what do we really know about it?" *Dalton Transactions*, vol. 38, pp. 10681–10689, 2009.

[7] I. Łakomska, H. Kooijman, A. L. Spek, W. Z. Shen, and J. Reedijk, "Mono- and dinuclear platinum(II) compounds with 5,7-dimethyl-1,2,4- triazolo[1,5-a]pyrimidine. Structure, cytotoxic activity and reaction with 5′-GMP," *Dalton Transactions*, no. 48, pp. 10736–10741, 2009.

[8] P. J. S. Miguel, M. Roitzsch, L. Yin et al., "On the many roles of NH3 ligands in mono- and multinuclear complexes of platinum," *Dalton Transactions*, no. 48, pp. 10774–10786, 2009.

[9] E. R. Jamieson and S. J. Lippard, "Structure, recognition, and processing of cisplatin-DNA adducts," *Chemical Reviews*, vol. 99, no. 9, pp. 2467–2498, 1999.

[10] J. Reedijk, "Why does cisplatin reach guanine-N7 with competing S-donor ligands available in the cell?" *Chemical Reviews*, vol. 99, no. 9, pp. 2499–2510, 1999.

[11] G. Cossa, L. Gatti, F. Zunino, and P. Perego, "Strategies to improve the efficacy of platinum compounds," *Current Medicinal Chemistry*, vol. 16, no. 19, pp. 2355–2365, 2009.

[12] E. Gabano, M. Ravera, and D. Osella, "The drug targeting and delivery approach applied to Pt-antitumour complexes. A coordination point of view," *Current Medicinal Chemistry*, vol. 16, no. 34, pp. 4544–4580, 2009.

[13] L. Kelland, "The resurgence of platinum-based cancer chemotherapy," *Nature Reviews Cancer*, vol. 7, no. 8, pp. 573–584, 2007.

[14] M. Campone, J. M. Rademaker-Lakhai, J. Bennouna et al., "Phase I and pharmacokinetic trial of AP5346, a DACH-platinum-polymer conjugate, administered weekly for three out of every 4 weeks to advanced solid tumor patients," *Cancer Chemotherapy and Pharmacology*, vol. 60, no. 4, pp. 523–533, 2007.

[15] C. N. Sternberg, P. Whelan, J. Hetherington et al., "Phase III trial of satraplatin, an oral platinum plus prednisone vs. prednisone alone in patients with hormone-refractory prostate cancer," *Oncology*, vol. 68, no. 1, pp. 2–9, 2005.

[16] M. D. Hall, H. R. Mellor, R. Callaghan, and T. W. Hambley, "Basis for design and development of platinum(IV) anticancer complexes," *Journal of Medicinal Chemistry*, vol. 50, no. 15, pp. 3403–3411, 2007.

[17] D. Lebwohl and R. Canetta, "Clinical development of platinum complexes in cancer therapy: an historical perspective and an update," *European Journal of Cancer*, vol. 34, no. 10, pp. 1522–1534, 1998.

[18] C. A. Rabik and M. E. Dolan, "Molecular mechanisms of resistance and toxicity associated with platinating agents," *Cancer Treatment Reviews*, vol. 33, no. 1, pp. 9–23, 2007.

[19] R. B. Weiss and M. C. Christian, "New cisplatin analogues in development: a review," *Drugs*, vol. 46, no. 3, pp. 360–377, 1993.

[20] E. Wong and C. M. Giandomenico, "Current status of platinum-based antitumor drugs," *Chemical Reviews*, vol. 99, pp. 2451–2466, 1999.

[21] G. Hata, Y. Kitano, T. Kaneko, H. Kawai, and M. Mutoh, "Synthesis, structure and antitumor of a water-soluble platinum complex, (1R, 3R, 4R, 5R)-(-)-Quinato(1R, 2R-cyclohexanediamine)platinum(II)," *Chemical and Pharmaceutical Bulletin*, vol. 40, pp. 1604–1605, 1992.

[22] H. Kawai, Y. Kitano, M. Mutoh, and G. Hata, "Synthesis, structure and antitumor activity of a new water-soluble platinum complex, (1R, 2R-Cyclohexanediamine-N, N')[2-hydroxy-4-oxo-2-pentenoato(2-)-O^2]platinum(II)," *Chemical and Pharmaceutical Bulletin*, vol. 41, pp. 357–361, 1993.

[23] D. L. Ma, T. Y. T. Shum, F. Zhang, C. M. Che, and M. Yang, "Water soluble luminescent platinum terpyridine complexes with glycosylated acetylide and arylacetylide ligands: photoluminescent properties and cytotoxicities," *Chemical Communications*, no. 37, pp. 4675–4677, 2005.

[24] D. D. N 'Da and E. W. Neuse, "Water-soluble macromolecular platinum conjugates derived from 1,2-dihydroxyl-functionalized carrier polymers," *Journal of Inorganic and Organometallic Polymers*, vol. 20, pp. 468–477, 2010.

[25] J. Bariyanga, M. T. Johnson, E. M. Mmutlane, and E. W. Neuse, "A water-soluble polyamide containing cis-dicarboxylato-chelated platinum(II)," *Journal of Inorganic and Organometallic Polymers*, vol. 15, no. 3, pp. 335–340, 2005.

[26] R. Song, Y. S. Kim, and Y. S. Sohn, "Synthesis and selective tumor targeting properties of water soluble porphyrin-Pt(II) conjugates," *Journal of Inorganic Biochemistry*, vol. 83, pp. 83–88, 2002.

[27] C. Lottner, R. Knuechel, G. Bernhardt, and H. Brunner, "Distribution and subcellular localization of a water-soluble hematoporphyrin-platinum(II) complex in human bladder cancer cells," *Cancer Letters*, vol. 215, no. 2, pp. 167–177, 2004.

[28] M. Benedetti, D. Antonucci, D. Migoni, V. M. Vecchio, C. Ducani, and F. P. Fanizzi, "Water-soluble organometallic analogues of oxaliplatin with cytotoxic and anticlonogenic activity," *ChemMedChem*, vol. 5, no. 1, pp. 46–51, 2010.

[29] G. Horvath, T. Premkumar, A. Boztas, E. Lee, S. Jon, and K. E. Geckeler, "Supramolecular nanoencapsulation as a tool: solubilization of the anticancer drug *trans*-Dichloro(dipyridine)platinum(II) by complexation with β-cyclodextrin," *Molecular Pharmaceutics*, vol. 5, pp. 358–363, 2008.

[30] A. Pasini and F. Zunino, "New cisplatin analogues-on the way to better antitumor agents," *Angewandte Chemie*, vol. 26, pp. 615–624, 1987.

[31] S. Moradell, J. Lorenzo, A. Rovira et al., "Water-soluble platinum(II) complexes of diamine chelating ligands bearing amino-acid type substituents: the effect of the linked amino acid and the diamine chelate ring size on antitumor activity, and interactions with 5'-GMP and DNA," *Journal of Inorganic Biochemistry*, vol. 98, pp. 1933–1946, 2004.

[32] C. G. Hartinger, A. A. Nazarov, S. M. Ashraf, P. J. Dyson, and B. K. Keppler, "Carbohydrate-metal complexes and their potential as anticancer agents," *Current Medicinal Chemistry*, vol. 15, pp. 2574–2591, 2008.

[33] Y. Chen, M. J. Heeg, P. G. Braunschweiger, W. Xie, and P. G. Wang, "A carbohydrate-linked cisplatin analogue having antitumor activity," *Angewandte Chemie*, vol. 38, pp. 1768–1769, 1999.

[34] Y. Chen, A. Janczuk, X. Chen, J. Wang, M. Ksebati, and P. G. Wang, "Expeditious syntheses of two carbohydrate-linked cisplatin analogs," *Carbohydrate Research*, vol. 337, no. 11, pp. 1043–1046, 2002.

[35] Y. Mikata, Y. Shinohara, K. Yoneda et al., "Unprecedented sugar-dependent in vivo antitumor activity of carbohydrate-pendant *cis*-diamminedichloroplatinum(II) complexes," *Bioorganic & Medicinal Chemistry Letters*, vol. 11, pp. 3045–3047, 2001.

[36] M. V. De Almeida, E. T. Cesar, A. P. S. Fontes, and E. De C. A. Felicio, "Synthesis of platinum complexes from sugar derivatives," *Journal of Carbohydrate Chemistry*, vol. 19, pp. 323–329, 2000.

[37] R. Srinivas, P. P. Karmali, D. Pramanik et al., "Cationic amphiphile with shikimic acid headgroup shows more systemic promise than its mannosyl analogue as dna vaccine carrier in dendritic cell based genetic immunization," *Journal of Medicinal Chemistry*, vol. 53, pp. 1387–1391, 2010.

[38] C. Schweinsberg, V. Maes, L. Brans et al., "Novel glycated [^{99m}Tc(CO)$_3$]-labeled bombesin analogues for improved targeting of gastrin-releasing peptide receptor-positive tumors," *Bioconjugate Chemistry*, vol. 19, pp. 2432–2439, 2008.

[39] V. Maes, E. Garcia-Garayoa, P. Blauenstein, and D. Tourwe, "Novel ^{99m}Tc-labeled neurotensin analogues with optimized biodistribution properties," *Journal of Medicinal Chemistry*, vol. 49, pp. 1833–1836, 2006.

[40] E. Garcia-Garayoa, C. Schweinsberg, V. Maes et al., "New [^{99m}Tc]bombesin analogues with improved biodistribution for targeting gastrin releasing-peptide receptor-positive tumors," *The Quarterly Journal of Nuclear Medicine and Molecular Imaging*, vol. 51, pp. 42–50, 2007.

[41] N. Farrell, J. D. Roberts, and M. P. Hacker, "Shikimic acid complexes of platinum. Preparation, reactivity, and antitumor activity of (R,R-1,2-diaminocyclohexane) bis(shikimato)platinum(II). Evidence for a novel rearrangement involving platinum-carbon bond formation," *Journal of Inorganic Biochemistry*, vol. 42, pp. 237–246, 1991.

[42] R. J. Holmes, M. J. McKeage, V. Murry, W. A. Denny, and W. D. McFadyen, "*cis*-Dichloroplatinum(II) complexes tethered to 9-aminoacridine-4-carboxamides: synthesis and action in resistant cell lines in vitro," *Journal of Inorganic Biochemistry*, vol. 85, pp. 209–217, 2001.

[43] H. H. Lee, B. D. Palmer, B. C. Baguley et al., "DNA-directed alkylating agents. 5. Acridinecarboxamide derivatives of (1,2-diaminoethane) dichloroplatinum (II)," *Journal of Medicinal Chemistry*, vol. 35, no. 16, pp. 2983–2987, 1992.

[44] S. Rajagopal and A. F. Spatola, "Mechanism of palladium-catalyzed transfer hydrogenolysis of aryl chlorides by formate salts," *The Journal of Organic Chemistry*, vol. 60, pp. 1347–1355, 1995.

[45] L. M. Sun, Z. L. Meng, and J. A. Guo, "Application of ammonium formate in catalytic transfer hydrogenation," *Chemical Reagents*, vol. 27, pp. 279–282, 2005.

[46] M. Tollabi, E. Framery, C. Goux-Henry, and D. Sinou, "Palladium-catalyzed asymmetric allylic alkylation using chiral glucosamine-based monophosphines," *Tetrahedron Asymmetry*, vol. 14, no. 21, pp. 3329–3333, 2003.

[47] J. V. D. adhav and F. P. Schmidtchen, "A novel synthesis of chiral guanidinium receptors and their use in unfolding the energetics of enantiorecognition of chiral carboxylates," *The Journal of Organic Chemistry*, vol. 73, pp. 1077–1087, 2008.

[48] M. L. Bowen, Z. -F. Chen, A. M. Roos et al., "Long-chain rhenium and technetium glucosamine conjugates," *Dalton Transactions*, vol. 38, pp. 9228–9236, 2009.

[49] Z. F. Chen, X. Y. Song, Y. Peng, Y. C. Liu, and H. Liang, "High cytotoxicity of dihalo-substituted 8-quinolinolato-lanthanides," *Dalton Transactions*, vol. 40, pp. 1684–1692, 2011.

[50] H. C. Clark, A. B. Goel, R. G. Goel, and S. Goel, "Preparation and characterization of platinum(II) complexes Pt(P-C)LX, where (P-C) is metalated tri-tert-butylphosphine," *Inorganic Chemistry*, vol. 19, pp. 3220–3225, 1980.

[51] J. Ruiz, V. Rodríguez, N. Cutillas, A. Espinosa, and M. J. Hannon, "Novel C,N-chelate platinum(II) antitumor complexes bearing a lipophilic ethisterone pendant," *Journal of Inorganic Biochemistry*, vol. 105, pp. 525–531, 2011.

[52] J. A. Platts, G. Ermondi, G. Caron et al., "Molecular and statistical modeling of reduction peak potential and lipophilicity of platinum(IV) complexes," *Journal of Biological Inorganic Chemistry*, vol. 16, no. 3, pp. 361–372, 2011.

[53] C. Francisco, S. Gama, F. Mendes et al., "Pt(ii) complexes with bidentate and tridentate pyrazolyl-containing chelators: synthesis, structural characterization and biological studies," *Dalton Transactions*, vol. 40, no. 21, pp. 5781–5792, 2011.

[54] M. R. Reithofer, A. K. Bytzek, S. M. Valiahdi et al., "Tuning of lipophilicity and cytotoxic potency by structural variation of anticancer platinum(IV) complexes," *Journal of Inorganic Biochemistry*, vol. 105, no. 1, pp. 46–51, 2011.

[55] B. Biersack, A. Dietrich, M. Zoldakova et al., "Lipophilic Pt(II) complexes with selective efficacy against cisplatin-resistant testicular cancer cells," *Journal of Inorganic Biochemistry*, vol. 105, pp. 1630–1637, 2011.

[56] I. Buß, D. Garmann, M. Galanski et al., "Enhancing lipophilicity as a strategy to overcome resistance against platinum complexes?" *Journal of Inorganic Biochemistry*, vol. 105, no. 5, pp. 709–717, 2011.

[57] I. V. Tetko, I. Jaroszewicz, J. A. Platts, and J. Kuduk-Jawworska, "Calculation of lipophilicity for Pt(II) complexes: experimental comparison of several methods," *Journal of Inorganic Biochemistry*, vol. 102, pp. 1424–1437, 2008.

[58] H. Varbanov, S. M. Valiahdi, A. A. Legin et al., "Synthesis and characterization of novel bis(carboxylato)dichloridobis(ethylamine)platinum(IV) complexes with higher cytotoxicity than cisplatin," *European Journal of Medicinal Chemistry*, vol. 46, pp. 5456–5464, 2011.

[59] A. M. J. Fichtinger-Schepman, J. L. VanderVeer, J. H. J. Denhartog, P. H. M. Lohman, and J. Reedijk, "Adducts of the antitumor drug cis-diamminedichloroplatinum(II) with DNA: formation, identification, and quantitation," *Biochemistry*, vol. 24, pp. 707–713, 1985.

[60] M. S. Robillard, M. Galanski, W. Zimmermann, B. K. Keppler, and J. Reedijk, "(Aminoethanol)dichloroplatinum(II) complexes: influence of the hydroxyethyl moiety on 5′-GMP and DNA binding, intramolecular stability, the partition coefficient and anticancer activity," *Journal of Inorganic Biochemistry*, vol. 88, pp. 254–259, 2002.

[61] D. Lemaire, M. H. Fouchet, and J. Kozelka, "Effect of platinum N7-binding to deoxyguanosine and deoxyadenosine on the H8 and H2 chemical shifts. A quantitative analysis," *Journal of Inorganic Biochemistry*, vol. 53, pp. 261–271, 1994.

[62] S. Moradell, J. Lorenzo, A. Rovira et al., "Water-soluble platinum(II) complexes of diamine chelating ligands bearing amino-acid type substituents: the effect of the linked amino acid and the diamine chelate ring size on antitumor activity, and interactions with 5′-GMP and DNA," *Journal of Inorganic Biochemistry*, vol. 98, pp. 1933–1946, 2004.

[63] F. J. Dijt, G. W. Canters, J. H. J. Den Hartog, and A. T. M. Reedijk, "Reaction products of cis-diammineplatinum(II) compounds with 5′-guanosine monophosphate characterized by high-frequency proton NMR," *Journal of the American Chemical Society*, vol. 106, pp. 3644–3647, 1984.

[64] T. Li, H. Lin, and Z. Guo, "Binuclear monofunctional platinum(II) complexes formed by hexaazamacrocyclic bisdien ligands: crystal structure, DNA binding and cytotoxicity studies," *Inorganica Chimica Acta*, vol. 362, pp. 967–974, 2009.

[65] S. K. Miller and L. G. Marzilli, "Interaction of platinum antitumor agents with guanine nucleosides and nucleotides. Platinum-195 and proton NMR spectroscopic characterization of compound III," *Inorganic Chemistry*, vol. 24, pp. 2421–2425, 1985.

[66] E. R. Jamieson and S. J. Lippard, "Structure, recognition, and processing of cisplatin–DNA adducts," *Chemical Reviews*, vol. 99, pp. 2467–2498, 1999.

19

Synthesis Characterization and DNA Interaction Studies of a New Zn(II) Complex Containing Different Dinitrogen Aromatic Ligands

Nahid Shahabadi and Somaye Mohammadi

Department of Inorganic Chemistry, Faculty of Chemistry, Razi University, Kermanshah 74155, Iran

Correspondence should be addressed to Nahid Shahabadi, nahidshahabadi@yahoo.com

Academic Editor: Imre Sovago

A mononuclear complex of Zn(II), [Zn(DIP)$_2$ (DMP)] (NO$_3$)$_2 \cdot$2H$_2$O in which DIP is 4,7-diphenyl-1,10-phenanthroline and DMP is 4,4′-dimethyl-2,2′-bipyridine has been prepared and characterized by ^{1}HNMR spectroscopy, FT-IR, UV-Vis and elemental analysis techniques. DNA-binding properties of the complex were studied using UV-vis spectra, circular dichroism (CD) spectra, fluorescence, cyclic voltammetry (CV), and viscosity measurements. The results indicate that this zinc(II) complex can intercalate into the stacked base pairs of DNA and compete with the strong intercalator ethidium bromide for the intercalative binding sites.

1. Introduction

In recent years, many researches [1–3] have been focused on interaction of small molecules with DNA. DNA is generally the primary intracellular target of anticancer drugs, so the interaction between small molecules and DNA can cause DNA damage in cancer cells, blocking the division of cancer cells and resulting in cell death [4, 5]. Small molecule can interact with DNA through the following three noncovalent modes: intercalation, groove binding, and external electrostatic effects. Among these interactions, intercalation is one of the most important DNA-binding modes, which is related to the antitumor activity of the compound. In this regard, mixed-ligand metal complexes were found to be particularly useful because of their potential to bind DNA *via* a multitude of interactions and to cleave the duplex by virtue of their intrinsic chemical, electrochemical, and photochemical reactivities [6, 7]. It has been reported that the intercalating ability of the complex was involved in the planarity of ligands, the coordination geometry, ligand donor atom type, and the metal ion type. Additionally, metal complexes with tris-dinitrogen ligands and their analogs have attracted much attention for the chiral recognition of DNA double helices with the enantiomeric complexes and

for the photochemical electron transfer reaction initiated by the complex bound to DNA [8–10]. A singular advantage in the use of these metallointercalators for such studies is that the ligands or the metal ion in them can be varied in an easily controlled manner to facilitate individual applications [11–13]. In this regards, binding of zinc(II) complexes to DNA has attracted much attention [14]. In this study, we investigated the mode of DNA binding of a new zinc(II) complex containing mix aromatic dinitrogen ligands. Many techniques have been used to investigate the interactions of complex with DNA. These include (i) molecular spectroscopy methods such as UV spectrophotometry [15], fluorescence [16], cyclic voltametry (CV), and circular dichroism (CD) spectropolarimetry [17] and (ii) dynamic viscosity measurements [18].

2. Experimental

All chemicals such as Zn(NO$_3$)$_2 \cdot$6H$_2$O, 4,7-diphenyl-1,10-phenanthroline (DIP), and 4,7-dimethyl-1,10-phenanthroline (DMP) were purchased from Merck, and Tris-HCl highly polymerized calf thymus DNA (CT-DNA) were purchased from Sigma Co. Experiments were carried out in

Synthesis Characterization and DNA Interaction Studies of a New Zn(II) Complex Containing Different Dinitrogen Aromatic Ligands

173

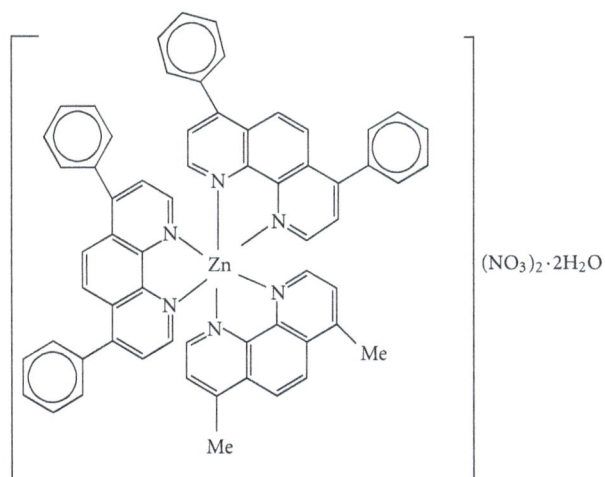

FIGURE 1: The structure of $[Zn(DIP)_2(DMP)](NO_3)_2 \cdot H_2O$ complex.

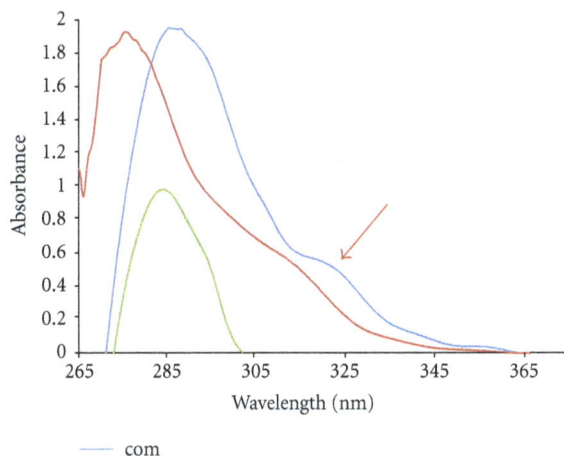

FIGURE 2: The UV-Vis spectra of the free ligands (DIP and DMP) and the zinc complex.

Tris-HCl buffer at pH = 7.0. A solution of calf thymus DNA gave a ratio of UV absorbance at 260 and 280 nm more than 1.8, indicating that DNA was sufficiently free from protein [19]. The stock solution of CT-DNA was prepared by dissolving of DNA in 10 mM of the Tris-HCl buffer at pH = 7.0. The DNA concentration (monomer units) of the stock solution (1×10^{-2} M per nucleotide) was determined by UV spectrophotometer, in properly diluted samples, using the molar absorption coefficient $6600 \, M^{-1} \, cm^{-1}$ at 258 nm. The stock solutions were stored at 4°C and used over no more than 4 days.

2.1. Synthesis of the $[Zn(DIP)_2(DMP)](NO_3)_2 \cdot 2H_2O$ Complex. The complex $[Zn(DIP)_2(DMP)](NO_3)_2 \cdot 2H_2O$ (Figure 1) was prepared by mixing the appropriate molar quantities of the ligands and the metal salt using the following procedure. A methanolic solution of $Zn(NO_3)_2 \cdot 6H_2O$ (0.297 g, 1 mmol) was stirred with a methanolic solution (4 mL) of the 4,7-dimethyl-1,10-phenanthroline (DMP) (0.21 g, 1 mmol) for *ca.* 1 h. To the above mixture, a methanolic solution (5 mL) of 4,7-diphenyl-1,10-phenanthroline (0.664 g, 0.2 mmol) was added in a 1 : 1 : 2 molar ratio, and the stirring was continued for *ca.* 1 h. The obtained solid product was filtered and washed with methanol. A bright pale brown precipitate was formed, which was filtered off and washed with ice-cold water and diethyl ether. Yield: 63%. Elemental analysis for $ZnC_{62}H_{44}N_6(NO_3)_2 \cdot 2H_2O$. Found (calculated): C, 65.8 (67.82); H, 4.0 (4.37); N, 11.0 (10.21). IR data (cm^{-1}): 1545 (ring), 1515 (C=C), 1429 (CCH), 840 (Phen), 729 (Phen), 1342 (Me), 486 (Zn–N); molar conductance, ($\Omega^{-1} \, cm^2 \, mol^{-1}$) in DMF: 121 (1 : 2 electrolyte). In the ^{1}H-NMR spectra of the Zn(II) complexes, the protons of the ligands are shifted downfield due to coordination to the metal ion.

3. Instrumentation

^{1}HNMR spectra were recorded using a Bruker Avance DPX200 MHz (4.7 Tesla) spectrometer with $CDCl_3$ as the

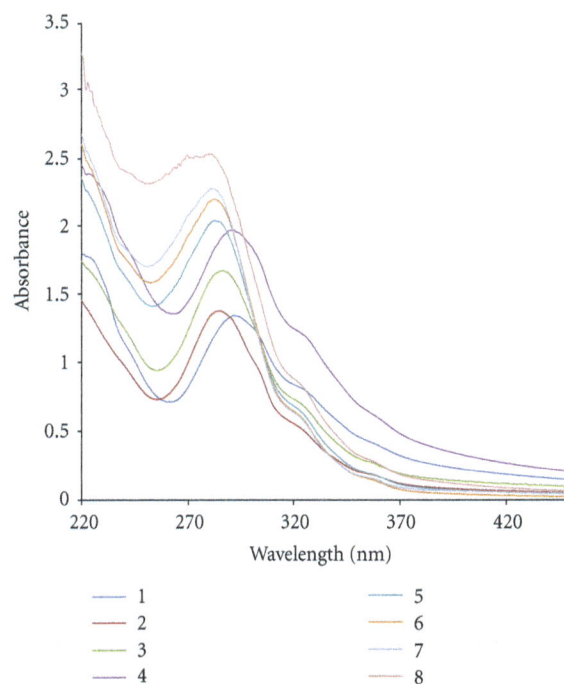

FIGURE 3: Absorption spectra of Zn(II) complex (2.9×10^{-5} M) in the absence and presence of increasing amounts of CT-DNA: $r_i =$ 0.0, 0.2, 0.5, 0.75, 1.5, 2, 3, 4.

solvent. The elemental analysis was performed using a Heraeus CHN elemental analyzer. Absorbance spectra were recorded using an HP spectrophotometer (Agilent 8453) equipped with a thermostated bath (Huber Polystat cc1). Absorption titration experiments were conducted by keeping the concentration of complex constant (2.9×10^{-5} M) while varying the DNA concentration from 0 to 1.9×10^{-4} M (r_i = [DNA]/[complex] = 0.0, 0.2, 0.5, 0.75, 1.5, 2, 3, 4). Absorbance values were recorded after each successive addition of DNA solution and equilibration (ca. 10 min).

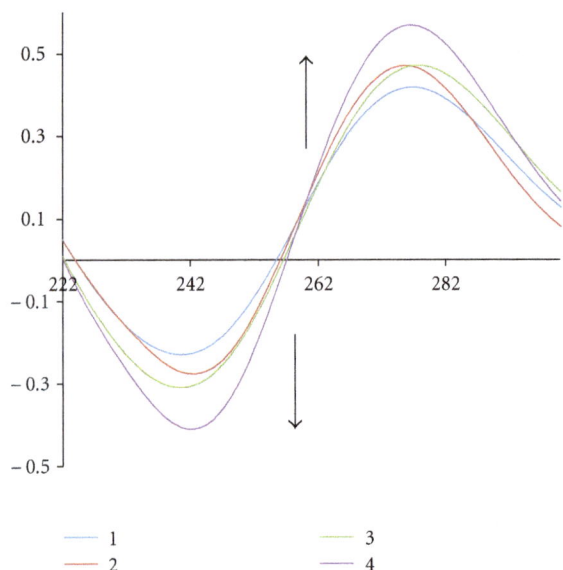

FIGURE 4: Circular dichroism spectra of CT-DNA (5.9×10^{-5} M) in Tris buffer (10 mM), in the presence of increasing amounts of the complex.

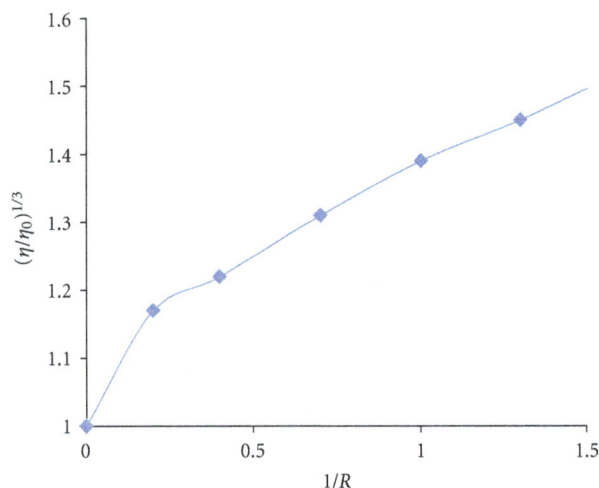

FIGURE 5: Effect of increasing amounts of Zn(II) complex on the viscosity of CT-DNA.

The data were then fitted to (1) to obtain intrinsic binding constant, K_b [20]:

$$\frac{[DNA]}{(\varepsilon_a - \varepsilon_f)} = \frac{[DNA]}{(\varepsilon_b - \varepsilon_f)} + \frac{1}{K_b(\varepsilon_b - \varepsilon_f)}, \qquad (1)$$

where ε_a, ε_f, and ε_b are the apparent, free, and bound complex extinction coefficients, respectively. In particular, ε_f was determined by a calibration curve of the isolated metal complex in aqueous solution; following Beer's law, ε_a was determined as the ratio between the measured absorbance and the Zn(II) complex concentration, $A_{obs}/[complex]$. A plot of $[DNA]/(\varepsilon_a - \varepsilon_f)$ versus $[DNA]$ gave a slope of $1/(\varepsilon_b - \varepsilon_f)$ and a Y intercept equal to $1/K_b(\varepsilon_b - \varepsilon_f)$; K_b is the ratio of the slope to the Y intercept.

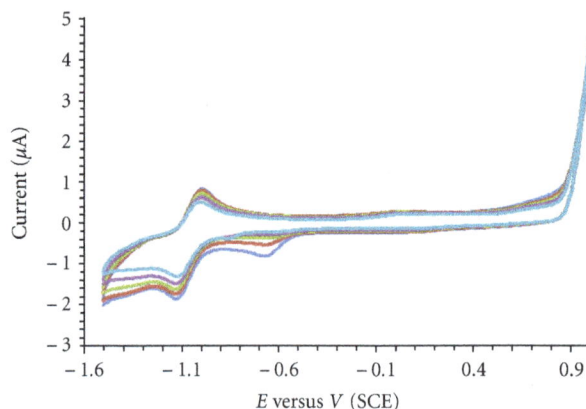

FIGURE 6: Cyclic voltammetry for the Zn(II) complex in the presence of different concentrations of CT-DNA.

CD measurements were recorded on a JASCO (J-810) spectropolarimeter, keeping the concentration of DNA constant (5.9×10^{-5} M) while varying the complex concentration ($r_i = [complex]/[DNA] = r_i = 0, 0.4, 0.5$).

Viscosity measurements were made using a viscosimeter (SCHOT AVS 450) maintained at $25.0 \pm 0.5°$C using a constant temperature bath. The DNA concentration was fixed at 5.9×10^{-5} M, and flow time was measured with a digital stopwatch. The mean values of three measurements were used to evaluate the viscosity g of the samples. The values for relative specific viscosity $(\eta/\eta_0)^{1/3}$, where g_0 and g are the specific viscosity contributions of DNA in the absence (η_0) and in the presence of the complex (η), were plotted against r_i ($r_i = [DNA]/[complex] = 0.0, 0.3, 0.6, 0.9, 1.2, 1.5, 2.0$).

All fluorescence measurements were carried out with a JASCO spectrofluorimeter (FP6200) by keeping concentration of complex constant while varying the DNA concentration from 0 to 36.9×10^{-5} ($r_i = [DNA]/[complex] = 0.0, 0.2, 0.4, 0.6, 0.8, 1$) at three different temperatures (298, 308, 318 K).

The cyclic voltammetric, linear sweep voltammetry, and differentials pulse voltammetry (DPV) measurements were performed using an AUTOLAB model (PG STAT C), with a three-electrode system: a 0.10 cm diameter Glassy carbon (GC) disc as working electrode, an Ag/AgCl electrode as reference electrode, and a Pt wire as counter electrode. Electrochemical experiments were carried out in a 25 mL voltammetric cell at room temperature. All potentials are referred to the Ag/AgCl reference. Their surfaces were freshly polished with 0.05 mm alumina prior to each experiment and were rinsed using double-distilled water between each polishing step. The supporting electrolyte was 0.01 M of Tris-HCl buffer solution (pH = 7.4) which was prepared with double-distilled water. Before experiments, the solution was deaerated via purging with pure nitrogen gas for 1 min, and during measurements a stream of nitrogen was passed over the solution. The current-potential curves and experimental data were recorded on software GPES [21].

Synthesis Characterization and DNA Interaction Studies of a New Zn(II) Complex Containing Different Dinitrogen Aromatic Ligands

175

4. Results and Discussion

4.1. Synthesis and Characterization of [Zn(DIP)₂(DMP)](NO₃)₂·2H₂O Complex.

4.1. Synthesis and Characterization of [Zn(DIP)₂(DMP)] (NO₃)₂·2H₂O Complex. The complex conforms to the formula $[Zn(DIP)_2(DMP)](NO_3)_2·2H_2O$ determined on the basis of elemental analysis. The IR spectrum of the complex was characterized by the appearance of a band at 486 cm⁻¹ due to the $\nu(Zn-N)$. The coordination of the nitrogen atoms is confirmed with the presence of this band. However, the broad band at 3400–3500 cm⁻¹ is assigned to $[\nu(H_2O)]$. ¹HNMR; in the aromatic region the signals at δ = 8.8, δ = 7.6, δ = 7.4, and δ = 7.2 were assigned to the protons of DIP ligand. The protons of DIP ligands are seen to be shifted ($\approx$−0.2 ppm) with corresponding free ligand and suggesting complexation. The UV-Vis spectra of the complex and the free ligands were recorded. The spectrum of the complex shows a d-d band at around 365 nm. This band may be assign to $2E_g$ to $2T_2g$ transitions, characteristics for a distorted octahedral structure. In addition, the intense higher energy bands at around 286 and 288 nm can be attributed to intraligand π-π^* transitions which shifted after complexation (Figure 2). Based on the elemental analysis, electronic, IR, and ¹HNMR data, the proposed structures of the complex are given in Figure 1.

4.2. DNA Interaction Studies

4.2.1. UV-Vis Spectroscopy.

4.2.1. UV-Vis Spectroscopy. Absorption spectroscopy is one of the most useful techniques to study the binding of any drug to DNA. The extent of hypochromism generally indicates the intercalative binding strength [22]. "Hyperchromic" and "hypochromic" effects are the spectra features of DNA concerning its double helical structure [23]. Hyperchromism has been observed for the interaction of many drugs with DNA [24]. The hyperchromic effect might be ascribed to external contact (electrostatic binding [25]) or to partial uncoiling of the helix structure of DNA, exposing more bases of the DNA [26]. Figure 3 illustrates the spectral changes in aqueous solution for the Zn(II) complex during titration with increasing amounts of CT-DNA at r_i values of 0, 0.05, 0.5, 1.5, 5. The initial spectrum is of a fixed concentration ($2.9 \times 10^{-5} M^{-1}$) of the free complex in the absence of CT-DNA. In the UV region, the intense absorption band at 290 nm is attributed to the interligand π-π^* transitions of the coordinated groups. The appreciable changes in the position of the maximum wavelength are observed on addition of CT-DNA (7 nm; blue shift), and the intensity of the band of the Zn(II) complex at 290 nm increases, resulting in hyperchromism. The hyperchromic effect may also be due to the electrostatic interaction between positively charged $[Zn(DIP)_2(DMP)]^{2+}$ complex unit and the negatively charged phosphate backbone at the periphery of the double-helix CT-DNA [27].

Structurally, intercalation to DNA may not be one of the binding modes, since the tris(bidentate) ligand strands wrap around the zinc center, possessing pseudo-threefold rotation axis that passes through the metal ions. Therefore, the complete intercalation of the ligands between a set of adjacent base pairs is sterically impossible, but some partial

TABLE 1: Dynamic enhancement, bimolecular enhancement and formation constants of the complex at different temperatures.

Temperature (K)	k_D	k_B	n	$k_f \times 10^5$
298	0.343	0.343×10^8	0.496	2.87
308	1.083	1.083×10^8	0.484	1.97
318	1.07	1.07×10^8	0.517	1.34

intercalation can be envisioned [28]. It should be pointed out that the absorbance of the shoulder at 326 nm was not changed upon addition of DNA, suggesting that metal-ligand supramolecular architecture was not significantly modified by binding.

The intrinsic binding constant, K_b, of the $[Zn(DIP)_2(DMP)]^{2+}$ complex was determined to be $(3.1 \pm 0.02) \times 10^6 M^{-1}$ using (1). Interestingly, the K_b values obtained for our Zn(II) complex sample are very much higher than those for the other known ordinary complexes containing 1,10-phenanthroline and/or modified phenanthroline ligands available in the literature [28]. This shows that comparatively our complex samples can bind very strongly with DNA. Viscosity and circular dichroism measurements could have been helpful to us to confirm the intercalative behavior.

4.2.2. Fluorescence Spectroscopy.

4.2.2. Fluorescence Spectroscopy. As the zinc complex is luminescent in the absence of DNA, it does show appreciable increase in emission upon addition of CT-DNA (Figure 4). This figure shows that a regular increase in the fluorescence intensity of the complex (1) with a shift of fluorescence emission maximum (414–416 nm) took place upon increasing the concentration of DNA at 25.0°C and at pH = 7.2. These fluorescence enhancements indicate that the complex interacted with DNA and the quantum efficiency of complex was increased. Like the quenching process, the enhancement constant can be obtained by the following [29]:

$$\frac{F_0}{F} = 1 - K_E[E]. \tag{2}$$

If a dynamic process is part of the enhancing mechanism, the above equation can be written as follows:

$$\frac{F_0}{F} = 1 - k_D[E] = 1 - K_B\tau_0, \tag{3}$$

where k_D is the dynamic enhancement constant (like a dynamic quenching constant), k_B is the bimolecular enhancement constant (like to a bimolecular quenching constant), and τ_0 is the lifetime of the fluorophore in the absence of the enhancer. The dynamic enhancement constants of the complex at different temperatures were calculated using (3) (Figure 4, Table 1). Since fluorescence lifetimes are typically near 10^{-8} s, the bimolecular enhancement constant (k_B) was calculated from $k_D = k_B\tau_0$ (Table 1).

By considering the equivalency of the bimolecular quenching and enhancement constants, it can be seen that the latter is greater than the largest possible value ($1.0 \times 10^{10} M^{-1}s^{-1}$) in aqueous medium. Thus, the fluorescence enhancement is not initiated by a dynamic process; it is suggested that a static process involves complex formation in the ground state [30].

4.2.3. Equilibrium Binding Titration.

The binding constant (K_f) and the binding stoichiometry (n) for the complex formation between the complex and DNA were measured using the following [30]:

$$\frac{\log(F - F_0)}{F} = \log K_f + n \log[\text{DNA}]. \tag{4}$$

Here F_0 and F are the fluorescence intensities of the fluorophore in the absence and in the presence of different concentrations of DNA, respectively. The linear equations of $\log(F - F_0)/F$ versus $\log[\text{DNA}]$ at different temperature are shown in Table 1. The values of K_f clearly underscore the remarkably high affinity of the complex to DNA.

4.2.4. Thermodynamic Parameters of DNA Binding.

To have a better understanding of thermodynamics of the reaction between the complex and DNA, it is useful to determine the contributions of enthalpy and entropy of the reaction. Therefore, the evaluation of formation constant for the complex-DNA at three different temperatures (298, 308, 318 K) allows determining the thermodynamic parameters such as enthalpy (ΔH) and entropy (ΔS) of complex-DNA formation by van't Hoff's equation and plotting $\log K_f$ versus $1/T$. The positive slope of the plot ($-\Delta H/R$, R is the gas constant) indicates that the reaction of DNA with the complex is exothermic and enthalpy favored. The ΔH and ΔS values of the complex-DNA were -179.948 ± 0.6 kJ/mol and -488.042 ± 2 J/mol K, respectively. In general, electrostatic interactions exhibit small enthalpy and positive entropy changes. Hydrophobic interactions are generally indicated by positive enthalpy and entropy changes. Hydrogen bonding and van der Waals interactions are usually characterized by negative standard enthalpies of interaction [30, 31]. From the thermodynamic data, it is quite clear that while complex formation is enthalpy favored, it is also entropy disfavored. Formation of the complex therefore results in a more ordered state, possibly due to the freezing (fixing) of the motional freedom of both the complex and DNA molecules. These thermodynamic data are lower than those of previously reported for an intercalator [32] and confirmed the non-intercalating interaction mode.

The analogous $[\text{Ru(phen)}_3]^{2+}$ [33, 34] which is suggested to bind electrostatically to DNA has been reported to exhibit a positive enthalpy change, but Mudasir et al. [35] found that the mode of binding of CT-DNA with $[\text{Fe(phen)(DIP)}_2]^{2+}$ ($\Delta H = -33.1$ kJ mol^{-1}) was intercalation. Furthermore, Chaires studies of the thermodynamic parameters of drug-DNA interactions revealed that for all the intercalators except actinomycin, the binding enthalpy changes are large and negative [36]. When cationic binding agent such as this complex binds to DNA, it is likely replace a countercation from the compact inner (stern) layer or the defuse outer layer surrounding DNA. This counter ion release process is believed to be nearly entirely entropic [37] and the signs of

ΔH and ΔS obtained for DNA binding of complex should be positive. Since the complex is also a cationic binding agent and its structure is similar to other complexes which were previously studied [35], it is expected that counter ion release, hydration, and hydrophobic interaction should occur in this complex. Evidently, the overall negative enthalpy and entropy changes are obtained for DNA binding of this complex. In addition to counter ion release, there should be another type of molecular interaction process in the DNA binding of the complex from which large negative enthalpy and entropy changes are produced. Intercalation of the phenyl group and partial phenanthroline ring of one DIP ligand in complex which involves π-stacking interaction between these planar substituents and DNA base pairs seems to be consistent with the enthalpy and entropy changes. Intercalation, in which two phenyl rings of the ligand are inserted into the DNA base pairs, may be less favorable, because the sheer size of the DIP ligand would preclude simultaneous intercalation of both phenyl rings as the diameter of the ligand is much larger than the width of base pairs. In addition, the crystal structure data of solid $[\text{Cu(bcp)}_2]\text{BF}_4$ (bcp = 2,9-dimethyl-4,7-diphenyl-1,10-phenanthroline) [38] complexes show that all the phenyl groups are skewed about 40 out of phenanthroline plane owing to steric effect between hydrogen atoms in the phenyl and phenanthroline rings. Intercalation of the entire DIP moiety thus would require that the phenyl groups rotate into the plane of the phenanthroline moiety in order to minimize stacking. This may produce unfavorable intermolecular interaction between the complex and DNA. In order to explain such steric hindrance, a model of partial intercalation of the phenanthroline ligand with only one phenyl substituent and partial phen ring inserted between the base pairs in the major groove (an asymmetric docking) has been proposed [39]. In this model, the other phenyl group and nonintercalating DIP ligands are aligned along the major groove and involved in hydrophobic nonstacking interactions with the base pairs along the helical major groove. This model does not require any rotation of the phenyl groups and takes into account the participation of the nonintercalated phenyl group and the other DIP ligands in supporting the binding. Furthermore, even when the model is not operating, intercalative binding of this complex could still be possible if the binding energy is large enough to overcome the barrier against rotation of the phenyl group into the plane.

4.2.5. Circular Dichroism Spectroscopy.

To establish in more detail whether binding of the complexes brings about any significant conformational change of the DNA double helix, CD spectra of CT-DNA were recorded at increasing complex/CT-DNA ratios. The observed CD spectrum of natural calf thymus DNA consists of a positive band at 275 nm (UV: $k$max, 260 nm) due to base stacking and a negative band at 245 nm due to helicity, which is characteristic of DNA in right-handed B form (Figure 4) [40]. The effect of the complex on the conformation of the secondary structure of DNA was studied by keeping the concentration of CT-DNA at 5.9×10^{-5} M while varying the concentration of complex in a buffer solution of 10 mM of Tris ($r_i = 0, 0.1, 0.2, 0.4, 0.6$). Spectrum of the CT-DNA and those with the additives were monitored from 215 to 800 nm. As shown in Figure 4, the CD spectrum of DNA exhibits a positive absorption at 277 nm due to the base stacking and a negative band at 240 nm due to the helicity of B-DNA. In the presence of the complex, both the positive and negative peak intensities of the CD spectra of DNA were increased. The changes in the CD spectra in the presence of the complex show stabilization of the right-handed B form of CT-DNA.

4.2.6. Viscosity Measurements.

The viscosity measurements of CT-DNA are regarded as the least ambiguous and the most critical tests of a binding model in solution in the absence of crystallographic structural data. A classical intercalation model demands that the DNA helix lengthens as base pairs are separated to accommodate the bound ligand, leading to the increase of DNA viscosity. In contrast, a partial, nonclassical intercalation of ligand could bend (or kink) the DNA helix, reducing its length and, concomitantly, its viscosity. In addition, complexes that bind exclusively in the DNA grooves by partial and/or nonclassical intercalation, under the same conditions, typically cause less pronounced (positive or negative) or no change in DNA solution viscosity [40, 41]. The values of relative specific viscosity $(\eta/\eta_0)^{1/3}$ (where η_0 and η are the specific viscosity contributions of DNA in the absence and in the presence of the Pt complex, resp.) were plotted against $1/R$ (R = [DNA]/[complex] (Figure 5). It is known that the groove binder like Hoechst 33258 does not cause an increase in the axial length of the DNA [42, 43] and therefore did not alter the relative viscosity. In contrast, cisplatin which is known to kink DNA through covalent binding, shortening the axial length of the double helix [44], caused a decrease in the relative viscosity of the solution. Partial intercalators also reduce the axial length observed as a reduction in relative viscosity, whereas the classical organic intercalators such as ethidium bromide increased the axial length of the DNA and it becomes more rigid [42, 43] resulting in an increase in the relative viscosity. Results confirm the sensitivity of viscosity measurements to the different modes of DNA binding. In this study, it was observed that increasing the complex concentration leaded to an increase of the DNA viscosity. Thus we may deduce that the complex, certainly, is a DNA intercalator. Since the interaction of this complex with DNA can make DNA longer, we would expect that the relative viscosity of DNA increases with a slope between 0 and 0.96 (a value measured for ethidium bromide) [45] if the intercalation of the complex was either only one interaction mode or much stronger than other interaction(s). But, in this study, the relative viscosity of DNA increase with a slope of 0.3 (Figure 5) and it is reasonably believed that may be other interaction(s) between DNA and the complex occurred. In addition, it should be noted that the DNA binding constant measured for this complex is ten orders of magnitude lower than those determined for ethidium bromide. Therefore, the greater increase in viscosity observed for ethidium bromide

compared to the Zn(II) complex is likely due to the lower binding constant of the latter to DNA. These results clearly show that the importance of using several techniques to ascertain intercalation [46].

4.2.7. Electrochemical Behaviour of the Complex in the Absence and Presence of DNA.

Recently, the electrochemical techniques extensively were used as a simple and rapid method to study DNA interaction with different compounds. The electrochemical behaviour of zinc is well known and was strongly influenced by the electrode material. A well-defined and sensitive peak was observed from the solutions of the complex with a GC electrode rather than the Pt one. Therefore a GC electrode was used in this investigation. When CT-DNA is added to a solution of complex both the anodic and cathodic peak current heights of the complex decreased in the same manner of increasing additions of DNA (Figure 6). Also during DNA addition the anodic peak potential (E_{pa}), cathodic peak potential (E_{pc}), and $E_{1/2}$ (calculated as the average of E_{pc} and E_{pa}) all showed positive shifts. These positive shifts are considered as evidences for intercalation of complex into the DNA, because this kind of interaction is due to hydrophobic interaction. From the other point of view, if a molecule binds electrostatically to the negatively charged deoxyribose-phosphate backbone of DNA, negative peak potential shifts should be detected. Therefore, the positive shift in the CV peak potentials of the complex is indicative of intercalative binding mode of the complex with DNA [47].

5. Conclusions

In summary, we have synthesized a new tris-chelate complex of Zn(II), $[Zn(DIP)_2(DMP)](NO_3)_2 \cdot 2H_2O$, which exhibits high binding affinity to CT-DNA. Different instrumental methods were used to finding the interaction mechanism. The following results supported the fact that the $[Zn(DIP)_2(DMP)](NO_3)_2 \cdot 2H_2O$ complex can bound to CT-DNA by the mode of partial intercalation and electrostatic binding.

(1) In absorption spectrum, the absorption intensity of the complex increased (hyperchromism) evidently after the addition of DNA, which indicated the interactions between DNA and the complex.

(2) The intrinsic binding constant ($K_b = (3.1 \pm 0.02) \times 10^6$ M^{-1}) is comparable to intercalative binding complexes.

(3) Fluorescence studies showed appreciable increase in the complex emission upon addition of DNA.

(4) The positive slope in van't Hoff plot indicated that the reaction of the Zn(II) complex and DNA was enthalpy favored ($\Delta H = -179.948 \pm 0.6$ kJ/mol).

(5) The changes in the CT-CD spectra of DNA in the presence of increasing amounts of the complex show stabilization of the right-handed B form of CT-DNA.

(6) The positive shift in the CV peak potentials of the complex is indicative of intercalative binding mode of the complex with DNA.

(7) Increase of the relative viscosity of CT-DNA in the presence of the complex showed that the intercalative binding must be predominant.

Acknowledgment

Financial support from the Razi University Research center is gratefully acknowledged.

References

[1] Y. M. Song, Q. Wu, P. J. Yang, N. N. Luan, L. F. Wang, and Y. M. Liu, "DNA Binding and cleavage activity of Ni(II) complex with all-*trans* retinoic acid," *Journal of Inorganic Biochemistry*, vol. 100, no. 10, pp. 1685–1691, 2006.

[2] M. Kožurková, D. Sabolová, L. Janovec et al., "Cytotoxic activity of proflavine diureas: synthesis, antitumor, evaluation and DNA binding properties of 1′,1″-(acridin-3,6-diyl)-3′,3″-dialkyldiureas," *Bioorganic and Medicinal Chemistry*, vol. 16, no. 7, pp. 3976–3984, 2008.

[3] C. Tan, J. Liu, L. Chen, S. Shi, and L. Ji, "Synthesis, structural characteristics, DNA binding properties and cytotoxicity studies of a series of Ru(III) complexes," *Journal of Inorganic Biochemistry*, vol. 102, no. 8, pp. 1644–1653, 2008.

[4] G. Zuber, J. C. Quada Jr., and S. M. Hecht, "Sequence selective cleavage of a DNA octanucleotide by chlorinated bithiazoles and bleomycins," *Journal of the American Chemical Society*, vol. 120, no. 36, pp. 9368–9369, 1998.

[5] S. M. Hecht, "Bleomycin: new perspectives on the mechanism of action," *Journal of Natural Products*, vol. 63, no. 1, pp. 158–168, 2000.

[6] T. Ghosh, B. G. Maiya, A. Samanta et al., "Mixed-ligand complexes of ruthenium(II) containing new photoactive or electroactive ligands: dynthesis, spectral characterization and DNA interactions," *Journal of Biological Inorganic Chemistry*, vol. 10, no. 5, pp. 496–508, 2005.

[7] F. Q. Liu, Q. X. Wang, K. Jiao, F. F. Jian, G. Y. Liu, and R. X. Li, "Synthesis, crystal structure, and DNA-binding properties of a new copper (II) complex containing mixed-ligands of 2,2′-bipyridine and *p*-methylbenzoate," *Inorganica Chimica Acta*, vol. 359, no. 5, pp. 1524–1530, 2006.

[8] C. Hiort, B. Norden, and A. Rodger, "Enantiopreferential DNA binding of [Ru(II)(1,10-phenanthroline)₃]²⁺ studied with linear and Circular Dichroism," *Journal of the American Chemical Society*, vol. 112, no. 5, pp. 1971–1982, 1990.

[9] J. K. Barton, J. M. Goldberg, C. V. Kumar, and N. J. Turro, "Binding modes and base specificity of tris(phenanthroline)ruthenium(II) enantiomers with nucleic acids: tuning the stereoselectivity," *Journal of the American Chemical Society*, vol. 108, no. 8, pp. 2081–2088, 1986.

[10] D. Campisi, T. Morii, and J. K. Barton, "Correlations of crystal structures of DNA oligonucleotides with enantioselective recognition by Rh(phen)₂phi³⁺: probes of DNA propeller twisting in solution," *Biochemistry*, vol. 33, no. 14, pp. 4130–4139, 1994.

[11] R. K. Egdal, A. Hazell, F. B. Larsen, C. J. McKenzie, and R. C. Scarrow, "A dihydroxo-bridged Fe(II)-Fe(III) complex: a new member of the diiron diamond core family," *Journal of the American Chemical Society*, vol. 125, no. 1, pp. 32–33, 2003.

[12] A. Silvestri, G. Barone, G. Ruisi, M. T. Lo Giudice, and S. Tumminello, "The interaction of native DNA with iron(III)-N,N′-ethylene-*bis*(salicylideneiminato)-chloride," *Journal of Inorganic Biochemistry*, vol. 98, no. 4, pp. 589–594, 2004.

[13] M. Navarro, E. J. Cisneros-Fajardo, A. Sierralta, M. Fernandez-Mestre, D. S. Arrieche, and E. Marchan, "Design of copper DNA intercalators with leishmanicidal activity," *Journal of Biological Inorganic Chemistry*, vol. 8, p. 401, 2003.

[14] X. Sheng, X. Guo, X. M. Lu et al., "DNA binding, cleavage, and cytotoxic activity of the preorganized dinuclear zinc(II) complex of triazacyclononane derivatives," *Bioconjugate Chemistry*, vol. 19, no. 2, pp. 490–498, 2008.

[15] T. Biver, F. Secco, M. R. Tinè, and M. Venturini, "Kinetics and equilibria for the formation of a new DNA metal-intercalator: the cyclic polyamine Neotrien/copper(II) complex," *Journal of Inorganic Biochemistry*, vol. 98, no. 1, pp. 33–40, 2004.

[16] C. Liu, J. Zhou, and H. Xu, "Interaction of the copper(II) macrocyclic complexes with DNA studied by fluorescence quenching of ethidium," *Journal of Inorganic Biochemistry*, vol. 71, no. 1-2, pp. 1–6, 1998.

[17] S. Mahadevan and M. Palaniandavar, "Spectroscopic and voltammetric studies of copper (II) complexes of *bis*(pyrid-2-yl)-di/trithia ligands bound to calf thymus DNA," *Inorganica Chimica Acta*, vol. 254, no. 2, pp. 291–302, 1997.

[18] C. A. Mitsopoulou, C. E. Dagas, and C. Makedonas, "Characterization and DNA interaction of the Pt(II)(pq)(bdt) complex: a theoretical and experimental research," *Inorganica Chimica Acta*, vol. 361, no. 7, pp. 1973–1982, 2008.

[19] S. D. Kennedy and R. G. Bryant, "Manganese-deoxyribonucleic acid binding modes. Nuclear magnetic relaxation dispersion results," *Biophysical Journal*, vol. 50, no. 4, pp. 669–676, 1986.

[20] A. Wolfe, G. H. Shimer, and T. Meehan, "Polycyclic aromatic hydrocarbons physically intercalate into duplex regions of denatured DNA," *Biochemistry*, vol. 26, no. 20, pp. 6392–6396, 1987.

[21] N. Shahabadi, S. Mohammadi, and R. Alizadeh, "DNA interaction studies of a new platinum(II) complex containing different aromatic dinitrogen ligands," *Bioinorganic Chemistry and Applications*, vol. 2011, Article ID 429241, 8 pages, 2011.

[22] J. B. Chaires, "Tris(phenanthroline)ruthenium(II) enantiomer interactions with DNA: mode and specificity of binding," *Biochemistry*, vol. 32, no. 10, pp. 2573–2584, 1993.

[23] N. Tian, Z. Y. Zhou, S. G. Sun, Y. Ding, and L. W. Zhong, "Synthesis of tetrahexahedral platinum nanocrystals with high-index facets and high electro-oxidation activity," *Science*, vol. 316, no. 5825, pp. 732–735, 2007.

[24] P. J. Cox, G. Psomas, and C. A. Bolos, "Characterization and DNA-interaction studies of 1,1-dicyano-2,2-ethylene dithiolate Ni(II) mixed-ligand complexes with 2-amino-5-methyl thiazole, 2-amino-2-thiazoline and imidazole. Crystal structure of [Ni(i-MNT)(2a-5mt)₂]," *Bioorganic and Medicinal Chemistry*, vol. 17, no. 16, pp. 6054–6062, 2009.

[25] R. F. Pasternack, E. J. Gibbs, and J. J. Villafranca, "Interactions of porphyrins with nucleic acids," *Biochemistry*, vol. 22, no. 10, pp. 2406–2414, 1983.

[26] G. Pratviel, J. Bernadou, and B. Meunier, "DNA and RNA cleavage by metal complexes," *Advances in Inorganic Chemistry*, vol. 45, pp. 251–312, 1998.

[27] N. Shahabadi, S. Kashanian, M. Mahdavi, and N. Sourinejad, "DNA interaction and DNA cleavage studies of a new platinum(II) complex containing aliphatic and aromatic dinitrogen ligands," *Bioinorganic Chemistry and Applications*, vol. 2011, Article ID 525794, 10 pages, 2011.

Synthesis Characterization and DNA Interaction Studies of a New Zn(II) Complex Containing Different Dinitrogen Aromatic Ligands

179

[28] R. S. Kumar, S. Arunachalam, V. S. Periasamy, C. P. Preethy, A. Riyasdeen, and M. A. Akbarsha, "Synthesis, DNA binding and antitumor activities of some novel polymer-cobalt(III) complexes containing 1,10-phenanthroline ligand," *Polyhedron*, vol. 27, no. 3, pp. 1111–1120, 2008.

[29] N. Shahabadi, S. Kashanian, and F. Darabi, "In Vitro study of DNA interaction with a water-soluble dinitrogen schiff base," *DNA and Cell Biology*, vol. 28, no. 11, pp. 589–596, 2009.

[30] N. Shahabadi and A. Fatahi, "Multispectroscopic DNA-binding studies of a tris-chelate nickel(II) complex containing 4,7-diphenyl 1,10-phenanthroline ligands," *Journal of Molecular Structure*, vol. 970, no. 1–3, pp. 90–95, 2010.

[31] S. Satyanarayana, J. C. Dabrowiak, and J. B. Chaires, "Tris(phenanthroline)ruthenium(II) enantiomer interactions with DNA: mode and specificity of binding," *Biochemistry*, vol. 32, no. 10, pp. 2573–2584, 1993.

[32] Mudasir, N. Yoshioka, and H. Inoue, "DNA binding of iron(II) mixed-ligand complexes containing 1,10- phenanthroline and 4,7-diphenyl-1,10-phenanthroline," *Journal of Inorganic Biochemistry*, vol. 77, no. 3-4, pp. 239–247, 1999.

[33] S. Kashanian, S. Askari, F. Ahmadi, K. Omidfar, S. Ghobadi, and F. A. Tarighat, "In vitro study of DNA interaction with clodinafop-propargyl herbicide," *DNA and Cell Biology*, vol. 27, no. 10, pp. 581–586, 2008.

[34] I. Haq, P. Lincoln, D. Suh, B. Norden, B. Z. Chowdhry, and J. B. Chaires, "Interaction of Δ- and Λ-[Ru(phen)$_2$DPPZ]$^{2+}$ with DNA: a calorimetric and equilibrium binding study," *Journal of the American Chemical Society*, vol. 117, no. 17, pp. 4788–4796, 1995.

[35] Mudasir, N. Yoshioka, and H. Inoue, "DNA binding of iron(II) mixed-ligand complexes containing 1, 10-phenanthroline and 4, 7-diphenyl-1, 10-phenanthroline," *Journal of Inorganic Biochemistry*, vol. 77, no. 3-4, pp. 239–247, 1999.

[36] A. Silvestri, G. Barone, G. Ruisi, D. Anselmo, S. Riela, and V. T. Liveri, "The interaction of native DNA with Zn(II) and Cu(II) complexes of 5-triethyl ammonium methyl salicylidene ortophenylendiimine," *Journal of Inorganic Biochemistry*, vol. 101, no. 5, pp. 841–848, 2007.

[37] M. T. Record, C. F. Anderson, and T. M. Lohman, "Thermodynamic analysis of ion effects on the binding and conformational equilibria of proteins and nucleic acids: the roles of ion association or release, screening, and ion effects on water activity," *Quarterly Reviews of Biophysics*, vol. 11, no. 2, pp. 103–178, 1978.

[38] F. K. Klemens, P. E. Fanwick, J. K. Bibler, and D. R. McMillin, "Crystal and molecular structure of [Cu(bcp)$_2$]BF$_4$·CH$_3$OH (bcp = 2,9-dimethyl-4,7-diphenyl-1,10-phenanthroline)," *Inorganic Chemistry*, vol. 28, no. 15, pp. 3076–3079, 1989.

[39] M. Cusumano, M. L. Di Pietro, and A. Giannetto, "Relationship between binding affinity for calf-thymus DNA of [Pt(2,2'-bpy)(n-Rpy)$_2$]$^{2+}$ (n =2,4) and basicity of coordinated pyridine," *Chemical Communications*, no. 22, pp. 2527–2528, 1996.

[40] S. Ramakrishnan and M. Palaniandavar, "Interaction of *rac*-[Cu(diimine)$_3$]$^{2+}$ and *rac*-[Zn(diimine)$_3$]$^{2+}$ complexes with CT DNA: effect of fluxional Cu(ii) geometry on DNA binding, ligand-promoted exciton coupling and prominent DNA cleavage," *Dalton Transactions*, no. 29, pp. 3866–3878, 2008.

[41] J. M. Kelly, A. B. Tossi, D. J. Mcconnell, and C. Ohui-gin, "A study of the interactions of some polypyridyl-ruthenium(II) complexes with DNA using fluorescence spectroscopy, topoisomerisation and thermal denaturation," *Nucleic Acids Research*, vol. 13, no. 17, pp. 6017–6034, 1985.

[42] E. C. Long and J. K. Barton, "DNA intercalation On demonstrating," *Accounts of Chemical Research*, vol. 23, pp. 271–273, 1990.

[43] S. W. Kaldor, B. A. Dressman, M. Hammond et al., "Isophthalic acid derivatives: amino acid surrogates for the inhibition of HIV-1 protease," *Bioorganic and Medicinal Chemistry Letters*, vol. 5, no. 7, pp. 721–726, 1995.

[44] L. Kapicak and E. J. Gabbay, "Effect of aromatic cations on the tertiary structure of deoxyribonucleic acid," *Journal of the American Chemical Society*, vol. 97, no. 2, pp. 403–408, 1975.

[45] F. H. Li, G. H. Zhao, H. X. Wu et al., "Synthesis, characterization and biological activity of lanthanum(III) complexes containing 2-methylene-1,10-phenanthroline units bridged by aliphatic diamines," *Journal of Inorganic Biochemistry*, vol. 100, no. 1, pp. 36–43, 2006.

[46] N. Shahabadi and L. Nemati, "DNA interaction studies of a platinum(II) complex containing l-histidine and 1,10-phenanthroline ligands," *DNA and Cell Biology*, vol. 31, no. 5, pp. 883–890, 2012.

[47] Y. Ni, D. Lin, and S. Kokot, "Synchronous fluorescence, UV-visible spectrophotometric, and voltammetric studies of the competitive interaction of *bis*(1,10-phenanthroline)copper (II) complex and neutral red with DNA," *Analytical Biochemistry*, vol. 352, no. 2, pp. 231–242, 2006.

20

Binuclear Cu(II) and Co(II) Complexes of Tridentate Heterocyclic Shiff Base Derived from Salicylaldehyde with 4-Aminoantipyrine

Omar Hamad Shihab Al-Obaidi

Chemistry Department, College of Education for Women, Al-Anbar University, Ramadi, Iraq

Correspondence should be addressed to Omar Hamad Shihab Al-Obaidi, dromaralobaidi@yahoo.com

Academic Editor: Nick Katsaros

New binuclear Co(II) and Co(II) complexes of ONO tridentate heterocyclic Schiff base derived from 4-aminoantipyrine with salicylaldehyde have been synthesized and characterized on the bases of elemental analysis, UV-Vis., FT-IR, and also by aid of molar conductivity measurements, magnetic measurements, and melting points. It has been found that the Schiff bases with Cu(II) or Co(II) ion forming binuclear complexes on (1 : 1) "metal : ligand" stoichiometry. The molar conductance measurements of the complexes in DMSO correspond to be nonelectrolytic nature for all prepared complexes. Distorted octahedral environment is suggested for metal complexes. A theoretical treatment of the formation of complexes in the gas phase was studied, and this was done by using the HyperChem-6 program for the molecular mechanics and semi-empirical calculations. The free ligand and its complexes have been tested for their antibacterial activities against two types of human pathogenic bacteria: the first type (*Staphylococcus aureus*) is Gram positive *and* the second type (*Escherichia coli*) is Gram negative (by using agar well diffusion method). Finally, it was found that compounds show different activity of inhibition on growth of the bacteria.

1. Introduction

Amino heterocyclic compounds containing two or more potential donor centers play an important role in the study of competitive reactivity of an bidentate ligand system [1]. Heterocyclic phenazone and their derivatives (4-aminoantipyrine) are known to act as bidentate or tridentate ligands when coordinated to metal ion [2].

Phenazone ligand can form mononuclear and binuclear complexes [3–5].

Transition metal complexes containing a salicylaldehyde are commonly found in biological media and play important roles in processes such as catalysis of drug interaction with biomolecules [6]. Phenazone Schiff base chemistry is less extensive, and our laboratory has been exploring this chemistry [7–10].

In this paper we are reporting the synthesis of the binuclear Co(II) and Cu(II) complexes of some heterocyclic Schiff base ligands (Figure 1) containing 4-aminoantipyrine. Spectral and magnetic studies have been used to characterize the structure of the complexes.

2. Experimental

2.1. Physical Measurements. A Fisher-100 infrared spectrophotometer was used to record the IR spectra as KBr and CsI disc, and UV/VIS spectra were measured by a HITACHI U-2000 spectrophotometer. Determination of all metals percentage by atomic absorption spectrophotometry on AA-680G (Shimadzu). Electrical conductance was measured on conductivity CDC304 (Jenway4070) melting points determined by an electric heated block apparatus (GallenKamp) and were uncorrected. Room temperature magnetic susceptibility measurements were carried out on a B.M 6 BRUKER type magnets, balance. Diamagnetic correction was done using pascal constants.

2.2. Materials. All the chemicals and solvents used for the synthesis were reagent grade and procured from (BDH chemicals or Sigma-Aldrich or Fluka). Metal salts were purchased from E. Merck and used as received. All solvents were dried and purified before used.

2.2.1. Preparation of the Schiff Base Ligands (L1 and L2). The ligands were prepared by condensation of 4-aminoantipyrine

Binuclear Cu(II) and Co(II) Complexes of Tridentate Heterocyclic Shiff Base Derived from Salicylaldehyde
with 4-Aminoantipyrine

181

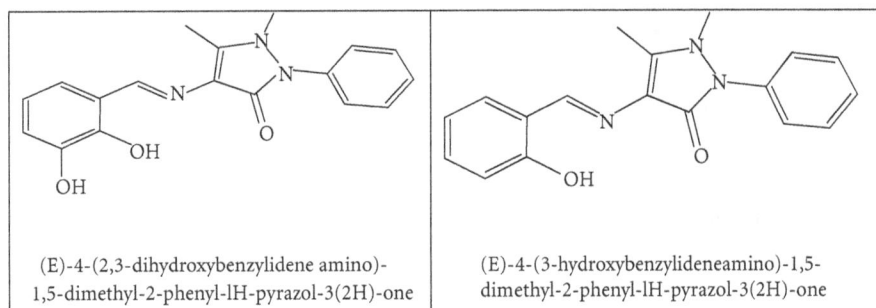

FIGURE 1: Structure of the ligands L1 and L2.

(E)-4-(2,3-dihydroxybenzylidene amino)-1,5-dimethyl-2-phenyl-1H-pyrazol-3(2H)-one

(E)-4-(3-hydroxybenzylideneamino)-1,5-dimethyl-2-phenyl-1H-pyrazol-3(2H)-one

TABLE 1: Analytical and physical data of all the complexes.

No.	Complexes/FW	Colour	ΔM (Ω^{-1} cm^2 mol^{-1}) In DMSO	M.P ($^\circ$C)	Yield (%)	Elemental analysis (% found) % Cal.				
						C	H	M	N	O
1	[Cu$_2$(L1)$_2$(H$_2$O)$_2$] C$_{36}$H$_{34}$Cu$_2$N$_6$O$_8$/805	Light orange	9	>240	75	53.66 (53.61)	4.25 (4.21)	15.77 (15.73)	10.43 (10.39)	15.88 (15.84)
2	[Cu$_2$(L2)$_2$(H$_2$O)$_4$] C$_{36}$H$_{40}$Cu$_2$N$_6$O$_8$/811	Orange	7	>240	65	53.26 (53.21)	4.97 (4.93)	15.65 (15.60)	10.35 (10.30)	15.77 (15.73)
3	[Co$_2$(L1)$_2$(H$_2$O)$_2$] C$_{36}$H$_{34}$Co$_2$N$_6$O$_8$/796	Light brown	5	>240	70	54.28 (54.23)	4.30 (4.26)	16.07 (16.02)	14.80 (14.76)	10.55 (10.50)
4	[Co$_2$(L2)$_2$(H$_2$O)$_4$] C$_{36}$H$_{40}$Co$_2$N$_6$O$_8$/802	Brown	8	>240	75	53.87 (53.83)	5.02 (4.98)	14.69 (14.65)	10.47 (10.41)	15.95 (15.90)

with salicylaldehyde in ethanol. This preparation was performed as cited in the literature [11].

The general structures of ligands obtained from chemical analysis and spectral methods are given in Figure 1. The full name of the ligand will be replaced with (L1 and L2) for the rest of this paper.

2.2.2. Preparation of the Binuclear Metal Complexes. 1.00 mm of the ligands were dissolved in 30 mL of ethanol, and solution of 1.00 mm of the metal salts [CuCl$_2 \cdot$4H$_2$O (0.20 g)] in 15 mL ethanol was added dropwise with continuous stirring. The mixture was stirred further for 2-3 h. at 80°C. The precipitated solid was then filtered off, washed with diethylether, followed by cold ethanol, and dried under vacuum. The same method was applied for the preparation of [CoCl$_2 \cdot$6H$_2$O] complexes by using the corresponding (L1 or L2) working in the same conditions with their respective molar ratio.

The physical properties of prepared complexes are listed in Table 1. The molar ratio of the complexes was determined according to the methods [12].

2.2.3. Study of Biological Activity for Ligands (L1 and L2) and Their Metal Complexes. The biological activity of the ligands and their metal complexes were studied against two selected types of bacteria which included *Escherichia coli* that are gram negative (−ve) and *Staphylococcus aureus* that are gram positive (+ve) to be cultivated and as control for the disc sensitivity test [13]. This method involves the exposure of the zone of inhibition toward the diffusion of microorganism on agar plat. The plates were incubated for (24 hours), at 37°C, and the zone of inhibition of bacteria growth around the disc was observed.

3. Result and Discussion

The Schiff bases ligands are soluble in common organic solvents. But its metal complexes are generally soluble in DMF and DMSO. The elemental analytical data of the complexes reveal that the compounds have "metal : ligand" an ion stoichiometry ratio of 1 : 1; the analytical data and other spectral analysis are in good agreement with the proposed stoichiometry of the complexes. The colors, yields, melting points, IR, and electronic absorption spectral data of all the compounds are presented in Table 2. The molar conductance of solutions of all the complexes in DMSO are in the range [5–8, 14] Ω^{-1} cm^2 mole^{-1} Table 1. These observations suggest that all the complexes are nonelectrolytes [15] in DMSO ($1 \ast 10^{-3}$ M) at room temperature. Polydentate complexes were obtained from 1 : 1 molar ratio reactions with metal ions and L1 and L2 ligands. The ligands L1 and L2 on reaction with Cu(II) and Co(II) salt yields complexes are corresponding to the formulas [Cu$_2$(L1)$_2$H$_2$O], [Co$_2$(L1)$_2$H$_2$O], [Cu$_2$(L2)$_2$(H$_2$O)$_4$], and [Co$_2$(L2)$_2$(H$_2$O)$_4$].

3.1. Infrared Spectral Study. The most important infrared spectral bands of the investigated metal complexes in the present article are summarized in Table 2. The free Schiff base ligands are characterized by strong band at 1690, 1625, and 1270 cm^{-1} for L1, 1720, 1625, and 1290 cm^{-1} for L2 which

TABLE 2: Characteristic IR and electronic spectral data of the metal complexes.

No.	Complexes	UV/ VIS λ max (cm^{-1})	μ_{eff}.BM	IR spectra cm^{-1}				
				$vC=O$	$vC=N$	$vC-O$	$vM-N$	$vM-O$
1	[Cu$_2$(L1)$_2$H$_2$O]	11085, 16595, 27990	0.9	1650 s	1600 m	1265 s	460 w	530 w
2	[Cu$_2$(L2)$_2$(H$_2$O)$_4$]	11080, 16590, 27995	1.69	1670 s	1610 m	1260 s	465 w	535 w
3	[Co$_2$(L1)$_2$H$_2$O]	10370, 14380, 18720, 35690	5.04	1665 s	1610 m	1270 s	470 w	520 w
4	[Co$_2$(L2)$_2$(H$_2$O)$_4$]	10375, 14385, 18725, 35695	4.90	1660 s	1590 m	1270 s	460 w	515 w

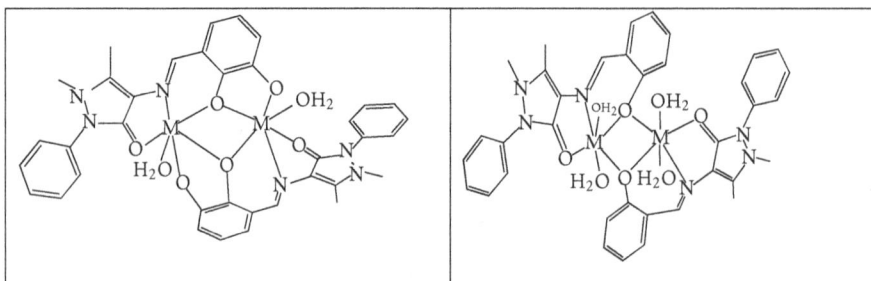

FIGURE 2: The proposed structure of the complexes where M = Cu(II) or Co(II)of the ligands L1 and L2.

may be ascribed to the stretching vibrations of C=O groups, C=N (imine) and C–O (phenolic) groups, respectively [7–10]. The band at 1630–1620 cm^{-1} due to the stretching mode of the C=N group in the spectrum of the free ligands shows a remarkable negative shift with splitting in the 1590–1610 cm^{-1} region in all the complexes spectra suggesting that the coordinating azomethine nitrogen atoms of the Schiff bases are involved in the complexes formation [7–10].

In the spectra of all binuclear complexes, the phenolic band at 1260–1270 cm^{-1} is shifted to lower frequency (10–30 cm^{-1}). It is suggested that the oxygen atom of this phenolic (C–O) group is bridged to the metal ions. An additional band at 1160 cm^{-1} suggests that water molecules are coordinated to metal ion [16, 17]. This band may be assigned to water molecule OH heterocyclic ring vibration at 1580–1200 cm^{-1}. Other band of M–O and M–N bands appear respectively at 535–515, 460–470 cm^{-1} Table 2.

3.2. Electronic Spectra and Magnetic Measurements.

The electronic spectra were recorded in DMSO. In the spectrum of the ligand, the bands in the 380–340 nm range are assigned to the n-π^* transitions of the azomethine group. During the formation of the complexes, these bands are shifted to lower wavelength, suggesting that the nitrogen atom of the azomethine group is coordinated to the metal ion. The values in the 325–245 nm range are attributed to the π-π^* transition of the aromatic rings. In the spectra of the complexes, these bands are shifted slightly to lower wavelength.

3.3. Cu(II) Complexes.

On the basis of the magnetic moment measurements, the Cu(II) complexes at room temperature probably have a binuclear structure with phenolic oxygen bridges. The magnetic moment lies in the range 0.9–1.69 B.M

(for per Cu^{+2}) Table 1, and this is abnormally small consistent with a dimeric structure [18, 19].

The electronic spectra of the copper complexes Table 1 recorded in DMSO supported a near octahedral geometry for them and support the proposal that H$_2$O groups are coordinated axially to Cu (II) ions [10]. The spectrum of the Cu(II) complexes exhibits absorption bands at 11085, 16595, and 27990 cm^{-1}.

These bands may be considered to the following three spin allowed [20] transitions: $^2B_{1g} \rightarrow {}^2A_{1g}$ ($dx^2 - y^2 \rightarrow dz^2$), $^2B_{1g} \rightarrow {}^2B_{2g}$ ($dx^2 - y^2 \rightarrow dzy$), and $^2B \rightarrow {}^2Eg$ ($dx^2 - y^2 \rightarrow dxy, dyz$), and these transitions suggest D$_{4h}$ symmetry. The energy level sequence will depend on the amount of tetragonal distortion due to ligand field and Jahn Tellar distortion effect.

3.4. Co(II) Complexes.

At room temperature the magnetic moment measurements of Co(II) complexes at 4.90–5.04 B.M correspond to three unpaired electrons, Table 1.

The electronic spectra of all the Co(II) complexes display absorption at 10370, 14380, 18720, and 356960 cm^{-1}; these bands may be assigned to the following transitions: $^4T_{1g}$ (F) $\rightarrow {}^4T_{2g}$ (v_1), $^4T_{1g} \rightarrow {}^4A_{2g}$ (v_2), and $^4T_{1g}$(F) $\rightarrow {}^4T_{2g}$ (P) (v_3), respectively [21]. It is difficult to give the assignments for the fourth band, and it may be due to chargetransfer. The position of electronic spectral bands indicates that these complexes have distorted octahedral geometry [17, 21, 22].

3.5. The Proposed Structure.

According to the results obtained from (ir, uv/vis, molar ratio, molar conductivity, and atomic absorption) measurements for the prepared complexes, the proposed molecular structure of the complexes has an octahedral structure as shown in Figure 2.

Binuclear Cu(II) and Co(II) Complexes of Tridentate Heterocyclic Shiff Base Derived from Salicylaldehyde with 4-Aminoantipyrine

183

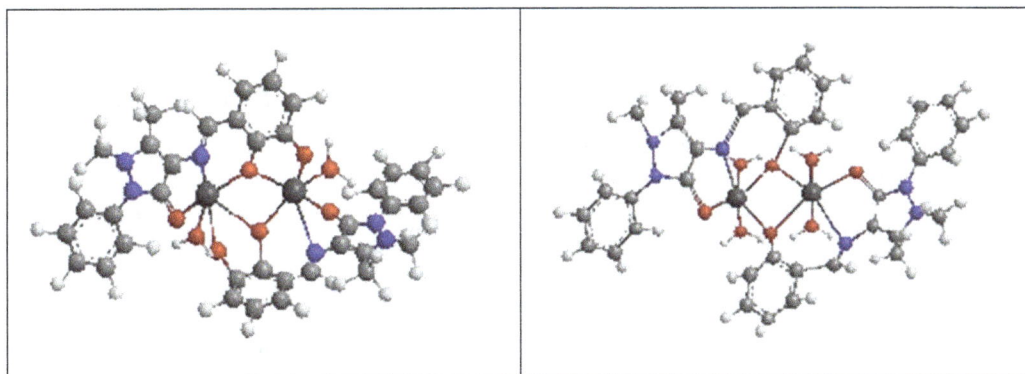

FIGURE 3: The optimized structural geometry of Cu(II) complexes.

TABLE 3: Structural parameters, bond length ($°A$), and angles ($°$) of the $[Cu_2(L2)_2(H_2O)_4]$ complex.

Parameters			
Bond lengths ($°A$)		Bond angles ($°$)	
		O(19)–Cu(50)–O(52)–H(85)	179.9998
Cu(136)–O(138)	1.5230	O(19)–Cu(50)–O(52)–H(86)	−0.0001
Cu(135)–O(137)	1.5230	O(24)–Cu(50)–O(52)–H(85)	174.1519
O(134)–Cu(135)	2.0400	O(24)–Cu(50)–O(52)–H(86)	−5.8480
O(129)–Cu(135)	2.3621	O(30)–Cu(50)–O(52)–H(85)	−40.8405
O(129)–Cu(136)	1.9129	O(30)–Cu(50)–O(52)–H(86)	139.1596
N(125)–Cu(136)	2.3774	N(39)–Cu(50)–O(52)–H(85)	9.7808
O(116)–Cu(136)	1.8092	N(39)–Cu(50)–O(52)–H(86)	−170.2191
O(110)–Cu(136)	1.3403	O(43)–Cu(50)–O(52)–H(85)	−3.4183
O(105)–Cu(136)	1.8419	O(43)–Cu(50)–O(52)–H(86)	176.5818
O(105)–Cu(135)	1.7708	O(6)–Cu(49)–O(51)–H(83)	−179.9999
N(101)–Cu(135)	1.3875	O(6)–Cu(49)–O(51)–H(84)	0.0011
O(92)–Cu(135)	1.5551	N(15)–Cu(49)–O(51)–H(83)	−122.7109
Cu(50)–O(52)	1.5228	N(15)–Cu(49)–O(51)–H(84)	57.2900
Cu(49)–O(51)	1.5227	O(19)–Cu(49)–O(51)–H(83)	41.5484
O(48)–Cu(49)	2.0398	O(19)–Cu(49)–O(51)–H(84)	−138.4507
C(47)–O(48)	1.4033	O(43)–Cu(49)–O(51)–H(83)	37.5946
C(46)–H(82)	1.1000	O(43)–Cu(49)–O(51)–H(84)	−142.4044
C(46)–C(47)	1.0149	O(48)–Cu(49)–O(51)–H(83)	37.5946
C(45)–H(81)	1.1000	O(48)–Cu(49)–O(51)–H(84)	142.4044
C(45)–C(46)	1.1328	C(47)–O(48)–Cu(49)–O(6)	−155.1509
C(44)–H(80)	1.1000	C(47)–O(48)–Cu(49)–N(15)	−34.7220
C(44)–C(45)	1.0148	C(47)–O(48)–Cu(49)–O(19)	55.8245
O(43)–Cu(49)	2.3617	C(47)–O(48)–Cu(49)–O(43)	52.8179
O(43)–Cu(50)	1.9130	C(47)–O(48)–Cu(49)–O(51)	−127.1821
O(30)–Cu(50)	1.8090	Cu(50)–O(43)–Cu(49)–O(48)	180.000
C(29)–C(37)	1.0468	Cu(50)–O(43)–Cu(49)–O(51)	180.000
C(28)–N(39)	1.0144	C(42)–O(43)–Cu(50)–O(19)	−150.107
C(28)–C(29)	1.1317	C(42)–O(43)–Cu(50)–O(24)	−143.209
C(27)–O(30)	1.1265	C(42)–O(43)–Cu(50)–O(30)	66.7882
C(27)–C(28)	0.9997	C(42)–O(43)–Cu(50)–N(39)	18.0306
N(26)–C(31)	0.9735	C(42)–O(43)–Cu(50)–O(52)	32.7885
N(26)–C(27)	1.0636	Cu(49)–O(43)–Cu(50)–O(19)	−2.8955
N(25)–C(38)	0.9597	Cu(49)–O(43)–Cu(50)–O(24)	4.0021
		Cu(49)–O(43)–Cu(50)–O(30)	−146.00

3.6. *Theoretical Study.* The ball and cylinders and some of selected structural parameters (bond length and angles) of the optimized geometries are shown in Figure 3 and Table 3.

As shown in this figure, there is no obvious trend for the variation of these parameters. The values of the bond length and angles of the optimized geometries are quite similar to the experimental results of the corresponding compounds.

3.7. *Antibacterial Studied.* The ligand and its transition metal complexes were evaluated against different species of bacteria

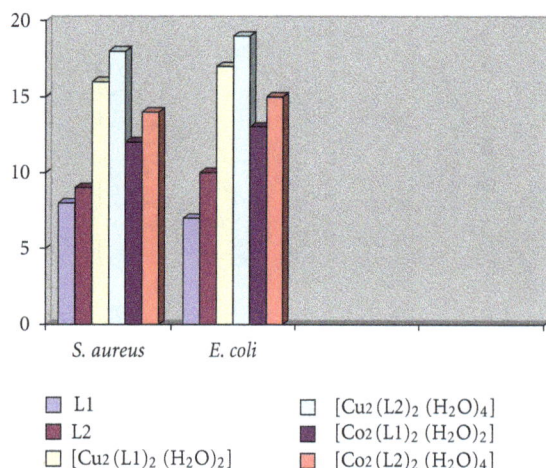

FIGURE 4: The effect of ligands and their metal complexes toward bacteria.

[23, 24]. The antibacterial action of the ligands and the complexes of Co(II) and Cu(II) was checked by the disc diffusion technique. This was done on *Staphylococcus aureus* (gram positive) and *Escherichia coli* (gram negative) bacteria at 25°C. The disc of whatman no. 4 filter paper having the diameter 6.00 mm was soaked in the solution of compounds in DMSO ($1.0 \, mg \, cm^{-1}$). After drying it was placed on nutrient agar plates. The inhibition area wasobserved after 48 hr. DMSO used as control Figure 4. The moderate effect was observed with Cu(II) complexes against *Staphylococcus aureus*, which is known as a resistant to most commercial antibiotic. The Schiff base ligands had more antibacterial activity than other ligands used, and this effect may be due to the presence of –ph, –OH and –N=C– groups which are electron releasing. The antibacterial results evidently showed that the activity of the ligand compounds became more pronounced when coordination to the metal ions.

References

[1] A. D. Garnovskii, "Ambident chelating ligands," *Zhurnal Neorganicheskoj Khimii*, vol. 43, no. 9, pp. 1491–1500, 1998.

[2] M. Sönmez, I. Berber, and E. Akbaş, "Synthesis, antibacterial and antifungal activity of some new pyridazinone metal complexes," *European Journal of Medicinal Chemistry*, vol. 41, no. 1, pp. 101–105, 2006.

[3] S. M. E. Khalil, H. S. Seleem, B. A. El-Shetary, and M. Shebl, "Mono- and bi-nuclear metal complexes of schiff-base hydrazone (ONN) derived from o-hydroxyacetophenone and 2-amino-4-hydrazino-6-methyl pyrimidine," *Journal of Coordination Chemistry*, vol. 55, no. 8, pp. 883–899, 2002.

[4] S. M. Saleem, S. Hussain, and Z. Nazir, "Gastric teratoma—a rare benign tumour of neonates," *Annals of Tropical Paediatrics*, vol. 23, no. 4, pp. 305–308, 2003.

[5] M. Weitzer and S. Brooker, "Bimetallic complexes of a structurally versatile pyridazine-containing Schiff-base macrocyclic ligand with pendant pyridine arms," *Dalton Transactions*, no. 14, pp. 2448–2454, 2005.

[6] P. O. Lumme and H. Knuuttila, "Studies on coordination compounds-VIII. Syntheses, structural, magnetic, spectral and thermal properties of some cobalt(II), nickel(II) and copper(II) complexes of 2-aminopyrimidine," *Polyhedron*, vol. 14, no. 12, pp. 1553–1563, 1995.

[7] M. Sönmez, A. Levent, and M. Şekerci, "Synthesis and characterization of Cu(II), Co(II), Ni(II), and Zn(II) complexes of a schiff base derived from 1-amino-5-benzoyl-4-phenyl-1H-pyrimidine-2-one and 3-Hydroxysalicylaldehyde," *Synthesis and Reactivity in Inorganic and Metal-Organic Chemistry*, vol. 33, no. 10, pp. 1747–1761, 2003.

[8] M. Sönmez, A. Levent, and M. Sekerci, "Synthesis, characterization, and thermal investigation of some metal complexes containing polydentate ONO-donor heterocyclic Schiff base ligand," *Koordinatsionnaya Khimiya*, vol. 30, no. 9, pp. 695–699, 2004.

[9] M. Sönmez, "Synthesis and spectroscopic studies of Cu(II), Co(II), Ni(II) and Zn(II) schiff base complexes from 1-amino-5-benzoyl-4-phenyl-1H pyrimidine-2-on with 2-hydroxynaphthaldehyde," *Polish Journal of Chemistry*, vol. 77, no. 4, pp. 397–402, 2003.

[10] M. Sönmez and M. Şekerci, "A new heterocyclic Schiff base and its metal complexes," *Synthesis and Reactivity in Inorganic and Metal-Organic Chemistry*, vol. 34, no. 3, pp. 489–502, 2004.

[11] E. Q. Gao, H. Y. Sun, D. Z. Liao, Z. H. Jiang, and S. P. Yan, "Synthesis of and magnetic interactions in binuclear Cu(II)-M(II) (M = Cu, Ni and Mn) complexes of macrocyclic oxamido ligands," *Polyhedron*, vol. 21, no. 4, pp. 359–364, 2002.

[12] H. H. Jaffé, "A reëxamination of the hammett equation," *Chemical Reviews*, vol. 53, no. 2, pp. 191–261, 1953.

[13] Z. H. Chohan and M. Praveen, "Biological role of anions (sulfate, nitrate, oxalate and acetate) on the antibacterial properties of cobalt(II) and nickel(II) complexes with pyrazinedicarboxaimide derived, furanyl and thienyl compounds," *Metal-Based Drugs*, vol. 6, no. 2, pp. 95–99, 1999.

[14] M. Sönmez and M. Sekerci, "complexes from 1-amino-5-benzoyl-4-phenyl-1H-pyrimidine-2-one with salicylaldehyde," *Polish Journal of Chemistry*, vol. 76, no. 7, pp. 907–914, 2002.

[15] W. J. Geary, "The use of conductivity measurements in organic solvents for the characterisation of coordination compounds," *Coordination Chemistry Reviews*, vol. 7, no. 1, pp. 81–122, 1971.

[16] K. Y. El-Baradie, "Mononuclear and binuclear Fe(III), Co(II), Ni(II), and Cu(II) complexes of 3,4′-dihydroxyazobenzene-3′,4-dicarboxylic acid and 3-carboxy-4-hydroxyphenylazo-3-carboxy-4-hydroxynaphthalene," *Monatshefte für Chemie*, vol. 136, no. 5, pp. 677–692, 2005.

[17] L. J. Bellamy, *The Infrared Spectra of Complex Molecules*, Chapman & Hall, London, UK, 1978.

[18] A. B. P. Lever, *Crystal Field Spectra. Inorganic Electronic Spectroscopy*, Academic Press, Amsterdam, The Netherlands, 1st edition, 1968.

[19] S. Chandra and K. Gupta, "Chromium(III), manganese(II), iron(III), cobalt(II), nickel(II) and copper(II) complexes with a pentadentate, 15-membered new macrocyclic ligand," *Transition Metal Chemistry*, vol. 27, no. 2, pp. 196–199, 2002.

[20] A. B. P. Lever, *Inorganic Electronic Spectroscopy*, Elsevier, Amsterdam, The Netherlands, 2nd edition, 1984.

[21] S. Chandra and K. Gupta, "Twelve-, fourteen- and sixteen-membered macrocyclic ligands and a study of the effect of ring size on ligand field strength," *Transition Metal Chemistry*, vol. 27, no. 3, pp. 329–332, 2002.

[22] K. Nakamoto, *Infrared Spectra of Inorganic and Coordination Compounds*, Wiley-Interscience, New York, NY, USA, 1970.

Binuclear Cu(II) and Co(II) Complexes of Tridentate Heterocyclic Shiff Base Derived from Salicylaldehyde with 4-Aminoantipyrine

185

[23] M. A. Pujar, B. S. Hadimani, S. Meenakumari, S. M. Gadded, and Y. F. Neelgund, *Current Science*, vol. 55, p. 353, 1986.

[24] S. Chandra and S. D. Sharma, "Template synthesis of copper(II) complexes of two twelve-membered tetradentate nitrogen donor macrocyclic ligands," *Journal of the Indian Chemical Society*, vol. 79, no. 6, pp. 495–497, 2002.

21

Synthesis, Crystal Structure, and DNA-Binding Studies of a Nickel(II) Complex with the Bis(2-benzimidazolymethyl)amine Ligand

Huilu Wu, Tao Sun, Ke Li, Bin Liu, Fan Kou, Fei Jia, Jingkun Yuan, and Ying Bai

School of Chemical and Biological Engineering, Lanzhou Jiaotong University, Lanzhou 730070, China

Correspondence should be addressed to Huilu Wu, wuhuilu@163.com

Academic Editor: Santiago Gómez-Ruiz

A V-shaped ligand Bis(2-benzimidazolymethyl)amine (bba) and its nickel(II) picrate (pic) complex, with composition $[Ni(bba)_2](pic)_2 \cdot 3MeOH$, have been synthesized and characterized on the basis of elemental analyses, molar conductivities, IR spectra, and UV/vis measurements. In the complex, the Ni(II) ion is six-coordinated with a N_2O_4 ligand set, resulting in a distorted octahedron coordination geometry. In addition, the DNA-binding properties of the Ni(II) complex have been investigated by electronic absorption, fluorescence, and viscosity measurements. The experimental results suggest that the nickel(II) complex binds to DNA by partial intercalation binding mode.

1. Introduction

Binding studies of small molecules to DNA are very important in the development of DNA molecular probes and new therapeutic reagents [1]. Transition metal complexes have attracted considerable attention as catalytic systems for use in the oxidation of organic compounds [2], probes in electron-transfer reactions involving metalloproteins [3], and intercalators with DNA [4]. Numerous biological experiments have demonstrated that DNA is the primary intracellular target of anticancer drugs; interaction between small molecules and DNA can cause damage in cancer cells, blocking the division and resulting in cell death [5–7].

Since the benzimidazole unit is the key-building block for a variety of compounds which have crucial roles in the functions of biologically important molecules, there is a constant and growing interest over the past few years for the synthesis and biological studies of benzimidazole derivatives [8–10]. Since the characterization of urease as a nickel enzyme in 1975, the knowledge of the role of nickel in bioinorganic chemistry has been rapidly expanding [11]. The interaction of Ni(II) complexes with DNA appears to be mainly dependent on the structure of the ligand exhibiting intercalative behavior [12–14].

In this context, we synthesized and characterized a novel Ni(II) complex. Moreover, we describe the interaction of the novel Ni(II) complex with DNA using electronic absorption and fluorescence spectroscopy and viscosity measurements.

2. Experimental

2.1. Materials and Methods. Calf thymus DNA (CT-DNA) and Ethidium bromide (EB) were purchased from Sigma Chemicals Co. (USA). All chemicals used were of analytical grade. All the experiments involving interaction of the ligand and the complexes with CT-DNA were carried out in doubly distilled water buffer containing 5 mM Tris and 50 mM NaCl and adjusted to pH 7.2 with hydrochloric acid. A solution of CT-DNA gave a ratio of UV absorbance at 260 and 280 nm of about 1.8–1.9, indicating that the CT-DNA was sufficiently free of protein [15]. The CT-DNA concentration per nucleotide was determined spectrophotometrically by employing an extinction coefficient of $6600 \, M^{-1} \, cm^{-1}$ at 260 nm [16].

Elemental analyses were performed on Carlo Erba 1106 elemental analyzer. The IR spectra were recorded in the $4000–400 \, cm^{-1}$ region with a Nicolet FT-VERTEX

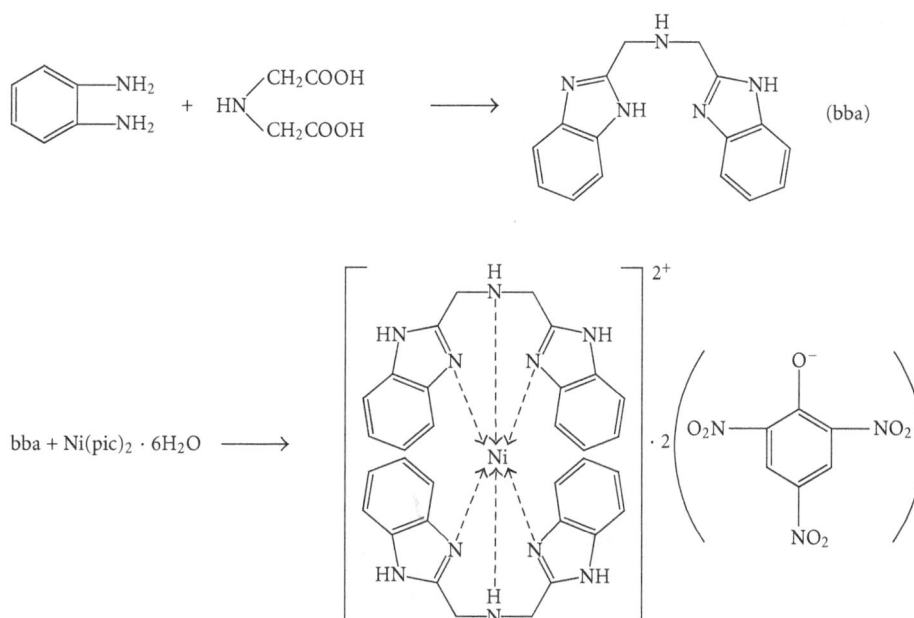

SCHEME 1: The synthesis of ligand bba and its Ni(II) complex.

70 spectrometer using KBr pellets. Electronic spectra were taken on a Lab-Tech UV Bluestar spectrophotometer. The fluorescence spectra were recorded on a 970-CRT spectrofluorophotometer. ^{1}HNMR spectra were obtained with a Mercury plus 400 MHz NMR spectrometer with TMS as internal standard and DMSO-d_6 as solvent. Electrolytic conductance measurements were made with a DDS-11A type conductivity bridge using a 10^{-3} mol·L^{-1} solution in DMF at room temperature.

2.2. Electronic Absorption Spectra.

Absorption titration experiment was performed with fixed concentrations of the complexes while gradually increasing concentration of CT-DNA. While measuring the absorption spectra, a proper amount of CT-DNA was added to both compound solution and the reference solution to eliminate the absorbance of CT-DNA itself. From the absorption titration data, the binding constant (K_b) was determined using [17]

$$\frac{[DNA]}{\varepsilon_a - \varepsilon_f} = \frac{[DNA]}{\varepsilon_b - \varepsilon_f} + \frac{1}{K_b(\varepsilon_b - \varepsilon_f)}, \quad (1)$$

where [DNA] is the concentration of DNA in base pairs, the apparent absorption coefficient, ε_a, ε_f, and ε_b correspond to $A_{obsd}/[M]$, the extinction coefficient of the free compounds and the extinction coefficient of the compound when fully bound to DNA, respectively. In plots of $[DNA]/(\varepsilon_a - \varepsilon_f)$ versus [DNA], K_b is given by the ratio of slope to the intercept.

2.3. Fluorescence Spectra.

EB emits intense fluorescence in the presence of CT-DNA due to its strong intercalation between the adjacent CT-DNA base pairs. It was previously reported that the enhanced fluorescence can be quenched by the addition of a second molecule [18]. The extent of fluorescence quenching of EB bound to CT-DNA can be used to determine the extent of binding between the second molecule and CT-DNA. The competitive binding experiments were carried out in the buffer by keeping [DNA]/[EB] = 1 and varying the concentrations of the compounds. The fluorescence spectra of EB were measured using an excitation wavelength of 520 nm and the emission range was set between 550 and 750 nm. The spectra were analyzed according to the classical Stern-Volmer equation [19],

$$\frac{I_0}{I} = 1 + K_{sv}[Q], \quad (2)$$

where I_0 and I are the fluorescence intensities at 599 nm in the absence and presence of the quencher, respectively, K_{sv} is the linear Stern-Volmer quenching constant, [Q] is the concentration of the quencher.

2.4. Viscosity Measurements.

Viscosity experiments were conducted on an Ubbelohde viscometer, immersed in a thermostated water-bath maintained at 25.0 ± 0.1°C. DNA samples approximately 200 bp in average length were prepared by sonicating in order to minimize complexities arising from DNA flexibility [20]. Titrations were performed for the compounds (3 mM), and each compound was introduced into the CT-DNA solution (50 μM) present in the viscometer. Data were presented as $(\eta - \eta_0)^{1/3}$ versus the ratio of the concentration of the compound to CT-DNA, where η is the viscosity of CT-DNA in the presence of the complex, and η_0 is the viscosity of CT-DNA alone. Viscosity values were calculated from the observed flow time of CT-DNA containing solutions corrected from the flow time of buffer alone (t_0), $\eta = (t - t_0)/t_0$.

2.5. Synthesis. The synthetic route for the ligand bba and its Ni(II) complex are shown in Scheme 1.

2.5.1. Bis(2-benzimidazolymethyl)amine (bba). The ligand bba was synthesized according to the procedure reported by Berends and Stephan [21]. The infrared spectra and UV spectra of the bba were almost consistent with the literature. Elemental analysis: $C_{16}H_{15}N_5$ (Mr = 277.33 g·mol^{-1}) calcd: C 69.30; H 5.45; N 25.26%; found: C 69.35; H 5.47; N 25.16%. IR (KBr, pellet, cm^{-1}): 1270s (ν_{C-N}), 1620s ($\nu_{C=N}$), UV-vis (λ, nm): 277, 283, ε_{277} = 5.99 × 10^2 L·mol^{-1}·cm^{-1}, ε_{283} = 5.73 × 10^2 L·mol^{-1}·cm^{-1}. ^{1}HNMR (DMSO-d_6, 300 MHz) δ: 12.3 (1H, N-H); 7.144 (m, 4H); 7.5 (d, 4H); 4.0 (s, 4H). Λ_M (DMF, 297 K): 1.29 S·cm^2·mol^{-1}.

2.5.2. [Ni(bba)$_2$](pic)$_2$·3MeOH. The ligand bba (0.4 mmol) and Ni(II) picrate (0.2 mmol) were dissolved in methanol (15 mL). A blue-green crystalline product which formed rapidly was filtered off, washed with methanol and absolute Et$_2$O, and dried in vacuo. The dried precipitate was dissolved in DMF resulting in a blue-green solution that was allowed to evaporate at room temperature. Blue-green crystals suitable for X-ray diffraction studies were obtained after one week. $C_{47}H_{36}N_{16}Ni\,O_{17}$ (Mr = 1155.63 g·mol^{-1}) calcd: C 48.85; H 3.14; N 19.39%; found: C 48.79; H 3.16; N 19.53%. IR (KBr, pellet, cm^{-1}): 1272s (ν_{C-N}), 1434 ($\nu_{C=N-C=C}$), 1487s ($\nu_{C=N}$), UV-vis (λ, nm): 275, 280, 407, ε_{275} = 6.55 × 10^2 L·mol^{-1}·cm^{-1}, ε_{280} = 6.50 × 10^2 L·mol^{-1}·cm^{-1}, ε_{407} = 7.99 × 10^2 L·mol^{-1}·cm^{-1}. Λ_M (DMF, 297 K): 128.5 S·cm^2·mol^{-1}.

2.6. Crystal Structure Determination. A suitable single crystal was mounted on a glass fiber and the intensity data were collected on a Bruker Smart CCD diffractometer with graphite-monochromated Mo Kα radiation (λ = 0.71073 Å) at 296 K. Data reduction and cell refinement were performed using the SMART and SAINT programs [22]. The structure was solved by direct methods and refined by full-matrix least squares against F^2 of data using SHELXTL software [23]. All H atoms were found in different electron maps and were subsequently refined in a riding-model approximation with C–H distances ranging from 0.95 to 0.99 Å. Basic crystal data, description of the diffraction experiment, and details of the structure refinement are given in Table 1.

3. Results and Discussion

The ligand bba and its Ni(II) complex are very stable in the air. They are remarkably soluble in polar solvents such as DMF, DMSO, and MeCN; slightly soluble in ethanol, methanol, ethyl acetate, and chloroform. The molar conductivities in DMF solution indicate that bba (1.29 S·cm^2·mol^{-1}) is nonelectrolyte compound and its Ni(II) complex is 1 : 2 electrolyte compound [24].

3.1. Spectral Characterization. In the bba ligand, a strong band is found at ca. 1270 cm^{-1} together along with a broad band at 1436 cm^{-1}. By analogy with the assigned bands of

TABLE 1: Crystallographic data and data collection parameters for the Ni(II) complex.

Complex	[Ni(bba)$_2$](pic)$_2$·3MeOH
Molecular formula	$C_{47}H_{36}N_{16}NiO_{17}$
Molecular weight	1155.63
Crystal system	Triclinic
Space group	P-1
a (Å)	10.4758 (9)
b (Å)	16.1097 (13)
c (Å)	17.2302 (14)
α (°)	107.5590 (10)
β (°)	107.5880 (10)
γ (°)	96.9150 (10)
V (Å^3)	2570.1 (4)
Z	2
ρ_{cald} (mg m^{-3})	1.493
Absorption coefficient (mm^{-1})	0.467
F (000)	1188
Crystal size (mm)	0.41 × 0.38 × 0.31
θ range for data collection (°)	2.04–25.00
h/k/l (max, min)	−12, 12/−16, 19/−20, 20
Reflections collected	18579
Independent reflections	8974 [R(int) = 0.0203]
Data/restraints/parameters	8974/6/746
Goodness-of-fit on F^2	1.097
Final R_1, wR_2 indices [$I > 2\sigma(I)$]	0.0383, 0.1135
R_1, wR_2 indices (all data)	0.0466, 0.1194
Largest differences peak and hole (eÅ^{-3})	0.734 and −0.384

imidazole, the former can be attributed to ν(C=N–C=C), while the latter can be attributed to ν(C=N) [25–27]. One of them shift to the higher frequency by around 41 cm^{-1} in the complex, which implies direct coordination of all three imine nitrogen atoms to metal ions. This is the preferred nitrogen atom for coordination as found for other metal complexes with benzimidazoles [28]. Information regarding the possible bonding modes of the picrate and benzimidazole rings may also be obtained from the IR spectra, such as 709, 744, 1272, 1363, 1434, 1487, and 1633 cm^{-1} [29]. This fact agrees with the result determined by X-ray diffraction.

DMF solutions of ligand bba and its complexes show, as expected, almost identical UV spectra. The UV bands of bba (275, 280 nm) are only marginally blue shifted (1-2 nm) in the complexes, which is clear evidence of C=N coordination to the metal ions center. The absorption bands are assigned to $\pi \rightarrow \pi^*$ (imidazole) transitions. The bands of picrate (407 nm) are assigned to $\pi \rightarrow \pi^*$ transitions.

3.2. Crystal Structure of [Ni(bba)$_2$](pic)$_2$·3MeOH. The molecular structure of the Ni(II) complex is shown in Figure 1, selected bond lengths and angles are summarized in Table 2. The Ni(II) atom is six-coordinate with a NiN$_4$O$_2$ environment. The bba ligand acts as a tridentate N-donor and O-donor. The coordination geometry of the Ni(II)

Synthesis, Crystal Structure, and DNA-Binding Studies of a Nickel(II) Complex with the Bis(2-benzimidazolymethyl) amine Ligand

189

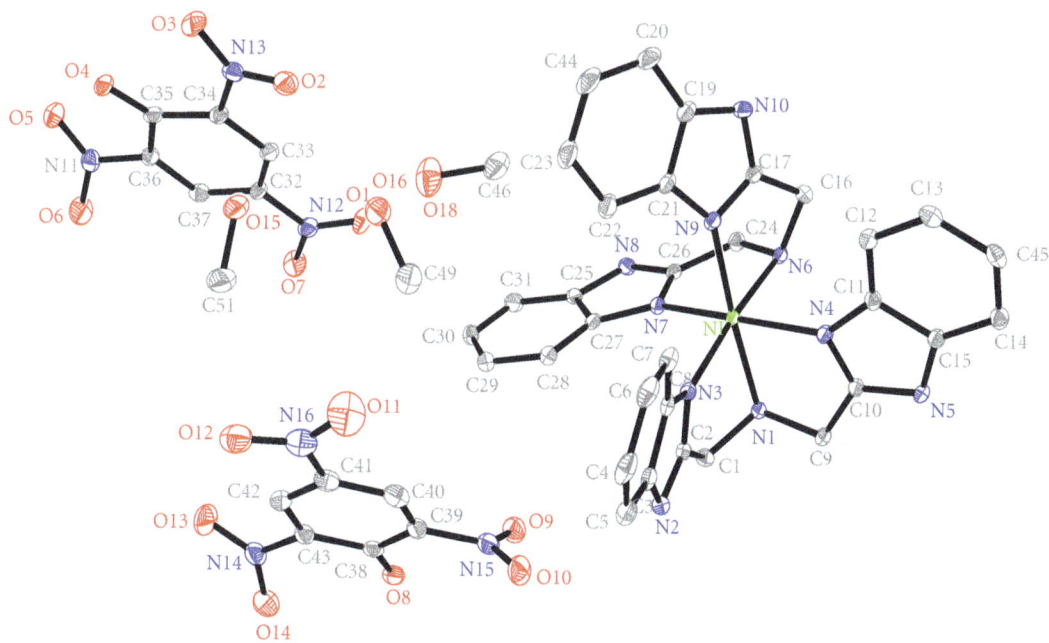

FIGURE 1: The molecular structure of the Ni(II) complex showing displacement ellipsoids at the 30% probability level. Hydrogen atoms have been omitted for clarity.

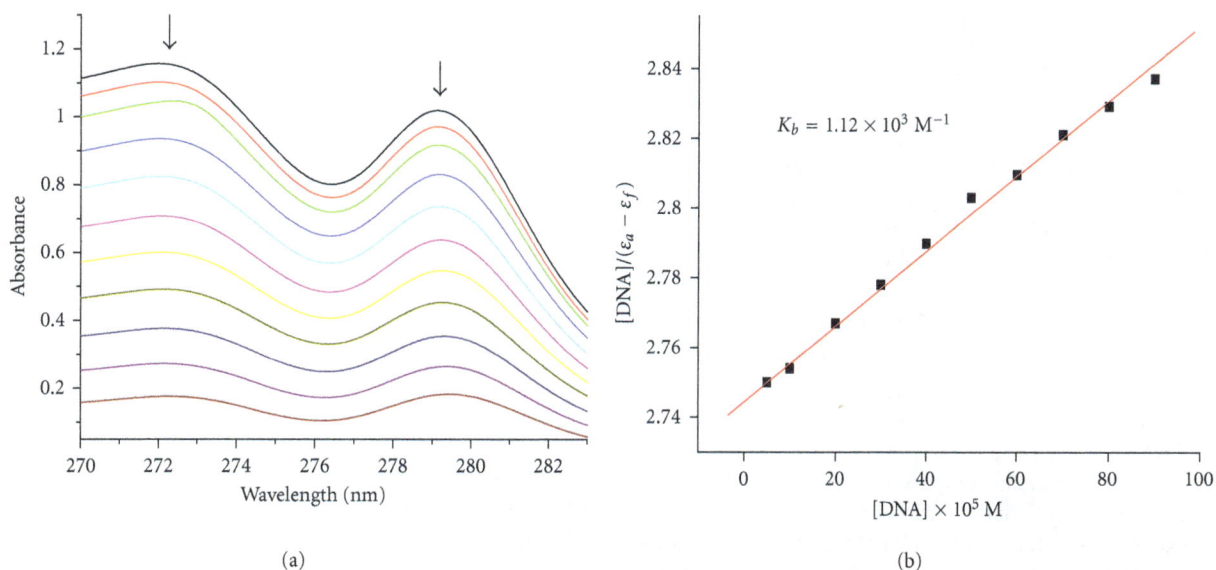

(a)

(b)

FIGURE 2: Electronic spectra of the Ni(II) complex (30 μM) in the presence of 0, 5, 10, 20, 30, 40, 50, 60, 70, 80, and 90 μL CT-DNA. [DNA] = 2.5 $\times 10^{-5}$ M. Arrow shows the absorbance changes upon increasing CT-DNA concentration. Plots of [DNA]/$(\varepsilon_a - \varepsilon_f)$ versus [DNA] for the titration of the Ni(II) complex with CT-DNA.

TABLE 2: Selected bond lengths (Å) and angles (deg) of the Ni(II) complex.

Bond lengths					
Ni–N(1)	2.1647 (19)	Ni–N(4)	2.0793(18)	Ni–N(6)	2.1788(19)
Ni–N(3)	2.0899 (19)	Ni–N(7)	2.0667 (18)	Ni–N(9)	2.0628 (19)
Bond angles					
N(1)–Ni–N(6)	94.12 (7)	N(9)–Ni–N(7)	173.40 (7)	N(9)–Ni–N(4)	98.11 (7)
N(3)–Ni–N(7)	173.40 (7)	N(3)–Ni–N(1)	79.29 (7)	N(7)–Ni–N(4)	166.55 (7)
N(9)–Ni–N(3)	107.52 (7)	N(7)–Ni–N(3)	98.95 (7)	N(3)–Ni–N(4)	89.98 (7)
N(1)–Ni–N(9)	173.19 (7)	N(7)–Ni–N(1)	90.23 (7)	N(1)–Ni–N(4)	81.52 (7)
N(9)–Ni–N(6)	79.07 (8)	N(7)–Ni–N(6)	81.11 (7)	N(4)–Ni–N(6)	88.87 (7)

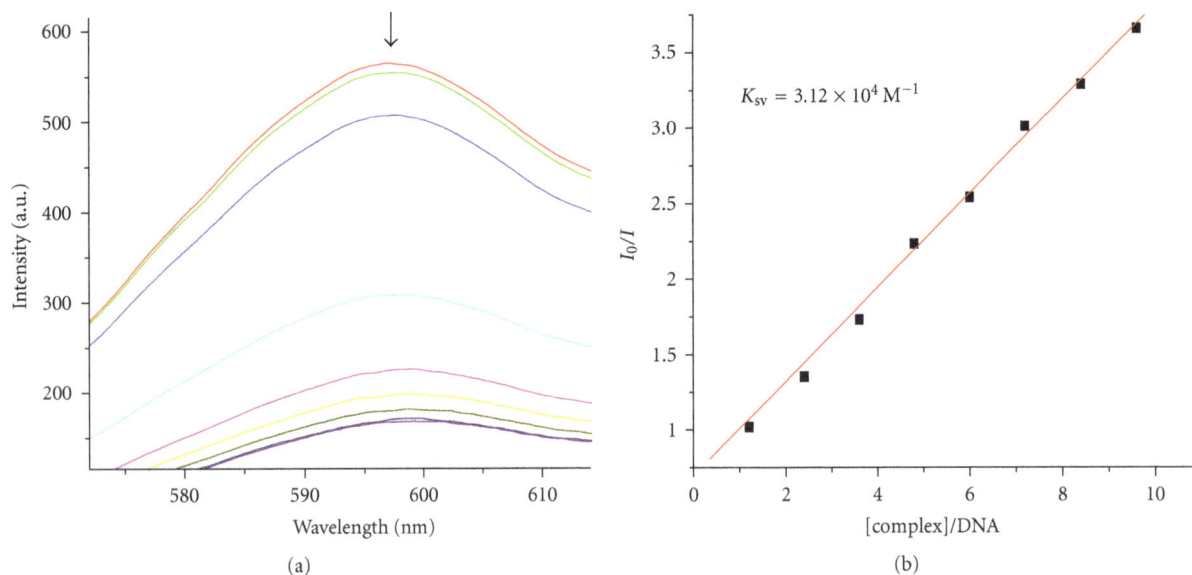

FIGURE 3: Emission spectra of EB bound to DNA in the presence of the complex. [Complex] $= 3 \times 10^{-3}$ M; $\lambda_{ex} = 520$ nm. The arrow shows the intensity changes upon increasing concentrations of the complex. Fluorescence quenching curves of EB bound to CT-DNA by the Ni(II) complex. (Plots of I_0/I versus [Complex]/DNA.).

may be best described as distorted octahedral with four coordination nitrogen atoms from an ideal equatorial plane. The maximum deviation (N9) from the plane containing these four N atoms is 0.764 Å. The bond average length between the Ni ion and the apical N atom (N1, N6) is 2.171 Å, which is about 0.097 Å longer than the bond average length between the Ni ion and four coordination N atoms from an equatorial plane. This geometry is assumed by the Ni(II) to relieve the steric crowding. Therefore, compared with a regular octahedron, it reflects a relatively distorted coordination octahedron around Ni(II).

3.3. Spectral Studies of the Interactions with DNA

3.3.1. Electronic Absorption Titration. Electronic absorption spectroscopy is universally employed to determine the binding characteristics of metal complexes with DNA [30–32]. The absorption spectra of the Ni(II) complex in the absence and presence of CT-DNA are given in Figure 2. There are two well-resolved bands at about 272, 278 nm for the complex. The λ for the ligand increases only from 272 to 273, and for the complex from 278 to 279 nm, a slight red shift about 1 nm under identical experimental conditions. The slight red shift suggests that the Ni(II) complex interacts with DNA [33].

The binding constant K_b for the complex has been determined from the plot of [DNA]/($\varepsilon_a - \varepsilon_f$) versus [DNA] and was found to be 1.12×10^3 M^{-1}. Compared with those of the so-called DNA-intercalative ruthenium complexes (1.1×10^4–4.8×10^4 M^{-1}) [34], the binding constants (K_b) of the Ni(II) complex suggest that the complex with DNA with an affinity is less than the classical intercalators.

3.3.2. Fluorescence Spectroscopic Studies. intensity in the EB-DNA adduct allows determination of the affinity of the

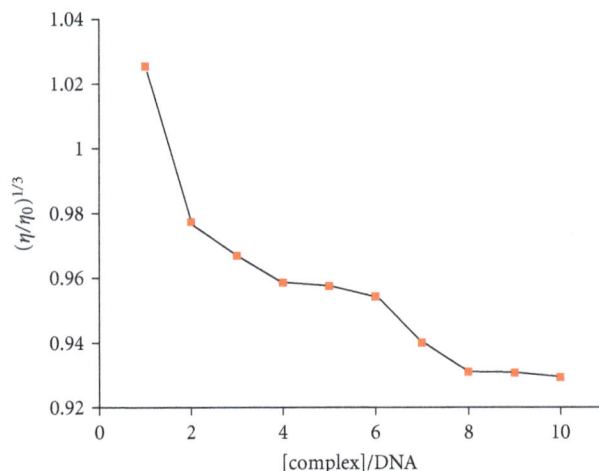

FIGURE 4: Effect of increasing amounts of the Ni(II) complex on the relative viscosity of CT-DNA at $25.0 \pm 0.1°$C.

complex for DNA, whatever the binding mode may be. If a complex can replace EB from DNA-bound EB, the fluorescence of the solution will be quenched due to the fact that free EB molecules are readily quenched by the surrounding water molecules [35]. For all the compounds, no emission was observed either alone or in the presence of CT-DNA in the buffer. The fluorescence quenching of EB bound to CT-DNA by the Ni(II) complex is shown in Figure 3. The quenching of EB bound to CT-DNA by the Ni(II) complex is in good agreement with the linear Stern-Volmer equation, which provides further evidence that the Ni(II) complex bind to DNA. The quenching plots illustrate that the quenching of EB bound to DNA by the complex is in good agreement with the linear Stern-Volmer equation, which also proves that the complex binds to DNA. The K_{sv}

Synthesis, Crystal Structure, and DNA-Binding Studies of a Nickel(II) Complex with the Bis(2-benzimidazolymethyl) amine Ligand

191

value for the Ni(II) complex is $3.12 \times 10^4 \, M^{-1}$. The data suggest that the Ni(II) complex interacts with DNA.

3.3.3. Viscosity Studies.

Optical photophysical techniques are widely used to study the binding model of the ligand, metal complexes, and DNA but not to give sufficient clues to support a binding model. Therefore, viscosity measurements were carried out to further clarify the interaction of metal complexes and DNA. Hydrodynamic measurements that are sensitive to the length change (i.e., viscosity and sedimentation) are regarded as the least ambiguous and the most critical tests of a binding model in solution in the absence of crystallographic structural data [15, 20]. A classical intercalative mode causes a significant increase in viscosity of DNA solution due to increase in separation of base pairs at intercalation sites and hence an increase in overall DNA length. By contrast, complexes that bind exclusively in the DNA grooves by partial and/or nonclassical intercalation, under the same conditions, typically cause less pronounced (positive or negative) or no change in DNA solution viscosity [20]. The values of $(\eta - \eta_0)^{1/3}$ were plotted against [compound]/[DNA] (Figure 4). For the Ni(II) complex, as increasing the amounts of compound, the viscosity of DNA decreases steadily. The decreased relative viscosity of DNA may be explained by a binding mode which produced bends or kinks in the DNA and thus reduced its effective length and concomitantly its viscosity. The results suggest that the Ni(II) complex may bind to DNA by partial intercalation.

4. Conclusions

In this paper, a new Ni(II) complex has been synthesized and characterized. Moreover, the DNA-binding properties of the Ni(II) complex were investigated by electronic absorption, fluorescence, and viscosity measurements. The experimental results indicate that the Ni(II) complex can bind to CT-DNA by partial intercalation mode. Information obtained from our study will be helpful to understand the mechanism of interactions of benzimidazoles and their complexes with nucleic acids and should be useful in the development of potential probes of DNA structure and conformation.

Appendix

Additional Data

CCDC 825141 contains the additional crystallographic data for this paper. These data can be obtained free of charge from The Cambridge Crystallographic Data Centre via http://www.ccdc.cam.ac.uk/data_request/cif.

Acknowledgments

The authors acknowledge the financial support and a grant from "Qing Lan" Talent Engineering Funds by Lanzhou Jiaotong University. The grant from "Long Yuan Qing Nian" of Gansu Province is also acknowledged.

References

[1] M. Mrksich and P. B. Dervan, "Antiparallel side-by-side heterodimer for sequence-specific recognition in the minor groove of DNA by a distamycin/1-methylimidazole-2-carboxamide-netropsin pair," *Journal of the American Chemical Society*, vol. 115, no. 7, pp. 2572–2576, 1993.

[2] C. Kokubo and T. Katsuki, "Highly enantioselective catalytic oxidation of alkyl aryl sulfides using Mn-salen catalyst," *Tetrahedron*, vol. 52, no. 44, pp. 13895–13900, 1996.

[3] S. Schoumacker, O. Hamelin, J. Pécaut, and M. Fontecave, "Catalytic asymmetric sulfoxidation by chiral manganese complexes: acetylacetonate anions as chirality switches," *Inorganic Chemistry*, vol. 42, no. 24, pp. 8110–8116, 2003.

[4] C. M. Dupureur and J. K. Barton, "Structural Studies of Λ-and Δ-[Ru(phen)$_2$dppz]$^{2+}$ Bound to d(GTCGAC)$_2$: characterization of Enantioselective Intercalation," *Inorganic Chemistry*, vol. 36, no. 1, pp. 33–43, 1997.

[5] C. Hemmert, M. Pitié, M. Renz, H. Gornitzka, S. Soulet, and B. Meunier, "Preparation, characterization and crystal structures of manganese(II), iron(III) and copper(II) complexes of the bis[di-1,1-(2-pyridyl)ethyl] amine (BDPEA) ligand; evaluation of their DNA cleavage activities," *Journal of Biological Inorganic Chemistry*, vol. 6, no. 1, pp. 14–22, 2001.

[6] V. S. Li, D. Choi, Z. Wang, L. S. Jimenez, M. Tang, and H. Kohn, "Role of the C-10 substituent in mitomycin C-1-DNA bonding," *Journal of the American Chemical Society*, vol. 118, no. 10, pp. 2326–2331, 1996.

[7] G. Zuber, J. C. Quada, and S. M. Hecht, "Sequence selective cleavage of a DNA octanucleotide by chlorinated bithiazoles and bleomycins," *Journal of the American Chemical Society*, vol. 120, no. 36, pp. 9368–9369, 1998.

[8] A. Gellis, H. Kovacic, N. Boufatah, and P. Vanelle, "Synthesis and cytotoxicity evaluation of some benzimidazole-4,7-diones as bioreductive anticancer agents," *European Journal of Medicinal Chemistry*, vol. 43, no. 9, pp. 1858–1864, 2008.

[9] Ö. Ö. Güven, T. Erdoğan, H. Göker, and S. Yildiz, "Synthesis and antimicrobial activity of some novel phenyl and benzimidazole substituted benzyl ethers," *Bioorganic and Medicinal Chemistry Letters*, vol. 17, no. 8, pp. 2233–2236, 2007.

[10] K. Kopańska, A. Najda, J. Zebrowska et al., "Synthesis and activity of 1*H*-benzimidazole and 1*H*-benzotriazole derivatives as inhibitors of Acanthamoeba castellanii," *Bioorganic and Medicinal Chemistry*, vol. 12, no. 10, pp. 2617–2624, 2004.

[11] K. C. Skyrianou, F. Perdih, I. Turel, D. P. Kessissoglou, and G. Psomas, "Nickel-quinolones interaction—part 2: interaction of nickel(II) with the antibacterial drug oxolinic acid," *Journal of Inorganic Biochemistry*, vol. 104, no. 2, pp. 161–170, 2010.

[12] K. C. Skyrianou, C. P. Raptopoulou, V. Psycharis, D. P. Kessissoglou, and G. Psomas, "Structure, cyclic voltammetry and DNA-binding properties of the bis(pyridine)bis(sparfloxacinato)nickel(II) complex," *Polyhedron*, vol. 28, no. 15, pp. 3265–3271, 2009.

[13] Y. Jin, M. A. Lewis, N. H. Gokhale, E. C. Long, and J. A. Cowan, "Influence of stereochemistry and redox potentials on the single- and double-strand DNA cleavage efficiency of Cu(II)-and Ni(II)·Lys-Gly-his-derived ATCUN metallopeptides," *Journal of the American Chemical Society*, vol. 129, no. 26, pp. 8353–8361, 2007.

[14] F. Bisceglie, M. Baldini, M. Belicchi-Ferrari et al., "Metal complexes of retinoid derivatives with antiproliferative activity: synthesis, characterization and DNA interaction studies," *European Journal of Medicinal Chemistry*, vol. 42, no. 5, pp. 627–634, 2007.

[15] J. B. Chaires, "Tris(phenanthroline)ruthenium(II) enantiomer interactions with DNA: mode and specificity of binding," *Biochemistry*, vol. 32, no. 10, pp. 2573–2584, 1993.

[16] J. Marmur, "A procedure for the isolation of deoxyribonucleic acid from microorganisms," *Methods in Enzymology*, vol. 6, pp. 726–738, 1963.

[17] A. Wolfe, G. H. Shimer, and T. Meehan, "Polycyclic aromatic hydrocarbons physically intercalate into duplex regions of denatured DNA," *Biochemistry*, vol. 26, no. 20, pp. 6392–6396, 1987.

[18] M. Chauhan, K. Banerjee, and F. Arjmand, "DNA binding studies of novel copper(II) complexes containing L-tryptophan as chiral auxiliary: in vitro antitumor activity of Cu-Sn$_2$ complex in human neuroblastoma cells," *Inorganic Chemistry*, vol. 46, no. 8, pp. 3072–3082, 2007.

[19] J. R. Lakowicz and G. Weber, "Quenching of fluorescence by oxygen. A probe for structural fluctuations in macromolecules," *Biochemistry*, vol. 12, no. 21, pp. 4161–4170, 1973.

[20] S. Satyanarayana, J. C. Dabrowiak, and J. B. Chaires, "Neither Δ- nor Λ-tris(phenanthroline)ruthenium(II) binds to DNA by classical intercalation," *Biochemistry*, vol. 31, no. 39, pp. 9319–9324, 1992.

[21] H. P. Berends and D. W. Stephan, "Copper(I) and copper(II) complexes of biologically relevant tridentate ligands," *Inorganica Chimica Acta*, vol. 93, no. 4, pp. 173–178, 1984.

[22] Bruker, *Smart Saint and Sadabs*, Bruker Axs, Inc., Madison, Wisc, USA, 2000.

[23] G. M. Sheldrick, *SHELXTL*, Siemmens Analytical X-Ray Instruments, Inc., Madison, Wisc, USA, 1996.

[24] W. J. Geary, "The use of conductivity measurements in organic solvents for the characterisation of coordination compounds," *Coordination Chemistry Reviews*, vol. 7, no. 1, pp. 81–122, 1971.

[25] C. Y. Su, B. S. Kang, C. X. Du, Q. C. Yang, and T. C. W. Mak, "Formation of mono-, bi-, tri-, and tetranuclear Ag(I) complexes of C_3-symmetric tripodal benzimidaxole ligands," *Inorganic Chemistry*, vol. 39, no. 21, pp. 4843–4849, 2000.

[26] R. J. Sundberg and R. B. Martin, "Interactions of histidine and other imidazole derivatives with transition metal ions in chemical and biological systems," *Chemical Reviews*, vol. 74, no. 4, pp. 471–517, 1974.

[27] V. McKee, M. Zvagulis, and C. A. Reed, "Further insight into magnetostructural correlations in binuclear copper(II) species related to methemocyanin: X-ray crystal structure of a 1,2-μ-nitrito complex," *Inorganic Chemistry*, vol. 24, no. 19, pp. 2914–2919, 1985.

[28] T. J. Lane, I. Nakagawa, J. L. Walter, and A. J. Kandathil, "Infrared investigation of certain imidazole derivatives and their metal chelates," *Inorganic Chemistry*, vol. 1, pp. 267–276, 1962.

[29] H. Wu, R. Yun, K. Li, K. Wang, X. Huang, and T. Sun, "Synthesis, crystal structure and spectra properties of the nickel (II) complex with 1,3-bis(1-benzylbenzimidazol2-yl)-2-oxopropane," *Synthesis and Reactivity in Inorganic, Metal-Organic and Nano-Metal Chemistry*, vol. 39, no. 9, pp. 614–617, 2009.

[30] H. Li, X. Y. Le, D. W. Pang, H. Deng, Z. H. Xu, and Z. H. Lin, "DNA-binding and cleavage studies of novel copper(II) complex with L-phenylalaninate and 1,4,8,9 tetra aza-triphenylene ligands," *Journal of Inorganic Biochemistry*, vol. 99, no. 11, pp. 2240–2247, 2005.

[31] V. G. Vaidyanathan and B. U. Nair, "Synthesis, characterization, and DNA binding studies of a chromium(III) complex containing a tridentate ligand," *European Journal of Inorganic Chemistry*, no. 19, pp. 3633–3638, 2003.

[32] V. G. Vaidyanathan and B. U. Nair, "Nucleobase oxidation of DNA by (terpyridyl)chromium(III) derivatives," *European Journal of Inorganic Chemistry*, no. 9, pp. 1840–1846, 2004.

[33] J. Liu, T. Zhang, T. Lu et al., "DNA-binding and cleavage studies of macrocyclic copper(II) complexes," *Journal of Inorganic Biochemistry*, vol. 91, no. 1, pp. 269–276, 2002.

[34] A. M. Pyle, J. P. Rehmann, R. Meshoyrer, C. V. Kumar, N. J. Turro, and J. K. Barton, "Mixed-ligand complexes of ruthenium(II): factors governing binding to DNA," *Journal of the American Chemical Society*, vol. 111, no. 8, pp. 3051–3058, 1989.

[35] B. C. Baguley and M. Le Bret, "Quenching of DNA-ethidium fluorescence by amsacrine and other antitumor agents: a possible electron-transfer effect," *Biochemistry*, vol. 23, no. 5, pp. 937–943, 1984.

22

Alkene Cleavage Catalysed by Heme and Nonheme Enzymes: Reaction Mechanisms and Biocatalytic Applications

Francesco G. Mutti

Department of Chemistry, Organic and Bioorganic Chemistry, University of Graz, Heinrichstrasse 28, 8010 Graz, Austria

Correspondence should be addressed to Francesco G. Mutti, francesco.mutti@uni-graz.at

Academic Editor: Ian Butler

The oxidative cleavage of alkenes is classically performed by chemical methods, although they display several drawbacks. Ozonolysis requires harsh conditions ($-78°C$, for a safe process) and reducing reagents in a molar amount, whereas the use of poisonous heavy metals such as Cr, Os, or Ru as catalysts is additionally plagued by low yield and selectivity. Conversely, heme and nonheme enzymes can catalyse the oxidative alkene cleavage at ambient temperature and atmospheric pressure in an aqueous buffer, showing excellent chemo- and regioselectivities in certain cases. This paper focuses on the alkene cleavage catalysed by iron cofactor-dependent enzymes encompassing the reaction mechanisms (in case where it is known) and the application of these enzymes in biocatalysis.

1. Introduction

The oxidative cleavage of alkenes is a widely employed method in synthetic chemistry, particularly to introduce oxygen functionalities into molecules, remove protecting groups, and degrade large molecules. Moreover, the synthesis of a large amount of bioactive compounds involves the alkene cleavage as a key step. Ozonolysis is the most employed chemical method for cleaving alkenes since it is considered the most efficient and cleanest. However, the ozonolysis requires harsh conditions such as low temperature (ca. $-78°C$), hence imposing the use of a special equipment (e.g., ozoniser) and reducing reagents in molar amounts during the workup [1]. Furthermore, safety hazards complicate this reaction on large scale, and serious accidents from explosion have been reported [2, 3]. Alternative protocols envisage the use of poisonous heavy metals such as Cr, Os, or Ru which are plagued by mediocre yields and selectivities [4–6]. Conversely, enzymes can activate the most innocuous oxidant, that is, molecular oxygen, and catalyse the alkene cleavage at ambient temperature and atmospheric pressure in aqueous buffer. Besides, in certain cases enzymes are capable to cleave olefinic functionalities in high chemo- and regioselective fashion allowing biocatalysis to compete with chemical methods [7–9].

Otherwise, the rising popularity of natural products during the last decade has triggered off remarkable research activities regarding the use of biocatalysis for the production of flavour compounds [10]. In fact, products derived from the bioprocess of natural substrates (i.e., using wild-type microorganisms or isolated enzymes thereof) are defined as natural. The tag natural was one of the main reasons for seeking biochemical routes to high-priced natural flavours such as vanillin, and nowadays biocatalysis constitutes a convenient alternative to synthesise them.

This paper focuses on the alkene cleavage catalysed by iron cofactor-dependent enzymes encompassing the reaction mechanisms (in case where it is known) and the application of these enzymes in biocatalysis. In the first part heme peroxidases are examined, for which the alkene cleavage is principally a promiscuous activity. In the second part heme and nonheme oxygenases are discussed, for which a more detailed survey concerning the reaction mechanisms is available in literature.

2. Alkene Cleavage by Heme Peroxidases

Peroxidases are ubiquitously found in microorganisms, plants, and animals; these enzymes are named after their natural sources such as horseradish peroxidase, lactoperoxidase,

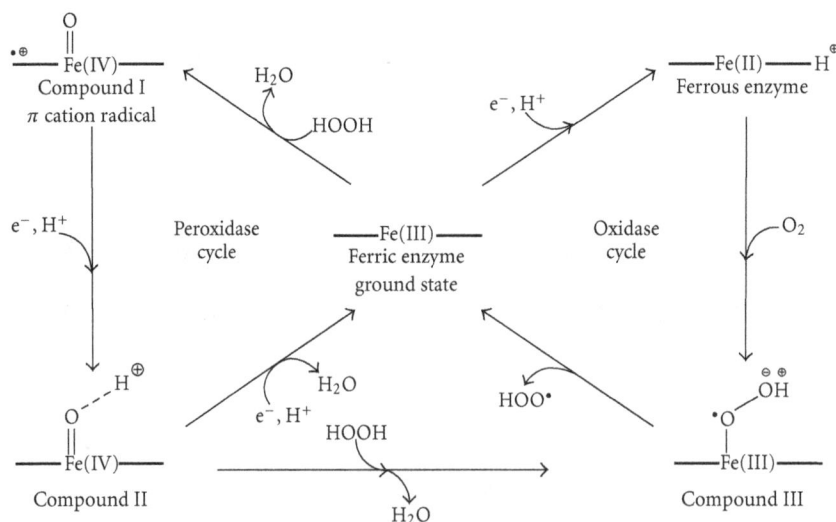

SCHEME 1: The five oxidation states of peroxidases.

myeloperoxidase or after their natural substrates such as cytochrome *c*, chloroperoxidase, and lignin peroxidase. The principally studied peroxidases are heme enzymes, hence possessing a ferric protoporphyrin IX (protoheme) as the prosthetic group [11]. Heme-containing enzymes participate in a strikingly diverse range of chemistry; yet all biological oxidation reactions catalysed by these enzymes involve very similar high oxidation state intermediates whose reactivity is modulated by the protein environment (Scheme 1) [12, 13].

Consequently, peroxidases are extremely promiscuous enzymes since they catalyse diverse chemical transformations such as peroxidase, peroxygenase, and oxidase reactions, acting on a vast array of substrates including phenols, aromatic and alkyl amines, NAD(P)H, H_2O_2, peracids, alkenes, thioethers, and aldehydes [11]. Particularly, it was shown that peroxidases catalyse the aerobic oxidation of aldehydes to yield carbonylic products via an enol tautomer, therefore postulating the formation of a dioxetane intermediate (α-oxidation) [14]. This mechanism somehow resembles the alkene cleaving activity, which has been observed as a side reaction in various peroxidases. Depending on the enzyme and the specific substrate involved, few speculative reaction mechanisms for the aerobic alkene cleavage of alkenes leading to aldehydes have been proposed, albeit a catalytic cycle has not been proved to date. A survey of alkene cleavage reactions catalysed by peroxidases is presented in the following sections.

2.1. Chloroperoxidase. Chloroperoxidase (CPO) was isolated from *Caldariomyces fumago* [15, 16], and it is one of the most versatile and promising heme enzymes for synthetic applications [17–20]. For instance, various transformations typical for catalases and cytochrome P-450 monooxygenases are also catalysed by CPO [21]. Furthermore, the enzyme catalyses halide-dependent as well as halide-independent reactions acting on various substrates and using hydrogen peroxide or organic peroxides as oxygen source [11]. Other studies have

shown that CPO catalyses the epoxidation of a number of functionalised or unfunctionalised olefins with high degree of enantio- and diastereoselectivity [22–24]. Epoxidation was often accompanied by the formation of aldehydes as well as by allylic hydroxylation [18]. Bougioukou and Smonou reported the alkene cleaving activity for the oxidation of conjugated dienoic esters employing CPO and *tert*-butyl-hydroperoxide (*t*BHP) as terminal oxidant [25, 26]. The reactions were carried out on alkenes with *cis*- and *trans*-configuration under anaerobic and aerobic conditions. In absence of molecular oxygen only two reactions occurred: the allylic oxidation and the epoxidation to give compounds **1** and **2**, respectively (Scheme 2). Both reactions proceeded with a high degree of regioselectivity, since the C=C bond proximal to the ester moiety was not converted. Surprisingly, the reaction performed under aerobic conditions furnished an additional aldehyde **3** as a minor product (13%) via the cleavage of the terminal C=C double bond. Also in this case, the alkene cleavage proceeded with perfect regioselectivity.

Moreover, the relative amounts of the products depended on the stereochemistry of the double bond, since the allylic aldehyde was the preferred product starting from *trans*-dienes. Nevertheless, the same reaction performed on methyl-(2Z,4Z)-hexadecanoate furnished the cleaved alkene as the main product (38%), followed by the epoxide (35%) and the allylic aldehyde (27%). The proposed explanation for the formation of all products, included 3 under aerobic but not anaerobic conditions, involves the formation of an intermediate radical cation (**4**). This is probably generated by a direct electron transfer from the substrate to the formally oxoiron (V) centre (compound I, see Scheme 1) formed in the CPO catalytic cycle [12, 27]. Alternatively a similarly generated *t*-BuOO• radical may act as a mediator, hence abstracting the electron from the π C=C bond of the substrate. Finally, the radical intermediate **4** can react with dioxygen, leading to the cleavage product via a dioxetane intermediate (**5**) (Scheme 3).

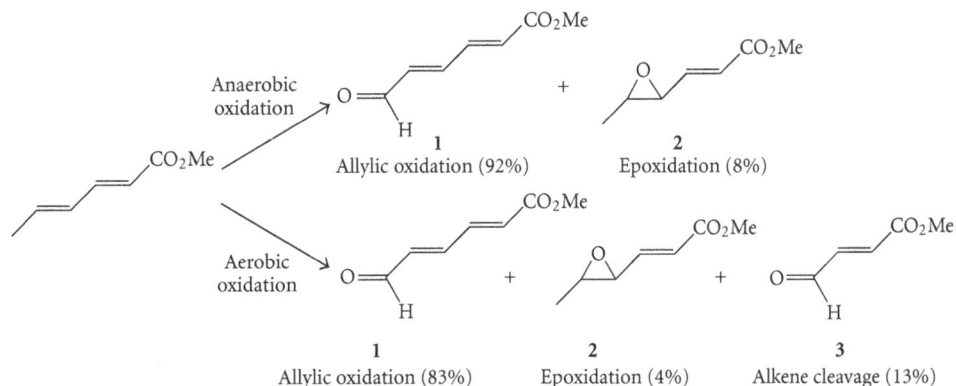

SCHEME 2: Aerobic and anaerobic CPO-catalysed oxidation of methyl-(2E,4E)-hexadienoate.

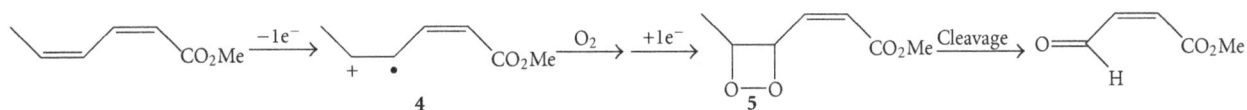

SCHEME 3: Proposed mechanism for the aerobic alkene cleavage catalysed by CPO.

2.2. Horseradish Peroxidase.

Horseradish peroxidase (HRP) is the most studied and well-characterised peroxidase, whose crystal structure [28] and catalytic pathway have been elucidated at high resolution [12]. HRP catalyses the oxidation of phenols, anilines, and a variety of other electron-rich compounds at the expense of H_2O_2 and alkyl hydroperoxides [11]. Ortiz De Montellano and Grab observed the oxidation of styrene to styrene oxide and benzaldehyde (molar ratio 4 : 1) upon incubating the substrates with HRP, H_2O_2, and 4-methylphenol as cooxidant [29]. By removing one of the compounds, styrene oxidation was not detected anymore. Besides, $^{18}O_2$ and $H_2^{18}O$ labelling studies have shown that the reaction mechanism for the formation of styrene oxide involves a radical intermediate which originates from the cooxidant. In the same work an explanation for the formation of benzaldehyde was not provided.

In another study, some HRP mutants showed an enhanced activity compared to the wild-type enzyme for the epoxidation of styrene as well as cis-β-methyl-styrene and trans-β-styrene in absence of any cooxidant [30]. Whilst a significant amount of benzaldehyde was produced, also in this case the catalytic cycle leading to this product was not elucidated. The increased activity stemmed probably from the improved accessibility of the substrate into the catalytic site, due to the replacement of the sterically hindering Phe-41 residue, located close to the heme centre, with a smaller amino acid such as leucine or threonine.

Oxidative cleavage of 3-methyl-indole and 3-ethyl-indole to the corresponding ring-opened ortho-acyl formanilides and oxindoles was carried out on a 50 mg scale using HRP under aerobic conditions [31]. Interestingly for a practical application, the molar ratio substrate HRP was 10000 : 1. The radical oxidation was initiated by the addition of a catalytic amount of H_2O_2, required to generate compounds

I and II (Scheme 1) from the resting state of the HRP; however, H_2O_2 concentration was kept below $30\,\mu M$ to avoid enzyme deactivation. Under aerobic conditions the main product was the carbonylic compound from the alkene cleavage, whereas under anaerobic conditions the radical intermediate completely polimerised. A mechanism for the aerobic oxidative cleavage of indole was finally proposed, involving hydrogen abstraction by HRP compound I or II (6), interaction with dioxygen to lead the hydroperoxide (7), and final rearrangement to afford the carbonylic product (8) (Scheme 4).

Mutti et al. have recently shown that some peroxidases (i.e., horseradish peroxidase, lignin peroxidase, and Coprinus cinereus peroxidase) catalyse the cleavage of a C=C double bond adjacent to an aromatic moiety for selected substrates at the expense of molecular oxygen and at an acidic pH (Scheme 5) [32]. Among the three active peroxidases, HRP turned out to be the most active when an equal concentration of enzyme was employed. A thorough study of the reaction showed that the highest activity was obtained at ambient temperature, at pH 2, and at 2 bars of pure dioxygen pressure. Addition of DMSO as cosolvent up to $15\%\,v\,v^{-1}$ increased the conversion, probably due to the improved solubility of the substrates in the aqueous reaction medium, while a further addition of DMSO led to a progressive decline of the enzymatic activity. Using trans-anethole as substrate (9) $(6\,g\,L^{-1})$ and HRP at low catalyst loading (3 mg, equal to 0.2–0.3 mol%), quantitative conversion was achieved within 24 h. The main product was para-anisaldehyde (11) (i.e., 92% chemoselectivity), whereas the side product accounted completely for the vicinal diol (12). The substrate spectrum was quite narrow, since only other two substrates, that is, isoeugenol (10) and indene (13), could be cleaved by HRP with 12% and 72% conversion, respectively.

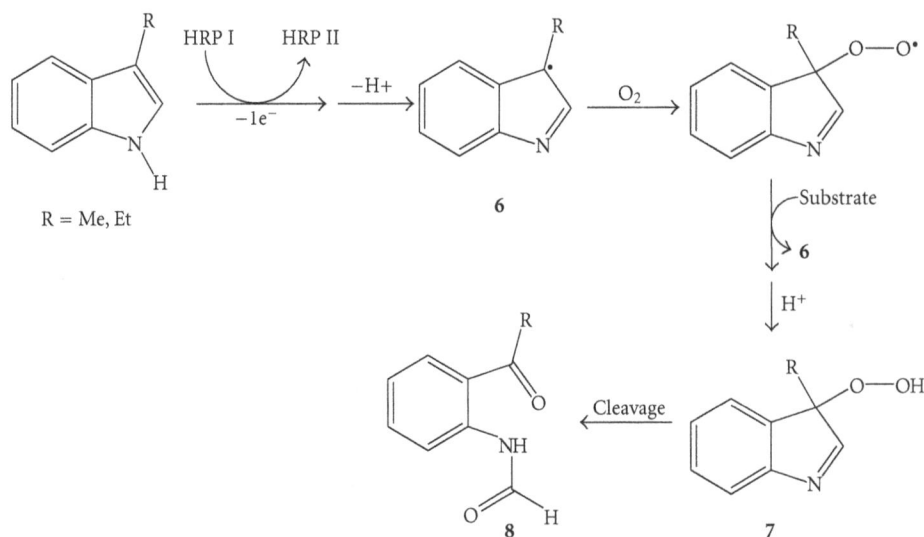

SCHEME 4: Proposed mechanism for the cleavage of indoles catalysed by HRP under aerobic conditions.

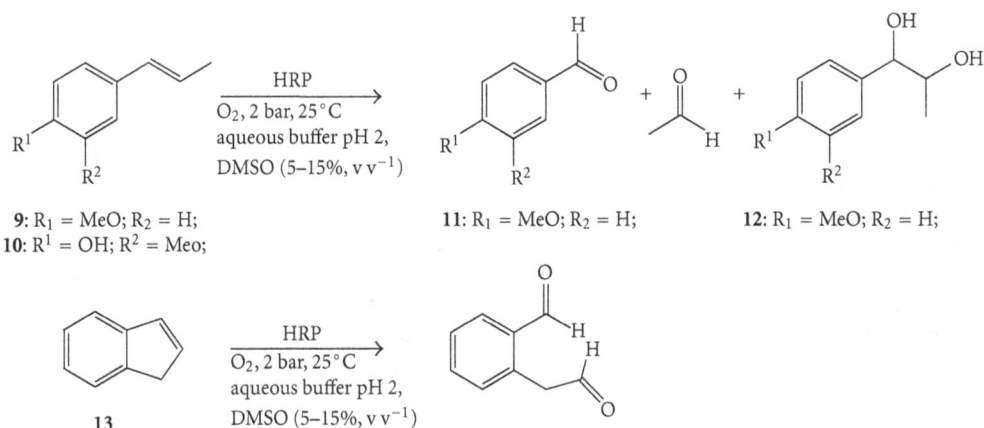

9: R₁ = MeO; R₂ = H;
10: R¹ = OH; R² = Meo;

11: R₁ = MeO; R₂ = H;

12: R₁ = MeO; R₂ = H;

13

SCHEME 5: Alkene cleavage of *trans*-anethole (**9**), isoeugenol (**10**), and indene (**13**) catalysed by HRP at 2-bar dioxygen pressure and ambient temperature.

2.3. Myeloperoxidase and Coprinus cinereus Peroxidase. Myeloperoxidase (MPO) and *Coprinus cinereus* peroxidase (CiP) catalyse the enantioselective epoxidation of styrene and a number of substituted derivates in moderate yield [33]. Additionally, as a consequence of the $C_\alpha=C_\beta$ double bond cleavage of variously substituted styrene precursors, both MPO and CiP form significant amounts of substituted benzaldehydes. The alkene cleavage is the most prominent reaction catalysed by CiP, whereas MPO forms a larger amount of epoxide. The reaction was performed with a continuous and controlled flow of H_2O_2 (1 μmol/h) to allow the formation of compound I and compound II (Scheme 1). The downside of employing such a low flow of oxidant is the long reaction time (16 h). The reaction mechanism of the alkene cleavage catalysed by CiP in presence of H_2O_2 as oxidant is different from the one previously described for CPO in presence of O_2. In fact, the addition of a cosubstrate is not required in the case of the alkene cleavage catalysed by CiP using H_2O_2. Styrene (1 mM) was converted to

styrene oxide (18%) and benzaldehyde (30%) employing CiP (20 μM). Conversely, MPO furnished the epoxide as main product (17%) with only traces of the other compounds (5%). Activated substituted styrenes bearing chlorogroups in *orto*-, *meta*- and *para*-position and *cis*-β-methyl-styrene were converted as well, yielding to generally higher amounts of aldehydes.

3. Alkene Cleavage by Heme and Nonheme Oxygenases

Iron cofactor-dependent oxygenases constitute a very heterogeneous group of enzymes. Each enzyme family activates dioxygen in different manner, and it was also postulated that the same enzyme may cleave olefinic functionalities using different mechanisms depending on the substrate (e.g., carotenoid cleavage oxygenases). This property has been already reviewed for other enzymes, and it was named catalytic enzyme promiscuity [34–36]. The alkene cleavage

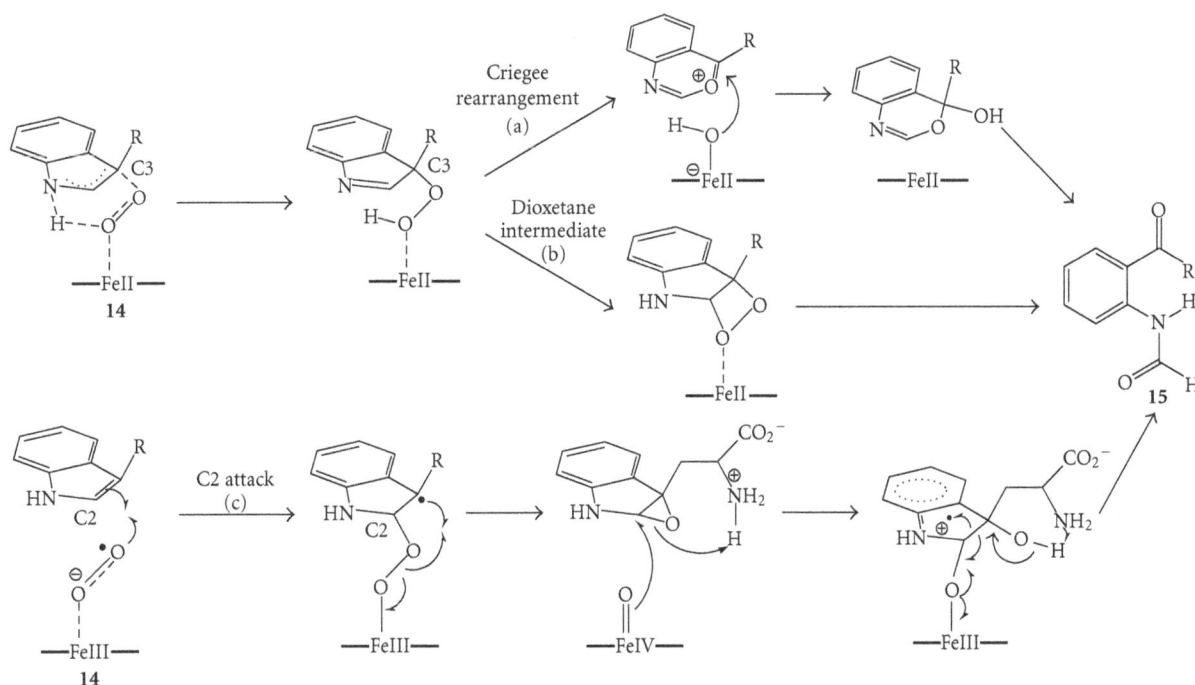

SCHEME 6: C=C double bond cleavage catalysed by tryptophan 2,3-dioxygenase (TDO). Historical perspective of the proposed reaction mechanisms: (a) C3-attack and Criegee rearrangement; (b) C3-attack and formation of dioxetane intermediate; (c) C2-attack via direct radical addition (current accepted mechanism).

is mainly a secondary activity for the peroxidases, whereas it is generally the natural and often unique activity for the oxygenases.

3.1. *Tryptophan 2,3-Dioxygenase and Indoleamine 2,3-Dioxygenase.* Tryptophan 2,3-dioxygenase (TDO) and indoleamine 2,3-dioxygenase (IDO) are unique heme dependent dioxygenases which cleave the C=C double bond of the pyrrole ring of the tryptophan (**14**) to afford N-formylkynurenine (**15**) [37]; hence, both oxygen atoms are incorporated from molecular oxygen. TDO and IDO catalyse essentially the same reaction, with the different denominations merely reflecting the wider substrate promiscuity of IDO. Various reaction mechanisms for this intriguing alkene cleavage have been proposed since the 1930; yet a conclusive study has not been published so far. The first proposal involved the deprotonation of the aminogroup of the indole ring by the highly conserved histidine residue present in the active site of the enzyme [38]. In contrast, later studies demonstrated that methyltryptophane could also be cleaved by the enzyme whilst at reduced reaction rate [39]. Moreover site direct mutagenesis of the histidine residue to an alanine or a serine led to variants which still cleaved the natural substrate [39–41]. These findings are in agreement with the well-documented chemistry of indoles [42] which do not react by base-catalysed reaction ($PK_{a,N-H} \approx 17$), thus ruling out the essential role of the histidine residue in a possible deprotonation step. The successive proposals encompassed the electrophilic addition of an activated Fe(II)-dioxo species to the C3 of the indole ring, followed by either Criegee-type rearrangement (Scheme 6, path (a)) or formation of a dioxetane intermediate (Scheme 6, path (b)). [39, 43]. On the contrary, the recent isolation and characterisation of a cyclic aminoacetal [44] (i.e., probably generated by the rearrangement of a 2-3-epoxide intermediate) coupled with novel Raman studies suggested a sequential insertion of oxygen [45]. A further computational study supported a mechanism whereby an Fe(III)-superoxo species (compound III, see Scheme 1) may be involved in a direct radical addition to the C2 of the indole ring, followed by the homolytic O–O cleavage and the formation of the 2-3 epoxide intermediate. In a second step, the Fe(IV)-oxo intermediate (compound II, see Scheme 1) might open the epoxide ring, attacking again the C2 of the indole ring, finally leading to the cleavage via an intramolecular rearrangement (Scheme 5, path (c)) [46]. It is noteworthy that the ultimate proposal does not require the initial deprotonation of the amino group of the indole.

3.2. *Catechol Dioxygenases.* The catechol dioxygenases catalyse the oxidative cleavage of catechols and substituted catechols (**16**) as a central step in the bacterial degradation of aromatic compounds [47]. Hayaishi et al. discovered two nonheme families of these dioxygenases, namely, the intradiol dioxygenases (e.g., catechol 1,2-dioxygenase also called pyrocatechase) and the extradiol dioxygenases (e.g., catechol 2,3-dioxygenase also called metapyrocatechase). The intradiol dioxygenases cleave the C=C double bond between the phenolic hydroxyl groups to yield muconic acid (**17**) [48], whereas the extradiol dioxygenases cleave the C=C double bond adjacent to the phenolic hydroxyl groups to yield 2-hydroxymuconaldehyde (**18**) [49]. In the initial step, the intradiol dioxygenase interacts with molecular oxygen

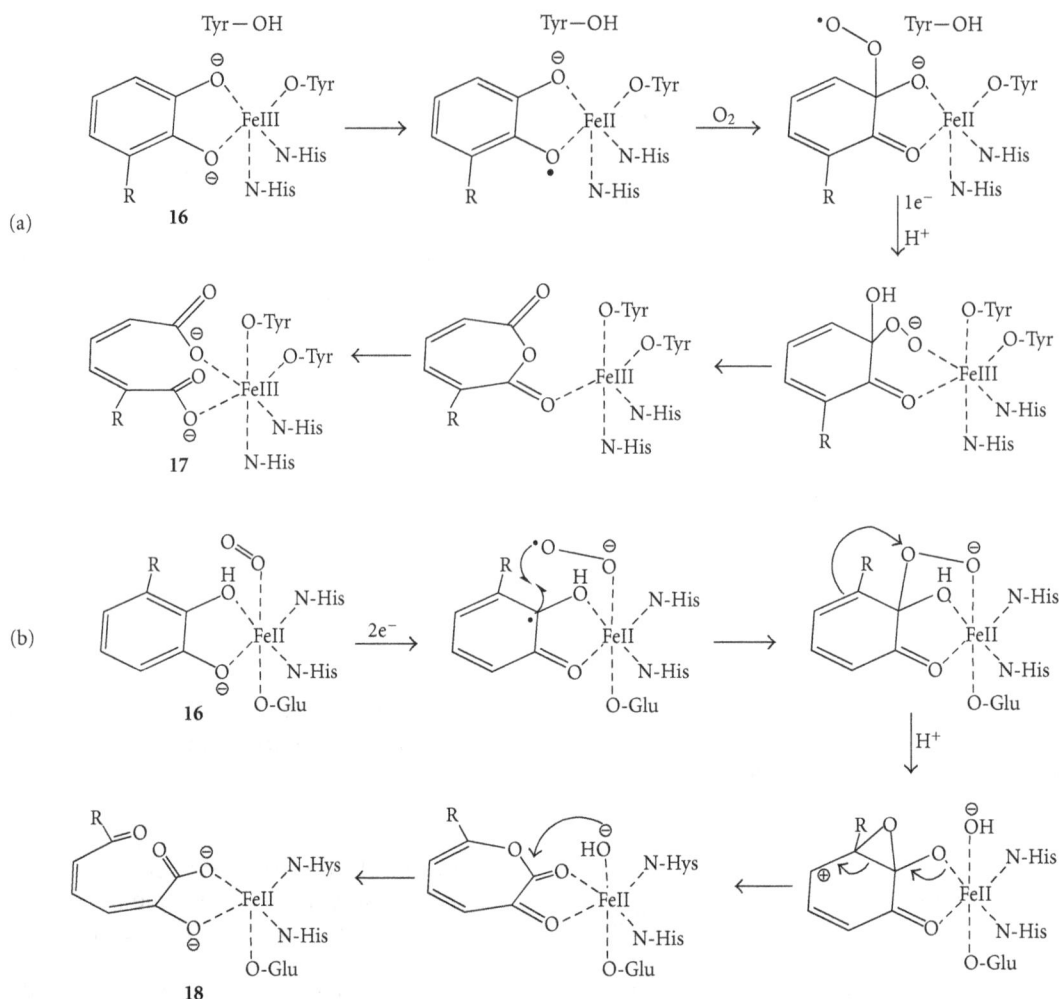

SCHEME 7: The mechanisms of the C=C double bond cleavage of catechols catalysed by (a) intradiol dioxygenase and (b) extradiol dioxygenase.

and probably generates an Fe(II)-semiquinone intermediate which then leads to Fe(III) species as confirmed by EPR studies. The successive steps, which end up to the C=C double bond cleavage of the aromatic ring, were previously believed to occur via a dioxetane intermediate; however, recent $^{18}O_2$ experiments supported an alternative mechanism involving a Criegee rearrangement to furnish an anhydride intermediate which then hydrolyses to muconic acid (Scheme 7(a)) [50]. Similarly, the mechanism of alkene cleavage catalysed by extradiol dioxygenase proceeds initially again through an Fe(II)-semiquinone complex, although then diverging to form an Fe(II) proximal hydroperoxide intermediate; the latter undergoes to Criegee rearrangement followed by hydrolysis to afford 2-hydroxymuconaldehyde (Scheme 7(b)). Several findings supported this mechanism: (i) the analysis of the product distribution from the reaction with substrates analogues carrying a cyclopropyl radical [51], (ii) UV-visible and Raman spectroscopic studies [52] and (iii) $^{18}O_2$ labelling studies [53].

3.3. Carotenoid Cleavage Oxygenases. Carotenoid cleavage oxygenases (CCOs) are widespread in bacteria, plants, and animals and catalyse the C=C double bond cleavage of carotenoids to give apocarotenoids [54, 55]. The family members require an Fe(II) centre which is bound to four highly conserved and catalytically essential histidine residues [56]. CCOs often exhibit substrate promiscuity, which probably contributes to the natural diversity of apocarotenoids and derivates, whilst often retaining a perfect regioselectivity (i.e., cleavage of a specific C=C double bond of the carotenoid chain). Thus, depending on the enzyme, the alkene cleavage can occur either at the central C=C double bond of the carotenoid substrates (i.e., C15=C15′ position, central cleavage) or at another position (i.e., excentric cleavage). In the literature, this enzyme family is usually referred to as carotenoid cleavage dioxygenases (CCD). However, in this paper, the broader definition as carotenoid cleavage oxygenases (CCOs) was adopted since the classification as mono- or dioxygenases is still subject of debate within the scientific community. A monooxygenase enzyme activates molecular oxygen to incorporate one oxygen atom into the substrate, whereas the second oxygen originates from a water molecule. Conversely, a dioxygenase enzyme incorporates the two oxygen atoms coming from one

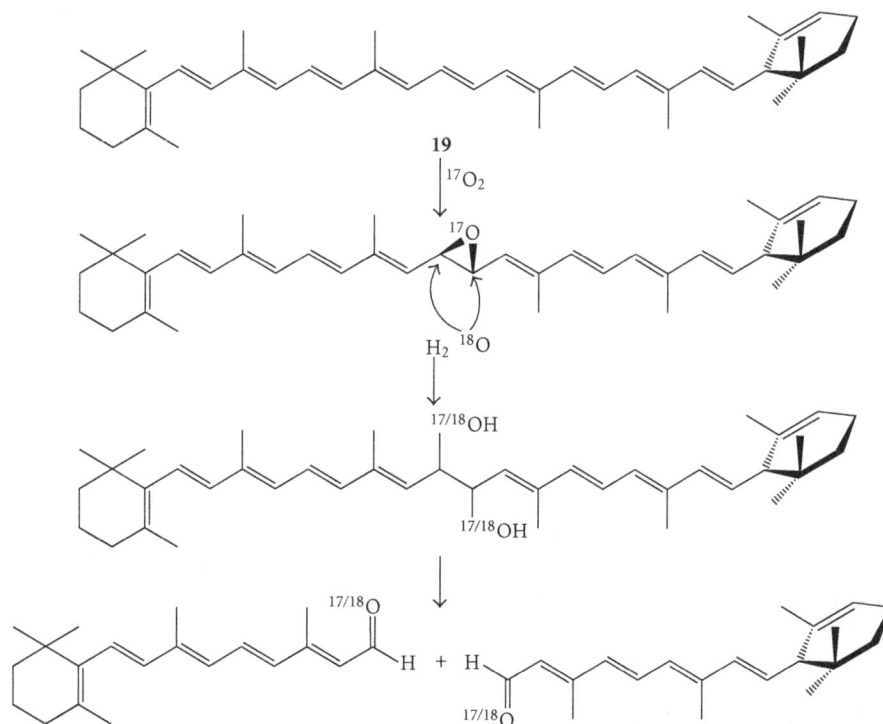

SCHEME 8: The proposed mechanism for the C15-C15′ double bond cleavage of β-carotene. Labelling experiment using $^{17}O_2$ and $^{18}H_2O$.

molecule of molecular oxygen. Distinguishing between these two mechanisms relies on experiments using isotopically labelled O_2 and H_2O, which pose a serious problem in the analysis of the product distribution due to the rapid exchange of the aldehydic oxygen with the water medium. The controversial case of assignment of the cleavage mechanism for β-15-15′-carotenoid cleavage oxygenase (β-CCO) exemplifies this issue. The β-CCO from rat liver and rat intestine was initially termed as "β-carotene 15,15′-dioxygenase" albeit an evidence concerning a dioxygenase mechanism was not previously reported [57]. Later Leuenberger et al. claimed that the cleavage occurs via a monooxygenase activity [58]. α-Carotene (19) was chosen as substrate for the labelling studies, since only the use of a nonsymmetrical carotenoid giving different aldehydes may provide an exact information on the origin of the oxygen atoms incorporation. With the aim to minimise the oxygen scrambling between the product aldehydes and the medium, the experiment was carried out combining the β-CCO with a horse liver alcohol dehydrogenase (HL-ADH) so that the generated aldehydes were in situ reduced to the corresponding alcohols. The experiment was performed employing labelled $^{17}O_2$ into labelled $H_2^{18}O$ as reaction medium and the obtained product distribution was apparently consistent with a monooxygenase pathway, since equal amounts of ^{17}O- and ^{18}O-labelled aldehydes were revealed (Scheme 8).

Despite the fact that β-CCO was reassigned as a β-carotene 15,15′-monooxygenase, several authors questioned about these results due to the long reaction time of the enzymatic reaction (7.5 h) and the dismutase activity of the HL-ADH, especially at increased level of NADH, which may

lead to an unspecific water-derived oxygen incorporation into retinol [59]. In another study the reaction mechanism of the carotenoid cleavage oxygenase from *Arabidopsis thaliana* (AtCCD1) was investigated [60]. AtCCD1 cleaves β-apo-8′-carotenal (20) as natural substrate at the 9,10 double bond to deliver one molecule of β-ionone (21) and a molecule of C_{17} dialdehyde (22), (Scheme 9).

In this case the very low rate for the exchange of oxygen atoms between the keto moiety of β-ionone and the medium coupled with a shorter reaction time (30 min) due to higher enzyme activity might provide an accurate evaluation of the reaction mechanism. When the experiment was performed in presence of $^{18}O_2$, the obtained β-ionone was 96% labelled, whereas in the experiment in presence of $^{18}H_2O$, the same product was completely unlabelled. The C_{17} dialdehyde underwent a partial oxygen exchange during the reaction time as supported by blank experiments; however, a significant fraction of C_{17} dialdehyde (27%) showed incorporation of one ^{18}O atom when the experiment was performed in $^{18}O_2$, hence supporting a dioxygenase mechanism. Furthermore, a recent computational study based on the crystal structure of AtCCD1 [61] estimated that the energy barrier for the formation of the epoxide intermediate (16.6 kcal mol^{-1}) is only slightly exceeding the one for the dioxetane intermediate (15.9 kcal mol^{-1}), probably due to the sterical effect of the Thr136 residue in the catalytic site [62]. Nevertheless a putative stilbene oxygenase sharing high sequence homology with the AtCCD1 was shown to cleave variously substituted stilbene derivatives via a monooxygenase mechanism (this enzyme will be discussed in the next session) [63]. Thus, subtle changes in the active

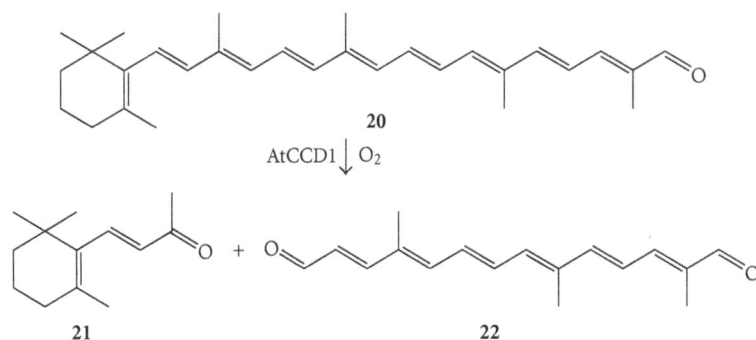

20

AtCCD1 $\downarrow$ O$_2$

21   **22**

SCHEME 9: The C9–C10 double bond cleavage catalysed by the carotenoid cleavage dioxygenase from *Arabidopsis thaliana* (AtCCD1).

site may favour one mechanism over the other or the same carotenoid cleavage oxygenase may display both reactivities depending on the substrate.

The products of the alkene cleavage of carotenoids and apocarotenoids constitute important natural flavours for the aroma industry; hence the possibility to exploit AtCCD1 in biocatalysis has attracted the interest of academia and industry. Schilling et al. reported an improved protocol for the expression of AtCCD1 in *E. coli* as a fusion protein with glutathione-S-transferase [64]. The recombinant enzyme showed higher level of heterologous expression as well as ameliorated biocatalytic performance. In the first report, an 18-fold increased activity in vitro was found when the enzymatic assay was conducted in a micellar dispersion using Triton X-100 as surfactant (substrate-surfactant, 0.008 ratio) and adding methanol as organic cosolvent (15% v v^{-1}). Due to improved solvation of the lipophilic substrate in the aqueous micellar medium, β-apo-8'-carotenal was cleaved to give β-ionone with high conversion (>90%) and perfect regioselectivity. In a more recent publication it was shown that the maximum activity for β-apo-8'-carotenal varied dependently on the surfactant employed [65]. Testing diverse apocarotenoids and carotenoids in combination with different surfactants demonstrated that the most suitable surfactant varied dependently on the lipophilicity of the substrate. Nevertheless, the substrate concentration currently applied is too low (less than 1 mM) to meet the requirement for a possible industrial application.

3.4. Stilbene-α-β-Oxygenase and Isoeugenol Oxygenase.
Kamoda et al. identified and purified four isoenzymes of the stilbene-α-β-oxygenase (also named lignostilbene-α-β-dioxygenase, LSD) from *Sphingomonas paucimobilis* TMY 1009 (previously named *Pseudomonas p.*). This enzyme family was arbitrarily classified as dioxygenase, albeit studies to shed light on the mechanism were never undertaken. Therefore, the definition as oxygenase was adopted in this paper. The four enzymes contain one equivalent of iron, and they are constituted by two subunits: isoenzyme I ($\alpha\alpha$), isoenzyme II ($\alpha\beta$), isoenzyme III ($\beta\beta$), and isoenzyme IV ($\gamma\gamma$) [66, 67]. The catalytic activity of all isoenzymes required a stilbene-type substrate possessing *trans*-configuration and bearing a hydroxylic substituent in *para*-position on the

aromatic ring. The four isoenzymes cleaved 4,4'-dihydroxy-3,3'-dimethoxystylbene (**23**) as well as 4,2'-dihydroxy-3,3'-dimethoxy-5'-(2''-carboxyvinyl)-stilbene (**24**), although showing different substrate specificities (Scheme 10(a)); the isoenzyme I was the most active enzyme with a preference for **23**, whereas the others cleaved preferentially substrate **24** [68]. Particularly, the isoenzyme I was stable at 50 °C and showed increased activity upon the addition of methanol as cosolvent (30% v v^{-1}) [69]. Despite the fact that the cleavage of compound **23** can furnish directly two molecules of vanillin, an important flavour and fragrance for the food and cosmetic industry, the enzymatic reaction was never exploited on a preparative scale. A recent survey of the bacterial genome sequence for carotenoid cleavage oxygenase homologues allowed to identify two putative stilbene oxygenases from *Novosphingobium aromaticivorans* DSM 12444 (NOV1 and NOV2) [63]. NOV1 and NOV2 cleaved selectively *trans*-stilbene-type substrates bearing a hydroxy- or a methoxy moiety in *para*-position of the phenyl ring such as rhapontigenin (**25**), resveratrol (**26**), rhaponticin (**27**), and piceatannol (**28**), (Scheme 10(b)). Interestingly, the two enzymes were not able to cleave carotenoids. Labelling studies using ^{18}O or ^{18}H$_2$O showed that NOV1 and NOV2 incorporated only one oxygen atom from molecular oxygen into the substrate. Thus the two putative stilbene oxygenases were classified as monooxygenases. It is interesting to note that AtCCD1 was classified as dioxygenase while NOV1 and NOV2 as monooxygenases in spite of high sequence homology.

In another study, a novel enzyme was unveiled from *Pseudomonas putida* IE27 when the microorganism was cultivated from soils containing isoeugenol as a sole carbon source [70]. The enzyme was purified and assayed for the catalytic reaction in vitro, demonstrating high activity for the alkene cleavage of isoeugenol (**29**) to yield vanillin (**30**) in strict presence of molecular oxygen (Scheme 11). Besides, 2-methoxy-4-vinylphenol was also a substrate albeit at a reduced reaction rate (i.e., two orders of magnitude). Hence the enzyme was named after its substrate as isoeugenol oxygenase. Interestingly the analysis of the amino acid sequence of the isoeugenol oxygenase revealed high homology with the previously mentioned stilbene-α-β-oxygenase from *Sphingomonas paucimobilis* TMY 1009 isoenzyme I and III

23: R^1 = OMe; R^2 = OH; R^3, R^4 = H
24: R^1 = CH=CH-COOH; R^2 = H; R^3 = OMe; R^4 = OH

25: R^1, R^2, R^5 = OH; R^3 = H, R^4 = OMe
26: R^1, R^2, R^4 = OH; R^3, R^5 = H
27: R^1 = O-Glu; R^2, R^5 = OH; R^3 = H; R^4 = OMe
28: R^1, R^2, R^4, R^5 = OH; R^3 = H

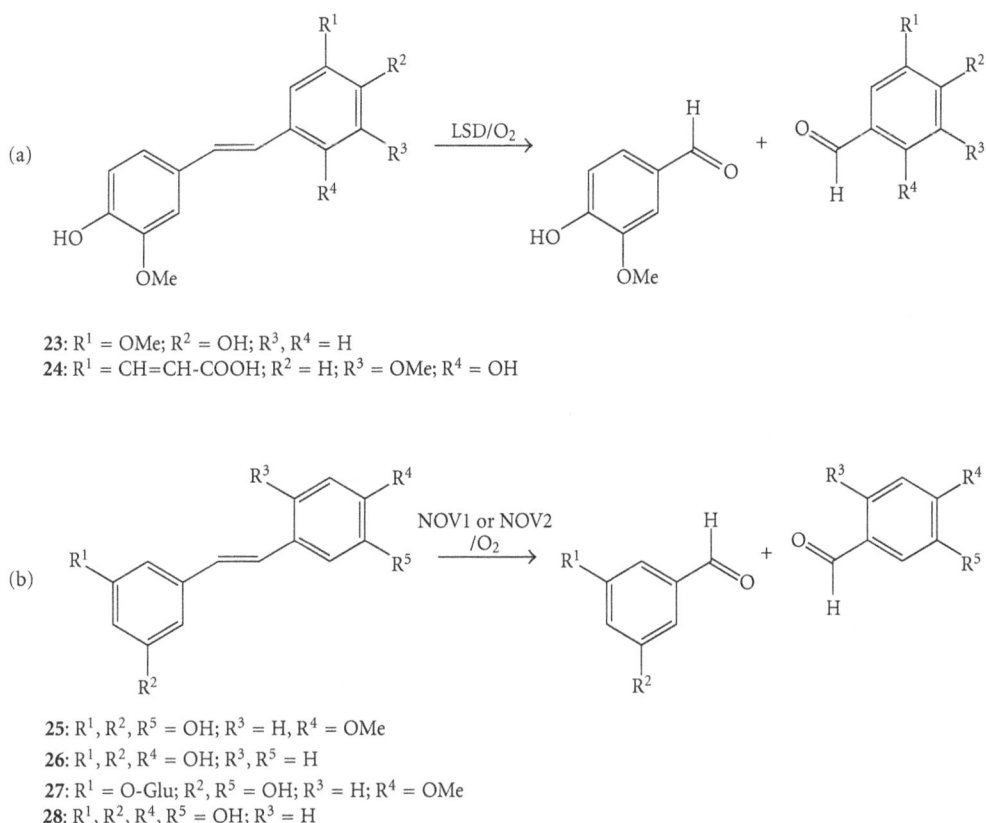

SCHEME 10: Alkene cleavage of stilbene derivates: (a) alkene cleavage catalysed by the isoenzymes of lignostilbene α,β-oxygenase from *Sphingomonas paucimobilis* TMY 1009; (b) alkene cleavage catalysed by stilbene monooxygenases from *Novosphingobium aromaticivorans* DSM 12444 NOV1 and NOV2.

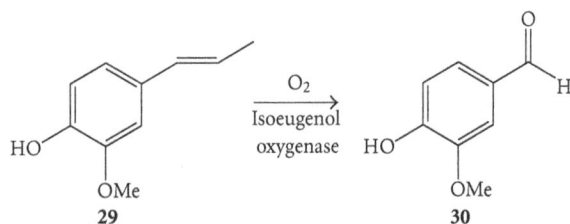

SCHEME 11: Alkene cleavage catalysed by isoeugenol oxygenase from *Pseudomonas putida* IE27.

(42% identity) and a putative stilbene oxygenase from *Novosphingobium aromaticivorans* DSM 12444 (40% identity). The alkene cleavage of isoeugenol was postulated to occur through a monooxygenase mechanism on the base that vanillin incorporated a labelled ^{18}O atom when the reaction was carried out either in labelled $^{18}O_2$ or $^{18}H_2O$. Nonetheless the monooxygenase mechanism cannot be considered as conclusive, due to the lack of data about isotopic product distribution; data about exchange of oxygen atoms from substrate to the aqueous medium were not reported as well.

3.5. Other Oxygenases. Other putative oxygenases have been isolated acting on various substrates. In this last paragraph, the most promising enzymes for biocatalytic applications will be examined. The enzymatic C=C double bond cleavage of natural rubber (i.e., poly-(*cis*-1,4-isoprene)) (31) and synthetic rubbers was observed using a purified protein from *Xanthomonas* sp., currently named as rubber oxygenase (RoxA), (Scheme 12) [71]. RoxA was characterised with the aid of UV-visible spectroscopy and gene sequence analysis, revealing two heme prosthetic groups located into protein scaffold and a conserved sequence motif which is a distinctive feature of the cytochrome *c* peroxidases. The enzymatic activity strictly necessitated molecular oxygen and completely ceased when heme inhibitors such as potassium cyanide and carbon monoxide were added into the reaction mixture, thus confirming the essential catalytic role of the metal centre. RoxA showed high regioselectivity since it was capable to cleave poly-(*cis*-1,4-isoprene) at regular intervals, principally cutting off three isoprene units per step (32).

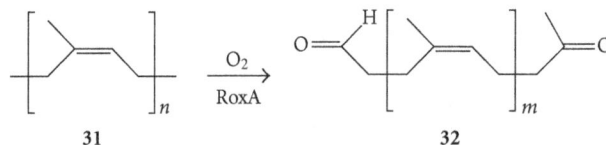

SCHEME 12: Alkene cleavage of natural rubber catalysed by the rubber oxygenase (RoxA) from *Xanthomonas sp*. The major product showed $m = 2$.

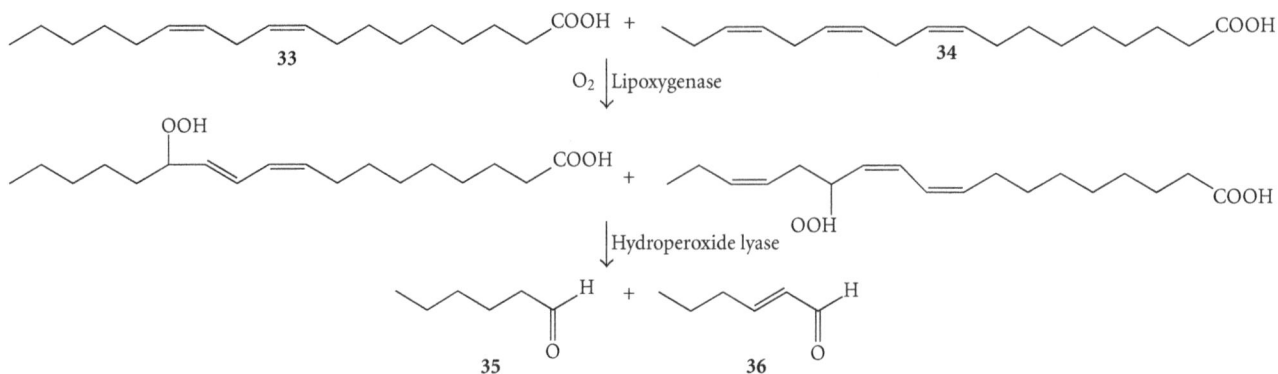

SCHEME 13: Formal alkene cleavage of polyunsaturated fatty acids combining a lipoxygenase and a hydroxyperoxide lyase.

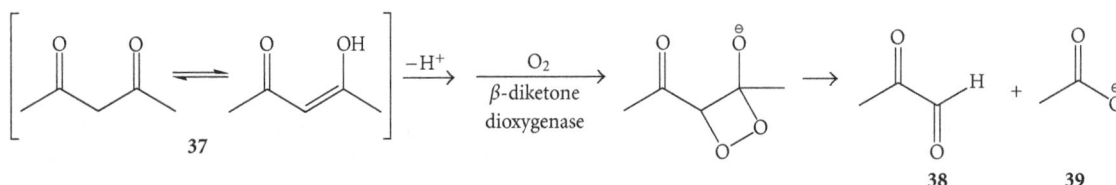

SCHEME 14: Alkene cleavage catalysed by β-diketone dioxygenase.

4. Conclusions and Outlook

A limited amount of major and minor degradation products was also detected, however always showing a carbonylic moiety at both terminal ends. Further labelling studies revealed RoxA to cleave following a dioxygenase mechanism [72].

Two enzymes can also cooperate to achieve the alkene cleavage. For instance, in the first step, the lipoxygenase from soybean flour activated molecular oxygen to enable the attack of the olefinic group. In the second step the hydroperoxide intermediate was cleaved by the fatty acid hydroperoxide lyase (HPO lyase) from green bell pepper. Thus, the formal two-step cleavage of a mixture of linoleic (**33**) and linolenic acids (**34**) to yield hexanal (**35**) and *trans*-hexenal (**36**), respectively, was carried out on a preparative scale with a productivity of ca. 300 mg L^{-1} (Scheme 13). The C6 aldehyde products are important flavours for the aroma industry [73].

Finally, even a formally single C–C bond can be cleaved following a dioxygenase mechanism. That is the case of the Fe(II)-dependent β-diketone dioxygenase from *Acinetobacter johnsonii* (Dke1). Dke1 cleaves the enol form of the 2,4-pentanedione (**37**) (plus related β-dicarbonyl compounds thereof), giving equimolar amounts of methylglyoxal (**38**) and acetate (**39**) and consuming only one equivalent of molecular oxygen (Scheme 14) [74].

While studying heme peroxidases to carry out chemical transformations such as the asymmetric epoxidation or the stereoselective mono- and dihydroxylation of unsaturated functionalities, the alkene cleavage has been revealed as a minor side reaction. In few cases, the alkene cleaving activity became the predominant one when the reaction conditions were properly adjusted (i.e., pH, dioxygen pressure, addition of cosolvents). Thus, this promiscuous activity can be potentially exploited in organic synthesis. In contrast, iron-dependent oxygenases often catalyse the alkene cleavage as the natural and unique reaction. Yet, the exploitation of these enzymes in organic synthesis is hampered by the limited solubility of apolar substrates in aqueous media.

On the other hand, especially in the case of heme peroxidases, a comprehensive understanding of the reaction mechanisms for the alkene cleavage necessitates more accurate studies. In fact, the proposed mechanisms are merely based on experimental observations, that is, analysis of the product distribution, requirement of dioxygen and cosubstrates, and so forth, or rely on studies of selective inhibition of the heme cofactor.

A detailed understanding of the alkene cleaving mechanism coupled with the advancement in gene cloning and protein engineering over the last decade will allow us to manipulate these enzymes to ameliorate their chemo- and regioselectivity and increase their tolerance to harsher reaction conditions. The design of improved variants could finally pave the way to the application of the enzymatic alkene cleavage on a large scale.

Acknowledgments

The author wishes to dedicate this paper to the memory of his former mentor Professor Michele Gullotti. He was an inspiring guide during the time of his doctoral studies and a dear friend.

References

[1] R. A. Berglund, *Encyclopedia of Reagents for Organic Synthesis*, vol. 6, Wiley, New York, NY, USA, 1995.

[2] K. Koike, G. Inoue, and T. Fukuda, "Explosion hazard of gaseous ozone," *Journal of Chemical Engineering of Japan*, vol. 32, no. 3, pp. 295–299, 1999.

[3] R. A. Ogle and J. L. Schumacher, "Investigation of an explosion and flash fire in a fixed bed reactor," *Process Safety Progress*, vol. 17, no. 2, pp. 127–133, 1998.

[4] J. March, *Advanced Organic Chemistry. Reactions, Mechanisms and Structures*, Wiley, New York, NY, USA, 4th edition, 1992.

[5] P. Y. Bruice, *Organic Chemistry. International Edition*, Pearson Education, Upper Saddle River, NJ, USA, 4th edition, 2004.

[6] R. U. Lemieux and E. Von Rudloff, "Periodate-permanganate oxidations: I. Oxidation of olefins," *Canadian Journal of Chemistry*, vol. 33, no. 11, pp. 1701–1709, 1955.

[7] C. E. Paul, A. Rajagopalan, I. Lavandera, V. Gotor-Fernandez, W. Kroutil, and V. Gotor, "Expanding the regioselective enzymatic repertoire: oxidative mono-cleavage of dialkenes catalyzed by *Trametes hirsuta*," *Chemical Communications*, vol. 48, no. 27, pp. 3303–3305, 2012.

[8] M. Lara, F. G. Mutti, S. M. Glueck, and W. Kroutil, "Biocatalytic cleavage of alkenes with O_2 and *Trametes hirsuta* G FCC 047," *European Journal of Organic Chemistry*, no. 21, pp. 3668–3672, 2008.

[9] M. Lara, F. G. Mutti, S. M. Glueck, and W. Kroutil, "Oxidative enzymatic alkene cleavage: indications for a nonclassical enzyme mechanism," *Journal of the American Chemical Society*, vol. 131, no. 15, pp. 5368–5369, 2009.

[10] J. Schrader, M. M. W. Etschmann, D. Sell, J. M. Hilmer, and J. Rabenhorst, "Applied biocatalysis for the synthesis of natural flavour compounds-current industrial processes and future prospects," *Biotechnology Letters*, vol. 26, no. 6, pp. 463–472, 2004.

[11] W. Adam, M. Lazarus, C. R. Saha-Moller et al., "Biotransformations with peroxidases," in *Advanced in Biochemical Engineering/Biotechnology*, T. Sheper, Ed., vol. 63, pp. 73–108, Springer-Verlag, Berlin, Germany, 1999.

[12] G. I. Berglund, G. H. Carlsson, A. T. Smith, H. Szöke, A. Henriksen, and J. Hajdu, "The catalytic pathway of horseradish peroxidase at high resolution," *Nature*, vol. 417, no. 6887, pp. 463–468, 2002.

[13] I. Schlichting, J. Berendzen, K. Chu et al., "The catalytic pathway of cytochrome P450cam at atomic resolution," *Science*, vol. 287, no. 5458, pp. 1615–1622, 2000.

[14] G. Cilento and W. Adam, "From free radicals to electronically excited species," *Free Radical Biology and Medicine*, vol. 19, no. 1, pp. 103–114, 1995.

[15] P. D. Shaw and L. P. Hager, "Biological Chlorination: VI: chloroperoxidase: a component of the β-ketoadipate chlorinase system," *Journal of Biological Chemistry*, vol. 236, no. 6, pp. 1626–1630, 1961.

[16] L. P. Hager, D. R. Morris, F. S. Brown, and H. Eberwein, "Chloroperoxidase. II. Utilization of halogen anions," *Journal of Biological Chemistry*, vol. 241, no. 8, pp. 1769–1777, 1966.

[17] J. A. Thomas, D. R. Morris, and L. P. Hager, "Chloroperoxidase. VII. Classical peroxidatic, catalatic, and halogenating forms of the enzyme," *Journal of Biological Chemistry*, vol. 245, no. 12, pp. 3129–3134, 1970.

[18] A. Zaks and D. R. Dodds, "Chloroperoxidase-catalyzed asymmetric oxidations: substrate specificity and mechanistic study," *Journal of the American Chemical Society*, vol. 117, no. 42, pp. 10419–10424, 1995.

[19] M. P. J. Van Deurzen, F. Van Rantwijk, and R. A. Sheldon, "Selective oxidations catalyzed by peroxidases," *Tetrahedron*, vol. 53, no. 39, pp. 13183–13220, 1997.

[20] S. Colonna, N. Gaggero, C. Richelmi, and P. Pasta, "Recent biotechnological developments in the use of peroxidases," *Trends in Biotechnology*, vol. 17, no. 4, pp. 163–168, 1999.

[21] S. R. Blanke and L. P. Hager, "Identification of the fifth axial heme ligand of chloroperoxidase," *Journal of Biological Chemistry*, vol. 263, no. 35, pp. 18739–18743, 1988.

[22] P. R. Ortiz De Montellano, Y. S. Choe, G. DePillis, and C. E. Catalano, "Structure-mechanism relationships in hemoproteins. Oxygenations catalyzed by chloroperoxidase and horseradish peroxidase," *Journal of Biological Chemistry*, vol. 262, no. 24, pp. 11641–11646, 1987.

[23] J. Geigert, T. D. Lee, D. J. Dalietos, D. S. Hirano, and S. L. Neidleman, "Epoxidation of alkenes by chloroperoxidase catalysis," *Biochemical and Biophysical Research Communications*, vol. 136, no. 2, pp. 778–782, 1986.

[24] E. J. Allain, L. P. Hager, L. Deng, and E. N. Jacobsen, "Highly enantioselective epoxidation of disubstituted alkenes with hydrogen peroxide catalyzed by chloroperoxidase," *Journal of the American Chemical Society*, vol. 115, no. 10, pp. 4415–4416, 1993.

[25] D. J. Bougioukou and I. Smonou, "Chloroperoxidase-catalyzed oxidation of conjugated dienoic esters," *Tetrahedron Letters*, vol. 43, no. 2, pp. 339–342, 2002.

[26] D. J. Bougioukou and I. Smonou, "Mixed peroxides from the chloroperoxidase-catalyzed oxidation of conjugated dienoic esters with a trisubstituted terminal double bond," *Tetrahedron Letters*, vol. 43, no. 25, pp. 4511–4514, 2002.

[27] W. Chamulitrat, N. Takahashi, and R. P. Mason, "Peroxyl, alkoxyl, and carbon-centered radical formation from organic hydroperoxides by chloroperoxidase," *Journal of Biological Chemistry*, vol. 264, no. 14, pp. 7889–7899, 1989.

[28] M. Gajhede, D. J. Schuller, A. Henriksen, A. T. Smith, and T. L. Poulos, "Crystal structure of horseradish peroxidase C at 2.15 Å resolution," *Nature Structural Biology*, vol. 4, no. 12, pp. 1032–1038, 1997.

[29] P. R. Ortiz De Montellano and L. A. Grab, "Cooxidation of styrene by horseradish peroxidase and phenols: a biochemical model for protein-mediated cooxidation," *Biochemistry*, vol. 26, no. 17, pp. 5310–5314, 1987.

[30] S.-I. Ozaki and P. R. Ortiz De Montellano, "Molecular engineering of horseradish peroxidase: thioether sulfoxidation and styrene epoxidation by Phe-41 leucine and threonine

mutants," *Journal of the American Chemical Society*, vol. 117, no. 27, pp. 7056–7064, 1995.

[31] K. Q. Ling and L. M. Sayre, "Horseradish peroxidase-mediated aerobic and anaerobic oxidations of 3-alkylindoles," *Bioorganic and Medicinal Chemistry*, vol. 13, no. 10, pp. 3543–3551, 2005.

[32] F. G. Mutti, M. Lara, M. Kroutil, and W. Kroutil, "Ostensible enzyme promiscuity: alkene cleavage by peroxidases," *Chemistry*, vol. 16, no. 47, pp. 14142–14148, 2010.

[33] A. Tuynman, J. L. Spelberg, I. M. Kooter, H. E. Schoemaker, and R. Wever, "Enantioselective epoxidation and carbon-carbon bond cleavage catalyzed by Coprinus cinereus peroxidase and myeloperoxidase," *Journal of Biological Chemistry*, vol. 275, no. 5, pp. 3025–3030, 2000.

[34] U. T. Bornscheuer and R. J. Kazlauskas, "Catalytic promiscuity in biocatalysis: using old enzymes to form new bonds and follow new pathways," *Angewandte Chemie*, vol. 43, no. 45, pp. 6032–6040, 2004.

[35] K. Hult and P. Berglund, "Enzyme promiscuity: mechanism and applications," *Trends in Biotechnology*, vol. 25, no. 5, pp. 231–238, 2007.

[36] P. J. O'Brien and D. Herschlag, "Catalytic promiscuity and the evolution of new enzymatic activities," *Chemistry and Biology*, vol. 6, no. 4, pp. R91–R105, 1999.

[37] M. Sono, M. P. Roach, E. D. Coulter, and J. H. Dawson, "Heme-containing oxygenases," *Chemical Reviews*, vol. 96, no. 7, pp. 2841–2887, 1996.

[38] S. G. Cady and M. Sono, "1-methyl-DL-tryptophan, β-(3-benzofuranyl)-DL-alanine (the oxygen analog of tryptophan), and β-[3-benzo(b)thienyl]-DL-alanine (the sulfur analog of tryptophan) are competitive inhibitors for indoleamine 2,3-dioxygenase," *Archives of Biochemistry and Biophysics*, vol. 291, no. 2, pp. 326–333, 1991.

[39] N. Chauhan, S. J. Thackray, S. A. Rafice et al., "Reassessment of the reaction mechanism in the heme dioxygenases," *Journal of the American Chemical Society*, vol. 131, no. 12, pp. 4186–4187, 2009.

[40] N. Chauhan, J. Basran, I. Efimov et al., "The role of serine 167 in human indoleamine 2,3-dioxygenase: a comparison with tryptophan 2,3-dioxygenase," *Biochemistry*, vol. 47, no. 16, pp. 4761–4769, 2008.

[41] S. J. Thackray, C. Bruckmann, J. L. R. Anderson et al., "Histidine 55 of tryptophan 2,3-dioxygenase is not an active site base but regulates catalysis by controlling substrate binding," *Biochemistry*, vol. 47, no. 40, pp. 10677–10684, 2008.

[42] G. Yagil, "The proton dissociation constant of pyrrole, indole and related compounds," *Tetrahedron*, vol. 23, no. 6, pp. 2855–2861, 1967.

[43] I. Efimov, J. Basran, S. J. Thackray, S. Handa, C. G. Mowat, and E. L. Raven, "Structure and reaction mechanism in the heme dioxygenases," *Biochemistry*, vol. 50, no. 14, pp. 2717–2724, 2011.

[44] J. Basran, I. Efimov, N. Chauhan et al., "The mechanism of formation of N-formylkynurenine by heme dioxygenases," *Journal of the American Chemical Society*, vol. 133, no. 40, pp. 16251–16257, 2011.

[45] A. Lewis-Ballester, D. Batabyal, T. Egawa et al., "Evidence for a ferryl intermediate in a heme-based dioxygenase," *Proceedings of the National Academy of Sciences of the United States of America*, vol. 106, no. 41, pp. 17371–17376, 2009.

[46] L. W. Chung, X. Li, H. Sugimoto, Y. Shiro, and K. Morokuma, "ONIOM study on a missing piece in our understanding of heme chemistry: bacterial tryptophan 2,3-dioxygenase with

dual oxidants," *Journal of the American Chemical Society*, vol. 132, no. 34, pp. 11993–12005, 2010.

[47] T. D. H. Bugg and C. J. Winfield, "Enzymatic cleavage of aromatic rings: mechanistic aspects of the catechol dioxygenases and later enzymes of bacterial oxidative cleavage pathways," *Natural Product Reports*, vol. 15, no. 5, pp. 513–530, 1998.

[48] O. Hayaishi, M. Katagiri, and S. Rothberg, "Mechanism of the pyrocatechase reaction," *Journal of the American Chemical Society*, vol. 77, no. 20, pp. 5450–5451, 1955.

[49] O. Hayaishi, "Crystalline oxygenases of pseudomonads," *Bacteriological Reviews*, vol. 30, no. 4, pp. 720–731, 1966.

[50] R. J. Mayer and L. Que, "^{18}O studies of pyrogallol cleavage by catechol 1,2-dioxygenase," *Journal of Biological Chemistry*, vol. 259, no. 21, pp. 13056–13060, 1984.

[51] E. L. Spence, G. J. Langley, and T. D. H. Bugg, "Cis-trans isomerization of a cyclopropyl radical trap catalyzed by extradiol catechol dioxygenases: evidence for a semiquinone intermediate," *Journal of the American Chemical Society*, vol. 118, no. 35, pp. 8336–8343, 1996.

[52] F. H. Vaillancourt, C. J. Barbosa, T. G. Spiro et al., "Definitive evidence for monoanionic binding of 2,3-dihydroxybiphenyl to 2,3-dihydroxybiphenyl 1,2-dioxygenase from UV resonance Raman spectroscopy, UV/Vis absorption spectroscopy, and crystallography," *Journal of the American Chemical Society*, vol. 124, no. 11, pp. 2485–2496, 2002.

[53] J. Sanvoisin, G. J. Langley, and T. D. H. Bugg, "Mechanism of extradiol catechol dioxygenases: evidence for a lactone intermediate in the 2,3-dihydroxyphenylpropionate 1,2-dioxygenase reaction," *Journal of the American Chemical Society*, vol. 117, no. 29, pp. 7836–7837, 1995.

[54] D. P. Kloer and G. E. Schulz, "Structural and biological aspects of carotenoid cleavage," *Cellular and Molecular Life Sciences*, vol. 63, no. 19-20, pp. 2291–2303, 2006.

[55] M. E. Auldridge, D. R. McCarty, and H. J. Klee, "Plant carotenoid cleavage oxygenases and their apocarotenoid products," *Current Opinion in Plant Biology*, vol. 9, no. 3, pp. 315–321, 2006.

[56] E. K. Marasco, K. Vay, and C. Schmidt-Dannert, "Identification of carotenoid cleavage dioxygenases from Nostoc sp. PCC 7120 with different cleavage activities," *Journal of Biological Chemistry*, vol. 281, no. 42, pp. 31583–31593, 2006.

[57] J. A. Olson and O. Hayaishi, "The enzymatic cleavage of beta-carotene into vitamin A by soluble enzymes of rat liver and intestine," *Proceedings of the National Academy of Sciences of the United States of America*, vol. 54, no. 5, pp. 1364–1370, 1965.

[58] M. G. Leuenberger, C. Engeloch-Jarret, and W. D. Woggon, "The reaction mechanism of the enzyme-catalyzed central cleavage of β-carotene to retinal," *Angewandte Chemie*, vol. 40, no. 14, pp. 2614–2617, 2001.

[59] A. During and E. H. Harrison, "Intestinal absorption and metabolism of carotenoids: insights from cell culture," *Archives of Biochemistry and Biophysics*, vol. 430, no. 1, pp. 77–88, 2004.

[60] H. Schmidt, R. Kurtzer, W. Eisenreich, and W. Schwab, "The carotenase AtCCD1 from Arabidopsis thaliana is a dioxygenase," *Journal of Biological Chemistry*, vol. 281, no. 15, pp. 9845–9851, 2006.

[61] D. P. Kloer, S. Ruch, S. Al-Babili, P. Beyer, and G. E. Schulz, "The structure of a retinal-forming carotenoid oxygenase," *Science*, vol. 308, no. 5719, pp. 267–269, 2005.

[62] T. Borowski, M. R. A. Blomberg, and P. E. M. Siegbahn, "Reaction mechanism of apocarotenoid oxygenase (ACO): a DFT study," *Chemistry*, vol. 14, no. 7, pp. 2264–2276, 2008.

[63] E. K. Marasco and C. Schmidt-Dannert, "Identification of bacterial carotenoid cleavage dioxygenase homologues that cleave the interphenyl α,β double bond of stilbene derivatives via a monooxygenase reaction," *ChemBioChem*, vol. 9, no. 9, pp. 1450–1461, 2008.

[64] M. Schilling, F. Patett, W. Schwab, and J. Schrader, "Influence of solubility-enhancing fusion proteins and organic solvents on the in vitro biocatalytic performance of the carotenoid cleavage dioxygenase AtCCD1 in a micellar reaction system," *Applied Microbiology and Biotechnology*, vol. 75, no. 4, pp. 829–836, 2007.

[65] C. Nacke and J. Schrader, "Micelle based delivery of carotenoid substrates for enzymatic conversion in aqueous media," *Journal of Molecular Catalysis B*, vol. 77, pp. 67–73, 2012.

[66] S. Kamoda, N. Habu, M. Samejima, and T. Yoshimoto, "Purification and some properties of lignostilbene-α,β-dioxygenase responsible for the $C(\alpha)$-$C(\beta)$ cleavage of a diarylpropane type lignin model compound from Pseudomonas sp. TMY1009," *Agricultural and Biological Chemistry*, vol. 53, no. 10, pp. 2757–2761, 1989.

[67] S. Kamoda, T. Terada, and Y. Saburi, "A common structure of substrate shared by lignostilbenedioxygenase isozymes from *Sphingomonas paucimobilis* TMY1009," *Bioscience, Biotechnology and Biochemistry*, vol. 67, no. 6, pp. 1394–1396, 2003.

[68] S. Kamoda and Y. Saburi, "Structural and enzymatical comparison of lignostilbene-alpha,beta-dioxygenase isozymes, I, II, and III, from Pseudomonas paucimobilis TMY1009," *Bioscience, Biotechnology, and Biochemistry*, vol. 57, no. 6, pp. 931–934, 1993.

[69] A. Makoto, A. Niwa, S. Kamoda, and Y. Saburi, "Reactivity and stability of Lignostilbene-α, β-dioxygenase-I in various pHs, temperatures, and in aqueous organic solvents," *Journal of Microbiology and Biotechnology*, vol. 11, no. 5, pp. 884–886, 2001.

[70] M. Yamada, Y. Okada, T. Yoshida, and T. Nagasawa, "Purification, characterization and gene cloning of isoeugenol-degrading enzyme from Pseudomonas putida IE27," *Archives of Microbiology*, vol. 187, no. 6, pp. 511–517, 2007.

[71] R. Braaz, P. Fischer, and D. Jendrossek, "Novel type of heme-dependent oxygenase catalyzes oxidative cleavage of rubber (poly-cis-1,4-isoprene)," *Applied and Environmental Microbiology*, vol. 70, no. 12, pp. 7388–7395, 2004.

[72] R. Braaz, W. Armbruster, and D. Jendrossek, "Heme-dependent rubber oxygenase RoxA of Xanthomonas sp. cleaves the carbon backbone of poly(cis-1,4-isoprene) by a dioxygenase mechanism," *Applied and Environmental Microbiology*, vol. 71, no. 5, pp. 2473–2478, 2005.

[73] G. Bourel, J. M. Nicaud, B. Nthangeni, P. Santiago-Gomez, J. M. Belin, and F. Husson, "Fatty acid hydroperoxide lyase of green bell pepper: cloning in Yarrowia lipolytica and biogenesis of volatile aldehydes," *Enzyme and Microbial Technology*, vol. 35, no. 4, pp. 293–299, 2004.

[74] G. D. Straganz, H. Hofer, W. Steiner, and B. Nidetzky, "Electronic substituent effects on the cleavage specificity of a nonheme Fe^{2+}-dependent β-diketone dioxygenase and their mechanistic implications," *Journal of the American Chemical Society*, vol. 126, no. 39, pp. 12202–12203, 2004.

Permissions

The contributors of this book come from diverse backgrounds, making this book a truly international effort. This book will bring forth new frontiers with its revolutionizing research information and detailed analysis of the nascent developments around the world.

We would like to thank all the contributing authors for lending their expertise to make the book truly unique. They have played a crucial role in the development of this book. Without their invaluable contributions this book wouldn't have been possible. They have made vital efforts to compile up to date information on the varied aspects of this subject to make this book a valuable addition to the collection of many professionals and students.

This book was conceptualized with the vision of imparting up-to-date information and advanced data in this field. To ensure the same, a matchless editorial board was set up. Every individual on the board went through rigorous rounds of assessment to prove their worth. After which they invested a large part of their time researching and compiling the most relevant data for our readers. Conferences and sessions were held from time to time between the editorial board and the contributing authors to present the data in the most comprehensible form. The editorial team has worked tirelessly to provide valuable and valid information to help people across the globe.

Every chapter published in this book has been scrutinized by our experts. Their significance has been extensively debated. The topics covered herein carry significant findings which will fuel the growth of the discipline. They may even be implemented as practical applications or may be referred to as a beginning point for another development. Chapters in this book were first published by Hindawi Publishing Corporation; hereby published with permission under the Creative Commons Attribution License or equivalent.

The editorial board has been involved in producing this book since its inception. They have spent rigorous hours researching and exploring the diverse topics which have resulted in the successful publishing of this book. They have passed on their knowledge of decades through this book. To expedite this challenging task, the publisher supported the team at every step. A small team of assistant editors was also appointed to further simplify the editing procedure and attain best results for the readers.

Our editorial team has been hand-picked from every corner of the world. Their multi-ethnicity adds dynamic inputs to the discussions which result in innovative outcomes. These outcomes are then further discussed with the researchers and contributors who give their valuable feedback and opinion regarding the same. The feedback is then collaborated with the researches and they are edited in a comprehensive manner to aid the understanding of the subject.

Apart from the editorial board, the designing team has also invested a significant amount of their time in understanding the subject and creating the most relevant covers. They scrutinized every image to scout for the most suitable representation of the subject and create an appropriate cover for the book.

The publishing team has been involved in this book since its early stages. They were actively engaged in every process, be it collecting the data, connecting with the contributors or procuring relevant information. The team has been an ardent support to the editorial, designing and production team. Their endless efforts to recruit the best for this project, has resulted in the accomplishment of this book. They are a veteran in the field of academics and their pool of knowledge is as vast as their experience in printing. Their expertise and guidance has proved useful at every step. Their uncompromising quality standards have made this book an exceptional effort. Their encouragement from time to time has been an inspiration for everyone.

The publisher and the editorial board hope that this book will prove to be a valuable piece of knowledge for researchers, students, practitioners and scholars across the globe.

List of Contributors

Jagvir Singh
Department of Chemistry, Meerut College, Meerut 250001, India

Prashant Singh
Department of Applied Chemistry, B.B.A. University, Lucknow 226025, India

Michele Iafisco and Ismaela Foltran
Dipartimento di Chimica "G. Ciamician," Alma Mater Studiorum Universit`a di Bologna, Via Selmi 2, 40126 Bologna, Italy
Dipartimento di Scienze Mediche, Universit`a del Piemonte Orientale, Via Solaroli 17, 28100 Novara, Italy

Norberto Roveri
Dipartimento di Chimica "G. Ciamician," Alma Mater Studiorum Universit`a di Bologna, Via Selmi 2, 40126 Bologna, Italy

Simona Sabbatini and Giorgio Tosi
Dipartimento di Scienze e Tecnologie Chimiche, Universit`a Politecnica delle Marche, Via Brecce Bianche, 60131 Ancona, Italy

Nahid Shahabadi, Zeinab Ghasemian and Saba Hadidi
Department of Inorganic Chemistry, Faculty of Chemistry, Razi University, Kermanshah 74155, Iran

Santiago Gómez-Ruiz
Departamento de Qu´ımica Inorganica y Anal₁tica, E.S.C.E.T., Universidad Rey Juan Carlos, 28933 Mostoles, Spain

Danijela Maksimović-Ivanić and Sanja Mijatović
Institute for Biological Research "Sinisa Stankovic", University of Belgrade, Boulevard of Despot Stefan 142, 11060 Belgrade, Serbia

Goran N. Kaluđerović
Institut fur Chemie, Martin-Luther-Universit"at Halle-Wittenberg, Kurt-Mothes-Straße 2, 06120 Halle, Germany

Vıctor M. Martınez-Juarez
Area Acad´emica de Medicina Veterinaria y Zootecnia, Instituto de Ciencias Agropecuarias, Universidad Aut´onoma del Estado de Hidalgo, Zona Universitaria, Rancho Universitario Km 1. C.P. 43600, Tulancingo de Bravo Hidalgo, Mexico

Juan F. Cárdenas-González, Marıa Eugenia Torre-Bouscoulet and Ismael Acosta-Rodríguez
Laboratorio de Micolog´ıa Experimental, Centro de Investigaci´on y de Estudios de Posgrado, Facultad de Ciencias Qu´ımicas, Universidad Aut´onoma de San Luis Potos´ı, Avenida Dr. Manuel Nava No. 6, Zona Universitaria, 78320 San Luis Potos´ı, SLP, Mexico

Lidi Gao, Yuichi Sato, Chong Li and Shuang Zhang
Graduate School of Science and Technology, Niigata University, Niigata 950-2181, Japan

Naoki Kano and Hiroshi Imaizumi
Department of Chemistry and Chemical Engineering, Faculty of Engineering, Niigata University, Niigata 950-2181, Japan

K. Kuroda and M. Okido
EcoTopia Science Institute, Nagoya University, Furo-cho, Chikusa, Nagoya 464-8603, Japan

S. A. Salman
EcoTopia Science Institute, Nagoya University, Furo-cho, Chikusa, Nagoya 464-8603, Japan
Graduate School of Engineering, Al-Azhar University, Nasr City, Cairo 11371, Egypt

Yunlan Li, Zhuyan Gao, Pu Guo and Qingshan Li
School of Pharmaceutical Science, Shanxi Medical University, Taiyuan 030001, China

Leonor Morgado, Joana M. Dantas and Carlos A. Salgueiro
Requimte-CQFB, Departamento de Qu´ımica, Faculdade de Ciˆencias e Tecnologia, Universidade Nova de Lisboa, Campus Caparica, 2829-516 Caparica, Portugal

Marta Bruix
Departamento de Espectroscop´ıa y Estructura Molecular, Instituto de Qu´ımica-F´ısica "Rocasolano", CSIC, Serrano 119, 28006 Madrid, Spain

Yuri Y. Londer
Biosciences Division, Argonne National Laboratory, Argonne, IL 60439, USA
New England Bio labs, 240 County Road, Ipswich, MA 01938, USA

Karin Sono, Diane Lye, Christine A.Moore, W. Christopher Boyd, Thomas A. Gorlin and Jason M. Belitsky
Department of Chemistry and Biochemistry, Oberlin College, 119 Woodland Street, Oberlin, OH 44074, USA

Ahmed E. Hannora
Faculty of Petroleum and Mining Engineering, Suez Canal University, Suez 43721, Egypt

Alexander S. Mukasyan
Department of Chemical and Biomolecular Engineering, University of Notre Dame, Notre Dame, IN 46556, USA

Zulkhair A. Mansurov
Institute of Combustion Problems, Almaty 050012, Kazakhstan

Yongli Zhang, Xiangsheng Wang, Wei Fang and Xiaoyan Cai
Department of Biology, School of Basic Courses, Guangdong Pharmaceutical University, Guangzhou, Guangdong 510006, China

Fujiang Chu
Guangdong Provincial Key Laboratory of Pharmaceutical Bioactive Substances, Guangzhou, Guangdong 510006, China

Xiangwen Liao and Jiazheng Lu
Chemistry Department, School of Pharmacy, Guangdong Pharmaceutical University, Guangzhou, Guangdong 510006, China

Samer M. Al-Hakami and Muataz Ali Atieh
Department of Chemical Engineering, KFUPM, Dhahran 31261, Saudi Arabia

Amjad B. Khalil
Department of Biology, KFUPM, Dhahran 31261, Saudi Arabia

Tahar Laoui
Department of Mechanical Engineering, KFUPM, Dhahran 31261, Saudi Arabia

Bassem A. Al-Maythalony, M. Monim-ul-Mehboob, Mohammed I. M. Wazeer and Anvarhusein A. Isab
Department of Chemistry, King Fahd University of Petroleum and Minerals, Dhahran 31261, Saudi Arabia

M. Nasiruzzaman Shaikh
Center of Research Excellence in Nanotechnology (CENT), King Fahd University of Petroleum and Minerals, Dhahran 31261, Saudi Arabia

Saleh Altuwaijri
Clinical Research Laboratory, Saad Research & Development Center, Saad Specialist Hospital, Al-Khobar 31952, Saudi Arabia

Dai Yamamoto and Ikki Kawai
Department of Materials Science and Engineering, Graduate School of Engineering, Nagoya University, Nagoya 464-8603, Japan

Kensuke Kuroda, Ryoichi Ichino and Masazumi Okido
EcoTopia Science Institute, Nagoya University, Nagoya 464-8603, Japan

Azusa Seki
Hamri Co., Ltd., Tokyo 110-0005, Japan

Ariadna Garza-Ortiz, Carlos Camacho-Camacho, Teresita Sainz-Espues, Irma Rojas-Oviedo and Luis Ral Gutiérrez-Lucas
Departamento de Sistemas Biologicos, Universidad Aut´onoma Metropolitana-Unidad Xochimilco, Calzada de Hueso 1100, Colonia Villa Quietud, 04960 Coyoac´an, D.F., Mexico

Atilano Gutierrez Carrillo and Marco A. Vera Ramirez
Departamento de Resonancia Magn´etica Nuclear, Universidad Aut´onoma Metropolitana, Unidad Iztapalapa, San Rafael Atlixco, No. 186, Col. Vicentina, 09340 Iztapalapa, D.F., Mexico

Peter A. Ajibade and Omoruyi G. Idemudia
Department of Chemistry, University of Fort Hare, Alice 5700, South Africa

Yan Peng, Min-Min Zhang, Zhen-Feng Chen, Kun Hu, Yan-Cheng Liu, Xia Chen and Hong Liang
State Key Laboratory Cultivation Base for the Chemistry and Molecular Engineering of Medicinal Resources, School of Chemistry and Chemical Engineering of Guangxi Normal University, Guilin 541004, China

Nahid Shahabadi and Somaye Mohammadi
Department of Inorganic Chemistry, Faculty of Chemistry, Razi University, Kermanshah 74155, Iran

Omar Hamad Shihab Al-Obaidi
Chemistry Department, College of Education for Women, Al-Anbar University, Ramadi, Iraq

Huilu Wu, Tao Sun, Ke Li, Bin Liu, Fan Kou, Fei Jia, Jingkun Yuan and Ying Bai
School of Chemical and Biological Engineering, Lanzhou Jiaotong University, Lanzhou 730070, China

Francesco G. Mutti
Department of Chemistry, Organic and Bioorganic Chemistry, University of Graz, Heinrichstrasse 28, 8010 Graz, Austria